VITA MATHEMATICA

HISTORICAL RESEARCH AND
INTEGRATION WITH TEACHING

The cover is the frontispiece of Scottish mathematician David Gregory's first edition of the known writings of Euclid—*Euclidis quae supersunt omnia* (Oxford, 1703). It illustrates a story that Vitruvius relates in Hellenized Latin in the preface to Book VI of *De architectura* (late 1st century B.C.). Vitruvius writes that after a shipwreck the Socratic philosopher Aristippus and his companions made it safely to the shore of Rhodes. It was in an area that they apparently did not know. When he saw geometric figures drawn in the sand, Aristippus yelled "Let us be of good hope, for I see the traces of men."

For more information on this episode, see pages 251 and 252 below.

The images on the cover and on pages 251 and 252 are courtesy of the Special Collections Division, USMA Library, West Point, NY.

©1996 by the Mathematical Association of America

ISBN 0-88385-097-4

Library of Congress Catalog Number 96-75298

Printed in the United States of America

Current Printing

10 9 8 7 6 5 4 3 2 1

VITA MATHEMATICA

HISTORICAL RESEARCH AND
INTEGRATION WITH TEACHING

RONALD CALINGER

Editor

Published by
THE MATHEMATICAL ASSOCIATION OF AMERICA

Dedicated to

Phillip S. Jones

Scholar-teacher, historian of mathematics, and colleague

Phil Jones has spent his academic career at the University of Michigan. He received his doctorate there in 1948 and rose from assistant professor that year, to associate professor in 1953, to profesor of mathematics and education in 1958. In 1982, the regents named Dr. Jones professor emeritus. He has enjoyed sabbaticals in Paris, Cambridge, and London. A leader in the mathematics education profession, he has served as president of the National Council of Teachers, a member of the Board of Governors of the Mathematical Association of America, and co-chair of the International Study Group on the Relations between History and Pedagogy of Mathematics.

His 120 books and articles have established Professor Jones as an international authority in his fields. Among these, he wrote *Mathematics, an Integrated Course* (4 books, 1965–1967), co-authored *The Historical Roots of Elementary mathematics* (1976), and co-edited *The Rhind Mathematical Papyrus* by Arnold Buffum Chace (2 volumes, 1979).

MAA Notes and Reports Series

The MAA Notes and Reports Series, started in 1982, addresses a broad range of topics and themes of interest to all who are involved with undergraduate mathematics. The volumes in this series are readable, informative, and useful, and help the mathematical community keep up with developments of importance to mathematics.

MAA Notes

39. Calculus: The Dynamics of Change, *CUPM Subcommittee on Calculus Reform and the First Two Years, A. Wayne Roberts,* Editor.

40. Vita Mathematica: Historical Research and Integration with Teaching, *Ronald Calinger,* Editor.

MAA Reports

1. A Curriculum in Flux: Mathematics at Two-Year Colleges, *Subcommittee on Mathematics Curriculum at Two-Year Colleges,* a joint committee of the MAA and the American Mathematical Association of Two-Year Colleges, *Ronald M. Davis,* Editor.

2. A Source Book for College Mathematics Teaching, *Committee on the Teaching of Undergraduate Mathematics, Alan H. Schoenfeld,* Editor.

3. A Call for Change: Recommendations for the Mathematical Preparation of Teachers of Mathematics, *Committee on the Mathematical Education of Teachers, James R. C. Leitzel,* Editor.

4. Library Recommendations for Undergraduate Mathematics, *CUPM ad hoc Subcommittee, Lynn A. Steen,* Editor.

5. Two-Year College Mathematics Library Recommendations, *CUPM ad hoc Subcommittee, Lynn A. Steen,* Editor.

6. Assessing Calculus Reform Efforts: A Report to the Community, *James R. C. Leitzel and Alan Tucker,* Editors.

7. Matching High School Preparation to College Needs: Prognostic and Diagnostic Testing, *John G. Harvey,* Editor and *Gail Burrill, John Dossey, John W. Kenelly, and Bert K. Waits,* Associate Editors.

These volumes can be ordered from:
MAA Service Center
P.O. Box 90973
Washington, DC 20036
1-800-331-1MAA   FAX: 1-301-206-9789

Foreword

This volume brings together papers by scholars from around the globe on the historiography and history of mathematics and their integration with mathematical pedagogy. These papers on selected critical ideas, methods, practices, and trends in each field are intended primarily for mathematics teachers, research mathematicians, historians of mathematics, and historians of science. The dual purposes of the volume are to advance leading scholarship in both fields and to expand the rich dialogue on their intersection occurring among these four main audiences. Its publication attests to the continuing encouragement of the Mathematical Association of America.

The historical articles are written at different levels of technical difficulty. Those on internal mathematical developments depend mainly on mathematical theory, fine-grained textual analysis, or linguistics, while examinations of the interacting social and intellectual context draw upon the history of ideas, history of science, social and institutional history, new categories from cultural history, prosopography, economics, philosophy, or religion. Chronologically arranged, the historical articles may be profitably read in different orders following the reader's interests. The articles on historiography and sources in Part One offer basic perspectives. The first identifies research trends in the history of mathematics; the second discusses the centrality of problems; and the third uses drama to recreate moments of mathematical discovery. In opening Part Three on pedagogy, Torkil Heiede explores the place of the history of mathematics in teaching mathematics and mistakes to be avoided.

The intersection of mathematics teaching and the history of mathematics[1] has had a long, fruitful tradition. Here are a few major examples beginning with the appeal to great books in teaching. In his 1725 tutorial for Leonhard Euler at the University of Basel, Johann Bernoulli assigned classics as well as leading books and articles of his time. Solving their challenging problems required Euler's full energy, but he was not exhausted by these since the readings kept him from going astray on methods of solution. The computational work so exhilarated Euler that he soon decided to become a mathematician. A century later the Norwegian Niels Abel in a marginal note in his unpublished notebook likewise urged the study of primary sources. Students who want to "make progress in mathematics," he wrote, "should study the masters and not the pupils."[2] At the Berlin Mathematics Seminar, Ernst Kummer and Karl Weierstrass emphasized the careful study of the classics together with research journals, such as Crelle's and Liouville's, as essential for gaining a clearer understanding of the foundations of mathematics and preparing to make original contributions.

Another way to utilize history in teaching mathematics has been to follow a quasi-historical rather than a strictly logical ordering of materials. French mathematician Alexis-Claude Clairaut had a mainly historical organization in his *Élemens de géométrie* (1741) and *Élemens d'algèbre* (1746). He sought thereby to have students rediscover the basics of geometry and algebra by themselves. By slowly increasing the level of difficulty in algebra, Clairaut attempted to reveal the necessity for symbolism and technique. A century later Englishman George Boole endeavored to enhance didactics and learning by employing an historical organization in his textbook *Differential Equations* (1859).

Contents

Part One of this volume addresses the historiography of mathematics and sources in three articles. David Rowe reviews continuing tensions between mathematical historians educated to be mathematicians and cultural historians prepared in the history of science or the history of ideas. He examines the heated debate over the origins of geometric algebra and considers the movement from unitary to pluralistic studies of themes and biographies in the history of mathematics. Evelyne Barbin, contending that problems are "the stamp of the true scientific mind," argues that the conceptual history of mathematics affects our epistemological concepts of mathematics and that historical perspective showing obstacles to the development of ideas may help teachers better understand students' errors. Seeking to bring out the emotions of the moment of discovery, Gavin Hitchcock has written two dialogues on the acceptance in Europe of decimal expansions of irrational numbers and of negative roots of equations.

Part Two offers a range of scholarship in the history of mathematics, including mathematical education, from antiquity to the present. The first section goes up to the scientific revolution. Jens Høyrup traces through Old Babylonian algebra, Greek metrical geometry, Islamic al-jabr, and Renaissance algebra a problem in mensurational algebra. Wilbur Knorr illustrates that the method of indivisibles was not new in the West with Cavalieri in the seventeenth century but occurs in Archimedes's *On Method.* Frank Swetz probes the indigenous development of computational methods and algebra in traditional China from ancient times, including the famous *Nine Chapters,* despite constraints posed by the humanist Confucian examination system. Victor Katz finds origins of combinatorics in third-century B. C. Palestine and India, and he traces its twelfth through fourteenth-century abstract development with proofs among Islamic and Hebrew scholars. Barnabus Hughes discovers in the thirteenth century the first correct algebraic solutions of new cubic equations in the Latin West.

The second section of Part Two deals chronologically with the modern period from the scientific revolution. Zarko Dadić investigates the beginnings of analytic geometry in the work of Marin Getaldić. Urging mathematics teachers not to undervalue diagrams and other visual modeling of statistical data in learning, John Fauvel brings out the power of Thomas Clarkson's subtle diagram that helped make the British reading public more socially aware in the late eighteenth century campaign to make the slave trade illegal. Demonstrating the diverse methods of problem solving, Judith Grabiner contrasts Maclaurin's study of the derivative and elliptic integrals using geometric curves with Lagrange's analysis, especially his use of Taylor's series with remainder to prove the fundamental theorem of calculus. Niels Jahnke examines the teaching of Euler's and Lagrange's algebraic analysis in nineteenth-century German gymnasia and its implications for modern curricular reform. Ronald Calinger delineates the intellectual and institutional origins of the Berlin Mathematics Seminar, which sought to engender original research and improve teaching. Calinger clarifies the roles of Ernst Kummer and Karl Weierstrass in founding it, and he describes its nature and research programs to 1883. Roger Cooke closely reconstructs the context within which Sof'ya Kovalevskaya, a student of Weierstrass, discovered the partial differential equations for the rotation of a rigid body, and adumbrates their influence in establishing the existence or nonexistence of analytic solutions of such equations to the present. Susann Hensel illuminates the raising of the level of mathematical instruction at German polytechnics in the 1890s. In those schools, tensions arose between purists stressing Weierstrassian rigor in analysis and increasing industrial and engineering pressures for more concrete problems and applications. Peggy Kidwell finds a series of geometrical models of polyhedra and surfaces intimately connected with reforms in mathematical pedagogy, such as those of Felix Klein, in nineteenth- and early twentieth-century America. Concluding this section, William Aspray, Andrew Goldstein, and Bernard Williams explore how the National Science Foundation nurtured the rise of theoretical computer science and engineering.

Part Three on the integration of history of mathematics with mathematics teaching begins with Torkil Heiede's discussion of reasons for the role of history of mathematics in mathematics education. Ubiratan d'Ambrosio next argues that ethnomathematics must be considered to have a complete picture of the development of mathematics. Frederick Rickey proposes a network for sharing information on history in teaching mathematics. Richard Laubenbacher and David Pengelley review their history course for mathematics teachers based on primary sources, and Israel Kleiner employs great quotations as a starting point in teaching. Michèle Grégoire shows how her mathematics class responded to the bicentennial celebration of the French Revolution, measuring an arc of meridian using eighteenth-century techniques, and Beatrice Lumpkin investigates Benjamin Banneker's work in colonial America on the method of false position in algebra. Karen Dee Michalowicz reexamines the life of Mary Boole and her thoughts on cooperative learning and fostering critical thinking and the discovery of patterns.

Peter Bero opens the final section on teaching calculus with a study of perceptions of the continuum among secondary school students. In tracing the evolution of the function concept from James Gregory to Georg Cantor, Manfred Kronfellner presents mathematics as a lively, developing science. Kronfellner elucidates preparatory formulas and graphs as they occurred in history and studies provisional definitions in the gaining of greater generality. This work provides insight into the stages students go through in learning calculus and possible errors that may appear. By introducing students of diverse background to Isaac Barrow's theorem relating tangent and area problems, Martin Flashman essentially presents a geometric version of the fundamental theorem of calculus and stimulates discussions of the algorithms and symbolism that Newton and Leibniz needed to invent their versions of calculus. To familiarize students with integration in finite terms and make them more aware of the excitement, skill, and theory involved in the advance of mathematics, Man-Keung Siu traces the development of integration in finite terms from Joseph Liouville's papers of the 1830s to its resurgence in the 1970s, when its coupling with computers kindled a rapid advance in symbolic integration. Combining history, sociology, and calculus, James Tattersall devises stimulating problems that relate to students' possible career goals. Applying the law of exponential growth and decay together with patterns of population increases through history, Tattersall approximates demographic growth.

HPM and Acknowledgements

These papers were selected and refined through a refereeing process from talks given in August 1992 at the Quadrennial Meeting of the International Study Group on the Relations between History and Pedagogy of Mathematics (HPM) at the University of Toronto and the Seventh International Congress on Mathematical Education (ICME) at Université Laval in Quebec City as well as a subsequent call to historians of mathematics for papers by the editor. HPM arranges conferences for mathematics teachers who wish to incorporate materials from the history of mathematics in the classroom to motivate students and deepen their understanding of mathematics, and it fosters related publications. Its conferences, which research mathematicians and historians also attend, address topics in the history of mathematics and pedagogical issues, the cultural contextualist approach to problem solving, and experiences with historical problems in the classroom. Dr. Florence Fasanelli of the Mathematical Association of America organized the Toronto and Quebec City meetings in 1992, and Professors Craig Fraser of Toronto and Israel Kleiner of York graciously hosted them.

HPM is one of four affiliated study groups of ICME, which the International Congress of Mathematical Instruction (ICMI) organized. From the time of Felix Klein as chair in 1898, ICMI has grown to include sixty member nations today. After Philip Jones organized with Leo Rogers the ICME 2 working group on the history of mathematics at Exeter in 1972, he began in 1983 the annual meetings of the Americas Section of HPM at the University of Michigan. The next year

HPM became an official associate at ICME 5. Professor Charles Jones first edited the quarterly HPM *Newsletter,* and Professor Victor Katz is an excellent successor.

I am indebted to Victor Katz for his extensive refereeing, to the many referees noted in individual papers, to Frederick Rickey for obtaining the cover picture, to Anita Solow and the copy editors for their thorough reading and editorial suggestions, and to the diligence, timely responses, and friendship of the authors. Finally, I am grateful for the unfailingly enthusiastic support from my wife Betty and our children John and Anne.

References

1. The history of mathematics has ancient origins. Studies in it stem from Eudemus of Rhodes, a student of Aristotle, in fourth-century B. C. Athens. His lost *History of Geometry* is chiefly known today because in his *Commentary on the First Book of Euclid's Elements,* the Neoplatonic philosopher Proclus (d. 485) condensed it. A critical history of mathematics did not emerge, however, until the French geometer Jean Étienne Montucla (d. 1799) composed his masterpiece, *Histoire de mathématiques* (1758) that replaced chronological organization with the narrative.

   For the Eudemian summary, see Ronald Calinger, editor and author of historical chapter introductions and biographies, *Classics of Mathematics* (Englewood Cliffs, NJ: Prentice Hall, 1995), pp. 48–49. For information on historiography before the twentieth century, see Dirk Struik, "The Historiography of Mathematics from Proclus to Cantor," *NTM Schriftenreihe für Geschichte der Naturwissenschaften, Technik, und Medizin* xvii (1980): 1–22 and Christoph J. Scriba, *et al,* "Historiographie et histoire des mathématiques," *Archives Internationale d'Histoire des Sciences* 42(1992): 1–144. For aims of the field, see the "opinionated introduction" to William Aspray and Philip Kitcher, eds., *History and Philosophy of Modern Mathematics* (Minneapolis: University of Minnesota Press, 1990), pp. 20–31; Saunders Mac Lane, "The Protean Character of Mathematics," in J. Echeverria, A. Ibarra, and T. Mormann, eds., *The Space of Mathematics: Philosophical, Epistemological, and Historical Explorations* (Berlin: Walter de Gruyter, 1992), pp. 1–13; Ivo Schneider, "The History of Mathematics: Aims, Results, and Future Prospects," in Sergei S. Demidov, *et al,* eds., *Amphora: Festschrift für Hans Wussing...* (Basel: Birkhäuser, 1992), pp. 619–31; and Hans Wussing, "Historiography of Mathematics: Aims, Methods, and Tasks," in William R. Woodward and Robert S. Cohen, eds., *World Views and Scientific Discipline Formation* (Dordrecht: Kluwer, 1991), pp. 63–73.

   For studies of the historiography of science, see Kostas Gavroglu, Jean Christianidis, and Efthymios Nicolaidis, eds., *Trends in the Historiography of Science* (Dordrecht: Kluwer Academic Publishers, 1993) and Rachel Laudan, "Histories of Science and Their Uses: A Review to 1913," *History of Sciences* xxxi (1993): 1–34. For an application of Thomas Kuhn's model to the history of mathematics, see Leo Corry, "Kuhnian Issues, Scientific Revolutions, and the History of Mathematics," *Studies in History and Philosophy of Science* 24 (1993): pp. 95–117. Arnold Thackray offers an extensive bibliography in "History of Science" in Paul Durbin, ed., *A Guide to the Culture of Science, Technology and Medicine* (New York: Free Press, 1980), pp. 3–69.

2. See Øystein Ore, *Niels Henrik Abel: Mathematician Extraordinary* (Minneapolis: University of Minnesota Press, 1957), p. 138 and Harold M. Edwards, "Read the Masters," in Lynn Arthur Steen, ed., *Mathematics Tomorrow* (New York: Springer-Verlag, 1981), pp. 105–110.

Contents

III. Integration of History with Mathematics Teaching: Fundamentals and Selected Cases

Origins and Teaching of Calculus

I

Historiography
and
Sources

New Trends and Old Images in the History of Mathematics

David E. Rowe
University of Mainz

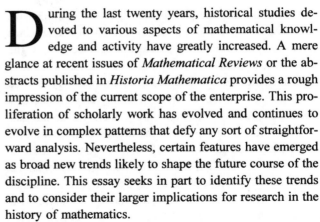

During the last twenty years, historical studies devoted to various aspects of mathematical knowledge and activity have greatly increased. A mere glance at recent issues of *Mathematical Reviews* or the abstracts published in *Historia Mathematica* provides a rough impression of the current scope of the enterprise. This proliferation of scholarly work has evolved and continues to evolve in complex patterns that defy any sort of straightforward analysis. Nevertheless, certain features have emerged as broad new trends likely to shape the future course of the discipline. This essay seeks in part to identify these trends and to consider their larger implications for research in the history of mathematics.

Rather than attempting to review all the major new directions in contemporary historical research, I have chosen instead to take a retrospective view of a constellation of themes and issues that, in my judgment, are of central importance for the history of mathematics taken as a whole. This set of issues bears upon the working assumptions of the discipline as well as upon philosophical views, value judgments, and methodological predilections that have played a role in historical research devoted to mathematics for more than a century. During the late nineteenth century, Moritz Cantor and Hieronymus Georg Zeuthen, two leading historians of mathematics of the era, pursued strikingly different research programs while maintaining a cordial personal relationship.[1] Unfortunately, such cordiality did not prevail in the 1970s when similar differences between historians regarding fundamental methodological issues created deep tensions that flared into open hostilities. Since that time, the warring parties have apparently left the field and retreated to their respective camps. The larger issues they have raised, however, continue to haunt the discipline of history of mathematics and, for that reason, deserve serious reexamination. Such a reappraisal constitutes a second goal of this paper.

Not surprisingly, the different viewpoints taken by historians on methodological issues have generally reflected the background and training of the individuals concerned. While these are obviously quite varied, for present purposes I wish to identify two principal camps: those who approach mathematics as historians of science, ideas, and institutions, on the one hand, and those who study the history of mathematics primarily from the standpoint of modern mathematicians, on the other. For brevity, I shall refer to the first group, which included M. Cantor, as cultural historians and the second, represented by H.G. Zeuthen, as mathematical historians.[2]

Since my own biases as an historian will be evident in what follows, and since most readers of this volume will be mathematicians and not historians, I wish to emphasize at the outset that, in my opinion, the history of mathematics should be approached from a variety of perspectives. At the same time, it should be recognized that historical studies in mathematics serve numerous different and very diverse constituencies. Indeed, the ever widening range of interests among scholars and students alike reflects one of the major trends now taking place in the history of mathematics. This trend is linked with a shift from a relatively narrow, Eurocentric vision of a monolithic body of mathematical knowledge to a broader, multi-layered picture of mathematical activity embedded in a rich variety of cultures and periods. A major challenge facing the history of mathematics as a discipline today will be to establish a constructive dialogue between the parties representing these differing interests. This survey hopes to contribute to such a dialogue not only by examining major disciplinary issues but also by indicating how influential mathematicians, historians, and philosophers have addressed them.

The Shift from Unity to Plurality

Beyond the exponential growth of literature in the history of mathematics, one of the most striking qualitative differences between the work done during the 1950s and 1960s and that of the ensuing period has been the trend away from a monolithic picture of mathematics to a more pluralistic outlook reflecting a broad range of diverse mathematical interests. This growing diversity of specialized interests helps account for the increased activity in the history of mathematics, much of which concerns areas that only recently entered the mainstream of literature published in major Western languages. For the purposes of this article, I shall concentrate on work published in the English language. Today, there exist good secondary accounts in English of mathematical developments in ancient China, India, medieval Islam, and other cultures.[3] In addition, there have been several recent studies of ethnomathematics, a subdiscipline devoted to the mathematics of pre-industrial, non-European cultures.[4]

To appreciate the contrast, consider the situation just twenty years ago. Earlier standard histories, classics like Thomas Heath's compendium study of Greek mathematics or Carl Boyer's sweeping account covering mathematics from its beginnings to the early decades of the twentieth century, presented a unified image of mathematics that nicely supplemented the material found in numerous textbooks on Western Civilization written primarily for North American undergraduate students.[5] These works exemplify the traditional cultural approach to the history of

mathematics championed by the German historian of mathematics, Moritz Cantor.[6]

Another influential, but strongly mathematical approach to the history of mathematics grew out of the collective efforts of the Bourbaki group.[7] Bourbaki's historical work reflected an image of mathematics closely tied to the group's mathematical interests—interests which sought to unearth the key structures and "mother structures" that reveal the underlying unity of *pure* mathematics. Thus, whereas the distinctly modern subjects of algebra and topology garnered considerable attention, less streamlined fields rooted in classical analysis, numerical methods, and statistics were either pushed to the periphery of Bourbaki's unified structural picture or simply never appeared in it at all.

This view of mathematics dominated in the 1960s, a situation that prompted Morris Kline's attempt to "set the record straight" in his *Mathematical Thought from Ancient to Modern Times,* published in 1972.[8] Whatever one may think of Kline's other works or his general views on the state of mathematics and mathematics education in the United States,[9] this impressive volume certainly gave the calculus, differential equations, and classical mechanics their rightful due and, at the same time, established Kline's credentials as a general authority on the history of mathematics from an internalist perspective.

Yet, whereas Kline extended the picture of mathematical knowledge and provided a more balanced account of its thematic development, his image of mathematics could hardly be described as radically new. In retrospect, this work appears much closer to its predecessors than to the studies that have recently come into vogue in the history of mathematics. Indeed, it may well mark the end of an era dominated by holistic images in the history of mathematics. Whereas Kline sought to present a comprehensive picture of the growth of mathematical knowledge, much subsequent work has cultivated a pluralist image of the history of mathematics which has found expression in numerous specialized studies devoted to the subject.

Consider, for example, some books that have since appeared on the history of probability and statistics, a huge area that receives practically no attention in Kline's book. Ian Hacking's *The Emergence of Probability* (1975) and *The Taming of Chance* (1990) deal with philosophical problems involving determinism and chance.[10] Donald MacKenzie's *Statistics in Britain, 1865–1930: The Social Construction of Scientific Knowledge* (1981) and Theodore Porter's *The Rise of Statistical Thinking, 1820–1900* (1986) both reflect the growing interest in sociology of science, a trend that began in the 1970s.[11] Lorraine Daston's

Classical Probability in the Enlightenment (1988) offers a sparkling account of how eighteenth-century thinkers came to grips with the theological implications of their probabilistic conceptions.[12] Finally, for more technical accounts one can turn to Anders Hald's *A History of Probability and Statistics and their Applications before 1750* (1990) and Stephen M. Stigler's *The History of Statistics— The Measurement of Uncertainty before 1900* (1986).[13] Despite their divergent approaches, all of the above works have received strong critical acclaim.[14]

These brief remarks should suffice to make the point. Here, in a single area of mathematics, one encounters not just a long list of substantial new works but a wide variety of styles, as the authors often take radically different tacks in dealing with their chosen subject matter. One encounters the same pluralist pattern in other areas and *genres* as well. Consider the following (again only partial) list of recent biographical studies of mathematicians: Michael Mahoney, *Pierre Fermat*; Joan Fisher Box, *R. A. Fisher*; Joseph Dauben, *Georg Cantor*; Thomas Hankins, *William Rowan Hamilton*; Richard S. Westfall, *Isaac Newton*; Ann Hibner Koblitz, *Sofia Kovalevskaya*; Andrew Hodges, *Alan Turing*; Jesper Lützen, *Joseph Liouville*; Pesi R. Masani, *Norbert Wiener*; William Aspray, *John von Neumann*; and Carol Parikh, *Oscar Zariski*.[15]

These studies reflect a broad range of themes and interests connected with the lives and work of prominent figures in the history of mathematics. For example, Dauben draws out the central role of philosophical ideas for Cantor's set theory, and Hankins does likewise for Hamilton's quaternions. In examining the creative achievements of Fermat and Newton, Mahoney and Westfall shed considerable light on the foremost methodological concerns of seventeenth-century mathematicians. In dealing with more recent subjects, Koblitz, Fisher, Hodges, Aspray, and Parikh focus more directly on the personal lives and professional roles of their respective protagonists, whereas Lützen and Masani strike a balance between the biographical and technical mathematical elements.

Noteworthy, too, are the rich primary source materials that inform these studies. Westfall's in many respects definitive biography of Newton drew heavily on the splendid edition of Newton's collected papers edited by Derek T. Whiteside.[16] The portraits of Cantor, Hamilton, Kovalevskaya, and Liouville drawn by Dauben, Hankins, Koblitz, and Lützen made essential use of unpublished European archival sources. Primary sources need not always be in written form. Oral interviews also greatly enriched the source material utilized in the biographies of Fischer, Turing, von Neumann, and Zariski published by Box, Hodges,

Aspray, and Parikh. Oral history has, in fact, emerged as a major subdiscipline in historical studies over the last three decades.

One could make similar observations about how other recent works have contributed to a new pluralist perspective. Much current research has taken the form of thematic studies devoted to specialized concepts or fields of mathematics. Among these there have been important contributions to axiomatic set theory (Gregory Moore's *Zermelo's Axiom of Choice: Its Origins, Development and Influence*)[17], discontinuous groups and automorphic functions (Jeremy Gray's *Linear Differential Equations and Group Theory from Riemann to Poincaré*), and manifold theory (Erhard Scholz's study)[18]. Thomas Hawkins's articles on the contributions of Sophus Lie, Wilhelm Killing, and others to the theory of Lie groups and Lie algebras have likewise shed considerable light on early developments in this central branch of modern mathematics.[19] Institutional histories that examine the structure and function of particular mathematical communities have also come to the fore recently, for example, Kurt Biermann's study of the Berlin school, Ivor Grattan-Guinness's analysis of Parisian mathematics from 1800-1840, and the account by Karen Parshall and David Rowe of the American research community in its infancy.[20] Perhaps the most striking evidence of the scope, diversity, and range of recent historical work can be found in the numerous volumes of encyclopedia articles, collected essays, conference and symposia proceedings, and *Festschriften* that have been published during the last decade, including the MAA volume edited by Esther Phillips that appeared in 1987.[21]

The above sampling suggests that the history of mathematics has entered a dynamic new phase of development. Yet, despite this virtual explosion of activity in the field, the present state of the discipline can hardly be described as unproblematic. One of its prime difficulties stems from the divergent interests represented by the two principal groups that have traditionally sustained it in the past, the cultural and mathematical historians. Some professional historians of mathematics have expressed concern that their field has become increasingly isolated from the spheres of interest that concern both of these other groups. Recently, Ivor Grattan-Guinness described the history of mathematics as "a classical example of a ghetto subject"[22] that has failed to attract serious attention from either mathematicians or historians of science. Whereas the latter group disdains technical mathematics, an attitude shared by most humanists within the population at large, Grattan-Guinness contends that those who feel otherwise, the mathematicians, "...usually view history as the record of a 'royal road to me'—that is, an account of how a particular modern theory arose out of older theories instead of an account of those older theories in their own right."[23]

William Aspray and Philip Kitcher sounded a somewhat similar note in their "Opinionated Introduction" to *History and Philosophy of Modern Mathematics,* a volume published in connection with a 1985 conference on history and philosphy of mathematics held at the University of Minnesota. Although they found that the historians and philosophers present at this conference had uncovered much common ground for future collaboration, they were "less sanguine about the prospects for dialogue among historians, philosophers, and mathematicians." One reason "was the tendency for some of the mathematicians present to dismiss the work of historians and philosophers as ignorant invasions of the mathematicians' professional turf."[24]

Another complaint leveled by historians against mathematicians is that they often display an almost cavalier attitude, if not outright indifference, toward historical research. Territorial rights may be one difficulty, but the real problems blocking fruitful mutual relations between mathematicians and historians lie deeper. In order to see what these are and how they operate in a concrete situation, I now turn to reconsider the debates of the 1970s in the field of ancient mathematics.

Obviously the following account does not offer an adequate synopsis of general historiographic trends in the area of ancient mathematics. Fortunately, however, there are a number of sources that provide overviews of major trends and issues bearing on this field. For a summary of major activity related to Greek mathematics up to 1984, the reader should consult Len Berggren's informative article "History of Greek Mathematics: A Survey of Recent Research" in *Historia Mathematica.*[25] The impressive achievements of Otto Neugebauer are nicely captured in Noel M. Swerdlow's memoir written for the American Philosophical Society.[26] Finally, for a general overview of research trends in the field of Mesopotamian mathematics, Jens Høyrup's recent article "Changing Trends in the Historiography of Mesopotamian Mathematics—An Insider's View" is warmly recommended.[27]

Algebra and its Role in Ancient Mathematics

In no area have the tensions between the two principal groups of practitioners, cultural and mathematical historians, been more apparent than in studies of ancient mathematics. Although signs of strong interpretive disagreement regarding the character of ancient Greek mathematics

stretch back to the 1930s, not until the 1970s did these latent tensions come to a head. As I shall indicate below, by then the debate had extended beyond ancient Greek to Babylonian mathematics, and the methodological issues carried clear historiographic implications for mathematics in general.

The standard interpretation of classical Greek mathematics up until the 1970s can be traced to the Danish algebraic geometer, H.G. Zeuthen.[28] Zeuthen contended that large portions of Euclid's *Elements*—Book II, the theorems on applications of areas in Books I and VI, and parts of the theory of incommensurable magnitudes in Book X—as well as many of the propositions found in Euclid's *Data* derived from an older, essentially algebraic corpus of knowledge. According to this view, Greek mathematicians drew upon this older tradition in developing a "geometric algebra" designed to replace the conventional Pythagorean approach for handling ratios and proportions of whole numbers. Following the discovery of incommensurable magnitudes during the fifth century B.C., Greek mathematicians realized that this purely number-theoretic approach could no longer serve as a tenable framework for geometry. This realization, in Zeuthen's estimation, led them to replace Pythagorean "pebble arithmetic" with a new kind of algebra in which general geometric magnitudes were represented by line segments. The product of two line segments, however, was no longer a line but rather was the rectangle formed by the two given sides, an operation employed over and again not only by Euclid, but by Archimedes and Apollonius as well.

This interpretive framework seemed to help explain some of the more perplexing portions of ancient Greek mathematics. In particular, it suggested a motivation for those parts of Euclid's *Elements*, such as most of Book II, which appeared practically devoid of any geometric content. The first ten propositions of Book II could now be seen as nothing more than a series of useful and familiar algebraic identities (such as Proposition II.4, which is mathematically equivalent to $(a + b)^2 = a^2 + 2ab + b^2$). More interesting still, other results, like Propositions II.11 and VI.27-29, could be interpreted as Greek methods for solving special types of quadratic equations by the techniques of geometric algebra.

This new angle on Greek mathematics soon became the standard interpretation and was adopted by such authorities as Thomas L. Heath and Paul Tannery.[29] The only substantial criticism appears to have come from Jacob Klein. In the 1930s, Klein published a lengthy study dealing with the evolution of Greek concepts of number and ratio from Plato to Diophantos, notions he found served as the point of departure for the conceptual innovations of later thinkers like François Viète, René Descartes, and Simon Stevin.[30] Klein's arguments, however, apparently fell on deaf ears until the late 1960s and early 1970s, perhaps because of the author's heavy philosophical and philological orientation.

In the meantime, a dramatic breakthrough occurred in historical research dealing with Babylonian mathematics.[31] Beginning in the late 1920s and early 1930s, Otto Neugenbauer and François Thureau-Dangin began to decode and study substantial numbers of cuneiform tablets, most stemming from the Old Babylonian period (2,000–1,600 B.C.). The results, first published by Neugebauer in the mid-1930s, proved nothing less than spectacular, as the analyses that accompanied these texts revealed sophisticated techniques for the manipulation of numerical relationships expressed by means of ideograms and sexagesimal numbers.[32] In particular, Neugebauer showed how Babylonian results could be verified by comparing them with modern formulations in algebra. Although he was cautious about overplaying the "algebraic" character of Babylonian mathematics (and generally placed such terms in quotation marks), in his more popular works Neugebauer tended to deemphasize such fine points. As Jens Høyrup has described the situation, Neugebauer's enthusiasm for this new enterprise apparently led to a slackening of his otherwise keen critical acumen, and in some of his analyses it becomes "impossible to distinguish *justification through* from *interpretation as* algebra."[33]

Once this distinction was blurred, momentum gained for a simple, but compelling, image of Babylonian mathematics that stressed its essentially algebraic character. In his *elogé* of Neugebauer, Noel Swerdlow has emphasized how his work reflected the deep tension that underlies the two principal directions of research in the history of mathematics:

At once a mathematician and cultural historian, Neugebauer was from the beginning aware of both interpretations and of the contradiction between them. Indeed, a notable tension between the analysis of culturally specific documents, whether the contents of a single clay tablet or scrap of papyrus or an entire Greek treatise, and the continuity and evolution of mathematical methods regardless of ages and cultures is characterisitic of all his work. And it was precisely out of this tension that was born the detailed and technical cross-cultural approach, in no way adequately described as the study of "transmission," that he applied more or less consistently to the

history of the exact sciences from the ancient Near East to the European Renaissance.

But if the truth be told, on a deeper level Neugebauer was always a mathematician first and foremost, who selected the subjects of his study and passed judgment on them, sometimes quite strongly, according to their mathematical interest.[34]

Neugebauer was operating in his dominant mode, that of the mathematical historian, when in 1936 he put together some of the essential pieces for a whole new picture of ancient mathematics. At its center stood the once mysterious, but never really controversial "geometric algebra" of the Greeks. Where did it come from? In Neugebauer's words:

The answer to the question what were the origins of the fundamental problem in all of geometric algebra [namely, the application of areas, as in Euclid II.44, VI.27-29] can today be given completely: they lie, on the one hand, in the demands of the Greeks to secure the general validity of their mathematics in the wake of the emergence of irrational magnitudes, on the other, in the resulting neccessity *to translate the results of the pre-Greek "algebraic" algebra as well.*

Once one has formulated the problem in this way, everything else is completely trivial and provides *the smooth connection between Babylonian algebra and the formulations of Euclid.*[35]

This view, which complemented Zeuthen's interpretation of Greek geometric algebra so nicely, quickly emerged as the standard position among historians of mathematics. Neugebauer reverted to a more cautious attitude in the six brilliant lectures he delivered at Cornell in 1949, later published under the title *The Exact Sciences in Antiquity*.[36] Others, however, took the accepted view on faith and proceeded to elaborate on it. Probably the single most influential work that followed in this direction was B. L. van der Waerden's *Science Awakening*, first published in Dutch, but which appeared in English translation in 1962.[37]

Not all made their peace with the new orthodoxy, however. Hungarian philologist-historian Arpád Szabó, for one, found the argument for such an early transmission of mathematical knowledge from Mesopotamia into Greece tenuous, and Michael Mahoney questioned the blanket characterization of Babylonian mathematics as algebraic.[38] These criticisms, however, seem not to have caused any substantial concerns at the time. Such polite skepticism did not survive for long, though. In 1975, the *Archive for History of Exact Sciences* published Sabetai Unguru's sweeping critique of the whole corpus of scholarship produced by leading historians of ancient mathematics, but taking van der Waerden's work especially hard to task.[39] Unguru directed much of his attack at the long-established theory of Greek geometric algebra, a notion he denounced as nothing but a sheer fantasy, a "monstrous, hybrid creature"[40] concocted by mathematicians who lacked any feel for history.

This frontal assault did not go unanswered and, fittingly, van der Waerden was the first to reply.[41] In doing so, he attempted to spell out clearly what he, Neugebauer, *et al.* meant by the word "algebra," since Unguru had so resolutely resisted the possibility of applying the term to either Greek or Babylonian mathematics. Unguru sought an opportunity to reply, but the *Archive's* editor, Clifford Truesdell, refused this request on the grounds that he did not wish to see his journal turn into a forum for disputes.[42] A year later, however, the journal carried a sharply-worded rejoinder to Unguru's position, written by the eminent Dutch mathematician, Hans Freudenthal, and shortly after that an open letter from the even more eminent Princeton mathematician, André Weil, containing a series of remarks that can only be described as venomous.[43] Its conclusion reads:

> when a discipline, intermediary in some sense between two already existing ones (say A and B) becomes newly established, this often makes room for the proliferation of parasites, equally ignorant of both A and B, who seek to thrive by intimating to practioners of A that they do not understand B, and vice-versa. We see this happening now, alas, in the history of mathematics. Let us try to stop the disease before it proves fatal.[44]

While he gave no indication of how widespread this "disease" had become, it should be noted that this was not the first time Weil had voiced profound misgivings regarding recent work of historians of mathematics. Five years earlier, he had published a vicious attack on Mahoney's biography of Fermat, clearly seeking to demonstrate that the author was simply incompetent.[45] Thus, the broader implications for history of mathematics as a discipline had been clearly drawn: "let no one unversed in modern mathematics enter these doors."

Throughout the three-year period that followed the publication of Unguru's 1975 article, Truesdell refused to allow him to answer his critics in the *Archive*. This forced Unguru to turn to the resources of *Isis*, the official journal of the History of Science Society, in order to defend his position. In replying to his critics, Unguru portrayed the two contrasting methodological positions in their sharpest possible relief. He began by asserting that

> most contemporary historians of mathematics, being mathematicians by training, assume tacitly or explicitly that mathematical ideas reside in the world of Platonic ideas where they wait patiently to be discovered by the genius of the working mathematician. ... Various forms of the same mathematical concept or operation are not considered merely mathematically equivalent but also historically equivalent.[46]

Citing Aristotle, Unguru argued that history is concerned with the "idiosyncratic rather than the nomothetic."[47] The historian aims to study "the event *qua* particular event," and to understand "each past event in its own right."[48] The same principle, he claimed, ought to apply for the history of mathematics, which should

> study ... the idiosyncratic aspects of the activity of mathematicians who themselves are engaged in the the study of the nomothetic, that is, of what is the case by law. If one is to write the history of mathematics, and not the mathematics of history, the writer must be careful not to substitute the nomothetic for the idiosyncratic, that is, not to deal with past mathematics as if mathematics had no past beyond trivial differences in the outward appearance of what is basically an unchangeable hard-core content.[49]

This 1979 article by Unguru was the last word, at least in print, dealing with these particular historiographic issues.[50] Not long before it appeared, however, André Weil spoke to his fellow mathematicians about mathematics and its history as well as the nature of the relationship between mathematicians and historians.

Bourbaki and the Platonists

In the wake of the debates over the notion of Greek geometric algebra, Weil drew some general conclusions from this episode pertaining to the "Why and How" as well as the "For Whom" of writing about the history of mathematics. In a 1978 lecture delivered at the International Congress of Mathematicians held in Helsinki, Weil indicated why the notion of a hidden algebraic agenda underlying Greek geometry posed no substantive difficulties for mathematicians.[51] Brushing the historiographic issues aside as trivial, Weil remarked:

> when quadratic equations, solved algebraically in cuneiform texts, surface again in Euclid, dressed

up in geometric garb without any geometric motivation at all, the mathematician will find it appropriate to describe the latter treatment as "geometric algebra" and will be inclined to assume some connection with Babylon, even in the absence of any concrete "historical" evidence. No one asks for documents to testify to the common origin of Greek, Russian and Sanskrit, or objects to their designation as indo-european [sic] languages.[52]

Weil went on to assert that only those possessing a knowledge of mathematics that extends "far beyond [the] ostensible subject matter"[53] of a given era can reach an in-depth understanding of the achievements of that period. With this kind of knowledge, one can obtain "a firm grasp on the ideas which are 'in the air', some of them flying all around us, some (to quote Plato) floating around in our own minds."[54] As a first example, Weil mentioned the theories of ratio and proportion presented in Book V (for general magnitudes) and Book VII (for natural numbers) in Euclid's *Elements*. Regarding this body of mathematics, Weil contended that "it is impossible for us to analyze properly the contents of Books V and VII of Euclid without the concept of group and even that of groups of operators, since the ratios of magnitudes are treated as a multiplicative group operating on the additive group of the magnitudes themselves."[55] Moreover, he suggested that this viewpoint not only clarifies the otherwise seemingly mysterious nature of the Euclidean presentation but it also enables one to follow easily "the line which leads directly from [Euclid] to Oresme and Chuquet, then to Neper and logarithms."[56]

Two points ought to be made here. First, the practical advantages of Weil's approach seem clear and indubitable. According to his perspective, the "true object"[57] of history of mathematics ought to be mathematical ideas, and since active mathematicians constitute the group most competent to judge such ideas, the issue "for whom" would appear settled, at least for Weil. Most working mathematicians, however, have neither the time nor the inclination to study dated work in the original. By training they have learned to study mathematical texts by skimming through abstracts, and, if actually forced to read a paper (usually a preprint), then they know how to get quickly to the heart of the matter. Confronted with this unpleasant task—for how many mathematicians particularly enjoy reading the terse, formalized, and largely impenetrable stuff that (at least until recently) has constituted the only acceptable form for mathematical writing?—the reluctant reader approaches the text in hand with a "what's really going on here?" frame of

mind. Call it a tool of the trade or simply a survival skill, this attitude nonetheless serves as the mathematicians' principal mode of interaction with contemporary mathematical literature. Whether they know any science or not, most mathematicians belong to a scientific and not to a humanistic culture.[58] (Obviously there are exceptions; Andr, Weil must be considered a full-fledged member of both.)

These circumstances clearly carry implications for the manner in which mathematicians approach a scholarly book or article dealing with the history of mathematics. For such readers—Weil's exclusive audience—the spokesman for Nicolas Bourbaki can offer a "royal road" into the past. Along the way, the traveler will be guided by a leading expert who knows which signposts need to be pointed out in conducting the journey. And if the signs themselves do not quite suffice, the guide will indicate what ought to be there, even if for some incidental reasons—call it historical accident—they cannot, or can no longer, literally be seen. What mathematician with even an inkling of interest in history would want to pass up a tour like this? Pity the poor chap who got stuck with an historian for a guide; the trip probably took twice as long to cover a tenth of the same terrain, and the guide's account mostly consisted of numerous boring and superfluous details.

No one, it seems to me, ought to dismiss the practical advantages of Weil's approach to history. True, it produces a highly rationalized picture of past achievements, not to mention a contracted image of how mathematics developed, but it also addresses the interests of the audience for which it is intended. Moreover, the historical studies of Weil, Dieudonn,, and others in the Bourbaki tradition represent serious and very valuable contributions to scholarship.[59] Clearly these works must be sharply differentiated from more popular ones, like E.T. Bell's *Men of Mathematics* or the biographies of Hilbert, Courant, and Neyman written by Constance Reid,[60] books which also fulfill an important purpose by illustrating the human side of mathematical activity.

Thus, my first point is that Weil's approach has definite merits which ought to be acknowledged even by those historians of mathematics whose work happens to be informed by radically different methodological assumptions. Still, the limitations of this purely mathematical approach—and this brings me to my second point—should be evident to those with some sensitivity for historical authenticity, that is, a critical interest in and respect for the integrity of past events. As anyone who has worked intensively with historical sources can testify, the more one learns about the past, the more complicated it appears.[61]

When Weil asserts that, with a little knowledge of group theory, the content of Euclidean proportion theory (and much else besides) becomes transparent, his goal is very different from the task of accounting for the complexities that appear in Books V and VII of Euclid's *Elements*. This body of mathematics poses subtle difficulties for our understanding of the ancient Greek notions of number, magnitude, and ratio, and their mutual relations, difficulties that continue to puzzle experts. Weil's approach seeks to pass over such complexities in order to make certain connections between the structures he finds in both ancient and modern mathematics. Why? As I have already indicated, if the answer to this were that he wants to meet the needs of his principal audience—mathematicians with an interest in history—then this would certainly be a reasonable justification for taking even such drastic shortcuts.

The problem, however, lies elsewhere and has little to do with finding fault in the historical writings of Weil, Dieudonn,, and others of similar persuasion. One can present the development of mathematical ideas as a steadily unfolding search for Platonic truths that transcend the particular cultural contexts in which these ideas arose, but only by discounting the rich variety of meanings that accompanied this work. Once mathematical knowledge is regarded as context-free and its "extra-mathematical" meanings are shorn away, it becomes far more accessible to us and yet drastically removed from the setting that originally prompted these earlier mathematical investigations. A purely Platonist approach thus ignores the specific intentions and motivations that have inspired mathematical creativity throughout time. Its fundamental assumption rests on an inherently timeless conception of mathematical ideas that somehow manifest themselves in different forms within various cultures. To maintain such a view implies, at the very least, a strong affirmation of the superiority of modern ideas over those of the past. For only from a modern, abstract perspective can one discover the sorts of structural similarities that Weil finds when he examines the proportion theory of Euclid's *Elements*. By interpreting past mathematical achievements as mere precursors to the mathematical ideas dominant in the 1950s and 1960s, Weil strongly suggests that only those who accept the latter canon as the one true perspective for mathematical knowledge can appreciate the "higher" truths our ancestors groped for but could never attain.

Historians and philosophers may be inclined to question whether mathematical concepts carry unambiguous meanings independent of the cultural setting that produced them, but André Weil is evidently not only persuaded of this, he also believes that with the help of modern algebra, et cetera, he and other talented mathematicians possess the keys needed to unlock many puzzling problems in the history of mathematics. It disturbs many historians of mathematics that such a prominent and influential mathematician should refuse to recognize the built-in limitations of this particularly abstract and ethereal brand of history. Sabetai Unguru has even asserted that this once dominant style in the history of mathematics should not be regarded as history at all. In his view, "...if scholars continue to neglect the peculiar specificities of a given mathematical culture, whether as a result of explicitly stated or implicitly taken-for-granted assumptions, then by definition their work is ahistorical and should be recognized as such by the community of historians."[62]

Thus, like Weil, Unguru appeared willing to entertain the possibility of simply drawing a sharp disciplinary line of demarcation between the cultural historians of mathematics (the "real" historians according to Unguru) and the mathematical historians. Luckily, no such barrier was officially erected, which is only to say that this open controversy subsided soon afterward. Clearly, such a radical severing of the discipline would have served no one's interests.

One could hope that a larger perspective on these historiographical matters might be gained by reaching a broad consensus about the nature of mathematical inquiry and mathematical knowledge as well as of historical knowledge. Just as mathematicians typically draw on thinking processes that enable them to find their way amid highly abstract theories and complicated formalisms, so, too, do historians rely on certain fundamental thought patterns which form part of a sensibility that instinctively "knows" what it means to think historically. What makes a good historian? While opinions about this obviously differ, most experts would certainly agree that an important attribute is a capacity to approach the past on its own terms, as something worth knowing about in and of itself.

This perspective was often absent in studies of the history of science prior to 1950. After the appearance of Herbert Butterfield's influential *The Origins of Modern Science*, however, which decried the evils of "Whig history," this principle has come to be regarded as axiomatic within the field.[63] Historians of mathematics have largely followed suit in this regard. Most do not view their primary task as showing how past achievements were absorbed into a more familiar body of modern knowledge that can be regarded as intrinsically better based on considerations of precision, generality, or elegance. Such issues, while perhaps of interest to mathematicians or historians of ideas, generally fall outside the domain of concerns that occupy specialists in the history of mathematics.

The very different perspective set forth by André Weil clearly reflects his deeply held views as one of the premier mathematicians of the century.[64] Weil's intolerance for other approaches to the history of mathematics surely stems from a Bourbaki tradition that has sought to enshrine a particular image of mathematics. Bourbaki's purist conception of mathematics, with its strong Platonistic overtones, has exerted, to be sure, a profound impact on twentieth–century mathematical thought. Modern mathematics would be simply unthinkable without Bourbaki. Yet, a sober appraisal of the strengths and weaknesses in the conception of mathematics championed by Bourbaki would seem vital if we are to reach a deeper understanding of the character of modern mathematics. And now that the Bourbaki phenomenon is receding into the past, a few historians are beginning the delicate task of separating myth from reality in the many lives of Nicolas Bourbaki.[65]

A central part of the Bourbaki legacy is an image of mathematical knowledge that has enjoyed widespread currency among mathematicians.[66] Bourbaki's leading members surely felt they had a solid grasp of what mathematics was about: the true essence of mathematics, that is, and not the dull and trivial forms of mathematics that occupy lesser lights, not to mention the messy results produced by mathematicians who work with computers and other machines. Philosophical nit-picking never interested Bourbaki much either, and in this regard André Weil was no exception.[67] In his Helsinki lecture, Weil merely let experience and authority speak for itself when it came to defining mathematical ideas. Declining "to consult outsiders" for advice (presumably he had historians in mind), he adapted his words to A.E. Housman's reply when the latter was asked to define poetry: "[the mathematician] may not be able to define what is a mathematical idea, but he likes to think that when he smells one he knows it."[68]

Perhaps the most disturbing consequence of Weil's pronouncements regarding what constitutes "good" history of mathematics is that his opinions reinforce and harden prevailing attitudes about history and historians within the mathematical community. The Platonist viewpoint Weil espouses has been the accepted orthodoxy for many years, not just among mathematicians but also in the philosophy of mathematics as well.[69] For mathematicians, it tends to translate into the notion that mathematical ideas lead lives of their own and therefore need only to be discovered. "To be a mathematician," according to the algebraic geometer David Mumford, "is to be an out-and-out Platonist" and he claimed further that:

the more you study mathematical constructions, the more you come to believe in their objective and prior existence. Mathematicians view themselves as explorers of a unique sort, explorers who seek to discover not just one accidental world into which they happen to be born, but the universal and unalterable truths of all worlds.[70]

Reuben Hersh has criticized the typical working mathematician for being a "Platonist on weekdays [when it matters] and a formalist on Sundays [when it does not]."[71] In other words, in reflecting about the ontological status of mathematical concepts, many mathematicians support the formalist position, first set forth in 1899 in David Hilbert's *Grundlagen der Geometrie*.[72] Having imbibed modern axiomatics, the "ideal mathematician"[73] repeats the standard catechism that "points," "lines," and "planes" do not really exist but rather form undefined elements in a system. One often says they are defined implicitily, meaning that their status is established solely in relation to the given axioms, which are conventions. But since philosophical reflections such as these seldom have any bearing on the conceptual world of the workaday mathematician, he or she often slips into a very different mode of thinking in which mathematical objects become real, sometimes even living, things (like functions that "live and breathe" in some abstract "space").

Even if a mathematician professes not to be Platonist, one should exercise considerable caution before accepting him at his word. Saunders Mac Lane, who insists that his views are "in sharp contrast to all variants of Platonism,"[74] refers to the variety pinpointed by Hersh and described above as "mythological Platonism."[75] He further suggests, in contrast to Mumford, that this type of Platonism need not be taken very seriously since the mathematicians who adopt it have little or no interest in philosophical matters anyway. In *Mathematics—Form and Function* (1986), Mac Lane argues that "a philosophy of Mathematics [significantly he spells mathematics with a capital "M" throughout this book] is not convincing unless it is founded on an examination of Mathematics itself."[76] For this reason, he dismisses the philosophical ideas of Ludwig Wittgenstein, and proceeds to offer his own account of what "Mathematics *is*"[77] (his italics). Although the picture Mac Lane portrays in his book contains much interesting mathematics, the image of mathematics he presents is static and frozen in time, as if what came before either did not matter or else had somehow found its niche in the present structure.

Thus, whereas Mac Lane claims to take a fresh new approach to the philosophy of mathematics, in effect he only props up the old Bourbaki image—the vision of Mathematics as essentially reducible to the forms and structures of modern pure mathematics—without appealing to an independent ontological Platonism. Taking this tack, the kinds of issues that would incorporate an historical or contextual dimension remain just as absent as before. Thus, it would seem that a very different kind of philosophical approach is needed in order to engender fruitful dialogue between historians and mathematicians. Happily, some recent trends in philosophy of mathematics suggest promising possibilities in this direction.

Recent Philosophical Trends

Perhaps the most important recent work aimed at breaking down rigid stereotypes about mathematics has been Philip J. Davis's and Reuben Hersh's delightful romp, *The Mathematical Experience.*[78] Written in a light, satirical style, this book has reached a wide audience that extends well beyond the mathematically literate. While underscoring the mythic elements "underneath the fig leaf," David and Hersh attempt to demystify mathematics by indicating how mathematical knowledge can be understood as a kind of intellectual activity.[79] To accomplish this, they pierce through the facade of jargon that separates the world of mental constructs employed by mathematicians from the mundane conditions that shape the actual context in which mathematical work is done.

Much of the inspiration that led to *The Mathematical Experience* can be traced to the work of the brilliant mathematician and pedagogue, George Polyá, and the closely related philosophical ideas of Imre Lakatos.[80] In his 1954 classic, *Induction and Analogy in Mathematics,* Polyá highlighted the role of "guesswork" in the process of mathematical discovery, an aspect sharply at odds with prevailing views regarding the nature of mathematical knowledge.[81] Polyá's empiricist approach was extended by Lakatos, who applied it not only to explain the process of discovery but mathematical arguments as well. Lakatos' famous *Proofs and Reputations,* written in the semi-popular style of a dialogue, appeared posthumously in 1976.[82] It has since spurred a major reorientation within the philosophy of mathematics.

The empiricist direction explored by Polyá and Lakatos, while still constituting a minority position, has gained considerable momentum among philosophers of mathematics during the last decade.[83] This shift stems in part from a growing sense that during the last fifty years the field had drifted too far from ideas and issues connected with mathematics. In an effort to begin reestablishing these

ties, Philip Kitcher, a leading exponent of this "maverick tradition," has challenged historians and philosophers of mathematics to begin examining more closely questions like: "How does mathematical knowledge grow? What is mathematical progress? What makes some mathematical ideas (or theories) better than others? What is mathematical explanation?"[84] These and similar questions cannot be answered simply in the abstract and may provide promising points of departure for collaborative studies between historians and philosophers of mathematics. Indeed, several signs of a convergence of interests between these two disciplines can already be discerned, beginning with the Aspray and Kitcher volume, *History and Philosophy of Modern Mathematics,* alluded to above.[85]

Another promising new avenue for research in the history of mathematics has been opened through the synthesizing work of Herbert Mehrtens, particularly his recent study *Moderne—Sprache—Mathematik.*[86] Inspired by a combination of interests ranging from social and cultural history to semiotics and the work of Michel Foucault, Mehrtens offers a global analysis of the foundations debates that lurked within the mathematical community for decades but shook it—particularly the German community—dramatically after World War I. Part of the novelty of Mehrtens's approach lies in the way he embeds the foundations debates in mathematics within the entire context of a European society straining to cope with the rapid changes of modernity. By placing the mathematical issues that divided antagonists like Hilbert and L.E.J. Brouwer within a larger cultural and disciplinary context, Mehrtens offers a provocative new interpretation of the deeper issues at stake that galvanized the rival parties in this conflict.

In one sense, Mehrtens's approach parallels that taken by Davis and Hersh in *The Mathematical Experience.* Both books engage the subtle myth-making process that seems endemic to the mental worlds inhabited by mathematicians since the ancient Pythagoreans. Part of Mehrtens's argument suggests that the cultural pessimism of Weimar Germany played a decisive role in the foundations debates and that the virulence engendered by the so-called foundations crisis had less to do with the ideas themselves than with the implications these carried for the conduct of research in mathematics. The ideas served as masks worn by the principal players in a long-standing power struggle that ended when Hilbert summarily dismissed Brouwer from the editorial board of *Mathematische Annalen.* Many readers will undoubtedly find Mehrtens's forays into psycho-sexual analyses of mathematical ideas far-fetched or his enthusiasm for semiotics problematic. Nevertheless, his book contains a wealth of new ideas that have not only generated considerable excitement (including agitation) among historians of mathematics but which also suggest promising avenues for future investigations.

Of course, this brief synopsis cannot possibly do justice either to the works discussed above or to the larger directions in research associated with them. My aim has been merely to provide a sense of the extent to which historical research has been steadily moving away from the cultivation of a monolithic image of mathematics as found in many well-known studies produced before 1970. In my estimation, the trend toward investigations informed by a variety of methodologies and set firmly in a socio-cultural context seems likely to grow stronger in the years ahead. Clearly, mathematicians will bring their own diverse interests and concerns to this endeavor, just as they have in the past, and it is critical that they do so. For to be successful the new history will have to be a common effort, a truly multi-disciplinary undertaking with the capability of tapping into the rich resources of human expertise that transcend the capacities of any one group.

Acknowledgements. The author wishes to thank Moritz Epple and Sabetai Unguru for their comments on an earlier draft of this paper and, especially, Ronald Calinger, for his numerous suggestions for improving both the style and content. Naturally, I take full responsibility for any errors of opinion or fact expressed herein.

References

1. For an interesting account of Zeuthen's historiographic orientation and the contrasting views of his contemporary and the other leading historian of the period, see Jesper Lützen and Walter Purkert, "Conflicting Tendencies in the Historiography of Mathematics, M. Cantor and H. G. Zeuthen," in Eberhard Knobloch and David E. Rowe, eds., *The History of Modern Mathematics,* vol. 3 (Boston: Academic Press, 1994), pp. 1–42.
2. Naturally, these categories overlook many important distinctions within these two camps. A more detailed analysis, for example, would have to take major trends in the history of science into account. This would include the influential studies of Alexandré Koyré and Thomas Kuhn which inspired much of the work done during the 1960s and 1970s. Another strong impulse can be traced to the French deconstructionists, Jacques Derrida and Michel Foucault. Many of the more recent trends in the history of science, however, appear to have affected the history of mathematics, at most, tangentially, for reasons that are too complex to be dealt with here. Dirk Struik's ever popular survey, *A Concise History of Mathematics,* 4th ed. (New York: Dover, 1987), reflects his Marxist orientation, but still falls comfortably within the mainstream Cantorian tradition of cultural history of mathematics. For a consideration of Kuhnian ideas in the history of mathematics, see Donald Gillies, ed., *Revolutions in Mathematics* (Oxford: Clarendon Press, 1992) and my review of the same in *Historia Mathematica,* 20(1993) 320–323.
3. For example, Li Yan and Du Shiran, *Chinese Mathematics: A Concise History,* trans. J.N. Crossley and A.W.-C. Lun (Oxford: Clarendon Press, 1987); John L. Berggren, *Episodes in the Mathematics of Medieval Islam* (New York: Springer-Verlag, 1986); and Michael P. Closs, ed., *Native American Mathematics* (Austin: University of Texas Press, 1986). For an overview, see Victor J. Katz's textbook, *A History of Mathematics* (New York: HarperCollins, 1993).
4. See, for example, Marcia Ascher, *Ethnomathematics: A Multicultural View of Mathematical Ideas* (Pacific Grove, CA: Brooks/Cole Publishing Co., 1991); and George Gheverghese, *The Crest of the Peacock: Non-European Roots of Mathematics* (New York: St. Martin's Press, 1991).
5. Thomas Heath, *A History of Greek Mathematics,* 2 vols. (Oxford: Clarendon Press, 1921); Carl. B. Boyer, *A History of Mathematics* (New York: John Wiley & Sons, 1968).
6. A still standard study is Moritz Cantor, *Vorlesungen über Geschichte der Mathematik,* 4 vols. (Leipzig: Teubner, 1880–1908).
7. Nicolas Bourbaki, *Éléments d'Histoire des Mathématiques* (Paris: Massou, 1984); and Jean Dieudonné, *Abrégé d'Histoire des Mathématiques, 1700–1900,* 2 vols. (Paris: Hermann, 1978).
8. Morris Kline, *Mathematical Thought from Ancient to Modern Times,* 2nd ed. (New York: Oxford University Press, 1991).
9. Other works include Morris Kline, *Mathematics in Western Culture* (New York: Oxford University Press, 1953); *Mathematics and the Physical World* (New York: Thomas Y. Crowall, 1959); *Mathematics, A Cultural Approach* (Reading, MA: Addison-Wesley, 1962).
10. Ian Hacking, *The Emergence of Probability* (Cambridge: Cambridge University Press, 1975); Ian Hacking, *The Taming of Chance* (Cambridge: Cambridge University Press, 1990).
11. Donald MacKenzie, *Statistics in Britain, 1865–1930: The Social Construction of Scientific Knowledge* (Edinburgh: Edinburgh University Press, 1981); Theodore M. Porter, *The Rise of Statistical Thinking, 1820–1900* (Princeton: Princeton University Press, 1986).
12. Lorraine Daston, *Classical Probability in the Enlightenment* (Princeton: Princeton University Press, 1988).
13. Anders Hald, *A History of Probability and Statistics and their Applications before 1750* (New York: John Wiley & Sons, 1990); Stephen M. Stigler, *The History of Statistics—The Measurement of Uncertainty before 1900* (Cambridge, MA: Harward University Press, 1986).
14. Ivo Schneider, Essay Review of Lorraine Daston, *Classical Probability in the Enlightenment,* Theodore M. Porter, *The Rise of Statistical Thinking, 1820–1900,* Stephen M. Stigler, *The History of Statistics—The Measurement of Uncertainty before 1900, Historia Mathematica,* 18 (1991) 67–74.
15. Michael Mahoney, *The Mathematical Career of Pierre de Fermat, 1601–1665* (Princeton: Princeton University Press, 1973); Joan Fisher Box, R. A. Fisher. The Life of a Scientist (New York: John Wiley & Sons, 1978); Joseph Dauben, *Georg Cantor: His Mathematics and Philosophy*

of the Infinite (Cambridge, MA: Harvard University Press, 1979); Thomas Hankins, *William Rowan Hamilton* (Baltimore: Johns Hopkins University Press, 1980); Richard S. Westfall, *Never at Rest. A Biography of Isaac Newton* (Cambridge: Cambridge University Press 1980); Ann Hibner Koblitz, *A Convergence of Lives: Sofia Kovalevskaia, Scinetist, Writer, Revolutionary* (Boston: Birkhäuser, 1983); Andrew Hodges, *Alan Turing: The Enigma* (New York: Simon & Schuster, 1983); Jesper Lützen, *Joseph Liouville 1809–1882: Master of Pure and Applied Mathematiques* (New York: Springer-Verlag, 1990); Pesi R. Masani, *Norbert Wiener 1894–1964*, (Vita Mathematica, vol. 5) (Basel: Birkhäuser, 1990); William Aspray, *John von Neumann and the Origins of Modern Computing* (Cambridge, MA: MIT Press, 1990); and Carol Parikh, *The Unreal Life of Oscar Zariski* (Boston: Academic Press, 1991).

16. Derek T. Whiteside, *The Mathematical Papers of Isaac Newton*, 8 vols. (Cambridge: Cambridge University Press, 1967–1981).

17. Gregory Moore *Zermelo's Axiom of Choice: Its Origins, Development and Influence* (New York: Springer-Verlag, 1982). For an assessment of its importance, see Thomas Drucker, Essay Review of Gregory H. Moore, *Zermelo's Axiom of Choice, Historia Mathematica*, 18 (1991) 364–369.

18. Jeremy Gray *Linear Differential Equations and Group Theory from Riemann to Poincaré* (Boston: Birkhäuser, 1986); Erhard Scholz, *Geschichte des Mannigfaltigkeitsbegriffs von Riemann bis Poincaré* (Basel: Birkhäuser Verlag, 1980).

19. See, for example, Thomas Hawkins, "Non-Euclidean Geometry and Weierstrassian Mathematics: The Background to Killing's Work on Lie Algebras," *Historia Mathematica* 7(1980) 289–342; "Wilhelm Killing and the Structure of Lie Algebras," *Archive for History of Exact Sciences* 26(1982) 127–192; "Line Geometry, Differential Equations, and the Birth of Lie's Theory of Groups," in David E. Rowe and John McCleary, eds., *The History of Modern Mathematics*, vol. 1, (Boston: Academic Press, 1988), pp. 275–327; "Jacobi and the Birth of Lie's Theory of Groups," *Archive for History of Exact Sciences* 42(1991) 187–278.

20. Kurt-R. Biermann, *Die Mathematik und ihre Dozenten an der Berliner Universität, 1810–1933* (Berlin: Akademie-Verlag, 1988); Ivor Grattan-Guinness, *Convolutions in French Mathematics, 1800–1840*, 3 vols. (Basel: Birkhäuser, 1990); Karen H. Parshall and David E. Rowe, *The Emergence of the American Mathematical Research Community, 1876–1900*, History of Mathematics, vol. 8 (Providence: American Mathematical Society, 1994).

21. See Esther R. Phillips, ed., *Studies in the History of Mathematics* (MAA Studies in Mathematics, vol. 26) (Washington, D.C.: Mathematical Association of America, 1987). Other volumes include William Aspray and Philip Kitcher, eds., *History and Philosophy of Modern Mathematics*, Minnesota Studies in the Philosophy of Science, vol. 11 (Minneapolis: University of Minnesota Press, 1988); David E. Rowe and John McCleary, eds., *The History of Modern Mathematics*, vols. 1–2, (Boston: Academic Press, 1988); Eberhard Knobloch and David E. Rowe, eds., *The History of Modern Mathematics*, vol. 3 (Boston: Academic Press, 1994); A. N. Kolmogorov and A. P. Yushkevich, eds., *Mathematics*

of the 19th Century. Mathematical Logic, Algebra, Number Theory, Probability (Basel: Birkhäuser Verlag, 1992); Ivor Grattan-Guinness, ed. *Companion Encyclopedia of the History and Philosophy of the Mathematical Sciences*, 2 vols. (London: Routledge, 1993); and Sergei S. Demidov, Menso Folkerts, David E. Rowe, and Christoph J. Scriba, eds., *Amphora. Festschrift für Hans Wussing zu seinem 65. Geburtstag* (Basel: Birkhäuser, 1992).

22. Ivor Grattan-Guinness, "Does the History of Science Treat of the History of Science? The Case of Mathematics," *History of Science* 28 (1990) 149–173, p. 158.

23. *Ibid.*, p. 157.

24. Aspray and Kitcher, *History and Philosophy of Modern Mathematics*, p. 51.

25. J. L. Berggren, "History of Greek Mathematics: A Survey of Recent Research," *Historia Mathematica*, 11 (1984) 394–410. Among the many important studies that have been recently published in this field are Wilbur R. Knorr, *The Evolution of the Euclidean Elements* (Dordrecht: Reidel, 1975); W. R. Knorr, *The Ancient Tradition of Geometric Problems* (Boston: Birkhäuser, 1986); Ian Mueller, *Philosophy of Mathematics and Deductive Structure in Euclid's Elements* (Cambridge, MA: MIT Press, 1981); David H. Fowler, *The Mathematics of Plato's Academy* (Oxford: Oxford University Press, 1986).

26. Noel M. Swerdlow, "Otto E. Neugebauer (26 May 1899—19 February 1990)," *Proceedings of the American Philosophical Society*, 137(1)(1993) 137–165.

27. Jens Høyrup, "Changing Trends in the Historiography of Mesopotamian Mathematics—An Insider's View," *Filosofi og Videnskabsteori pa Roskilde Universitetscenter* 3(3)(1991).

28. See, for example, H. G. Zeuthen, *Geschichte der Mathematik im Altertum und Mittelalter* (Copenhagen: Verlag von Andr. Fred. Höst & Sön, 1896), pp. 44–53.

29. See Thomas L. Heath, *The Thirteen Books of Euclid's Elements*, 3 vols. (New York: Dover Publications, Inc.), vol. 1, pp. 346–347, 372–374. Heath, *History*, vol. 1, pp. 379–380, 394–396. Paul Tannery, *Mémoires Scientifiques*, vol. 1, *Sciences Exactes dans L'Antiquité*, ed. J.-L. Heiberg and H.G. Zeuthen (Paris: Gauthier-Villars, 1912).

30. Jacob Klein, "Die griechische Logistik und die Entstehung der Algebra," *Quellen und Studien zur Geschichte der Mathematik, Astronomie und Physik*, Abteilung B: *Studien* 3, fasc. 1 (Berlin: Julius Springer, 1934) (Part I): 18–105; fasc. 2 (Berlin: Julius Springer, 1936) (Part II): 122–235. This study appeared thirty years later in English translation as Jacob Klein, *Greek Mathematical Thought and the Origin of Algebra*, trans. Eva Brann (Cambridge, MA: The MIT Press, 1968).

31. See Swerdlow, "Otto Neugebauer," pp. 143–148, and Høyrup, "Changing Trends," pp. 2–14.

32. See Otto Neugebauer, *Mathematische Keilschrift-texte. I-III, Quellen und Studien zur Geschichte der Mathematik, Astronomie und Physik*, Abteilung A: *Quellen*. 3. Band, erster-dritter Teil (Berlin: Julius Springer, 1935, 1935, 1937), reprinted (Berlin: Springer-Verlag, 1973).

33. Høyrup, "Changing Trends," p. 9. Høyrup noted further that one finds the same stance in such works as Francois Thureau-

Dangin, *Textes mathématiques babyloniens* (Leiden: Brill, 1938).

34. Swerdlow, "Otto Neugebauer," pp. 141–142.

35. Otto Neugebauer, "Zur geometrischen Algebra," *Quellen und Studien zur Geschichte der Mathematik, Astronomie und Physik*, Abt. B3: 245–259, p. 250. My translation, Neugebauer's italics.

36. Otto Neugebauer, *The Exact Sciences in Antiquity*, 2nd ed. (Providence: Brown University Press, 1957), reprinted (New York: Dover, 1969), pp. 146-152.

37. B.L. van der Waerden, *Science Awakening*, 2nd ed. (Groningen: Noordhoff, 1962).

38. Arpád Szabó, *Anfänge der griechischen Mathematik* (München: R. Oldenbourg, 1969), pp. 455 ff.; Michael Mahoney, "Babylonian Algebra: Form vs. Content," *Studies in History and Philosophy of Science*, 1 (1970–1971) 369–380.

39. Sabetai Unguru, "On the Need to Rewrite the History of Greek Mathematics," *Archive for History of Exact Sciences*, 15 (1975) 67–114.

40. *Ibid.*, p. 77.

41. B.L. van der Waerden, "Defence of a 'Shocking' Point of View," *Archive for History of Exact Sciences*, 15 (1976) 199–210. Later, van der Waerden even professed to find some merit in Unguru's position (B. L. van der Waerden, *Geometry and Algebra in Ancient Civilizations* (Berlin: Springer-Verlag, 1983)). The approach taken in this work, however, is far more speculative in its use of free-wheeling mathematical arguments than was *Science Awakening*.

42. Unguru and I spoke about this just before the appearance of his article in *Isis* (see note 46 below). He also verified the accuracy of this statement when he read an earlier draft of this paper.

43. Hans Freudenthal, "What is Algebra and What has it Been in History?" *Archive for History of Exact Sciences*, 16 (1977) 189–200; André Weil, "Who Betrayed Euclid?" *Archive for History of Exact Sciences* 19(1978)91–93; reprinted in Weil's *Collected Papers*, 3:431–433.

44. *Ibid.*, p. 433.

45. André Weil, "Review of Michael S. Mahoney, *The Mathematical Career of Pierre de Fermat*," *Bulletin of the American Mathematical Society* 79 (1973): 1138–1149; reprinted in Weil's *Collected Papers*, 3 266–278.

46. Sabetai Unguru, "History of Ancient Mathematics. Some Reflections on the State of the Art," *Isis* 70 (1979) 555–565, p. 555.

47. *Ibid.*, p. 556.

48. *Ibid.*, p. 562.

49. *Ibid.*, p. 563.

50. A more mathematically oriented critique of geometric algebra appeared shortly afterward in Sabetai Unguru and David E. Rowe, "Does the Quadratic Equation have Greek Roots? A Study of 'Geometric Algebra,' 'Application of Areas,' and Related Problems," *Libertas Mathematica*, 1(1981) 1–49 (Part 1); 2(1982) 1–62 (Part 2).

51. André Weil, "The History of Mathematics: Why and How," *Proceedings of the International Congress of Mathematicians, Helsinki 1978*, 2 vols. (Helsinki: Academia Scientiarum Fennica, 1980), 1: 227–236; reprinted in André Weil, *Oeuvres Scientifique/Collected Papers*, 3 vols. (New York:

Springer-Verlag, 1979), 3:434-442. The views Weil presented in this lecture were scrutinized in considerable detail in Joseph W. Dauben, "Mathematics: An Historian's Perspective," *Philosophy and the History of Science: A Taiwanese Journal*, 2(1)(1993) 1–21.

52. *Ibid.*, p. 435.

53. *Ibid.*, p. 438.

54. *Ibid.*

55. *Ibid.*, p. 439.

56. *Ibid.*

57. *Ibid.*

58. The clashing of the two cultures has been a major theme of discussion for cultural critics and historians since the publication of C. P. Snow's *The Two Cultures: And a Second Look. An Expanded Version of the Two Cultures and the Scientific Revolution* (London: Cambridge University Press, 1964). Its relevance for the history of mathematics, while obvious even from the present sketchy account, has not received sufficient attention.

59. Jean Dieudonné, *A History of Algebraic and Differential Topology, 1900–1960* (Boston: Birkhäuser, 1989). Jean Dieudonné, *History of Algebraic Geometry*, trans. Judith D. Sally (Monterey, Calif.: Wadsworth, 1985). Jean Dieudonné, *Mathematics—The Music of Reason* (Berlin: Springer-Verlag, 1992).

60. E.T. Bell, *Men of Mathematics* (New York: Simon & Schuster, 1937); Constance Reid, *Hilbert* (New York: Springer-Verlag, 1970); *Courant in Göttingen and New York: The Story of an Improbable Mathematician* (New York: Springer-Verlag, 1976); *Neyman—From Life* (New York: Springer-Verlag, 1982). The appeal of popular works about mathematics as opposed to serious historical scholarship is undeniable and, probably, inevitable. See, for example, John Ewing, "Essay Review of D.E. Rowe and J. McCleary, eds., *The History of Modern Mathematics*, vols. 1–2," *Historia Mathematica* 19(1992) 93–98.

61. In a somewhat related vein, Richard Westfall remarked, after working twenty years on his monumental biography of Isaac Newton, that "the more I have studied him, the more Newton has receded from me." Richard S. Westfall, *Never at Rest. A Biography of Isaac Newton* (Cambridge: Cambridge University Press 1980), p. ix.

62. S. Unguru, "History of Ancient Mathematics," pp. 555–556.

63. Herbert Butterfield, *The Origins of Modern Science*, 2nd ed. (New York: Macmillan, 1957).

64. See André Weil, *The Apprenticeship of a Mathematician* (Basel: Birkhäuser, 1992).

65. Liliane Beaulieu, *Bourbaki. Une histoire du groupe de mathématiciens francais et de ses travaux, 1934–1944*, 2 vols. PhD dissertation, Université de Montréal, 1989; Leo Corry, *The Origins of Category Theory as a Mathematical Discipline* (Hebrew), PhD dissertation, Tel-Aviv University, 1990.

66. The flavor of Bourbaki's image of mathematics can readily be seen in the titles of Dieudonn,'s final reflections, the French version based on C.G.J. Jacobi's famous phrase, Jean Dieudonné, *Pour l'honneur de l'esprit humain* (Paris: Hachette, 1987), and the English translation, *Mathematics—The*

Music of Reason (Heidelberg: Springer-Verlag, 1992), based on a saying of J.J. Sylvester.

67. For an account of the philosophical orientation of Bourbaki, see J. Fang, *Bourbaki. Towards a Philosophy of Modern Mathematics,* vol. 1 (Hauppauge, N.Y.: Paideia Press, 1970).

68. André Weil, "The History of Mathematics: Why and How," p. 437.

69. For a survey of major trands in the philosophy of mathematics over the course of this century, see Aspray and Kitcher, pp. 3–19.

70. David Mumford, "Forward for Non-Mathematicians" in Carol Parikh, *The Unreal Life of Oscar Zariski* (Boston: Academic Press, 1991), p. xvii.

71. Reuben Hersh, "Some Proposals for Reviving the Philosophy of Mathematics," *Advances in Mathematics* 31 (1979) 31–50.

72. This classic went through ten editions, seven during Hilbert's lifetime. See David Hilbert, *The Foundations of Geometry,* 2nd. English ed. based on the 10th revised and enlarged German ed. of Paul Bernays, trans. Leo Unger (Chicago: Open Court, 1971). One must be cautious in identifying Hilbert as a doctrinaire formalist. See, for example, my introduction to David Hilbert, *Natur und mathematisches Erkennen* (Basel: Birkhäuser, 1992).

73. The "Ideal Mathematician" was created by Philip J. Davis and Reuben Hersh, *The Mathematical Experience* (Boston: Birkhäuser Verlag, 1981), pp. 34–43.

74. Saunders Mac Lane, *Mathematics—Form and Function* (New York: Springer-Verlag, 1986), p. 447.

75. *Ibid.,* p. 448.

76. *Ibid.,* p. 1.

77. *Ibid.*

78. Philip J. Davis and Reuben Hersh, *The Mathematical Experience* (Boston: Birkhäuser Verlag, 1981).

79. *Ibid.,* pp. 89–112.

80. *Ibid.,* pp. 272–297, 345–362.

81. George Polyá, *Induction and Analogy in Mathematics* (Princeton: Princeton University Press, 1954).

82. Imre Lakatos, *Proofs and Refutations* (Cambridge: Cambridge University Press, 1976).

83. For a survey of these developments, see Knut Radbruch, "Philosophische Spuren in Geschichte und Didaktik der Mathematik," *Mathematische Semesterberichte* 40(1) (1993): 1–28. See further, Philip Kitcher, *The Nature of Mathematical Knowledge* (New York: Oxford University Press, 1983); and Thomas Tymoczko, ed. *New Directions in the Philosophy of Mathematics* (Boston: Birkhäuser, 1986).

84. Aspray and Kitcher, p. 17.

85. Aside from the Aspray and Kitcher volume, the reader should consult Javier Echeverria, Andoni Ibarra, and Thomas Mormann, eds. *The Space of Mathematics. Philosophical, Epistemological, and Historical Explanations* (Berlin: Walter de Gruyter, 1992); L. Boi, D. Flament, and J.-M. Salanskis, eds. *1830–1930: A Century of Geometry. Epistemology, History and Mathematics* (Heidelberg: Springer-Verlag, 1992); Donald Gillies, ed., *Revolutions in Mathematics* (Oxford: Clarendon Press, 1992).

86. Herbert Mehrtens, *Moderne—Sprache—Mathematik: eine Geschichte des Streits um die Grundlagen der Disziplin und des Subjekts formaler Systeme* (Frankfurt am Main: Suhrkamp, 1990).

The Role of Problems in the History and Teaching of Mathematics

Evelyne Barbin
University of Paris–North

Pour un esprit scientifique, toute connaissance est réponse à une question. S'il n'y a pas eu de question, il ne peut y avoir de connaissance scientifique. Rien ne va de soi. Rien n'est donné. Tout est construit.
—Gaston Bachelard

[For the scientific mind, all knowledge is a response to a question. If there had not been any questions, it would not have been possible to have scientific knowledge. Nothing comes of itself. Nothing is given. Everything is constructed.]

In what way may the history of mathematics influence mathematics education? Teachers of mathematics who at some point in their career become interested in the history of their subject often report that the understanding they gain influences their teaching or, at the very least, the way in which they perceive mathematics education. These teachers may or may not choose to introduce an historical perspective into their teaching, and they may or may not give their pupils historical texts to read. But they say that they have a different view about the errors their pupils make, and they have a better understanding of certain remarks their pupils make, and are better able to respond to them. They also pay more attention to the different stages that have to be passed theough in acquiring mathematical knowledge and, in particular, to those obstacles that must be overcome on the way.

To give an answer to the opening question, I believe that a study of the history of mathematics profoundly changes the epistemological concepts of mathematical knowledge: that the introduction of the history of mathematics will transform the practice of teaching mathematics.

Let me deal here with one of the main contributions of the history of mathematics according to this point of view: namely the importance of the role of problems in the historical construction of mathematical knowledge. As I often tell my students, who are future teachers of mathematics; there would never have been a construction of mathematical knowledge, if there had been no problems to solve.

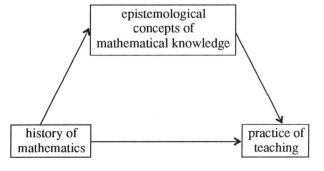

Epistemological Concepts and the Role of Problems

Epistemologist Gaston Bachelard has written: "It is precisely this notion of problem that is the stamp of the true scientific mind, all knowledge is a response to a question."[2] All knowledge is a response to a question: that is to say, the concepts and theories of mathematics exist as instruments for answering questions: they are tools for solving problems.[3]

The history of mathematics shows that mathematical concepts are indeed constructed, modified, and extended in order to solve problems. Problems come, as much into the birth of concepts, as into the different meanings attached to concepts as tools for the resolution of problems. This emphasis upon the role of problems in the historical construction of knowledge can lead to a new way of conceiving history. During the 1910s, a philosopher Léon Brunschvicg revitalized the history of mathematics by proposing a conceptual history of mathematics, that is, a history of those concepts which described how mathematics functions.[4] An alternative way of writing a history of mathematics is a history of problems.[5]

This emphasis upon problems corresponds equally to certain epistemological concepts of mathematical knowledge. Taken as a whole, there are two ways of thinking about mathematical knowledge: either as product or as process.[6] Thinking about mathematical knowledge as product means being concerned with the results and the structure of that knowledge, that is to say, with mathematical discourse. Thinking about mathematics as process means being concerned with mathematical activity. A history of mathematics centered on problems brings to the fore the process of the construction and rectification of knowledge arising out of the activity of problem solving.

On the other hand, the role accorded to problems also agrees with a constructivist view of mathematical knowledge. According to this view, mathematical objects are neither discoveries of what is present in reality, nor are they invented to be applied to that reality: they are constructed

and themselves construct a reality.[7] The construction of mathematical objects takes place at the same time as the construction of a reality,[8] the purpose of which is to make a problem intelligible or to solve it. We now have the schema above.

In order to illustrate these remarks, let us consider two examples, the concept of angle and the concept of the curve.

The Concept of Angle: Its Birth and Significance[9]

We know that the Ionian Greeks of the 6th century B.C. knew how to measure the distance of a boat at sea. They faced the problem of measuring an inaccessible distance: clearly it is not possible to take a stick as a unit of measure and lay it out across the surface of the water.

How, then, did the Ionians set about the task? By climbing to the top of a tower at the water's edge, it is possible to sight a boat with a quadrant. A quadrant consists of a quarter of a cricle and a moving part which can be used as a sight. The quadrant can then be turned towards the land and, using the same sight, directed towards a point on the ground whose distance from the tower can be determined.

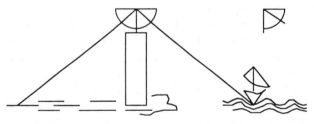

FIGURE 1

Along with the practical problem with which we started, we can draw a geometric figure (Figure 2). The effect of this figure is to schematise the problem situation

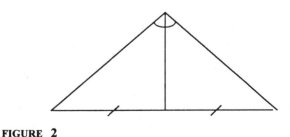

FIGURE 2

and it allows us to comprehend it: the line segments being the visual rays, without thickness, and the angles corresponding to the sights. Assuming that equal sights produce equal distances, we can argue from the figure and state a proposition about angles in a triangle. Thus we proceed to a geometric thought which structures the reality, which we can then comprehend. We have, at one and the same time, the construction of geometric objects, the construction of a reality, and the genesis of a reasoned argument. Geometric concepts and reasoning have taken place within the same problematic situation.

The concept of angle arises with other meanings in Euclid's *Elements*. Book I, Definition 8, reads: "A plane angle is the inclination to one another of two lines in a plane which meet one another and do not lie in a straight line." This definition is not operational, that is to say, Euclid does not use it in his proofs. In particular, in comparing angles, Euclid does not compare inclinations.

In order to compare angles Euclid enclosed them within triangles. Book I, Proposition 4 is as much a proposition about the equality of angles as it is a theorem about the equality of triangles. It asserts that if two triangles ABC and DEF have angles A and D equal with $DE = AB$ and $AC = DF$, then $EF = BC$ and the triangles are equal[10] (Figure 3). In other words, the equality of two angles, has been changed into the equality of two line segments.

The construct "angle in a triangle" allows Euclid to prove a number of propositions, such as, the exterior angle of a triangle is less than each of the interior opposite angles. But in order to prove that this angle equals the sum of the two interior opposite angles, Euclid needs to appeal to another construct: that of "angles and parallel lines." In the proof of Book I, proposition 32, he draws CE parallel to AB and concludes by using the already established results of equal-alternate angles and corresponding angles[11] (Figure 4).

When comparing angles inside circles, Euclid introduces a further idea of "angles in a circle." In proposition 26 of Book III, he proves that "in equal circles equal an-

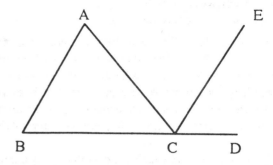

FIGURE 4

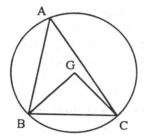
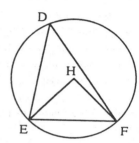

FIGURE 5

gles stand on equal circumferences, whether they stand at the centers or at the circumferences" (Figure 5). The subsequent proposition presents the converse of this result.[12]

In this way different problems lead to new meanings of angle. An angle is not only an inclination: as different geometric configurations are introduced, so the angle becomes associated with other concepts involving new relationships which, in turn, enrich its sense.

But this interpretation of *The Elements* does not claim in any way that Euclid's work was guided by problems concerning angles, nor does it claim that the work placed any value on the range of interpretations of angle in the resolution of these problems. In this respect, it is interesting to compare *The Elements* with those other *Elements of Geometry,* by Antoine Arnauld and Alexis Clairaut, which appeared in the seventeenth and eighteenth centuries. Doing so enables us to show the relationship between the importance these authors attached to the solution of problems with certain of their epistemological conceptions.

Three Epistemological Conceptions of Geometric Knowledge

In Euclid's *Elements,* as stated before, the definition of angle is not operational: according to the Aristotelian conception, the definition is simply given in order to say what the object is. Proofs which are produced subsequently have the object of making explicit the absolute, universal, and neces-

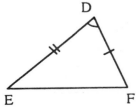
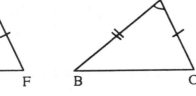

FIGURE 3

sary character of the propositions. The axiomatic-deductive system responds to that requirement. The reader cannot but be convinced of the truth of the propositions, which Euclid lines up one after the other. The order of the propositions is, therefore, a consequence of the deductive procedure.

It is precisely this order that Arnauld criticises in his *Nouveax éléments de géométrie* (1667). In the preface he writes

> les Elements d'Euclide [sont] tellement confus et brouillés, que bien loin de pouvoir donner á l'esprit l'idée et le goût du véritable ordre, ils ne pouvaient au contraire que l'accoutumer au désordre et á la confusion. [Euclid's *Elements* are so confused and muddled, that far from conveying to the mind an idea and taste for true order, on the contrary, they only make the mind used to disorder and confusion.][13]

Why is the deductive order adopted by Euclid judged to be so confused? And what is this 'true order'? Arnauld explains that natural order opposes the cutting up of triangles, which are composite figures, in order to prove results concerning straight lines, whcih are simple objects. The same objection applies to angles, Arnauld does not describe the angle until he reaches Book VIII, which is entirely dedicated to the subject. Book VIII presents a 'true method' for comparing angles.

Arnauld defines the angle as a portion of surface "determined by the proportional part of a circumference whose center is the point where these lines meet"[14] (Figure 6). He goes on to explain that there are four ways of measuring an angle: the arc AC, the sinus DE, and the base DF (Figure 7). The arc is "la mesure naturelle et vraie," the true and natural measure, whereas the base is "la mesure la plus imparfaite," the least perfect measure. And yet it is the base that Euclid chose to use in his Book I.

Arnauld finally presents a fifth way of comparing the magnitudes of angles by considering angles made by lines between parallel lines, and he states that all oblique lines between two parallel lines make equal alternate angles (Figure 8). This property allows him to prove an angle property, namely that "[the sum of] any angle plus the two angles made by its sides on the base is equal to two right angles."[15] If AB and AC are the sides of an angle and BC its base, the result can be deduced by drawing a line through A, parallel to BC (Figure 9).

For each of these angle measures, Arnauld identifies those problems that can be solved using that particular concept of angle. Consequently, Arnauld's chapter on angles is not so much a catalogue of propositions as a method

for comparing angles and solving problems. Readers will know how they learn: they will become enlightened. Arnauld explains why in particular the arc is the only "mesure vraie et naturelle" and why the base is "la plus imparfaite" measure of angle. In his work, the order of propositions is not governed by the demands of a deductive system. The propositions appear according to the way in which they aid understanding: the order is methodological.[16]

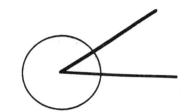

FIGURE 6

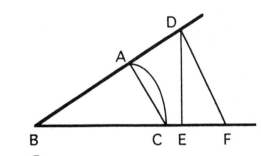

FIGURE 7

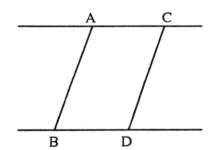

FIGURE 8

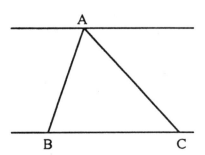

FIGURE 9

The desire to enlighten the reader also motivated Clairaut when he published his *Eléments de géométrie* almost a century later. Clairaut wrote in his preface that he wished "intéresser er éclairer les començants" (to interest and enlighten beginners].[17] He did not wish them to have to read theorems, "Those propositions where it is shown that such and such is true without showing how it came to be discovered." He wanted to present exploration and invention, and to this end he wished "to continually occupy his readers with the solution of problems." Clairaut proposed a "géométrie problématisée," that is, a geometry in which the geometric facts made sense because they were tools to be used in solving problems. Thus, readers would both learn, and learn how and why they learned; they would be able "very easily to acquire the spirit of invention."

In contrast to Euclid, who gives a long list of definitions at the beginning of each book of *The Elements,* Clairaut only introduces concepts as and when they are necessary to solve a problem. Whereas the imperative in dictating the order for introducing propositions in Euclid is the deductive order, with Clairaut it is the chosen problem: in the following case, the measurement of land. His order of work is the order of inventions.

Following this approach, the figure of the triangle arises when it occurs in a situation where one has to measure a rectilinear surface $ABCDE$, which is itself, perhaps, an approximation to a curved line which bounds a field (Figure 10). In this case, "it can be seen that, despite the infinite variety of rectilinear figures, they can all be measured in the same way, by dividing them up into figures of three sides, which are commonly called triangles."[18]

The concept of angle is introduced when a difficulty arises in measuring triangles: "But in the space $ABCDE$ some obstacle occurs, an elevation, for example, or a wood, a lake, etc. which prevents one from setting out the needed lines. What should then be done? What method should then be followed in order to remedy the inconvenience of the land?"[19] (See Figure 11.) Here is a problem of inaccessible distance. It leads us to have recourse to "equal and similar triangles. Suppose that we can only measure two sides of a triangle ABC, sides AB and BC, for example. If DE equals BC and DF equals BA, we cannot know in which position to place one of the two sides in relation to the other (Figure 12). The answer to the problem lies in the use of the angle: "DE is to slant in same way in relation to DE, as AB does to BC; or, to express it in the same way as the geometers, the angle FDE is to be given the same amount of opening as the angle ABC." Clairaut defines an angle

as "l'inclinaison d'une ligne sur une autre" [the inclination of one line upon another].[20]

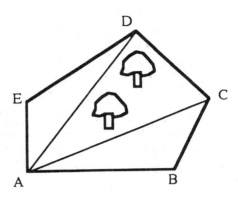

FIGURE 10

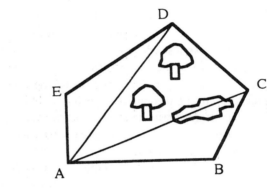

FIGURE 11

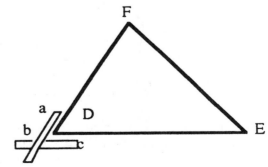

FIGURE 12

For Clairaut, knowledge implies the process by which it is known. Why should some fact become the object of geometrical research? How may the geometer arrive at the truth? Why and how, for example, can the geometer know that the sum of the angles of a triangle is two right angles? For Clairaut, geometrical knowledge is a means by which problems can be solved: his readers learn about a problem in such a way that it leads them to ask themselves questions about the angle sum of a triangle (the why), and they learn about those investigations that lead to the construction of a line parallel to one of the sides of the triangle (the how). They are both interested and enlightened. The proposition concerning the angle sum of a triangle, arises as the way of solving the following problem: to find a simple and effective way of ensuring that the measurements of the three angles of a triangle are exact. To establish the proof, Clairaut makes the side *BC* turn until it reaches the position where the line becomes parallel to *AC* (Figure 13).

In the three works just considered, the role accorded to problems can be set in the context of the way in which geometrical knowledge is perceived. For Euclid, Arnauld, and Clairaut respectively, theoretical mathematical knowledge is a product of incontestable truth, a method dictated by the requirement of developing understanding, and a process of constructing knowledge.

The history of the concept of the curve offers another example of the role that problems play in the transformation of a concept.[21] From ancient Greek geometry through to the birth of infinitesimal calculus, the concept of the curve changed from being thought of as geometric and static to being thought of as the trajectory of a moving point, or represented by an equation, or as an infinitely sided polygon. I shall consider each of these in turn.

The Curve as Viewed by the Ancients

In Greek geometry, with the exception of the Archimedean spiral, curves are defined as objects which are geometric and static. For example, the parabola is defined as the intersection of a cone with a plane. Curves such as the quadratrix, obtained by movement, and introduced to solve problems such as the quadrature of the circle or the trisection of an angle, are excluded from geometry proper.[22] Proofs of theorems which concern geometric curves are also geometric. Motion is not allowed in proofs. So, proofs about tangents to circles in Euclid's *Elements* or about tangents to spirals in Archimedes' treatise *On Spirals* proceed by the method of *reductio ad absurdum*. An argument by contradiction has the advantage of avoiding any reference to the infinite.

In *Elements* III, 16, for example, Euclid proves that "the straight line drawn at right angles to the diameter of a circle from its extremity will fall outside the circle."[23] To do this, he supposes that the straight line *AE* falls inside the circle, like *AC*. This implies that angles *DAC* and *DCA* are equal and so both right angles, which contradicts the fact that the sum of the angles of a triangle is two right angles (Figure 14). Similarly, he says, "we can prove that neither will it fall on the circumference," that is, no part of the circumference may be considered as part of a straight line.

For the geometers of the seventeenth century there was a major objection to proofs offered by Ancients: the proofs assumed that the result to be proved had already been discovered by some other means. Since the proofs proceeded by the method of contradiction, heuristic process was required on the part of the geometer. Furthermore, the proofs were inconvenient on account of their length and because particular methods were needed for each curve. Seventeenth century geometers set out to find "méthodes de

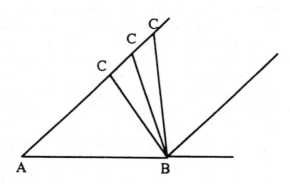

FIGURE 13

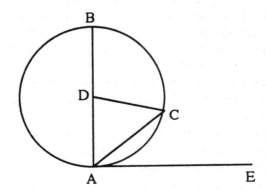

FIGURE 14

découverte," that is, general procedures that would allow them to find tangents to all curves. The problem of finding tangents led geometers to develop new conceptions of the curve.

The Curve as the Trajectory of a Moving Point

The curve, from the static object of Greek geometry, became a dynamic concept in seventeenth century mathematics. This transformation can be seen in the study of one of the major problems that would lead, at the end of the century, to the birth of infinitesimal calculus: namely, the study of motion.[24] In the fourth day of his *Dialogue Concerning Two New Sciences* of 1638, Galileo found the trajectory of a cannon ball.[25] Assuming a vacuum, the trajectory is a parabola. The parabola has now changed: it is no longer the intersection of a cone with a plane, but has become the trajectory of a point under the constraint of motions in two directions.[26] To establish his result Galileo found the composition of two motions. Gilles Roberval used the same approach to find tangents of curves.

Roberval's method for finding tangents dates from the 1630s. During this period geometers were trying to determine the nature of the cycloid, a problem proposed by Marin Mersenne. Not until 1693 was the method published under the title *Observations sur la composition des mouvements et sure le moyen de trouver les touchantes aux lignes courbes*. Roberval's method rests on the following "principle of invention":

> La direction du mouvement d'un point qui décrit une ligne courbe, est la touchante de la ligne courbe en chaque position de ce point-là. (The direction of the movement of a point that describes a curve is the tangent to the curve at each position of the said point.)[27]

The general method for finding the tangent at a point is the following: "by the specific properties of the curve (which you are given), examine the different motions of the point which describe it in the neighborhood where you wish to draw the tangent; find the single motion of which these motions are the components and you will have the tangent to the curve."[28]

Using this method, Roberval explains how to find the tangent at the point F of an ellipse with foci A, B (Figure 15). He writes:

> Since the point F moves away from one of the points A, B to the extent that it moves closer to the other, and the two directions of these movements are the directions BFC and BFD, I only have to divide one of the two angles AFC or

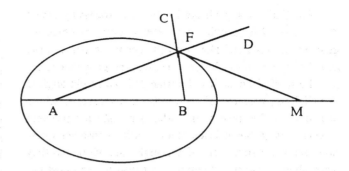

FIGURE 15

BFD into two equal parts by the line IFM, and that will be the tangent to the ellipse.[29]

Roberval's treatise relates a number of ideas: the concept of the curve, the trajectory of a moving point, the concept of the tangent, the direction of movement, and a mechanism for proof.

The Curve as an Equation

In *La géométrie* (1637) Descartes explains how to reduce solving problems in geometry to solving equations. In this way he demonstrates a general method for the problem of finding normals to curves, that is, perpendiculars to tangents. Descartes writes:

> This is why I have said all that is required for the elements of curved lines, since I have already given the general method for drawing those straight lines which fall at right angles to the curve at any of its points one may care to choose. And I dare to say that this is the Problem that is the mose useful and the most general, not only that I know, but that I may ever desire to know in Geometry.[30]

The method Descartes gives in *La géométrie* consists in considering a circle with its center on an axis of coordinates and passing through two points C and E of the curve. He makes these points of intersection coincide. Descartes studied the case where the curve is an ellipse: in which case the coordinates solve a second degree equation. The case where the equation has a double solution is used to obtain the coordinates of the point P where the circle "touches" the curve at a single point. For, writes Descartes, "the more these two points C and E come closer together, the less is the difference between the two roots, and finally they are exactly the same, they are both joined together as one."[31]

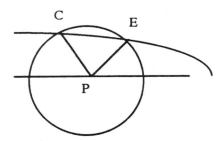

FIGURE 16

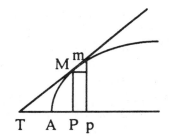

FIGURE 17

The tangent to the circle is already known: it will also be the tangent to the curve where the curve and circle "touch" (Figure 16).

Descartes' method can be used for all curves which can be associated with an algebraic equation, those curves which Descartes called geometric. Here again, a number of ideas are related: the use of an equation to stand for a curve, the concept of tangent, the limit position where an equation has a double solution, and an algebraic method of proof.

The Curve as a Polygon with an Infinite Number of Sides

The differential calculus developed by Gottfried Leibniz was popularized by the Marquis de L'Hospital's (1696) treatise, based on Johann Bernoulli's lectures, and entitled *Analyse des infiniment petits pour l'intelligence des lignes courbes*. This calculus is based upon a consideration of "the infinitely small part by which a quantity is continually increased or diminished, called the Difference." The difference of a quantity x is denoted by dx. L'Hospital requires that a curve "be considered as the assemblage of an infinity of straight lines, each infinitely small, or as a polygon having an infinite number of sides."[32] Recall that Euclid, on the contrary, took great care to prove that a line and a curve could not coincide.

In his treatise L'Hospital explains how the differential calculus may be used to determine the tangents to curves (Figure 17):

> Having drawn the appliqué [ordinate] MP, and supposing that the straight line MT that meets the diameter at T is the required tangent, imagine another appliqué mp infinitely close to the first and a small straight line MR parallel to AP. And by letting the given AP, PM be x and y; (whence pp or $MR = dx$, and $Rm = dy$) the similar triangles mRM and MPT im-

ply $mR/RM = MP/PT$ (whence $dy/dx = y/PT$). Now, by use of differences from the given equation, the value of dx can be found in terms of dy, which being multiplied by y and divided by dy, will give the value of the subtangent PT in terms which are entirely known and free of differences, by which means the desired tangent MT may be found.[33]

His method consists of considering a segment of a line, Mm, the same as part of the curve. Again, we have a number of ideas which are associated: the concept of the curve, the concept of the tangent, and the method of proof.

Despite their differences, the methods of Roberval, Descartes, and L'Hospital have this in common, they attempt to solve a common problem: the search for a general method of finding tangents to a curve. The problem has led to new ways of defining both curve and tangent and has given rise to new procedures of proof.

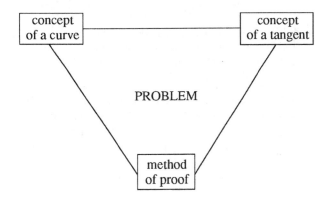

In addition, the problem of investigating tangents is itself linked with other problems, such as those concerning motion or optics. Problems other than those associated with tangents also played an important role in changing the concept of a curve.[34]

In conclusion, I should like to return to my opening remarks concerning the relation between history and mathematics education. If you should ask a French *lycéen* (age 15–18 years) the meaning of an increasing function, it would not be unsual to receive the reply: "a function whose derivative is positive." Here is an example of one of the perverse effects of education, namely, that answers are given to questions that have not been asked. The hsitory of mathematics shows us that questions must come first, and it is through questions that we make sense of mathematical concepts. It is essential that our education take account of this epistemological conception if our pupils are to understand better what is meant by mathematical activity.

References

1. This paper was originally read to the Congress: "History and Pedagogy of Mathematics," University of Toronto, 14 August 1992.
2. Gaston Bachelard, *La formation de l'esprit scientifique,* 5th ed., Paris: Vrin, 1967), p. 14.
3. See Bernard Charlot, "L'épistémologie implicite des pratiques mathématiques," in Bkouche, Charlot, and Rouche, *Faire des mathématiques, Le plaisir de sens* (Paris: Armand Colin, 1991).
4. Léon Brunschvig, *Les étapes de la philosophie mathématique* (1912: reprint Paris, Blanchard, 1981).
5. This corresponds, for example, to the work being prepared by the Groupe inter-IREM, *Histoire des Problèmes, Histoire des Mathématiques,*presented in both English and French versions in pre-published form to the Toronto Congress.
6. See Evelyne Barbin, "Les *Eléments de géométrie* de Clairaut," *Repères IREM,* (July 1991, 4, 119–113).
7. See Evelyne Barbin, "La démonstration mathématique: épistémologie, histoire et enseignement," Rosmorduc (ed.), *Actes des 2éme journées Paul Langevin* (Brest: University of Brest, 1991).
8. See Ferdinand Gonseth, *Les Mathématiques et la réalité,* (1936: new ed. Paris, Blanchard, 1974).
9. This section recapitulates Barbin, "La démonstration mathématique: épistémologie, histoire et enseignement," op. cit.
10. T. L. Heath, *The Thirteen Books of Euclid's Elements,* 3 vols., 2nd ed. (Cambridge: Cambridge University Press, 1925: reprint New York, Dover, 1956), vol. 1, p. 247.
11. *Ibid.,* pp. 316–317.
12. *Ibid.,* vol. 2, pp. 56–59.
13. Antoine Arnauld, *Nouveaux éléments de géométrie* (Paris: Savreux, 1667), Preface.
14. *Ibid.,* p. 142.
15. *Ibid.,* p. 154.
16. See Evelyne Barbin, "La démonstration mathématique...," op. cit.
17. Alexis Clairaut, *Eléments de géométrie,* (Paris, 1765: reprint Laval, Siloë, 1986), Preface.
18. *Ibid.,* p. 14.
19. *Ibid.,* p. 27.
20. *Ibid.,* p. 29.
21. See also Evelyne Barbin and Gilles Itard, "Le courbe et le droit," in Groupe inter-IREM *Histoire de problèmes, Histoire des mathématiques,* (Paris: Ellipses, 1993).
22. See Joëlle Delattre and Rudolf Bkouche, "Pourquoi la régle et le compas?" in Groupe inter-IREM *Histoire de problèmes, Histoire des mathématiques,* op. cit.
23. Heath, op. cit., vol. 2, p. 37.
24. See Morris Kline, *Mathematical Thought from Ancient to Modern Times,* (New York: Oxford University Press, 1972).
25. Galileo, *Discours concernant deux sciences nouvelles,* trans. Clavelin, (Paris, 1638: Armand Colin, 1970).
26. Evelyne Barbin and Michèle Cholière, "La trajectoire des projectiles de Tartaglia à Galilée" in *Mathématiques, Arts et techniques au XVIIème siècle* (Le Mans: Université du Maine, 1987).
27. Gilles Roberval, "Observations sur la composition des mouvements et sur le moyen de trouver les touchantes aux ligne courbes," *Recueil de l'Académie des Sciences, tome IV,* (Paris: Académie des Sciences, 1693), p. 24.
28. *Ibid.,* p. 25.
29. *Ibid.,* p. 32.
30. René Descartes, *La géométrie,* (1637: reprint Paris, Christian David, 1705), p. 63.
31. *Ibid.,* pp. 69–71.
32. L'Hospital, *Analyse des infiniment petits pour l'intelligence des lignes courbes,* (Paris, 1696: Montalent, 1716).
33. *Ibid.,* p. 11.
34. See Evelyne Barbin and Gilles Itard, "Le courbe et le droit," op. cit.

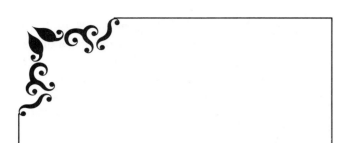

Dramatizing the Birth and Adventures of Mathematical Concepts: Two Dialogues

Gavin Hitchcock
University of Zimbabwe

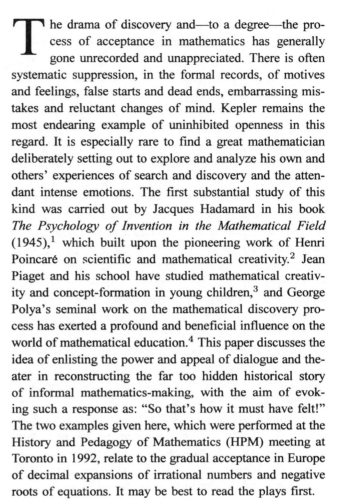

The drama of discovery and—to a degree—the process of acceptance in mathematics has generally gone unrecorded and unappreciated. There is often systematic suppression, in the formal records, of motives and feelings, false starts and dead ends, embarrassing mistakes and reluctant changes of mind. Kepler remains the most endearing example of uninhibited openness in this regard. It is especially rare to find a great mathematician deliberately setting out to explore and analyze his own and others' experiences of search and discovery and the attendant intense emotions. The first substantial study of this kind was carried out by Jacques Hadamard in his book *The Psychology of Invention in the Mathematical Field* (1945),[1] which built upon the pioneering work of Henri Poincaré on scientific and mathematical creativity.[2] Jean Piaget and his school have studied mathematical creativity and concept-formation in young children,[3] and George Polya's seminal work on the mathematical discovery process has exerted a profound and beneficial influence on the world of mathematical education.[4] This paper discusses the idea of enlisting the power and appeal of dialogue and theater in reconstructing the far too hidden historical story of informal mathematics-making, with the aim of evoking such a response as: "So that's how it must have felt!" The two examples given here, which were performed at the History and Pedagogy of Mathematics (HPM) meeting at Toronto in 1992, relate to the gradual acceptance in Europe of decimal expansions of irrational numbers and negative roots of equations. It may be best to read the plays first.

Mathematics-in-the-making

"I feel as if I'm watching the fourteen hundreds—but watching it *now!*" (statement overheard in the lunch queue at HPM Toronto meeting). This experience—however it is achieved—of re-living with the mathematical trail-blazers their struggles, excitements, tensions, delights and victories, is an elusive but treasured goal of all advocates of history in the teaching of mathematics. The aim is similar to Bruno Latour's "first rule of method":[5] to watch the black boxes being closed and thereby distinguish the contradictory explanations of the closure. There is generally a striking difference between the procedures and justifications uttered while the closure is taking place and those given when finished. We take "black box" in mathematics to mean a set of routine techniques and implicitly accepted principles. Two examples are decimal real number representation, and quadratic equation solving, and our two dialogues propose to give an experience of eavesdropping on the closing of these respective black boxes. The

mathematical *Janus* (like Latour's scientific *Janus*) has two faces: the hoary, all-wise, forbidding, granite face of ready-made mathematics (just learn definitions, theorems, and formulas ... when things are well-defined and understood within a nice theory, then they can be fruitfully applied); and the eager, curious, fluid face of mathematics-in-the-making (open questions, guesses, false leads, mistakes, debates about meanings and boundaries ... when things are bearing fruit in solving problems, then they start becoming accepted, defined and incorporated within nice theories). Following Latour further, we take a "black-box" statement $A = B$ (think of an assertion about decimal representation, or quadratic equation solving), frequently encountered by mathematics pupils as a completed, dry-packaged dish for instant, obedient consumption and application, if not outright force-feeding. Adding inverted commas "$A = B$," we give it a new status as a claim within some context of human discourse. Then we draw a speech-balloon around it, that is, we personalize it by putting it in the mouth of an identifiable person; and we steadily sketch in the wider context of social, cultural, and intellectual interactions, and represent the community of controversy, communication, and competition, engaged in the process of creative ferment through which "$A = B$" matured in language, symbolism, theory, and paradigm. How can this reconstruction of the historical and intellectual drama be achieved most vividly and most accurately? Our claim is that the devices of dialogue and theater furnish an answer to the first, at least.

Why dialogue/conversation form?

It is a very old discovery—exploited by Plato, Robert Recorde, Bernard Fontenelle, Galileo Galilei, and others, up to Imre Lakatos—that these devices provide excellent means of communication, not only of abstract philosophical ideas but also of technical and scientific matters. Why not attempt to harness this special appeal and power to combat false stereotypes of mathematics, to recover the exhilaration of the quest and the sense of wonder at newly-derived concepts, and so to help the student of mathematics to share something of the grand adventure which is so much a feature of living, growing mathematics? We are increasingly aware of the importance and delights of going back to the primary sources, but there are major problems in their effective critical use, as shown by the work on the scientific foundations for modern history from Leopold von Ranke to Anthony Grafton. numbered

- Sources may be opaque, if not inaccessible, for many.
- Records are fragmentary—much of the drama of creation goes unreported in the public arena, and the puzzle is fundamentally incomplete.
- Recorded fragments are scattered and isolated from their influences and neighbours in the dialectical process—their relationships are obscure.
- Records can be misleading if not downright deceitful, as regards the discovery process. Peter Medawar has said that it is no use looking to scientific papers—in this century at least—if one wants to find out what scientists actually do, for the papers "not merely conceal, but actively misrepresent the reasoning that goes into the work they describe ... Only additional unstudied evidence will do and that means listening at a keyhole."[6] Or being a "fly on the wall." Nicholas Broad and Anthony Wade describe vividly how the highly stylized nature of scientific reporting leads to a profoundly anti-historical portrayal of science;[7] and Thomas Kuhn claims that "the depreciation of historical fact is deeply, and probably functionally, ingrained in the ideology of the scientific profession."[8]

Thus, the challenge comes: use dialogue to help fill in the gaps and tell a more authentic story—to present concepts *in statu nascendi*, in the freshness of their becoming, and to reconstruct or reincarnate the spirited concatenation of ideas in time; employ the art of the dramatist to interpret and relate historical fragments, weaving them together into a seamless and coherent whole; capture with the natural flow of a good conversation or story-line something of the complex historical dialectic which lies behind the official facade—and spice the story with the essence of human genius and human frailties, emotions, and idiosyncracies. Authenticity is not won cheaply, and professional historians will look askance at uncritical extrapolations of an unruly imagination. Four characteristics seem important for our enterprise (the present author makes no claim to have fully met these ideals in the dialogues presented here!):

- A wide perspective on various personalities, communities, periods, and areas of mathematics involved.
- Careful interpretation of micro-studies against the background of large-scale movements.
- An integrated multidisciplinary approach: cognitive, psychological, social, ethnic, rhetorical, and philosophical, as well as mathematical and historical.
- Support from meticulous historiographical research which includes the "informal sector" of mathematical records: correspondence and diaries, as well

as mathematical perceptions scattered through non-mathematical literature.

For revealing excavation of this kind into the vast substratum of informal mathematical intercourse and creativity, the imagination needs to be well-disciplined and methodologically well-prepared. The enterprise may appear daunting—especially to creative spirits in whom the gifts of dramatist and story-teller do not sit comfortably with the demands of scholarship in the history of mathematics. But, in an age which spawns University classes for whom the idea that "joy" might have characterized the creation of mathematics evokes incredulity and derision, these gifts are important. Let them be encouraged wherever possible, that the dry bones may be clothed with flesh and live! It is likely that the scholar will benefit as much as the student.

Varieties of Dialogue

Several authors have attempted to use dialogue as a means of displaying the nature of mathematical exploration and creativity. Alfred Renyi's three wonderful dialogues (featuring Socrates, Archimedes, and Galileo, respectively) set out to be "lively and vivid" and "comprehensible to non-specialists," and yet present problems "in their full complexity"—an achievement possible only (Rényi felt) by means of dialogue form.[9] He aimed to emulate the original writers' styles, preserving historical fidelity and "avoiding anachronism as far as possible," always assuming some poetic licence.

Imre Lakatos clearly separated the "concept story" and the "human story," telling the latter as a footnoted counterpoint to the conversations between his characters (named alpha, beta, gamma, etc.) embodying the former.[10] He set out to explore the *logic of discovery,* emphasizing the problem-situation, the groping for definition, the flawed proofs and half-formed concepts, and the provocative or embarrassing counterexamples, that all play their role in what Lakatos summarized as "the incessant improvement of guesses by the logic of proofs and refutations." Lakatos succeeded in providing a splendid representation of the dialectic of ideas, but his work makes few allowances for the faint-hearted and is aimed at philosophers rather than children. Arend Heyting had previously used the device of de-personalized and de-historicized representative characters (more suggestively named Class, Form, Int, Letter, Prag, Sign) in his *Disputation,* a discussion of intuitionism, formalism, and other approaches to the philosophy of mathematics.[11]

Donald Knuth and Rolf Grunseit have also made important contributions. In a bold and effective attempt to communicate the experience of creating new mathematics, Knuth wrote his dialogue as he was "actually doing the research" himself—redeveloping for himself John Horton Conway's recent approach to numbers.[12] By recording false starts and frustrations as well as good ideas, Knuth's aims were to give "a reasonably faithful portrayal of the important principles, techniques, joys, passions, and philosophy of mathematics;" to show how "math can be taken out of the classroom and into life;" and to "communicate the techniques of mathematical research." Rolf Grunseit has written a number of stories and plays, designed primarily for classroom use with younger pupils.[13] While occasionally risking offence to historians' sense of correctness, these have potential for opening hearts to the love of mathematics. Such isolated pioneering efforts deserve to be applauded and encouraged; it would be good to have a booklet of similar short and accessible items as a teachers' resource.

At HPM Toronto 1992, Robert Brabanec described the fruitful incorporation of spontaneous role-playing (for example, Newton and Leibniz) in his History of Mathematics classes at Wheaton College; and Jan van Maanen spoke of the effectiveness of challenging students to write their own dialogues. Clearly, there are differing ways in which dialogue can be used, with quite distinct aims. The historiographical, pedagogical, and artistic priorities will be mixed in varied degrees, and the inevitable tensions should not be underestimated. The historian, who wishes to avoid anachronisms and cut through pretensions, will be at odds with the educator who insists on avoiding opaqueness and keeping it relevant and snappy! Meanwhile, the true dramatist, with an eye to audience appeal and entertainment value (even Knuth wrote his dialogue "mostly for fun"), will be concerned with character development, humor, irony, dramatic tension and flow, and the digestibility of speeches.

Of the two extracts presented below, the first is a dialogue more oriented towards pedagogical ends, taking more liberties with historical characters than the second, which stays closer to primary source material. The protagonists in the first dialogue, Michael Stifel and Simon Stevin, may not have met in the flesh, but they are, as it were, summoned, as real, colourful, historical characters and convenient representatives of differing positions, to appear on behalf of ideas which did interact historically. With Rényi, I have " felt free to attribute to them views and ideas at which they may have arrived, particularly if these were logical developments of such ideas with which they were definitely familiar."

It is important that any presentation of such an imaginative reconstruction be accompanied by a disclaimer on the historical accuracy of the characters' lines, and a clear statement of its purpose to stimulate the learning and enjoyment of mathematics. The second extract (which is part of Scene 5 in a dramatization of the rise of the negative numbers whose synopsis appears in Gavin Hitchcock[14]) in attempting to be more authentic on the historical moment, uses period expressions and language which sounds to the modern ear formal, even pompous. Such archaisms, like science fiction, will make demands upon the listener, but (apart from any intrinsic charm) may encourage the suspension of modern preconceptions and thought-patterns, and so assist unencumbered imaginative entry to that remote world.

Play 1 : A Cloud of Infinity
Facing up to the Irrationals

Characters:

MICHAEL STIFEL (c.1487–1567).   Born in Esslingen, Germany, he entered the local Augustinian monastery as a young man, to become a monk and then a priest. He was one of the earliest followers of the fiery reformer Martin Luther, falling foul of his superiors, over what he felt were abuses in the practice of receiving 'indulgences' for granting absolution from sin—and also over his composing a song in honour of Luther! He had to flee, and took refuge in Luther's house in Wittenberg.

As a Lutheran pastor henceforth, Stifel's life was turbulent, but he did settle down, thanks to the influence of Luther and Philip Melanchthon, to take a Master's degree at the University of Wittenberg and write a very influential algebra textbook, *Arithmetica Integra* (1544); and eventually he became a Professor at the University of Jena. He was forever engaging in acrimonious theological disputes and his extreme passion for numerology, or number mysticism, led him to associate the name of Pope Leo X with the number 666, and so prove him to be the "beast" of the Book of Revelation. Stifel also contrived to predict, using Biblical numbers, that the end of the world would come at 8 o'clock on 18 October 1533. His friend Luther forgave him (referring to his "little temptation"), but begged him to stop spreading his fantastic notions abroad.

In spite of these activities, Michael Stifel is recognised as the greatest mathematician in sixteenth-century Germany. He was a pioneer of algebraic symbolism, and gave general methods for solving equations and for computing roots. Although he struggled to come to terms with the nature of irrational numbers, he was very good at manipulating them in surd form, and he was one of the first Europeans to face up to negative numbers, cheerfully extending his use of exponents to include observations akin to saying $2^{-3} = 1/8$.

SIMON STEVIN (1548–1620).   Born in Bruges, he became a central figure in the economic and cultural renaissance of the northern Netherlands, writing many books on mathematics, as well as being a brilliant practical inventor and engineer. He has been called (by George Sarton) "perhaps the most original man of science of the second half of the sixteenth century"; he was certainly the most influential mathematician of the century in the Netherlands.

Stevin is justly celebrated for giving the first systematic exposition of decimal fractions in his *De Thiende* (1585), which became widely known as *La Disme (or the Art of Tenths): Teaching how all Computations that are met in Business may be performed by Integers alone without the aid of Fractions.* Perhaps because of the unifying influence of this approach, and perhaps because he was such a very practical man, Stevin boldly stated his credo about the reality of "any number which turns up": "there are no absurd, irrational, irregular, inexplicable or surd numbers." He also made significant contributions to statics, hydrostatics, and navigation; among his inventions were a wind-driven drainage mill and a wind-driven carriage which could carry twenty-eight passengers.

Stevin became quartermaster of the army of the States of the Netherlands and a powerful official of Maurice of Nassau, Prince of Orange. He was famous for his work on fortifications and military engineering.

Background to the Play

Simon Stevin was not the first to extend the Hindu-Arabic decimal place-value numeral system to fractions; this was done earlier by al-Kashî (died c. 1436), Christoph Rudolff (c. 1500–1545), who used a vertical stroke as "decimal point" in his compound interest tables, and François Viète (1540–1603). But Stevin gave the first clear and popular exposition, which was sound yet practical, persuasive in tone, and ultimately very influential. He recognized that, in decimal expansion, "we might continue this indefinitely," and that, in a number, like $4/3$, "it appears that the quotient will be infinitely many threes with always one third in addition." But his practical nature was quite undaunted, for "in business one does not take account of the thousandth part of a maille or a grain"; and such as the great Ptolemy "did not make up their tables with the utmost accuracy that could be reached," for, in view of the purpose of the tables, "approximation is more useful than perfection."

Originating with Greek and Hindu mathematicians, "incommensurable magnitudes" and associated "irrational numbers"—such as the square root of 2 (arising from the ratio of the diagonal of a square to its side and admitting only approximations by whole number ratios)—were introduced to Mediaeval Europe largely through Arabic writings. The intriguing "pi" also, arising as the ratio of the circumference of a circle to its diameter, had long offered a challenge to mathematicians to approximate it ever more closely, or to construct straight line segments in the same ratio (equivalent to constructing a square equal in area to the given circle—known as "squaring the circle"). The great Hebrew Rabbi Maimonides (1135–1204) may stand as representative of the long interim period between Archimedes's time and Stevin's time: for Maimonides, the essence of pi had to remain unknown and could be conceived only as approximated by ratios such as 22/7.

Approximations to pi carried out to increasing numbers of decimal places became an industry about the time of Stevin's work on decimals. François Viète, in France, reached nine places; Ludolph van Ceulen, in Germany, computed pi to thirty-five places—and had the fruit of his labours appear on his tombstone! It was not until 1766 that pi was finally proved to be irrational (by Lambert), and the inconstructibility of pi (hence impossibility of squaring the circle) had to wait a century longer for proof (Lindemann 1882).

The Italians, Luca Pacioli (1445–1517) and Girolamo Cardano (who published his *Ars Magna* in 1545—the year after Michael Stifel published his *Arithmetica Integra*) used irrational numbers freely and classified them into types. But they remained uneasy about their status as numbers. Even Pascal, Barrow, and Newton, a century later, could not believe in these numbers as independent entities: their existence was tied up with the geometrical magnitudes they represented. Stevin, and later Wallis (in his *Algebra* of 1685), were among the first to fully accept irrationals into the family of legitimate numbers.

About fifteen years after the appearance of Cardano's *Ars Magna*, Rafael Bombelli developed practical ways of approximating irrationals (using continued fractions), and when publishing his work twelve years later referred to "young Stifel, a German," among the worthy algebraists of his time (although Stifel was then dead five years, and was well over sixty years old when his *Arithmetica* was published!). Perhaps this is an indication of how Stifel's eager mind and original work affected his contemporaries. Bombelli (like Cardano) describes how he blundered, al-

most by chance, into a new kind of number (later known as complex numbers). His specimens were the cube roots of {2 plus-or-minus the square root of minus 121}, which by "a wild thought" he guessed as {2 plus-or-minus the square root of minus 1}, and found to work like magic in the solution of the cubic equation $x^3 = 15x+4$. He was sufficiently daring to follow up his "sophistry" with a formal calculus of operations with these numbers, but, in spite of this imaginative breakthrough, most mathematicians (including Stevin) found the new numbers suspect, calling them impossible, absurd, fictitious, meaningless, imaginary, or just (like Stevin) uncertain.

The air of mystery and suspicion surrounding complex numbers took nearly three centuries to be dispelled. Both complex and irrational numbers had to wait for the nineteenth century to receive a firm logical foundation—and they were soon joined by the previously unthinkable "infinite numbers" of Georg Cantor (1845-1918), with an arithmetic all of their own.

A Cloud of Infinity

STIFEL [*looking at the decimal expansion of* $\sqrt{2}$: 1.4142⋯, *or* 1 ⓪ 4 ① 1 ② 4 ③ 2 ④ ..., *in Stevin's own notation*[15]]

But Meneer Stevin, it is always running away from you! How can you call something a true number which is in perpetual retreat from your advances?[16]

STEVIN Oh—I can approach as near to $\sqrt{2}$ as the problem requires. [17] Just give me time and I will have my runaway number trapped in an interval as small as you wish.

STIFEL But you will never finish! You will never catch it—you can never have it precisely![18] For $\sqrt{2}$ is an irrational number—you and I both know that; we have a simple proof, which the Greeks knew as well. I feel compelled to deny that such a badly-behaved creature is a number worthy of the name.[19]

STEVIN [*laughing*] Irrational! What kind of a name is that? We must really think of a more honorable name for this intriguing tribe of numbers—they deserve better of us! I say there are no absurd, unthinkable, or unreasonable numbers.[20] When a number turns up, let us face it, and embrace it with our minds!

Come now, Herr Stifel, don't you agree that there must be something inside all those intervals-within-intervals? Look, here is something else you would certainly call a true number:

[*writes* $1/3 = 0.3333\ldots$, or 3 ① 3 ② 3 ③ $3\frac{1}{3}$ ④].
Yet we never finish chasing that one either: there will be infinitely many threes, with always one third remainder.[21]

STIFEL At least with that number, I can see that the sequence has only threes; and so also for:

[*writes* $1/7$ = $0.142857\ldots$,   or 1 ① 4 ② 2 ③ 8 ④ 5 ⑤ 7 ⑥ $\ldots$],

the whole procession recurring again and again. It is periodic. I know the pattern—I could communicate it to you. But– God help us!– there is no pattern, no predictability at all, with that procession of numbers in pursuit of $\sqrt{2}$. This $\sqrt{2}$ is an incorrigible fugitive:[22] there is no discipline—no regimentation about it. It's like a deserter from your Dutch Army!

STEVIN Well, let me see … how about this?

[*writes* $0.1010010001\cdots$, or 1 ① 1 ③ 1 ⑥ 1 ⑩ $\ldots$].

Can you see a pattern? Is this a more acceptable recruit for our fine, well-disciplined army of true numbers?

STIFEL Hmmm …I must say, I am tempted to be more hospitable to this fellow … but no! Pattern there may be; but you can never write down with finality—with complete precision—what the number really is. You have given me instructions for finding each successive interval, but the number itself is clearly irrational—I can never look it in the face and say: Ah! So that is who you are! Whereas with the decimal enumeration of $1/3$, or $1/7$, or any other rational number, I know that I can apprehend each precisely in itself:[23] for such enumerations (and they will all be recognizable as terminating or periodic) are really just whole number ratios. But, as for this fellow [*gestures towards Stevin's new number*], it is impossible to call a number real which is so peculiar as to be devoid of precision, and have no known proportion to any real number.[24]

STEVIN Now, Herr Stifel, let me ask you—why do you think you can look these so-called rational numbers in the face? What is, shall we say [*writes* $237/459$] in itself? Is it not also an instruction? It commands me, "divide 237 units into 459 parts," or else, "divide unity into 459 parts and measure off 237 of them." Are you going to be suspicious of the poor number and keep it waiting out in the cold until you have actually completed carrying out the operation?

STIFEL Only the practical mind of a military engineer would conceive of a pure mathematical proportion in such an operational and utilitarian manner! But let us be practical by all means. The thing that frightens me about this monster that you would recruit as a true number, is that your instructions for apprehending it can never be fully carried out. The procedure for approaching it may be clear, but it goes on for ever and ever. That monstrous number lies hidden in a kind of cloud of infinity.[25] And the ratio between such a number and a rational number will be just as obscure as the ratio of an infinite number to a finite number.[26] Therefore, just as we can agree that an infinite number is not a number, so also an irrational number is not a number at all.[27]

STEVIN [*laughing*] I would have thought your years in the monastery would have accustomed you to the serene contemplation of infinity! What have these Lutheran reformers done to you?

STIFEL [*fervently*] I think that Martin Luther and Philip Melanchthon have taught me to be less mystical and less blindly submissive to authority; to be more honest, more logical. Faith and reason are not opposed—

STEVIN [*interrupting*] Well, well!—let us see what in the name of reason you mean by the pure mathematical proportion one-third, or one-to-three. First of all, you keep on insisting on proportion and ratio in the style of our mathematical ancestors, the Greeks. But we have become accustomed, with long use, to think of one-third as a thing-in-itself, just as you said—an individual rational number, not merely a ratio. There have been races who struggled to go further and conceive of two-thirds or three-quarters as things-in-themselves. But we have gone much further, and there is still further to go. Tell me, now: will you agree that $2 : 6$, or $11 : 33$, or $15 : 45$, are essentially the same thing?

STIFEL The same thing? You mean the same number—one-third? Yes, I will accept that.

STEVIN Then you must admit that one-third is just a name for the infinite class of pairs of numbers in a certain relationship. And yet you claim to be able to apprehend this number precisely in itself!

STIFEL You are confusing me! I think it is safest to say that one-third is the number we designate to measure what we get when we divide the unit length into three parts.

STEVIN [*shaking his head in mock sorrow*] Ah! Back to good old geometrical figures for refuge! When will we give due honour to our great science of pure number—our Arithmetic? But never mind—I will come with you and say that $\sqrt{2}$ is the number I designate to measure the diagonal of the unit square—and I can construct that even more easily than you can divide your unit length into three parts. Your $1/3$ is approached by a unique procession of decimal intervals. Each lies entirely within its forerunner, and each happens to be labelled "3." In the same way, my $\sqrt{2}$ calls forth its own procession of intervals: the first (after the unit) labelled "4," the second "1," and so forth. And it will be the same with any point you care to pick on the line.

STIFEL Surely not every point can be measured by a number? I have never believed that all lengths would submit to being captured by numbers.[28] For example, take this perplexing ratio between the circumference and diameter of a circle—that which some call "pi." Everybody who has ever tried to calculate it has used polygons to approximate the circle. Now, since a circle is an infinitely-many-sided polygon, this so-called number requires an infinite calculation. This is impossible in finite time, so I conclude that it is ridiculous to say this number really exists.[29]

STEVIN Unless we extend our hospitality to include infinite decimals as numbers—just as we were moved to entertain the idea of $\sqrt{2}$ as a number because it serves to measure the diagonal of the square. You yourself have been so bold (perhaps I should say rash) as to do business in your book with such a fine fellow as:

[*writes* $\sqrt[3]{(\sqrt{20} - \sqrt{10}) + (\sqrt{8} - \sqrt{2})}$].[30]

May not these new decimal numbers be just what we have need of? If only they can succeed in breaking down our prejudices and winning our confidence!

It is common for authors of books on Arithmetic to deal with numbers like $\sqrt{8}$ and similar ones, to which they are then so discourteous as to call them absurd, irrational, irregular, inexplicable, or just surd numbers. That literally means deaf numbers! Now I deny that any number which turns up deserves such a name—and when I open the door to infinite decimals I will defend myself and my guests with the same challenge: By what reason will the opposition prove them unreasonable?[31]

STIFEL Reason, Meneer Stevin, may be impotent to justify its dissatisfaction—but it still demands to be satisfied! Even that wild new conscript of yours [*gestures towards it*] has some order and decency—some reasonable regimentation about it. But as for this $\sqrt{2}$, or pi, if I dare to subject it to your numeration:[32]

[*writes* $pi = 3.14159...$, or $pi = 3\ ⓪\ 1\ ①\ 4\ ②\ 1\ ③\ 5\ ④\ 9\ ⑤...$],

where there is no pattern, no specification at all—if these are really true numbers, then it is only because they are determined as lengths, or ratios of lengths.

STEVIN Ah! Poor Master Arithmetic! Will he ever free himself of the apron strings of Mistress Geometry?

But look—neither of us has any doubt that we have at least one sure algorithm for calculating the fifth, or the

twentieth, or even the hundredth decimal place, given time. Isn't this adequate numerical specification? True—we may not know these numbers now, but we could find them if required. They do exist! Therefore $\sqrt{2}$ and pi do exist as decimals—as does my, uhh, wild conscript number. I am glad you seem more favourably disposed towards him!

STIFEL I hate to disappoint you again, but, while I may be prepared to accept it as potentially fully specified in the sense you describe, I am very uncomfortable about it, because I cannot see any hope of constructing it precisely as a length. What point on the line does it correspond to? Unless we can determine that, we surely have no logical foundation, nor practical means, for calculating with it.

STEVIN Never mind about exact calculations—I just want to persuade you that these numbers are there, calling for us to engage with them: each unambiguously specified and infinitely closely approaching some point.[33] Distinct decimals will be found eventually in distinct intervals at the stage of enumeration where the numbers first differ—hence they will approach distinct points.

STIFEL But what are these points—how do we construct them?

STEVIN Herr Stifel, in this matter we have to pass by intellect beyond practical geometrical experience. And impractical mathematicians like you can take Euclid as their model, as in so much else—for he shows the way in the recognition of incommensurable magnitudes. You see, because of the fallibility of our eyes and hands (which cannot perfectly see and divide), we would judge in the end that our decimal subdivisions all converge to a point after a finite number of steps. But also—even supposing it were possible for us to apply our process of subdivision several hundred thousand times, and continue that for several thousands of years, still (if the number is irrational) one would labour eternally, always remaining ignorant of what could still happen at the end. This way of thinking is therefore not legitimate—it's intellectually possible but practically impossible.[34]

STIFEL Now you are talking just like me! But where does Euclid come into it? As I understand him, Euclid denied, in his treatment of Book 10, that irrationals are true numbers.[35]

STEVIN Ah! Euclid knew very well that what really happens in nature cannot be understood in this spatial, geometrical way by the senses—but he did not throw up his hands in despair and give up trying to understand incommensurable magnitudes. Rather, he showed how to deal with them by means of numbers. There lies the answer: trust the numbers! If we use decimal numeration of an irrational number, then, just as with incommensurables, we can infinitely closely approach the true value. But we never arrive, else the irrational would be rational, and the incommensurable commensurable![36]

STIFEL I wonder what Euclid would say about your infinite decimals! Would he be so easy as to how these elusive points can be constructed with a compass? I am still uncomfortable . . .

STEVIN Herr Stifel, why don't we agree to accept the fundamental principle, that if we are to have a science of measurement which will be adequate for all geometric figures, then we must conscript and marshal our numbers so that they are arrayed to cover the line entirely—each point determining a number? Thus a number will be precisely determined by a sequence of intervals of our decimal subdivision method—no matter how we specify the sequence.

STIFEL I am impressed by the force of your military metaphor, Meneer Stevin. It would be good to know that there are sufficient numbers to capture the whole line. . . But then—if I am compelled to accept this great battalion of decimals as true numbers—then . . . I can conceive of all kinds of new ways of specifying numbers! The finite and the periodic ones will represent the rational numbers—oh! my apologies! We should not call them by that name—I mean the fractions! And all the others—the infinite ones, the chaotic or random ones—

STEVIN The wild ones!

STIFEL —these are what we call the irrational numbers. Heaven help us! They are by far the most numerous! Why—the rationals begin to appear like stars against the terrifying black vastness of these irrational numbers. And we have no precise means of calculating with them, and no idea how to construct them geometrically . . . My head is beginning to ache—I think I know how the ancient Pythagoreans felt, when they first came face to face with incommensurable magnitudes. It makes me want to run away!

STEVIN [*laughing*] Back to the monastery? Nowhere else is safe, it seems; being a mathematician is as dangerous as being a Lutheran preacher! Come, come, my friend, take courage! Fortunately for us, the rational numbers are densely distributed on the line. Like stars in the milky way, there is always another between any two of them. By our decimal enumeration method we can march forward as near to the real point as any problem requires. That means the

rationals are quite sufficient for all practical matters. In business no one takes account of one thousandth of a grain; and, similarly, the greatest Geometers and Arithmeticians, in even their most important calculations, have omitted insignificant remainders. Ptolemy and Tehan de Montroyal, for instance, did not make up their tables with the utmost possible accuracy: for they knew, keeping in view the purpose of these tables, that approximation is more useful than perfection.[37]

Herr Stifel, we shall never need that hundredth decimal place of the number pi!

STIFEL I am surprised at you, Meneer Stevin: that kind of talk may be all right for businessmen and engineers. But we mathematicians have always insisted that we are talking about an ideal world, where points are perfect points, lines have no width, and lengths of lines are precisely given. Why else do we speak about proving things?

If I have been persuaded to accept $\sqrt{2}$, pi, and the whole tribe of your infinite decimals, it is purely on these grounds: they do appear to correspond perfectly to definite points on the line—even if we do not yet know how to construct these points. And, in working with geometrical figures, whenever rational numbers desert us, certain of these irrational numbers come riding to our rescue, and prove precisely what rational numbers were not able to prove.

That has been my experience. And I am thus moved to admit that, since these representative irrational numbers have such very real, certain, and reliable effects in the ideal world of geometry, they deserve to be accepted and treated as true numbers.[38]

And if some, why not all? They may present themselves in the intimidating dress of infinite decimal enumeration—but you have held out to me the ravishing prospect of an ideal correspondence between numbers and points. I feel compelled to swallow what my intuition revolts against!

STEVIN Perhaps you should rather say these irrational numbers are swallowing you—overwhelming you by the force of their authentic mathematical personalities! Here lies Michael Stifel, Master of Arts, Minister of the Divine Word,[39] and Mathematician—engulfed by the cavalry brigade of the irrational numbers charging to his rescue out of the "cloud of infinity" he so feared!

STIFEL Yes—although it feels more like a vast horde of devils charging at me out of the smoke of the infinite abyss! Perhaps they are angels after all . . . This "ideal" world of ours is proving to be a great deal stranger than we ever imagined it could be.

STEVIN That is true—and it may turn out to be stranger yet. Forgive me—but I cannot resist placing before you the prospect of something even more devilish than irrationality. You raised the issue earlier in tones of alarm and horror: what if there are, among these irrational numbers, some that we cannot in principle ever construct with compasses?

STIFEL Meneer Stevin, I am almost beyond the point where I can be shocked now! Once, I would have rejected the notion as utterly preposterous; even now I am tempted to demand that such a number be instantly and dishonourably discharged from service! But, woe is me! I am forced by your arguments today to concede that I might be thus banishing a real point from the line!

And moreover, I will confess that I have for some time nursed dark, secret suspicions about the ratio pi. I have begun to wonder whether this army of people all over Europe, trying to square the circle, is not engaging in a futile exercise.[40] It may be that they cannot succeed in principle—that the number pi is not constructible after all. If so, then it is worse than irrational—it is positively indecent!

STEVIN And maybe pi can never be found as the root of any equation—as $\sqrt{2}$ is the root of $x^2 = 2$. Is it not conceivable (if we have the courage to face it) that there is a great alien tribe of such numbers, transcending both our geometric and algebraic procedures—an even deeper blackness in that black sky of irrationals?

STIFEL God help us! There are mathematicians who have the audacity to propose using what they call "imaginary" numbers;[41] I don't know which is worse!

STEVIN They will be talking about using infinite numbers next! As far as I'm concerned, these matters are a waste of time. I will leave them to you, and to others who take pleasure in that sort of thing. They can make of them whatever they choose.[42]

STIFEL But a short time ago you were proclaiming that there is no such thing as an absurd number! Are there degrees of absurdity which make even Simon Stevin avert his eyes?

STEVIN Signor Bombelli may be one of the great Arithmeticians of our time, but as far as I can see, these imaginary numbers of his have not been derived by any definite rule—they have no logic to give them clear definition, and they certainly do not arise naturally out of my decimal enumeration process. His method therefore seems unworthy of a place amongst legitimate propositions.[43]

STIFEL But it succeeds in solving his equation! However impossible these numbers seem, they are surely practical enough to earn your approval.

STEVIN It was lucky chance! For such a solution, Fortune deserves as much credit as Bombelli![44] Anyway, speaking for myself, there are quite enough legitimate things—even infinitely many—to work on, without getting carried away by obscure and dubious matters like these.[45] I must get back to business; if the wind is up, we are going to test my new drainage mill. Farewell!

[EXIT]

STIFEL [*to audience*] I regarded irrational numbers as very dubious things, and never imagined I would find myself drawn into admitting them as true numbers...God alone knows what strange and wonderful beasts we mathematicians will be entertaining in the future!

[CURTAIN]

THE END

Play 2: Gross Inhibitions in England
Multiple and negative roots of equations

Characters:

WILLIAM FREND (1757–1841). Aged 70, Fellow of Jesus College, Cambridge. He was previously mathematics Tutor to the University until he was dismissed in 1788, for propagating heretical theological views. For similar offences in 1793 he was put on trial in the University Vice-Chancellor's Court, and banished from University and College residence when he refused to retract. He has practised as an Actuary for the Rock Life Assurance Company since then. With his henchman Francis Masères, (lawyer and constitutionalist), lately deceased (1731-1824), Frend has fought a bitter, rearguard action against the evils of "symbolical algebra," "fictitious" and "imaginary" numbers, priding himself on being a "pure arithmetician" and a "noted oppugner of all that distinguishes Algebra from Arithmetic."

GEORGE PEACOCK (1791-1858). Aged 36, highly respected Cambridge mathematician. He is a founding member of the Cambridge Analytical Society, whose aim is to revive British mathematics out of its isolationist stagnation—in particular, by publicizing and making available the work of continental mathematicians.

AUGUSTUS DE MORGAN (1806-1871). Aged 21, talented student of Peacock. He is just graduating from Cambridge with a mathematics degree, and is soon to be appointed Professor of Mathematics at University College, London, at the young age of 22. He will eventually marry Frend's daughter Sophia Elizabeth, in spite of profound differences with his prospective father-in-law.

Background to the Play

This takes the form of a prologue, put in the mouth of Jean d'Alembert, French mathematician, and one of the major contributors to the "Encyclopédie ou Dictionnaire Raisonné des Sciences, des Arts, et des Métiers," 1751-1780. He is sitting at his desk, reading out of one of its 28 large volumes; it is 1783, the year of his death.

D'ALEMBERT " ... the algebraic rules of operations with negative numbers are generally admitted by everyone, and acknowledged as exact, whatever idea we may have about these quantities ... " [*laughs*]

Whatever idea we may have about these quantities! When I wrote that entry, I wished to express my impatience with those who were detaining our poor negative numbers for interminable philosophical interrogation, instead of permitting them to get on with the work they are so perfectly fitted to perform! What a business these negative numbers have had trying to persuade us that they should be treated as worthy numbers in their own right! Even irrational numbers have had it easier. A plague on the philosophers and their scruples! But even practical mathematicians like Girolamo Cardano had uneasy consciences. They called them names like "absurd numbers," "fictitious numbers," "invented numbers." Luca Pacioli and others found it necessary to deal with quadratic equations in six different forms—each with its own rule for solution. What a cumbersome bag of tricks! And no European mathematician, right through the sixteenth and seventeenth centuries, could bring himself to accept negative numbers as roots of equations. Blaise Pascal derided those who thought of taking four away from nothing and getting minus four; even René Descartes called his negative numbers "false roots," but he showed the beginnings of acceptance because they turned out to have a use.

Now—there is the secret: in spite of all these scruples, the negative numbers won for themselves more and more recognition over two centuries, because they proved themselves useful. They worked! Especially in Mechanics, when that great science came to the forefront. And no doubt the development of Analytic Geometry was a help.

It is astounding, in view of these long European reservations, to think that 600 years ago the Hindu mathematician Bhaskara was writing about positive and negative roots, and getting 50 and minus 5 formally as solutions for the equation $x^2 - 45x = 250$. He did say that the second root should not be used as "people do not approve of negative solutions." But those Hindus certainly recognized the role that negatives could play in representing debts, and, as early as the seventh century, Brahmagupta was calculating freely with negative numbers and giving rules for operating with them. Even the man we call the father of algebra—that great Greek Diophantus, in the third century—does not appear to have recognized that any of his equations might have multiple roots, let alone negative roots!

[*reads*] " ... the algebraic rules of operations with negative numbers are generally admitted by everyone ... "

A thousand years after Brahmagupta, we in Europe finally got there. I wrote that at a time when the philosophical quibblers were at last beginning to subside under the great evidences of success of the new Analysis. It took a long time for some people to learn that we have entered the Age of Mechanics; and outside Mechanics there is no sense or use of Algebra to speak of, hence quibbles about the nature of negative numbers in themselves, without any reference to any physical quantity, are unprofitable. It is enough that they work consistently in the great house of Mechanics!

Generally, the doubting mathematicians seem to have taken my injunction to heart: "Go on!" I said, "Faith will follow!" I do not speak, however, for the English. They remain, regrettably, in a past age of mathematics. They have reverted to studying their hero Newton in place of Nature. There is, not surprisingly, a rearguard action over there: the opposition to negative numbers is kicking still!

Gross Inhibitions

FREND [*writes where necessary*][46] A problem correctly posed will always lead to one true root of the appropriate equation. Here, I will give an example of how second-order equations ought to be proposed: Let

$$x + \sqrt{5x + 10} = 8$$

therefore

$$\sqrt{5x + 10} = 8 - x$$

(and it is quite clear by the conditions of the proposed problem that x is less than 8). Therefore, squaring:

$$5x + 10 = 64 - 16x + x^2,$$

$$5x + 16x - x^2 = 64 - 10$$

therefore

$$21x - x^2 = 54$$

(which is the equation some would foolishly assert to be the basic equation of the problem). Completing the square:

$$\frac{441}{4} - 21x + x^2 = \frac{441}{4} - 54$$

$$= \frac{441 - 216}{4} = \frac{225}{4}$$

$$\sqrt{\frac{441}{4} - 21x + x^2} = \sqrt{\frac{225}{4}},$$

$$\frac{21}{2} - x = \frac{15}{2}$$

$$x = \frac{21}{2} - \frac{15}{2} = \frac{6}{2} = 3,$$

(and in this case the root $x - 21/2$ cannot be used, for it is known that x is less than 8).

DE MORGAN [*writes where necessary*] But if you do use the root $x - 21/2 = 15/2$, you would produce a perfectly acceptable positive root $x = 18$. Why should that not constitute a valid alternative solution?

FREND It certainly does satisfy the equation $21x - x^2 = 54$ [*gestures towards it*]; but it does not satisfy my original equation, where the nature of the square root to be taken has been correctly defined. Nature herself is never ambiguous; no practical problem of science or commerce, correctly posed, will ever lead to more than one solution.

PEACOCK[47] I do not think you are right there. What about this problem? "I sold a horse for 24 pounds; and by so doing, lost as much percentage as the horse cost me: it is required to find the prime cost of the horse." Is this a fairly posed practical problem?

FREND It would appear so, indeed.

PEACOCK[48] Then let us take required quantity x; this is the cost, and I sold it for 24, so that the loss will be given by

$x - 24 = \dfrac{x}{100} \cdot x$, the percentage loss being given to be x.
Therefore

$$100x - x^2 = 2400$$

$$2500 - 100x + x^2 = 100$$

$$x^2 - 100x + 2500 = 100.$$

Therefore (and I willingly neglect the negative value of the square root!),

$$\begin{cases} 50 - x = 10, & x = 40 \\ x - 50 = 10, & x = 60 \end{cases}$$

and both solutions satisfy the conditions of the problem proposed.

FREND [*long pause*] That merely shows the problem to be indeterminate. Any well-posed, determinate problem will never lead to more than one solution.

PEACOCK [*laughing*] I think, Mr. Frend, that you are fully aware of the consequences of admitting the truth of this theory of the composition of equations, as far as your system is concerned, and I must allow that you and Baron Masères have struggled against it with considerable ingenuity.[48]. But now that you are reduced to empty tautology for an answer, you should admit defeat.

FREND I will admit the possibility of multiple positive roots: all will then be equally congruous to the problem, indicating that the problem is indeterminate. But I contend that these roots might be obtained by arithmetical Algebra alone, without trifling with those other useless fictions which can lead to no practical solution. Their only use is to show the problem to be impossible.[49]

PEACOCK What makes a root congruous to a problem? May it not be merely that it meets your preconceived specifications?

FREND It must be a real (positive) root—that is all! Such roots will always lead to practical solutions. Other so-called roots will not.

PEACOCK[50] But it would be very easy to propose a problem leading to an equation with a single, real positive root, which is yet not congruous to the problem proposed.

FREND Let me see it!

PEACOCK[51] If it was proposed: "to find a number, the double of whose square exceeds three times the number itself by 5," then we shall find 5/2 and −1 for the roots of the resulting equation.

$$2x^2 - 3x = 5$$

$$x^2 - \frac{3}{2}x = \frac{5}{2}$$

$$\left(x - \frac{3}{4}\right)^2 = \frac{49}{16}$$

$$\begin{cases} x - \dfrac{3}{4} = \dfrac{7}{4}, & x = \dfrac{5}{2} \\ \dfrac{3}{4} - x = \dfrac{7}{4}, & x = -1 \end{cases}$$

Now, you are forced to admit the existence of the root 5/2, and yet it is not a solution, if by number, in the proposed problem, be meant a whole positive number. Thus, incongruous and real, as well as negative and impossible roots, may equally indicate the impossibility of the problem proposed.

FREND But the root yielded by the Algebra shows that you have imposed an artificial restriction on the conditions of the problem: you should not only be looking for whole numbers—you should be prepared to admit the existence of a number such as 5/2 as a genuine solution to your problem.

PEACOCK You have been delivered into my hands, Mr. Frend! You accept this root 5/2 (your own principles force you to admit it), which is incongruous in terms of the expressed conditions of my problem. Because your Algebra presents you with a certain root, you would have me remove certain artificial restrictions I have short-sightedly imposed on my problem! Why not, then, be done with your own artificial restrictions, and admit the existence of other roots—whether negative or impossible – to which the algebraical solution of the problem might lead, even though they admit of no very direct interpretation in terms of the expressed conditions of the problem?[52]

Others, long before you, would have excluded any non-whole number solution, out of hand. You have merely drawn the unforgiving line elsewhere. It is not true congruity with real problems and real practical solutions that is the guiding principle of your Algebra, Sir, but congruity with the preconceptions of your mind.

DE MORGAN It does appear more and more clearly to me, that the prevalent gross inhibitions about negative numbers as roots of equations, and the persistent inability to form a distinct conception of them as real solutions of problems, are the aftermath of historical conditioning by the very forms in which we have perceived, proposed and expressed our problems. Perhaps in the future, under the influence of a new generation of problems, these numbers will admit of a more direct interpretation, and acquire thereby a more distinct meaning.

Their orienting role in Geometry and Mechanics seems well-defined enough, but I must confess the major function of negative numbers in my own experience has been indistinguishable from that described by Mr. Frend: their use is to show the problem to be ill-posed! Whether it is the imaginary expression: square root of minus a, or the negative expression: minus b, they have this resemblance, that either of them occurring as the solution of a problem, immediately disposes me to search for some inconsistency or absurdity. If I am to be honest, I have to admit that, as far as real meaning is concerned, I usually treat them both as equally imaginary, since for most practical purposes, zero minus a is as inconceivable as the square root of minus a.[53]

FREND Ha! Honesty is teaching you sense at last, young Augustus!

DE MORGAN[54] In either case, however, I recognize that, if I am brave enough to interpret these imaginary numbers in an appropriate way, often a marvellously rational answer to the problem may be given.

PEACOCK Confessions of a cautious mathematician! Can you give an example of what you mean?

DE MORGAN[55] Here is an illustration. I like telling people that I will be x years old in the year x squared, and then challenging them to find the year of my birth. Observing from my youthful appearance, that each of the years in question almost certainly occurs within the 19th Century, it is not hard to calculate that I will be 43 in the year 1849 (which is the square of 43), so that I was born in the year 1806!

$$(43)^2 = 1849, \quad 1849 - 43 = 1806$$

But someone, taking the problem as a purely algebraical one, might suggest that I be minus 42 in the year 1764 (which is the square of minus 42). Now, my first response is to protest, and tell him not to indulge in such folly but to confine himself to legitimate numbers. However, this solution does, wonderfully, also yield the correct year of my birth!

$$(-42)^2 = 1764, \quad 1764 - (-42) = 1806$$

PEACOCK There is a fine vindication of the legitimacy and power of algebra!

FREND [*with triumphant sneer*] Until some joker of a symbolical algebraist observes that his perverted algebra allows De Morgan to be minus 43 in the year 1849 (which is the square, [*sarcastically*] as we all know, of minus 43); and thus, led by his donkey's nose into futility and folly, he finds 1892 to be the auspicious year of the future birth (or should I say reincarnation?) of the honourable Augustus de Morgan.

$$(-43)^2 = 1849, \quad 1849 - (-43) = 1892$$

DE MORGAN Yes—I do admit it!—my original problem may be ill-posed. I shall have to be careful in future to add the facts that I am neither ghost nor Hindu.

PEACOCK Nor Pythagorean!

DE MORGAN[56] But here is a more subtle illustration of the same situation: Suppose that a father is 56 years old, and his son is 29; it is required to find how long it will be before the father is twice as old as the son. I naturally solve the equation:

$$56 + x = 2(29 + x), \quad \text{obtaining } x = -2 \text{ years!}$$

My immediate intuitive reaction is to recoil from this absurdity! I tell myself that something is wrong with my framing of the problem, and I replace x by $-x$, and solve the new equation:

$$56 - x = 2(29 - x), \quad \text{now obtaining } x = 2 \text{ years.}$$

The problem, I conclude, had too great a limitation in its expression—it should have demanded how many years will elapse before, or, maybe, have elapsed since the father is (or was) twice as old as his son.

PEACOCK A very clumsy and ambiguous expression! How do we decide which it is in advance—for we have to form an equation before we can solve it?

DE MORGAN[57] Yes—it will be necessary to suppose the event either past or future, and on that supposition form an equation. If we choose the wrong one, this error will be pointed out by the negative form of the result. In this case the negative result shows us we have made a mistake in reducing the problem to an equation.[58]

FREND There is hope for you yet, boy! You are now treating these negatives with the disrespect they deserve.

DE MORGAN[59] Oh, but I must point out again that, whichever supposition we made, if the result is interpreted in the right way, we can give a perfectly rational and acceptable answer to the problem.

PEACOCK For a moment, De Morgan, you appeared to be speaking almost as a convert of Mr. Frend's and Baron Masères' algebraical reformation.[60] I think most of us mathematicians spend a great part of our time with our feet in two worlds—the old and the new. Making the transition is exceedingly difficult. I am sure you are correct about the conditioning effects of practical problems: like myriad ministering angels bearing us aloft, as we devote ourselves to the foremost problems of our day as well as the apparently trivial wayside problems, it is they whose inexorable influence will ultimately carry us home.

It is therefore not surprising to me that the practical Hindus, as concerned with everyday problems about taxation and debts and interest as they were with the distant stars, should have conceived these ideas of the negative numbers and zero, as well as their marvellous decimal number system—ideas which never entered into the abstract speculations of all the great philosophers of ancient Greece!

No, Mr. Frend, Europe owes more to those mathematicians—dismissed by you as Oriental magicians obsessed with poverty and debts!—than it is willing to acknowledge, and also to those diligent Arab translators and mathematicians through whom so many ideas were preserved, refined, and transmitted into the European heritage. And how inadequately we celebrate that practical and widely-travelled Italian, Leonardo of Pisa (whom we all know as Fibonacci)! He wrote his famous Book of Counting at the outset of the 13th Century, in order that the Latin race might, as he put it, no longer be deficient in that knowledge of Eastern numeration, so superior to the burdensome Roman system then still used in his country. Was it not in that book, the fountain of the Mathematical Renaissance of the West, that he wrote that he conceived of a negative root as representing a loss instead of a gain? And is it not a fair supposition that the powerful Italian Medici family came so to dominate the commercial life of Europe, partly because they could keep accounts and balance their books more proficiently than others?[61]

Thus were sown the seeds which germinated and grew when the climate was favourable, to be shaped into the weapons of that Revolution in the Sciences, initiated by Galileo and Descartes. Yes—the influence of practical problems, whether of scientific or commercial origin, cannot be overestimated. But we must not neglect the equally powerful conditioning effects of a coherent theoretical system of symbols and operations: if you, Mr Frend, and Baron Masères, and the other authors of this misguided attempt at algebraical reform, had been better acquainted with the more modern results of the science, you would have felt in your bones the total inadequacy of the very limited science of arithmetical algebra to replace it.[62]

[CURTAIN]

THE END

Endnotes

1. Jacques Hadamard, *The Psychology of Invention in the Mathematical Field*. (Princeton, N.J. : Princeton University Press, 1945).
2. Henri Poincaré, *The Foundations of Science*. (New York : Science Press, 1946), and "Mathematical Creation," in *The World of Mathematics*, Vol. 4, ed. James R. Newman. (New York : Simon and Schuster, 1956), pp. 2039-2050. The first is a reprint in one volume of *Science and Hypothesis, The Value of Science*, and *Science and Method*, published in many editions in Paris in the early years of this century. The second is a lecture Poincaré delivered before the Psychological Society of Paris.
3. Jean Piaget, *The Child's Conception of Number*, tr. Caleb Gattegno and F.M. Hodgson. (London : Routledge and Kegan Paul, 1952); Jean Piaget, B. Inhelder and A. Szeminska, *The Child's Conception of Geometry*. (London : Routledge and Kegan Paul, 1960).
4. George Polyá, *Mathematics and Plausible Reasoning*. (Princeton, N.J. : Princeton University Press, 1954). See also Godfrey H. Hardy, *A Mathematician's Apology*. (Cambridge: Cambridge University Press, 1940), Hermann Weyl, "The Mathematical Way of Thinking," in *Science* 92, 1940, pp. 437–446, and John von Neumann, "The Mathematician," in Newman, *The World of Mathematics*, Vol. 4, pp. 2051–2063.
5. Bruno Latour, *Science in Action*. (Cambridge, Massachusetts : Harvard University Press, 1987).
6. Peter Medawar, *The Art of the Soluble : Creativity and Originality in Science*. (Harmondsworth, Middlesex : Penguin Books, 1969), p. 169.
7. William Broad and Nicholas Wade, *Betrayers of the Truth*. (New York : Simon and Schuster, 1982), p. 128.
8. Thomas Kuhn, *The Structure of Scientific Revolutions*. (Chicago : University of Chicago Press, 1970), 2nd ed.
9. Alfréd Rényi, *Dialogues in Mathematics*. (San Francisco : Holden Day, 1967).
10. Imre Lakatos, *Proofs and Refutations : the Logic of Mathematical Discovery*. (Cambridge: Cambridge University Press, 1976).
11. Arend Heyting, "Disputation," in *Philosophy of Mathematics : Selected Readings*, ed. Paul Benacerraf and Hilary Putnam. (Oxford : Basil Blackwell, 1964).
12. Donald E. Knuth, *Surreal Numbers*. (Reading, Massachusetts : Addison-Wesley, 1974).
13. Rolf Grunseit, *Thales of Miletus, The Genesis of Conics, Gauss*, etc., plays and stories available from the author, 4 Sandridge St., Bondi 2026, N.S.W., Australia. For discussion, evaluation and classroom trial, see Barry J. Fraser and Anthony J. Koop, *An Evaluation of some Motivational Materials related to the History of Mathematics*, Research Report, School of Education, Macquarie University, 1977; also Barry J. Fraser and Anthony J. Koop, " Changes in Affective and Cognitive Outcomes among Students using a Mathematical Play." *School Science and Mathematics*, 1981, vol. 81, pp. 55–59.
14. Gavin Hitchcock, "The 'Grand Entertainment' : Dramatising the Birth and Development of Mathematical Concepts," *For the Learning of Mathematics*, 1992, vol. 12 no. 1, pp. 21–27.
15. Simon Stevin, *De Thiende*. (Leiden, 1585); also *Disme, the Art of Tenths*, tr. Robert Norton. (London, 1608), *The Principal Works of Simon Stevin*, vol. II, ed. D.J. Struik. (Amsterdam, 1955–1968), and D.J. Struik, *A Source Book in Mathematics, 1200–1800*. (Cambridge, Massachusetts : Harvard University Press, 1969). We will refer to the *Principal Works* subsequently as *PW*.
16. Michael Stifel, *Arithmetica Integra*, Book II. (Nuremburg, 1544), p. 103. Translations of selected passages from the *Arithmetica* in Morris Kline, *Mathematical Thought from Ancient to Modern Times*. (New York : Oxford University Press, 1972), p. 251, and John N. Crossley, *The Emergence of Number*. (Singapore : World Scientific, 1987).
17. Stevin, *Disme*, in *PW*, p. 407. He does not write explicitly about the decimal expansion of the square root of 2, but refers in this manner to the expansion of 4/3.
18. Stifel, *Arithmetica*, p. 103.
19. ibid.
20. Stevin, *PW*, p. 532.
21. Stevin, *Disme*, in *PW*, p. 407.
22. Stifel, *Arithmetica*, p. 103r.
23. ibid.
24. ibid.
25. ibid.
26. ibid.
27. ibid.
28. Stifel, *Arithmetica*, quoted in Crossley, *Emergence*.
29. ibid.
30. Stifel, *Arithmetica*, p. 111.
31. Stevin, *PW*, p. 532 (basis for the paragraph).
32. Stifel, *Arithmetica*, p. 103, uses this phrase but not with particular reference to pi.

33. Simon Stevin, *L'Arithmetique* (Leiden, 1585, reprint 1635), p. 215 r., quoted in Crossley, *Emergence*, p. 126.
34. Stevin, *L'Arithmetique*, p. 215 (basis for the paragraph).
35. Stifel, *Arithmetica*, p. 104.
36. Stevin, *L'Arithmetique*, p. 215.
37. Stevin, *Disme*, in *PW*, p. 407.
38. Stifel, *Arithmetica*, p. 103 (basis for the previous three sentences).
39. The title given to Stifel in the register of the University of Jena is: "senex, artium Magister et minister verbi divini."
40. Stifel "dabbled in circle-squaring, but came to the conclusion that attempts were futile" (*Dictionary of Scientific Biography*).
41. Rafael Bombelli, *L'Algebra*. (Feltrinelli, 1966); first draft was composed about 1560, first published 1572.
42. Stevin, *PW*, pp. 619–620.
43. ibid.
44. ibid.
45. ibid.
46. William Frend, *The Principles of Algebra*. (London, 1796), p. 111.
47. George Peacock, "Report on the Recent Progress and Present State of Certain Branches of Analysis," *Report of the Third Meeting of the British Association for the Advancement of Science held at Cambridge in 1833*. (London : John Murray, 1834), p. 190.
48. Peacock, "Report," p. 191.
49. Frend, *Algebra*, p. 111.
50. Peacock, "Report," p. 191.
51. Peacock, "Report," p. 192.
52. ibid.
53. Augustus De Morgan, *On the Study and Difficulties of Mathematics*. (London : Society for the Diffusion of Useful Knowledge, 1831).
54. ibid.
55. De Morgan was fond of posing this riddle concerning the year of his birth.
56. De Morgan, *Study and Difficulties*.
57. ibid.
58. Jean d'Alembert, in his article under the heading "Negative" in the *Encyclopédie* (ed. Denis Diderot et al, 1751–65), says something similar: "Thus negative quantities really indicate positive quantities in calculations, but which one has supposed in a false position. The sign '−' which is found before a quantity serves to redress and correct an error that has been made in the hypothesis . . . "
59. De Morgan, *Study and Difficulties*.
60. Francis Masères, *Dissertation on the Use of the Negative Sign in Algebra*. (London, 1759).
61. For De Morgan's own writings on this period, see "On Leonardo da Vinci's use of + and −," *Philosophical Magazine*, (3) 20 (1842), pp. 135–137; and "On the early history of + and −," (4) 30 (1865), p. 376.
62. Peacock, "Report," p. 192.

II

Historical Studies

From Antiquity to the
Scientific Revolution

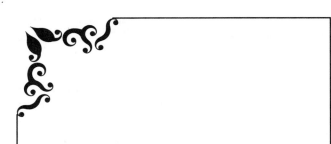

The Four Sides and the Area: Oblique Light on the Prehistory of Algebra

Jens Høyrup
University of Roskilde

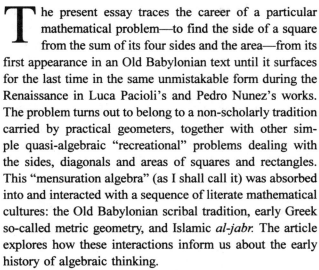

The present essay traces the career of a particular mathematical problem—to find the side of a square from the sum of its four sides and the area—from its first appearance in an Old Babylonian text until it surfaces for the last time in the same unmistakable form during the Renaissance in Luca Pacioli's and Pedro Nunez's works. The problem turns out to belong to a non-scholarly tradition carried by practical geometers, together with other simple quasi-algebraic "recreational" problems dealing with the sides, diagonals and areas of squares and rectangles. This "mensuration algebra" (as I shall call it) was absorbed into and interacted with a sequence of literate mathematical cultures: the Old Babylonian scribal tradition, early Greek so-called metric geometry, and Islamic *al-jabr*. The article explores how these interactions inform us about the early history of algebraic thinking.

As far as possible I have referred for detailed documentation to earlier publications, in particular to my analysis of Babylonian "algebra" and its reflections in later traditions. In cases where documentation is not discussed in depth elsewhere I have still tried to be concise, but nonetheless felt obliged to present at least an outline of the full argument.

I. An Old Babylonian "square problem"

A famous cuneiform mathematical text (BM 13901)[1] contains as its No. 23 the following problem

> In a surface, the f[o]u[r fronts and the surf]ace
> I have accumulated, 41′40″. 4, the f[ou]r fronts,
> yo[u inscr]ibe. The igi of 4 is 15′. 15′ to 41′40″
> [you r]aise: 10′25″ you inscribe. 1, the projec-
> tion, you append: 1°10′25″ makes 1°5′ equilat-
> eral. 1, the projection, which you have appended,
> you tear out: 5′ to two you repeat: 10′ nindan
> confronts itself.

The text was written in the Old Babylonian period, that is, between 2000 B.C. and 1600 B.C., and probably during the eighteenth century B.C.. Originally, it appears to have contained 24 problems of apparently algebraic character dealing with one or more squares and their sides. In its present state, the tablet is damaged, though most problems can be safely reconstructed.

The translation is meant to render the terminology as precisely as possible, and follows principles which I have developed for the translation of Babylonian "algebra".[2] In the present context, only a few words' explanation can be made. Numbers, first of all, are rendered in the degree-minute-second notation, which means that $1°10′25″$ is to be read $1 + \frac{10}{60} + \frac{25}{60 \cdot 60}$. (One should remember that the original

45

text contains no indicators of absolute order of magnitude, merely the sequence 1 10 25.) "Accumulating" (Akkadian *kamārum*[3]) is a genuine addition of numbers, where both addends lose their identity and merge into a sum; as here, it may be used for additions with no concrete interpretation (length plus area). "Appending" (*waṣābum*), on the other hand, is a concrete additive operation, where one entity (as an example one may think of one's own bank account) is augmented by another (the interest of the year—actually labelled "the appended" in Akkadian) without changing its identity (it remains *my* account). Appending possesses an inverse operation "tearing out" (*nasāhum*); the other ("comparative") subtractive operation "*a* exceeds *b* by *x*" (*a eli b x iter*) is used only for concretely meaningful comparisons, and is thus not a real inverse of "accumulating."

The "igi" of a number n is its reciprocal as listed in a table of reciprocals. When having to divide by n, the Babylonians would multiply by igi n, using an operation labelled "raising" (*našûm*)—probably best explained as "calculation [of something] by means of multiplication"; other multiplicative operations are "*a* steps of *b*" (*b* a-rà *a*), designating the multiplication of a number by a number in a multiplication table; "repeating to n" (*ina n ēṣēpum*), which is indeed an n-fold concrete repetition; and "making a and b hold each other" (the most plausible reading of *a ù b šutakūlum*), which means arranging the lines [with lengths] a and b as sides of a rectangle [whose area will then be $a \cdot b$]. A variant of the latter operation is "making a confront itself" (*a šutamhurum*), which means making a the side of a square. The reverse of the latter operation is to find out what "makes [the area] B equilateral" (*B* íb-si$_8$), that is, what length a will be the side if B is formed as a square (arithmetically: $a = \sqrt{B}$). The "projection" (*waṣītum*) 1, finally, is a line segment of length 1 which, projecting orthogonally from another line segment [with the length] a, transforms it into a rectangle [with the area] $1 \cdot a = a$. Lengths are measured in the unit nindan (1 nindan = 6 m) and areas in sar (= nindan^2)

With this is mind, we can understand the text. The first line tells that we are dealing with a surface (details in the grammar seem to suggest *a field*). The sum of the measuring numbers for *the four sides* (not just *four times* the side) and the area is $41'40''$. In modern notation, if s is the length of the side, this corresponds to the equation $s^2 + 4s = 41'40''$, which is the reason that this and similar Babylonian problems are generally regarded as algebra. The second line prepares a division by 4, which takes place in line 3; in our equation, this division would express itself in a transformation into $(s/2)^2 + 1 \cdot s = 41'40''/4 = 10'25''$. The addition of 1 in line 4 would tell *us* that $(s/2)^2 + 2 \cdot 1 \cdot$

$(s/2) + 1 = 1°10'25''$; finding the equilateral corresponds to the transformation $s/2 + 1 = \sqrt{1°10'25''} = 1°5'$, leading us to the further conclusion that $s/2 = 5'$—and finally $s = 10'$.

The numerical steps of the solution are thus meaningful when seen in the perspective of symbolic algebra, yet the use of the term "projection" (and the addition of a mere "1" instead of "1^2" in line 4, which is an otherwise compulsory Babylonian practice) tells us that the Babylonian calculator operated in a very different representation—see Figure 1: Each of the four sides was thought of as provided with a projection (that is, a "projecting width") 1[4] and thus represented by a rectangle $s \times 1$; the surface was a square $s \times s$; and the sum was hence represented by a cross-shaped configuration. When the Babylonian scribe divided by 4 in lines 2–3, what he did was to single out one fourth of this configuration, for example, the gnomon in the upper left corner. The addition of "1 the projection" calls for a general commentary: *We* think of a square as *being* (for instance) 4 square feet and *having* the side 2 feet (knowing that, strictly speaking, the square is a complex configuration which can equally well be characterized by any of these parameters). The Babylonians thought of the square as *being* 2 feet and *having* an area 4 square feet.[5] Appending "1 the projection" thus means fitting in the square contained by the gnomon, each of whose sides is indeed the projection. Thereby the gnomon is completed as a square with known area $1 + 10'25'' = 1°10'25''$, which is "made equilateral" by $\sqrt{1°10'25''} = 1°5'$. From this, the projection (this time, according to *our* distinction, viewed as the side of the completing square) is torn out, leaving $5'$ as the width of the gnomon leg. "Repeating" this to two, that is, uniting it with its mirror image, produces the side of the original square, that which "confronts itself."

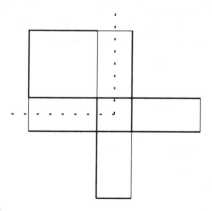

FIGURE 1
The procedure of BM 13901, No. 23.

This "cut-and-paste procedure" is "naive" in the sense that everything can be "seen" immediately to be correct. (Whenever the word is used in the following, it is to be read in this technical sense and never as "gullible.") There is no attempt to prove, for example, that the gnomon *is* a rectangular gnomon and contains precisely a square; such "critical" reflection (in a quasi-Kantian sense) had to wait until Euclid. But the procedure *can* be seen to be correct (and can be transformed into a "critical" proof without difficulty), and is thus justification and algorithm in one (as is the stepwise transformation of a modern algebraic equation). It is also analytical in the sense that the unknown side is treated as if it were known until it can be isolated from the complex relation in which it is entangled. If algebra is understood primarily as the application of analysis (as François Viète would have it), the method is clearly algebraic in nature. But if algebra is a science of *number* (or, post-Noether, generalized number) by means of abstract symbols, the Old Babylonian "algebra of measurable line segments" is *not* algebra. This proviso should be kept in mind in the following when I drop the quotes for reasons of stylistic simplicity, speaking simply of *Babylonian algebra.*

Many features of the present problem are shared by the Old Babylonian algebra texts in general: The distinction between two additive operations—that is, operations which when translated into modern equations become additions; the analogous distinction between two different subtractive and no less than four different multiplicative operations; and the use of naive cut-and-paste geometry in procedures which are their own immediate justification. Other features, however, single out the problem of "the four sides and the area" as a remarkable exception.

If by Q we designate the quadratic area and by s the corresponding side (Q_i and s_i, $i = 1, 2, \ldots$ when several squares are involved); by $_4s$ "the four" sides of a square); if $\square(a)$ stands for the area of the square on the line segment a and $\square(a, b)$ for that of the rectangle "held" by a and b, the tablet contains the following problems ($n^\prime$ stands for $n \cdot 60^1$):

1. $Q + s = 45'$
2. $Q - s = 14`30$
3. $Q - \frac{1}{3}Q + \frac{1}{3}s = 20'$
4. $Q - \frac{1}{3}Q + s = 4`46°40'$
5. $Q + s + \frac{1}{3}s = 55'$
6. $Q + \frac{2}{3}s = 35'$
7. $11Q + 7s = 6°15'$
8. $Q_1 + Q_2 = 21'40''$, $s_1 + s_2 = 50'$ (reconstructed)
9. $Q_1 + Q_2 = 21'40''$, $s_2 = s_1 + 10'$
10. $Q_1 + Q_2 = 21°15'$, $s_2 = s_1 - \frac{1}{7}s_1$
11. $Q_1 + Q_2 = 28°15'$, $s_2 = s_1 + \frac{1}{7}s_1$
12. $Q_1 + Q_2 = 21'40''$, $\square(s_1, s_2) = 10'$
13. $Q_1 + Q_2 = 28'20''$, $s_2 = \frac{1}{4}s_1$
14. $Q_1 + Q_2 = 25'25''$, $s_2 = \frac{2}{3}s_1 + 5'$
15. $Q_1 + Q_2 + Q_3 + Q_4 = 27'5''$, $(s_2, s_3, s_4) = (\frac{2}{3}, \frac{1}{2}, \frac{1}{3})s_1$
16. $Q - \frac{1}{3}s = 5'$
17. $Q_1 + Q_2 + Q_3 = 10`12°45'$, $s_2 = \frac{1}{7}s_1$, $s_3 = \frac{1}{7}s_2$
18. $Q_1 + Q_2 + Q_3 = 23'20''$, $s_2 = s_1 + 10'$, $s_3 = s_2 + 10'$
19. $Q_1 + Q_2 + \square(s_1 - s_2) = 23'20''$, $s_1 + s_2 = 50'$
20. [missing]
21. [missing]
22. [missing]
23. $_4s + Q = 41'40''$
24. $Q_1 + Q_2 + Q_3 = 29'10''$, $s_2 = \frac{2}{3}s_1 + 5'$, $s_3 = \frac{1}{2}s_2 + 2'30''$

We observe that No. 23 is the only problem referring to "the four" sides of a square. It is also the only problem mentioning the sides before the area. It is certainly not the only normalized mixed second-degree problem dealing with a single square, but all the others refer to a general method (in semi-modern terms: halving the number of sides, squaring this half, etc.). In geometric terms, a sides are expressed as $\square(a, s)$; this rectangle is bisected, and the total area $Q + 2\square(\frac{1}{2}a, s)$ is transformed into a gnomon which is then completed; etc.—see Figure 2. The procedure of No. 23, on the other hand, depends critically on the number 4; already at this point we may observe that this use of an amazing and elegant but non-generalizable solution makes the problem look more like a riddle than like a normal piece of mathematics (Babylonian or modern); so does, in fact, the presence of precisely *those four* sides which really belong to the square, instead of an arbitrary (and thus virtually general) multiple.

Other differences are no less striking. All remaining problems tell that they deal with squares by using the term which at one time designates the quadratic configuration *and* the length of the side; No. 23 is alone in stating at first that it deals with "a surface" or (probably) "a field." It is also alone in using the term translated here as "front" (*pūtum*), an Akkadian term corresponding to Sumerian sag, the "width" of a rectangle. In normal algebraic problems the Sumerian term is compulsory; the use of a word belonging

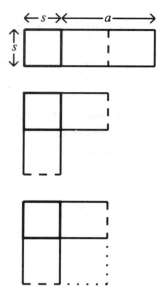

FIGURE 2
The "normal" procedure of BM 13901 for the solution of $Q + \alpha s = C$.

to the spoken vocabulary of surveyors indicates that we are supposed to think of a real piece of land.

Even the solution is uncommon. Other problems of the tablet dealing with a single square have the side equal to $30'$ (or 30), except for one case of $20'$. These are indeed the standard values of square sides in Old Babylonian algebra problems, which may have to do with the roundness of these numbers in the sexagesimal place value notation used in mathematics teaching ($30' = \frac{1}{2}$, $20' = \frac{1}{3}$).[6] All other cases where $10'$ is found are caused by the use of other favorites (ratios 4 and 7, differences $10'$ and $5'$). Only (23) (at least among those problems which are conserved) is constructed from the side $10'$ as a *deliberate* choice. And only No. 23 tells the unit of the result, as if it were to be entered into a cadastral or similar document (cf. note 6).

The final puzzling feature does not concern the problem itself but its place: Apart from No. 16 (which can be suspected of having been displaced), problems of the type $\alpha Q \pm \beta s = C$ occur in the beginning of the tablet, and the neighbours of No. 23 are considerably more complex. It seems as if the difference in method as reflected in the contrast between Figure 1 and Figure 2 was understood as a difference between mathematical genres.

II. The Proofs of al-jabr

No other Babylonian mathematical tablet contains a problem involving "the four" sides of a square or making use of

the peculiar method of Figure 1. In order to find parallels we have to make a jump to the early ninth century CE.

This was the moment when the Khalif al-Ma'mūn asked al-Khwārizmī to put together a treatise covering those parts of the field *al-jabr wa'l-muqābalah* that were either "brilliant" (laṭī f) or practically useful.[7] Al-Khwārizmī is thus not to be considered the inventor of *al-jabr* (Latinized as *algebra*), and, as we can read in a treatise by the slightly later Thābit ibn Qurrah[8], it was practiced by a group of "*al-jabr* people," evidently some kind of professional calculators. Yet within another generation or two, Abū Kāmil would regard it as al-Khwārizmī's discipline—and al-Khwārizmī appears indeed (together with his contemporary ibn Turk, from whose work only a fragment is extant) to have reshaped the discipline, in particular the treatment of second-degree problems, which was its core.[9]

The problem which we translate as $x^2 + 10x = 39$ would be formulated as follows by the *al-jabr* people: *A treasure together with 10 roots equals 39 dirhems.* Fundamentally, the problem thus tells that an unknown amount of money (the "treasure" or māl—more precisely "property") together with 10 times its [square] root (jadr) equals 39 dirhems (strictly speaking, the correct translation is hence $y + 10\sqrt{y} = 39$). They would find the root by an unexplained rule: You halve [the number of] roots (which gives 5), multiply it by itself (25), add this to the dirhems (64), take the root (8), and subtract the half of the [number of] roots. Thus the root is 3, and the treasure is 9.

This rule is given by al-Khwārizmī and repeated by Thābit ibn Qurrah. It can safely be assumed to belong to the inherited lore of the group. Al-Khwārizmī's most important innovation was to give a geometrical proof that the traditional rule (and the corresponding rules for the cases *treasure and number equal roots* and *roots and number*

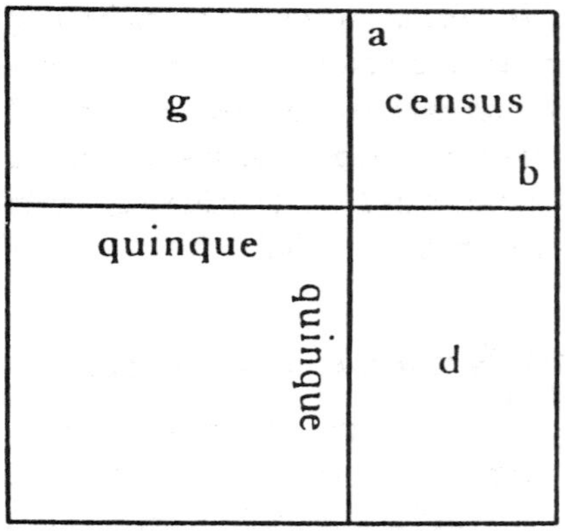

	a
g	census
	b
quinque	
quinque	d

FIGURE 3
Al-Khwārizmī's second proof. From B. B. Hughes, "Gerard of Cremona's Translation of al-Khwārizmī's *Al-Jabr*," p. 238.

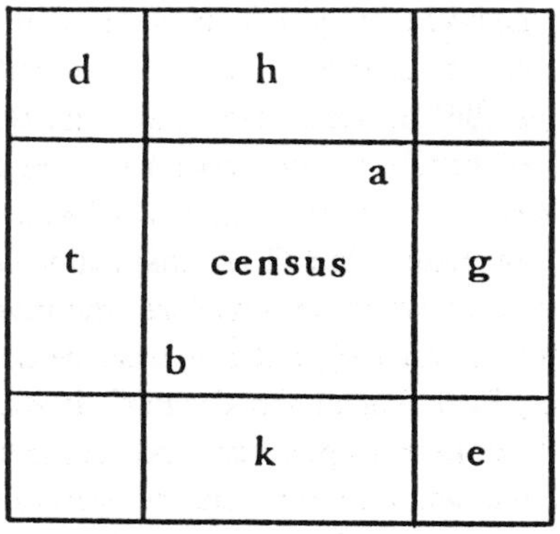

FIGURE 4

Al-Khwārizmī's first proof. From B. B. Hughes, "Gerard of Cremona's Translation of al-Khwārizmī's *Al-Jabr*," p. 237.

equal treasure) was correct. As in the Greek texts translated by al-Khwārizmī's colleagues at the Baghdad court, points and areas are labelled by letters in these proofs. In essence, however, they differ from the cut-and-paste proofs which we have encountered above only by being more precisely argued and hence less naive.

For the case *the treasure together with 10 roots equals 39 dirhems,* two different proofs are given. The second corresponds directly to the rule, and is made on a diagram similar to Figure 2 (see Figure 3, which renders Gherardo of Cremona's translation). The first corresponds to a procedure that differs from the one whose correctness is to be proved: 10 is divided by 4 ($2\frac{1}{2}$), squared ($6\frac{1}{4}$), multiplied by 4 (25), and added to 39. The diagram (see Figure 4) corresponds to that of Figure 1. There is no reason *within* al-Khwārizmī's text to bring a diagram so obviously at odds with what is to be proved—elsewhere, he confesses no particular infatuation with symmetry. If the diagram is there, it must be because it comes first to his mind, or because he expects it to come first to the reader's mind. It must hence be supposed to have been familiar either to al-Khwārizmī or to his "model reader"–not from the *al-jabr* but from some other tradition. (It is indeed also more naive in style than the following proofs.)

III. Abū Bakr's "mensuration algebra"

This conjecture is confirmed by another treatise, a *Liber mensurationum* written by one unidentified Abū Bakr. According to terminological criteria the work would be grossly contemporary with al-Khwārizmī's.[10] No manuscript of the Arabic text is known, but a careful Latin translation was made by Gherardo of Cremona.[11] Moreover, as we shall

see, Leonardo Fibonacci has used the work in his *Pratica geometrie.*

Formally, the work deals with practical geometry, and some of it really does. Thus, in the beginning of the first chapter it tells how, given the side of a square, the area and the diagonal can be calculated. Then, however, Abū Bakr goes on with "brilliant" problems of no or scarce practical interest and mostly asking for some kind of algebraic treatment; all in all, the initial chapter (on squares) contains the following problems:

1. $s = 10$: Q?
2. $s = 10$: d?
3. $s + Q = 110$: s?
4. $_4s + Q = 140$: s_u?
5. $Q - s = 90$: s?
6. $Q - _4s = 60$: s_u?
7. $_4s = \frac{2}{5} \cdot Q$: s_u?
8. $_4s = Q$: s_u?
9. $_4s - Q = 3$: s_u? (Both solutions are given.)
10. $d = \sqrt{200}$; s?
11. $d = \sqrt{200}$; Q?
12. $_4s + Q = 60$: s_u?
13. $Q - 3s = 18$: s?
14. $_4s = \frac{3}{8} \cdot Q$: s_u?[12]
15. $Q/d = 7\frac{1}{2}$: s_u?
16. $d - s = 4$: s?
17. $d - s = 5$ (No question, refers to the previous case.)
18. $d = s_u + 4$: s? (No reference is made to No. 16.)
19. $Q/d = 7\frac{1}{14}$: s?, d?

Here, Q again denotes the area and s the side of the square, d is the diagonal, $_4s$ stands for "[the sum of] its four sides" (or merely "its sides," meaning the same), and s_u for "each of its sides." (Below, A shall be used about the area of a rectangle, and ℓ_1 and $-\ell_2$ about its sides.) The next chapters (rectangles regarded as "quadrates longer on one side," and rhombi) are similarly weighted toward algebraic problems; only then come chapters dominated by genuine geometrical calculation (and clearly related to the Alexandrian/Heronian tradition). In order to possess a name for this particular kind of quasi-algebra I shall speak about "mensuration algebra"—dropping again the quotes in the following for stylistic reasons, even though the objections to this characterization of the technique as algebra *tout court* are even stronger than in the case of the scribe school discipline (cf. note 22).

Returning to the chapter on the square, we observe, first, that "the four sides and the area" turns up as No. 4, and again with a different numerical parameter as No. 12—the sides being once more mentioned first (in the *Liber mensurationum* this is the common usage); second, that all problems involving sides except No. 13 deal with *the* side or *the four* sides; later on, the sides of rectangles also invariable turn up in geometrically meaningful company— the shorter or the longer side alone, these two together, or all four together (similarly, also the diagonals of rhombi); third, that the standard square has a side equal to 10, the only real exceptions being Nos. 8–9 and Nos. 12–13.[13]

Abū Bakr solves many of the quasi-algebraic problems in what he regards as two different ways. One of these receives no special label and can thus be identified as a standard method, the method habitually belonging with the tradition of mensuration algebra as he knew it. The other is *al-jabr* (*aliabra* in Gherardo's translation). A literal translation of Nos. 3, 4, and 6 will serve as illustration:

3.   And if he [a "somebody" presented in No. 1] has said to you: I have aggregated the side and the area, and what resulted was 110. How much is then each side?

The working in this will be that you take the half of the side as the half and multiply it by itself, and one fourth results; this then add to 110, and it will be $110\frac{1}{4}$, whose root you then take, which is $10\frac{1}{2}$, from which you subtract the half, and 10 remain which is the side. Understand!

There is also another way for this according to al-jabr, which is that you posit the side as *a thing* and multiply it by itself, and what results will be *the treasure* which will be the area. This you thus add to the side according to what you have posited, and what results will be a treasure and a thing which equal 110. Do thus what you were told above in *al-jabr,* which is that you halve the thing and multiply it by itself, and what results you add to 110, and you take the root of the sum, and subtract from it the half of the root. Actually, what remains will be the side.

4.   And if he has said: I have aggregated its four sides and its area, and what resulted was 140, then how much is each side?

The working in this will be that you halve the sides which will be two, thus multiply this by itself and 4 results, which you add to 1 < 40 and what results will be 1 > 44, whose root you take

which is 12, from which you subtract the half of 4, what thus remains is the side which is 10.

6.   And if he has said: I subtracted its sides from its area and 60 have remained, how much thus is each side?

In this the working will be that you halve the sides which will be two. This you thus multiply by itself and add it to 60, and take the root of the sum which is 8, to this you thus add half the number of sides, and what results will be 10 which is the side.

But its working according *al-jabr* is that you posit the side as a thing, which you multiply by itself, and *a treasure* results which is the area. From this then subtract its four sides, which are 4 things; thus remains *a treasure* minus 4 *things* which equals 60, restore thus and oppose, that is that you restore the treasure by the 4 things that were subtracted, and join them to 60, and you will thus have a treasure which equals 4 things and 4 dragmas. Do thus what you were told above in the sixth question [of *al-jabr*], that is that you halve the roots and multiply them by themselves and join them to the number and take its root, and what results will be that which is 8. To this you then join the half and 10 results, which will be the side.

This piece of text calls for a number of commentaries. First we observe that the numerical steps of the basic and the *al-jabr* methods coincide (which is actually noticed by Abū Bakr, as can be seen by his identification "that which is 8" in No. 6). The difference between the two methods must thus depend on something else (even though, in certain other problems, the two also differ numerically).

Al-jabr is evidently the technique explained by al-Khwārizmī, and Abū Bakr's treatise on mensuration must have been produced as a companion piece to an explanation of *al-jabr*—though not to al-Khwārizmī's treatise but to something of more archaic style. This appears from certain terminological peculiarities: more precisely, from the use of the terms "restoration" (Arabic *al-jabr*) and "opposition" (Arabic al-muqābalah), precisely the ones that had given the technique its name.

Al-Khwārizmī uses "restoration" exclusively about the elimination of a subtractive term, in the way it is employed in Abū Bakr's No. 6; the elimination of a coefficient by division is termed differently, without distinction between coefficients larger than and smaller than 1.[14] In Abū Bakr's *al-jabr* expositions, "a treasure minus 4 things" is "restored" as "one treasure" by the addition of 4 things, and "one fourth of a treasure" is "restored" through the multiplication by 4 (in No. 55). In Abū Bakr's usage (which is confirmed in the standard treatment of No. 4, and again in the genuine geometrical part of the treatise, in Nos. 67, 100, and 102), restoration thus repairs any deficiency, whether subtractive or partitive. (On one occasion it even repairs an excess by subtracting it, viz in No. 55.)

"Opposition" as used by al-Khwārizmī is the converse of his restoration, the subtraction of an addend on both sides of an equation. In the *Liber mensurationum,* the meaning once again is less specific and mostly different. Where al-Khwārizmī has the recurrent phrase "restore, and add" (the restoration being the elimination of a subtractive term $-t$ on one side of the equation, and the addition the concomitant addition of an additive term t on the other), Abū Bakr has "restore, and oppose" (Nos. 5, 6, 9, etc.);[15] in one place (No. 22), the term covers an al-Khwārizmī an opposition; and repeatedly, when an entity A is "opposed with" or "by" another entity B, the meaning is that the equation $A = B$ is formed (most clearly in Nos. 41, 48, 49, and 50, but also in Nos. 7, 24, 25, 31, and elsewhere). Summed up in one concept, "opposition" means "putting on the opposite side," either in an already existing equation or by establishing an equation.[16]

Abū Bakr is not alone in not complying with the usage which was canonized thanks to the fame of al-Khwārizmī's treatise. Even al-Karajī, though he *defines* the terms as does al-Khwārizmī, uses "opposition" in Abū Bakr's way.[17] There can be little doubt that Abū Bakr's loose parlance is original and al-Khwārizmī's stricter usage an innovation, in all probability an intentional and premeditated innovation: the natural trend for the terminology of a mathematical culture undergoing a process of dynamic maturation (as that of ninth to tenth-century Islam) is to increase its precision and stringency, not to abandon its accuracy. Abū Bakr's *al-jabr* is thus pre-al-Khwārizmī an, if not *necessarily* by date then at least in substance and style (but, given the triumph of al-Khwārizmī's *Algebra,* it cannot then be too much later).

So much for the *al-jabr* method. Returning to the standard method, we remember that it did not (or did not always) differ from *al-jabr* in its numerical steps. Nonetheless it was regarded as something different by Abū Bakr. Why?

A first observation to make is the care with which the *al-jabar* sections explain that the treasure represents the area of the square, and the root (or "the thing," which is used in the same sense until standard equations are derived)[18] its side. The implication is that treasure and root/thing are not in themselves understood geometrically

but as numbers. The basic method may then differ from *al-jabar* precisely by referring directly to the geometric method.

This conjecture is confirmed by several further observations. One concerns the word "understand" (*intellige* in the Latin text), whose occurrences are scattered throughout the work, in somewhat varying contexts. On two occasions, the word stands as an exhortation to penetrate a deliberately opaque and superfluously intricate computation and to grasp why it works after all (Nos. 50 and 74). In a number of questions concerned with genuine geometrical computation it asks the disciple to look at or understand from actually appearing diagrams why the computation is correct (a square with diagonal in No. 2; an isosceles trapezium in No. 78; etc.). This recalls another Gherardian translation from an Arabic text, according to which the Indians "possess no demonstration [for a particular construction] but only the device *intellige ergo*"—where indeed Indian geometrical texts have the phrase *nyāsa,* "one draws" (etc.) followed by a diagram when they want to illustrate a rule, algorithm or algebraic identity which has just been stated.[19] Finally, the word is used repeatedly as in No. 3, that is, after the presentation the standard solution (but not the *al-jabar* solution) of a quasi-algebraic problem. Even though no diagrams are given on these occasions in Gherardo's version, the parallel to the real geometric problems suggest that here too the exhortation may have referred originally to understanding through a diagram—in No. 3 to a diagram similar to Figure 2.

Significantly, some of the solutions which carry the "understand" are termed in a way which shows that the original constitutive geometrical entities are thought of all the way through. One instance is No. 43, dealing with a rectangle (a "quadrate longer on one side") and indeed a rectangular version of "the four sides and the area":

> If indeed he has said to you: I have aggregated its four sides and the area, and what resulted was 76; and one side exceeds the other by two. How much thus is each side?
>
> The way to find this will be that you multiply the increase of one side over the other, always [that is, whatever the actual excess] by 2, and what results will be 4. Therefore subtract this from 76, and 72 will remain. Next aggregate the number of sides of the quadrate, which is 4, and join it to the increase of one side over the other, and what results will be 6. Thus take its half, which is 3, and multiply this by itself, and 9 results, which you join to the 72, and 81 results. Then take its root, which is 9, and subtract from

it the half of 6, which is 3, and the shorter side will remain, which is 6. To this then add 2, and the longer side will be 8. Understand.

The way according to *al-jabar*, however, ...

The numerical steps can be explained in several ways. Algebraically, we may call the width z, and the length thus $z + 2$; proceeding mechanically from here we get Abū Bakr's *al-jabar* procedure. Or we may call the two sides x and y ($x = y + 2$), and observe that the area plus the sides is then $x \cdot y + 2x + 2y = x \cdot y + 4y + 2 \cdot 2 = (x + 4) \cdot y + 4$; if $X = x + 4$, we therefore have $X \cdot y = 76 - 4 = 72$, $X = y + (2 + 4) = y + 6$. The problem has thus been reduced to finding the sides of a rectangle whose area is $76 - 4 = 72$ (4 being 2 *the excess* times *invariably* 2), and whose length exceeds the width by $2 + 4$ (4 being the number of sides). This interpretation makes sense not only of the numbers but also of most of the words of the text—including the use of the identity-conserving "joining" of 4 to the excess, since the result is still an excess. (As the Old Babylonian texts, Abū Bakr distinguishes between additions, even if less sharply.)

Still, some formulations remain unexplained, and x's and y's are anyhow anachronistic. The second interpretation therefore has to be reinterpreted itself in order to become relevant. This is done in Figure 5: Initially, the sides are thought of as provided with the standard width 1 (the "projection" of our Old Babylonian texts).[20] The excesses are cut off, after which the sides are "aggregated," and collectively "joined" to the excess. The rest goes as in Figure 2: The excess of the rectangle over the square is bisected, and a gnomon is formed, to which the quadratic complement is "joined," etc.

That the text refers to something more than mere numbers is confirmed by the recurrent phrase "what results/remains will be . . . ." The *al-jabr* sections (where we have the advantage of knowing what goes on) demonstrate that the phrase is no mere stylistic whim. Here the phrase also turns up time and again–but never in places where "what results" is nothing but the outcome of a computation. Instead of "what remains will be 72," such passages simply tell that "72 results." Invariably, "what results" is either a composite algebraic expression or equation, or a *something* which is identified with *something different*— as in the end of No. 3, where the numerical outcome of the algorithm is told to be the side, and again toward the end of No. 6.

Even within the descriptions of the standard method, we therefore have to read the phrase "what results will be a" as "the thing which results will have the numerical value a." But since it is never explained (as done in the

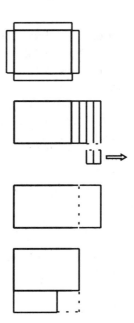

FIGURE 5
Liber mensurationum, the procedure of No. 43.

al-jabar sections) that something different represents the geometrical entities that the problems deal with, then the "things" whose existence is presupposed must be geometrical entities, derived by means of geometrical operations from the entities referred to in the statement. In No. 43, "the thing that is 4" will hence be the piece which is removed from the two rectangles representing the lengths–that is, the small square that is eliminated in the second step in Figure 5; and "the thing that is 6" will be the excess of the new length over the width.

No. 38—a kind of rectangular counterpart of No. 1— may be even more elucidating, because the solution builds on a fallacy which turns out to make excellent sense in a diagram:

> If indeed he has said to you: I have aggregated its longer and shorter sides and the area, and what resulted was 62; and the longer side exceeds the shorter by two. How much, then, is each side?
>
> The way to find this will be that you subtract 2 from 62, and 60 remains, then add 2 to the half of the number of sides, and 4 results. Join this to 60, and 64 results. Thus take its root, which is 8. This, in fact, is the longer side. And if you want the shorter, subtract 2 from 8, and 6 remains, which is the shorter side.

Figure 6 shows what goes on: We start as before, but this time, taking advantage of the coincidence between the number of sides involved and the excess (and thereby depriv-

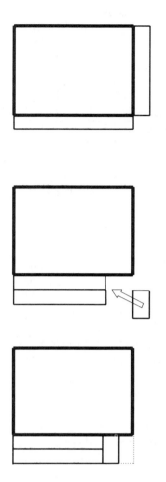

FIGURE 6
Liber mensurationum, the procedure of No. 38.

that the two together contain the completing square (the number of sides translated into 'projections")—that is, in a geometrical representation drawn or imagined in more or less correct proportions. All in all we may confidently conclude that Abū Bakr's standard method was based on geometrical operations—and that at least the method used in the problems translated above was in naive cut-and-paste style.[21] Moreover, the geometrical operations concern the very entities which define the problems[22]—and these, as pointed out in passing above, are always geometrically meaningful. They do not involve entities like αQ or βs (or $\gamma \ell_1 - \delta \ell_2$) but instead: the single area; the side, both sides, or all four sides; the two diagonals of a rhombus; etc.

The geometrical technique of Abū Bakr's mensuration algebra recalls what one encounters in Old Babylonian texts, and "the four sides and the area" certainly recalls BM 13901, No. 23. No surviving Babylonian problem possesses precisely the structure of Abū Bakr's Nos. 38 and 43, but one text (also belonging to the early phase of the development of Old Babylonian algebra) contains a close parallel, which happens also to make use of a trick for its solution which corresponds to a change of variable: AO 8862 No. 1.[23] Here, in symbolic translation, $x \cdot y + (y - x) = 3`3°$, $y + x = 27$; by addition, $x \cdot y + 2y = (x + 2) \cdot y = 3`30°$ or $X \cdot y = 3`30°$, $y + X = 27 + 2 = 29$.

Several other similarities between the Old Babylonian corpus and the standard part of Abū Bakr's quasi-algebraic problems can be enumerated: in particular, certain shared characteristic methods; furthermore, a highly systematic and rather intricate shift between past and present tense and between the first, second, and third grammatical person (there is also one significant though only partial divergence in this domain, which we shall discuss below). We may thus safely conclude that the two kinds of quasi-algebra are somehow connected. *How* they are connected is a question to which we shall return.

IV. Twelfth- and thirteenth-century evidence

First, however, we shall look at two later authors who still draw on the same tradition: Abraham Bar Ḥiyya—better known as Savasorda, from a twisted pronunciation of his court title—and Leonardo Fibonacci.

Savasorda's early twelfth-century *Ḥibbur ha-mešiḥah we'tišboret (Collection on Mensuration and Partition),* translated into Latin by Plato of Tivoli as *Liber embadorum (Book of Areas)*[24], has its main emphasis on genuine geometrical computation, in clear contrast to Abū Bakr's work. Equally in contrast to Abū Bakr, Savasorda also draws on the *Elements,* first in the initial chapter, where he copies

ing the solution of any general validity), we produce the gnomon by moving the width to a position along the length and splitting off the excess from the length. The gnomon is completed as a square by fitting in the loose end of the length together with another piece (with width 1 and length) equal to "the half of the number of sides" (that is, equal to the number of sides actually involved). The area of the completed square being 64, its side (which equals the length according to the diagram) is 8.

The correct solution of No. 43 might in principle have been obtained by means other than the use of a diagram (there are always many ways to obtain a correct result), even though it seems difficult to explain the precise phrasing without the geometrical cut-and-paste interpretation. The lapses of No. 38, on the other hand, can have resulted meaningfully only from a representation where it goes without saying, firstly that the excess of length over width equals the number of sides involved, and secondly the definitions from *Elements* I and VII and a number of theorems, and later in the work in a number of proofs. At one point (chapter 2, part 1, §7), however, he tells that, before going on with triangles and with those quadrangles whose treatment presuppose triangulation, he will present some problems "so that by solving them, with God's assistance you may prove yourself a keen and swift enquirer." First come some problems concerning squares:

§8. $s = 10$, d?
§9. $d = \sqrt{200}$; s?
§10. $Q - {}_4s = 21$, Q? s?
§11. $Q + {}_4s = 77$, Q? s?
§12. ${}_4s - Q = 3$, s_u? (Both solutions are given).

Without doubt Savasorda has borrowed this sequence of problems, and no doubt it is related to what we encountered in the *Liber mensurationum*. It is uncertain, however, and rather implausible that he used Abū Bakr's manual directly. If he had done so and then made the present meagre selection, changing furthermore the order in §§9–11 and the value of the unknown in §§10–11, it does not seem likely that he would keep §12 unchanged. (Comparison between the treatments of rectangles in the two treatises supports this conclusion.) That the side of §§10–11 is precisely 7 is also in itself noteworthy, as possibly related to the crude approximation that was behind Abū Bakr's Nos. 16 and 18 (side 10 and diagonal 14).

Abū Bakr's standard method appeared to be a geometrical cut-and-paste procedure referring to geometrical diagrams, but at least Gherardo's translation brings no diagrams beyond those that show the square, the rectangle, the rhombus (etc.) with which the problems deal. Savasorda's manual does contain diagrams demonstrating the correctness of his solutions. (On the other hand, Savasorda provides no *al-jabr* solutions.)[25] Formally, however, these refer to the Euclidean theorems which are reported in the introduction. It is therefore possible that they have been associated afresh with the traditional problems by some editor (Savasorda or a predecessor) with Euclidean schooling or familiar with Thābit ibn Qurrah's *Verification of the Problems of Algebra through Geometrical Demonstrations* (which proves the correctness of the standard algorithms of "the *al-jabr* people" for the solution of mixed second-degree problems by means of *Elements* II.5–6 in a way which is very similar to Savasorda's). It could also be, however, that this editor simply reformulated a number of traditional and still current naive geometrical procedures in Euclidean style–this would be quite easy, since the Euclidean theorems in question look precisely like "critical" recastings of a naive cut-and-paste inheritance. (Compare,

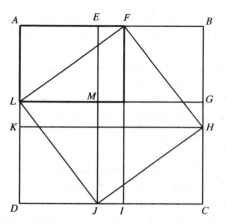

FIGURE 7
The naive diagram showing that $a^2 \pm 2A = (\ell_1 \pm \ell_2)^2$ in a rectangle

for instance, *Elements* II.6 with Figure 2; the argument is specified below, see p. ???.) In other words, it is possible but not sure that Savasorda's diagrams descend directly from the procedures traditionally connected with his quasi-algebraic problems.[26]

Leonardo Fibonacci wrote his *Pratica geometrie* (see note 4) in 1220, and certainly drew on many sources. As Maximilian Curtze pointed out in the critical notes to the Liber embadorum, Savasorda is one of them. The whole structure of the work indicates that Leonardo has read the Liber embadorum. Quite a few of the shared features, however, derive not from direct borrowing but from one or more shared sources.

This regards precisely the group of problems which concerns us here. As pointed out by Curtze, Savasorda's §§8–12 recur in the *Pratica*. Their order, however, has been changed, as have some of the parameters ($+n$ counts lines from the top, $-n$ from the bottom of the page).

p. 58^{+6}. $s = 10$, d?
p. 58^{-3}. $d = \sqrt{200}$; s?
p. 59^{+5}. $Q + {}_4s = 140$, Q? s?
p. 59^{-15}. $Q - {}_4s = 77$, Q? s?
p. 60^{+10}. ${}_4s - Q = 3$, s_u? (Both solutions are given).

The formulations, furthermore, are wholly different from Savasorda's, even though at other places (for example, when Abū Bakr's No. 38 is reproduced—cf. below) the phrases of a source are taken over without any change beyond grammatical polishing. Most decisive, however, is that several of Leonardo's deviations from Savasorda agree with the "background tradition" as we know it from Abū Bakr. Like the latter in Gherardo's translation, Leonardo refers to *quatuor eius latera,* while Savasorda takes away *omnium suorum laterum in unam summan collectum*; and, like Abū Bakr, Leonardo's side in the problem $Q+_4 s = A$ is 10.[27]

There can be no doubt that Leonardo had Gherardo's version of the *Liber mensurationum* (in full or in excerpt) on his desk while writing parts of the *Pratica*. A striking proof is provided by the problem dealt with from p. 66[−13] onward, which coincides with Abū Bakr's No. 38 (see above, p. ???):[28]

> Again, the two sides with the expanse amount to 62; and the larger side exceeds the smaller by two. How much then is each single side?
>
> The way to find this will be that you subtract 2 from 62, and 60 remain, then add 2 to the half of the sides, and 4 result. Join this to 60, and 64 result. Thus take their root, which is 8. That, in fact, is the longer side. And if you want the shorter, subtract 2 from 8, and 6 remain, that is the shorter side. For example: posit the smaller side as a thing, then the larger will be a thing and two dragmas. From the multiplication of this shorter side by the longer results the expanse. Therefore multiply the thing, that is the smaller side, by the thing and by two dragmas, and you will have a treasure and two roots as the expanse; which, if you add to them the two sides, namely 2 roots and 2 dragmas, will be a treasure and 4 roots and 2 dragmas, which equal 62 dragmas. Remove 2 dragmas in each place, and a treasure and 4 roots remain, which equal 60, and so on.

We see that the statement differs from Abū Bakr's—among other things, Leonardo speaks here about the "larger" and "smaller" side, where Abū Bakr/Gherardo has "longer" and "shorter." In the end, Leonardo gives a solution by means of *al-jabr* (which he seems to regard as an explanation, even though completion of the *al-jabr* procedure would highlight the fallacy),[29] where Abū Bakr has none in this particular problem. In the description of the standard procedure, however, all he has done is to change the grammatical number, considering "60" etc. as plurals and not singulars.

In other places, Leonardo has geometrical proofs, some of them similar to those of Savasorda. We may look at Leonardo's treatment of "the four sides and the area" (p. 59[+5]):

> And if the surface and its four sides make 140, and you want to separate the sides from the surface. Let a quadrate *ezit* be put together, and the rectangular surface *ae* added to it. And let *ai* prolong the straight line *it*, and be prolong the straight line *ez*; and let each of the straight

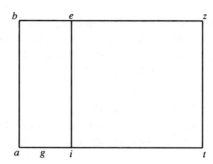

FIGURE 8
Leonardo's diagram for "the area and its four sides make 140."

lines *be* and *ai* be 4 because of the number of the sides of the quadrate; because the surface *ae* equals four sides of the quadrate *et*, since the side *ei* of the latter is one of the sides of the surface *ae*; and the surface *et* contains indeed the expanse of the quadrate *zi*, and [not] its four sides. Therefore the surface *za* is 140; and that is what we have said, namely that the treasure with four roots equal 140; and the treasure is the quadrate *et*, and its four roots are the surface *ae*. Divide indeed the straight line *ai* in two equals at the point *g*; and because the line *ti* is added to the line *ai*, then the rectangular surface *it* on *at* with the square on the line *gi* will be equal to the quadrate on the line *gt*. But the surface *it* on *at* is as the surface *zt* on *at*, since it is equal to *tz*. Thus the surface *zt* on *at* with the square on the line *gi* equals the square on the line *gt*. But *zt* on *at* is the surface *za*, which is 140. Which, when the square on the line *gi*, namely 4, is added to them, give 144 as the quadrate on the line *gt*; therefore *gt* is 12, namely the root of 144. Therefore, if *gi*, namely 2, are dropped from *gt*, remains it as 10, which is the side of the quadrate *et*; whose expanse, namely 100, if its four sides are added, which are 40, will be 140, as claimed. And like this is done in all questions in which a number equals one square and roots, namely that to this number is added the square of the half of the roots, and the root of the sum is found; from which the half of the posited roots is removed, and the root of the treasure which is asked for will remain; which when multiplied by itself makes the treasure. For example: 133 dragmas equal one treasure and twelve roots. Therefore, if we add the square on the half of the roots, namely 36, to 133, they will make 169; when 6,

56   VITA MATHEMATICA

namely the half of the roots, is subtracted from its root, namely from 13, 7 will remain as root of the treasure asked for; and the treasure will be 49.

The geometrical proof is similar to Savasorda's (and Thābit's), and the same observations could be made. The treatment of the problem "the two sides with the expanse amount to 62 ..." (above, p. ???) supports the conclusion that Leonardo has no direct access to the naive procedures which had still been known to al-Khwārizmī and Abū Bakr. It is also characteristic that Leonardo gives only an *al-jabr* treatment of the "four sides and rectangular area" (Abū Bakr's No 43, where the naive procedures were most clearly reflected in the phrasing—see above, p. ???).

This would go by itself if Leonardo's only windows on the tradition were Savasorda and Abū Bakr/Gherardo. Plausibly, however, he has known also at least one other version of Abū Bakr's manual or a close relative of this work. Gherardo, indeed, had worked on a defective manuscript, as revealed by certain corrupt passages and by references backward to problems which in the actual manuscript come later. Among the seemingly corrupt passages is the solution of problem No. 14, "I have aggregated the four sides [of a square], and they are 3/8 of its area." At the corresponding place, Leonardo has "the four sides and 3/8 of the expanse equal $77\frac{1}{2}$." It is unlikely that Leonardo (who was a fairly systematic writer) should have produced this problem in order to repair the defect in Gherardo's version, since the problem is preceded by $_4s = \frac{2}{9}Q$, and followed by $_4s = Q$ and $_4s = 2Q$. It is also remarkable that Leonardo this time mentions the sides before the area, as done by Abū Bakr and in our Old Babylonian tablet. In the preceding treatment of the problem "sides plus area equal 140," Leonardo has indeed normalized the order of the members; there is certainly no reason to expect that he would innovate in this respect when repeating an inherited problem and return to the ancestral idiom when inserting a problem of his own making. The problem will hence have been borrowed, if not from a different version of the *Liber mensurationum,* then from its closest kin.

Savasorda, Gherardo and Leonardo have thus been in touch with at least three different versions of the quasi-algebraic tradition to which the problem of "the four sides and the area" belongs. (As we shall see below, Pacioli seems to use material stemming from a fourth version.) All these versions, however, appear to have lost contact with the original naive-geometric techniques, replacing (or possibly recasting) those proofs which allowed that with corresponding propositions from the *Elements* II, and handing down those solutions which did not allow such Euclidiza-

tion (like Abū Bakr's Nos. 38 and 43) without geometrical support (which explains why Leonardo gave up in front of No. 38, cf. above, note 29).[30]

The transformation of the tradition between Abū Bakr's and Leonardo's time, and its gradual assimilation to an increasingly geometrized *al-jabr* tradition, are shown also by another feature. Abū Bakr, as we remember, took great care to distinguish the "standard procedure" from the *al-jabr* method, and to explain how "the treasure" of the latter represented the area of the square (etc.). Savasorda, as we saw, was even more respectful of the geometrical tradition, and does not mention the *al-jabr* tradition (which would anyhow, one may presume, not have been be very informative for his intended public); his only algebraic theory is borrowed from the *Elements* II. Leonardo, as we see, and, as it is made even more explicit in the beginning of the section on quadrilaterals (pp. 56f), has abolished the distinction completely. Where al-Khwārizmī tells number to fall into three classes, *roots, treasures,* and simple numbers without any reference to either[31], Leonardo tells the three natures of numbers and their fractions to be *roots of squares; squares;* and simple numbers: this in spite of obvious al-Khwārizmī an inspiration for the passage in question (revealed by characteristic phrases borrowed from Gherardo's translation of al-Khwārizmī).

Savasorda's and Leonardo's texts thus tell us two things: first, that the tradition carrying the problem about "the four sides and the area" was still present in their world; second, that it had been reduced to a shadow; after having served al-Khwārizmī's coordination of *al-jabr* with geometry, and after centuries of coexistence with the Euclidization of applied geometry, it had no mathematical standing of its own, and it survived only as a collection of venerated problems. As Gherardo must somehow have tried to express when translating Abū Bakr's *al-jabr* as *aliabra, algebra* had come to encompass much more than the purely numerical technique of the pre-al-Khwārizmīan *al-jabr* people.

V. Reconstructing the process

In the closing section we shall consider the end of the disintegration process. Since, however, the forces at work in this phase differ from those which shaped the earlier development, it may be convenient to discuss first what we can learn about the prehistory of algebra by following the career of "the four sides and the area" and its cognates from the cradle through the High Middle Ages. This we shall do, on one hand by summing up and connecting observations which were already made above, on the other by drawing new conclusions.

The first question concerns precisely the cradle. Our earliest encounter with the tradition and the characteristic problem embodying it was in an odd corner of an Old Babylonian mathematical scribe school text. Several features of the formulation of the problem, however, hinted at real surveying practice—and our next encounter with the problem was in an Islamic handbook concerned with that very practice. Is it likely that a problem created within the tradition of scribe school algebra but dressed as a real problem for surveyors would be adopted by these together with a narrow selection of other problems and continued as a tradition of mensuration algebra, while the main body of Old Babylonian algebra would remain the exclusive property of the scribe school and die with it? Or should we rather expect the scholar-scribes to be the debtors?

The question is a variant of a traditional problem of folklorists: are folktales *gesunkenes Kulturgut,* as the Romanticists believed, or not? Are folktales the remnants of myths and high-level literature, or are myths created on the basis of folk tale motifs? In the final instance: is genuine culture produced by prophets, priests and scholars alone, and the low culture of other strata merely derivative, misconstrued, and defective?

Several observations speak decisively against the hypothesis of a scribe school descent, and in favour of an origin of the mensuration algebra among practical geometers. One of these is the length of the side of the Old Babylonian version of "the four sides and the area." As in Abū Bakr's and Leonardo's corresponding problem, it is ten—but ten minutes. Now, 10 is an obvious choice in any culture using a decadic number system; $10'$, however, is not—neither *a priori* nor according to the Old Babylonian tablets. Indeed, 10 in any order of sexagesimal magnitude (including $10°$) would be an untypical side length in any Old Babylonian text. It is highly improbable (to say the least) that the queer problem should have been invented within the scribe school and been constructed around the anomalous value of the unknown side, and then taken over by people who by accident could correct $10'$ (which they would see as $1/6$) into the obvious value 10. The scribe school mathematician, however, if borrowing a problem with the parameter 10, could reasonably be expected to locate this number in his habitual order of magnitude, which in the tablet in question is that of minutes.

Another observation has to do with the topic and general character of the problem. As already hinted at, the combination of the geometrically meaningful (*all four* sides of a square field) with the practically meaningless (which

practitioner ever knew the sum of the sides and the area without first knowing them separately?) gives the problem the character of a bizarre *riddle.* Such riddles, when mathematical, are known as recreational problems. In pre-Modern times, they were transmitted within environments of mathematical practitioners, where they served the purpose told by Savasorda: "that by solving them, with God's assistance you may prove yourself a keen and swift enquirer"; or, in another formulation taken from a Carolingian problem collection (I quote the puzzle in full):

> A paterfamilias had a distance from one house of his to another of 30 leagues, and a camel which was to carry from one of the houses to the other 90 measures of grain in three turns. For each league, the camel would always eat 1 measure. Tell me, whoever is worth anything, how many measures were left.[32]

In other words, these problems—which according to their dress belong within the domain of the practitioners in question (surveyors and caravan traders, respectively) but which are more complex or more bizarre than the problems solved in everyday practice—serve to train the mental agility and enhance the professional self-esteem of the members of the craft (whence the term "brilliant" used by al-Khwārizmī to characterize the useless second-degree part of *al-jabr*—see above, p. ???).[33] Invariably, they have something stunning in their formulation: unless a clever trick is applied (an intermediate stop), the camel will eat *exactly* everything; in another widespread problem, 100 monetary units will buy *exactly* 100 animals; repeated doublings run to 30 or 64, because this fits the days of the month or the cases of a board game; etc.[34]

The topic—the real sides of a real field; the striking parameter—exactly all four sides; and the solution by means of a doubly weird trick—quadripartition and quadratic completion: all three features indicate that "the four sides and the area" was hatched not in a scribe school but in a non-scholastic environment of practical geometers.

A third observation allows us to locate this environment tentatively in time and space. As stated above (p. ???), Abū Bakr's discourse is astonishingly close to what we find in Old Babylonian school texts. There is one exception to this rule, however. Abū Bakr always has a hypothetical "somebody" posing the question (in the first person singular, past tense). Old Babylonian texts, instead, start directly with the question (as in BM 13901, No. 23), implying that it is the teacher who asks. One group of texts, however, starts its problems with the familiar "if somebody has asked." These texts come from Tell Ḥarmal and Tell Dhiba'i, both in the Kingdom of Ešnunna, and belong to the earliest eighteenth century B.C.[35] Ešnunna is an early focus for that Akkadian scribal culture which arose around the mid-Old Babylonian period: late nineteenth century Ešnunna produced the first law code in Akkadian, half a century in advance of the Codex Hammurapi. Since algebra is an Akkadian genre with no identified Sumerian antecedent, Ešnunna may thus be the location where the recreational lore of Akkadian-speaking practical geometers was adopted into the curriculum of the Akkadian scribal school.

An Akkadian origin fits the side of our square field. Akkadian, as Arabic (and as the likely intermediate carrier language of our tradition, Aramaic), is a Semitic language and has a decadic number system. It also fits the name "Akkadian method" given to the quadratic completion in a late Old Babylonian mathematical text; it agrees with the observation made by Robert Whiting that the problems contained in a school text from the Old Akkadian period (the 22nd century B.C.) dealing with area measurement are so much facilitated by familiarity with the geometric-"algebraic" rule $(R-r)^2 = R^2 - 2Rr + r^2$ that this rule is likely to have been presupposed; and it matches the presence of a tablet with a bisected trapezium (another favorite problem following our tradition until Abū Bakr and Leonardo) in an Old Akkadian temple.[36] It looks as if already the Old Akkadian scribe school had adopted part of the recreational lore of the Akkadian surveyors, but that the strictly utilitarian neo-Sumerian school (21st century B.C.) did not transmit it.[37]

Since there is, anyhow, close affinity between the Old Babylonian scribe school algebra and the tradition of mensuration algebra, it is reasonable to assume the former to have developed from the adoption of the latter under the fecundating influence of the systematic spirit of the school. The quadratic completion, originally another weird trick comparable to the quadripartition and the intermediate stop, may have been the cornerstone on which the whole stupendous edifice of Old Babylonian algebra was erected.

The overlap between the algebra of the scribe school and that of the *Liber mensurationum* (and other post-Babylonian sources) allows us to draw up a list of problems which can be ascribed with some confidence to the mensuration algebra of the early Old Babylonian epoch. Of course (sticking to the symbols introduced on p. ???), $s + Q = \alpha$ and $_4s + Q = \beta$ (we may even be confident that $\alpha = 110$, $\beta = 140$); probably also problems with differences (area minus side(s), and side(s) minus area) and questions about the diagonal when the side is given, and vice versa. For rectangles, furthermore, $A = \alpha$, $\ell_1 \pm \ell_2 = \beta$; $A + (\ell_1 \pm \ell_2) = \alpha$, $\ell_1 \mp \ell_2 = \beta$; $A = \alpha$, $d = \beta$ (this lat-

ter problem is found on the Tell Dhiba'i-tablet). Highly likely is also the presence of problems dealing with several squares, at least $Q_1 \pm Q_2 = \alpha$, $s_1 \pm s_2 = \beta$ (a partial alternative, less plausible however, is the presence of the rectangle problems $\ell_1 \pm \ell_2 = \alpha$, $d = \beta$).[38] Rhombi and right triangles (both of which are used as pretexts for the formulation of quasi-algebraic problems in the *Liber mensurationum*) seem to be beyond the horizon, as is anything involving non-right triangles.

Old Babylonian scribal algebra developed into a sophisticated discipline, but most of its higher achievements were lost when the Old Babylonian era was interrupted by conquest and social breakdown after 1600 B.C., at which occasion the scribe school also disappeared. The late Babylonian period, in particular in the Seleucid era (from 300 B.C. onwards), produced a certain revival of algebraic activity, it is true; discontinuity in the use of Sumerian word signs demonstrate, however, that much of the transmission had taken place outside the scribal environment, and that a readoption of material from the mensuration algebra tradition occurred.

In the meantime, it appears that new problem types had been invented or imported into this tradition. The most systematic Seleucid treatment of second-degree problems is found on the tablet BM 34568.[39] All problems except two deal with rectangles, where various combinations of sides, diagonal and area are given.[40] With a single exception, the rectangle problems recur in the *Liber mensurationum* (at times with other parameters); moreover, the exception ($\ell_1 + D$ and $\ell_2 + d$ given) is not really one, since Abū Bakr's No. 36 ($\ell_1 + d$ and $\ell_1 - \ell_2$ given) is reduced to the Seleucid problem and then solved in the same way.

Interestingly, the only rectangle problem dealing with a diagonal of whose presence in the early mensuration algebra we are sure (*viz* $A = \alpha$, $d = \beta$, found in the Tell Dhiba'i tablet) is absent from the Seleucid anthology. Also interesting is one of the two problems in the tablet which does not consider rectangles. It deals with a reed leaning against a wall, and is equivalent to the rectangle problem $d - \ell_1 = \alpha$, $\ell_2 = \beta$ (Abū Bakr's No. 31). Nothing with the same mathematical substance is found in the Old Babylonian corpus. The dress, on the other hand, is familiar, but originally it covered a problem translatable into the much more trivial $d = \alpha$, $\ell_1 = \alpha - \beta$.

On the whole, the Seleucid tablet thus looks like a listing of *new* problems; the reed problem may be meant to demonstrate how this fascinating new wine could be poured into an old cherished bottle, thereby lending new quality to both. In any case, and quite in contradiction to the traditional view, the tablet demonstrates the discontinuity of Babylonian mathematics in spite of apparent continuity.[41]

Also at variance with widespread convictions, but the other way round, is the perspective we get on the core of *Elements* II if we correlate propositions 1 to 10 of the Euclidean work with what we have come to know about mensuration algebra.[42] Postponing for a moment propositions 1 to 3, the rest can be seen as quasi-Kantian critiques of the familiar procedures: Proposition 4 is used, e.g, by Leonardo when he finds the sum of the sides of a rectangle from the diagonal and the area, while Savasorda (proceeding like the Tell Dhiba'i text) finds their difference via Proposition 7.[43] Proposition 6 explains the solution of all problems $Q \pm \alpha s = \beta$ (including "the four sides and the square") and $A = \alpha, \ell_1 - \ell_2 = \beta$ (and Leonardo quotes it on these occasions). Proposition 5 has a similar relation to rectangular problems $A = \alpha, \ell_1 + \ell_2 = \beta$ and to $\alpha s - Q = \beta$ (again noticed by Leonardo). Proposition 7, beyond the use made of it by Savasorda, explains the rule which seemed to be presupposed already in an Old Akkadian school text (see above, p. ???). Proposition 8 does not seem to enter any problem directly which we have discussed so far; but it may be connected to the configuration of "four sides and area" (showing that, if we add the four sides to a square $\square(s)$, we do not get a square $\square(s+2)$—instead, we have to add the four sides of the average square $\square(s+1)$; conversely it can be linked with the concentric inscription of one square into another (also familiar from Old Babylonian practical geometry). Propositions 9 and 10, finally, which like Proposition 8 serve nowhere else in the *Elements* (and which must therefore have been supposed to possess a value of their own),[44] solve the problems where the sum of two square areas and either the sum or the difference between their sides are known[45]. (Leonardo also makes appeal to Proposition 10 a couple of times.)

The proofs of Propositions 9 and 10 are obviously of the Greek and not the naive type. The others, however, fall into two sections, of which the second is in essence a cut-and-paste proof, and the first explains why the various constituents of the diagram are really squares, rectangles, etc. Section 1, we may say, takes care that the subsequent cut-and-paste section is not naive.

Propositions 1 to 3 have a similar function. Proposition 1 is a general "critique of mensurational reason," justifying the cutting and pasting of rectangles; Propositions 2 and 3 apply this insight to the particular situations where sides (provided with a "projection," it goes by itself) are added to or subtracted from a square.

Elements II.1–10, we may hence conclude, is closely connected to the cut-and-paste mensurational algebra and is

precisely, as formulated above, *a critique*. We may observe, furthermore, that the whole group of propositions points back to the stock of problems and procedures which seems to have been present in Old Babylonian times. There is no trace of the new problem types from the Seleucid tablet.

Arguments can be given that the kind of area geometry which was canonized in *Elements* II was developed in the fifth century B.C. in connection with a theoretical investigation inspired by surveyors' geometry and algebra.[46] If this is really so, then there is some reason to believe that the new problems reached or arose in the Near Eastern and Mediterranean world after 500 B.C., but before 200 B.C.. We may think either of the contacts resulting from Alexander's conquests, or of the general establishment of cultural interaction along the Silk Road.[47]

It may be added that the small group of second-degree problems in Diophantos's *Arithmetica* I also refer to what appears to be the original core of the mensuration algebra: a rectangle with given area and given sum of (Proposition 27) or difference between (Proposition 30) the sides; and two squares with given sum of the sides and given sum of (Proposition 29) or difference between (Proposition 29) the areas.

The next occasion on which the tradition of mensuration algebra turns up in familiar sources is at its encounter with the numerical *al-jabr* practice, and when al-Khwārizmī draws upon its cut-and-paste technique in order to demonstrate the correctness of the *al-jabr* calculations. These geometrical proofs were already discussed above and need not be taken up again. Only one observation should be added: when teaching the addition and subtraction of binomials involving roots, al-Khwārizmī's standard exemplification of the root—that is, we must presume, the first square root which his reader is expected to recognize as not reducible to a number—is $\sqrt{200}$, the diagonal of our familiar 10×10-square. Unless this concurrence is purely accidental (which is not likely—see also note 13 on the possibility to distinguish chronological strata in the mensuration tradition by means of changing approximations to this length), the practice from which al-Khwārizmī borrowed his proofs thus appears to have been fairly well known.

Mensuration algebra did not disappear as an independent tradition after al-Khwārizmī's integration of its methods with *al-jabr*. As we have seen, at least three or four different versions could be found in the Islamic world in the twelfth and thirteenth century. But as we have also seen, it had lost its *raison-d'être* as a separate mathematical tradition. In this as in other fields, Islamic mathematics initiated an integration of theoretical and practitioners' mathematics which was, in the Modern epoch, to transform the latter enterprise into *applied* [theoretical] *mathematics*. Gherardo, as a faithful translator, would still render Abū Bakr's sharp distinction between (geometrical) standard method and (numerical) *al-jabr*. Leonardo the mathematician, however, did not see the point, or saw no point in doing so.

VI. The End of a Tradition

However much the tradition of mensuration algebra had become superfluous from a theoretical point of view, it did not die easily in Christian Europe once it had been adopted. Thus, in the geometrical part of his *Summa de arithmetica*, Luca Pacioli tells that

> even though rather much has been said about the rule of algebra in the part on arithmetic: none the less, something must be said about it here.[48]

What needs to be said turns out to be precisely what Leonardo tells in his *Pratica geometrie*. The treatment is so close to Leonardo that misprints in Pacioli's lettering of diagrams can be corrected from Leonardo's text. (This was how I stumbled upon the affinity between the texts.) But there are certain puzzling exceptions to his faithfulness: thus Leonardo, as we remember, did not speak about "the four sides and the area" but about "the area and its four sides" making up 140. Pacioli, however, returns to the original pattern. Since this pattern was as foreign to Renaissance algebra as to Old Babylonian algebra, Pacioli can not be expected to have reinvented the ancestral formula on his own: it must have been around. As it has sometimes been suspected, Italian Late Medieval algebra, however much it was indebted to Leonardo, must have received impulses from the Islamic world through supplementary channels.[49]

The last appearance of the set of problems once belonging to the tradition of mensuration algebra is in Pedro Nunez *Libro de algebra en arithmetica y geometria* from 1567 (at least the last which I know about—but my reading of Renaissance sources is far from complete). Part III, chapter 7 has the heading "About the practice of algebra in geometrical cases or examples, and firstly about squares".[50] It is obvious that Nunez has profited much from Pacioli, as also told in his concluding address to the reader (fol. 323^V). In our now customary abbreviations, the examples about squares are the following:

1. $s = 3$: Q?
2. $Q = \alpha$: s?
3. $s = 3$: d?
4. $d = 6$: s?
5. $d + s = 6$: d? s?

6. $d \cdot s = 10$: d? s?
7. $d - s = 3$: d? s?
8. $s \cdot (d - s) = 15$: s? d?
9. $d \cdot (d - s) = 14$: s? d?
10. $s + Q = 90$: s? Q?
11. $d + Q = 12$: Q? s?
12. $s + d + Q = 37$: s? d? Q?
13. $Q \cdot s = 10$: s? Q?
14. $d \cdot Q = 12$: s? Q?

These translations are misleading insofar as they conceal the real format of the examples. This format follows that of the Euclidean *Data* (and of Jordanus de Nemore's *De numeris datis*)—for instance, No. 11 tells that "if the diameter and the area of the square together are known, then each is known separately." Only afterwards the numerical example is introduced. In this respect, the text is thus developing toward *theory*. It has also dropped the opaque solutions by unexplained numerical algorithms (the rudiments of naive cut-and-paste procedures), and starts directly with the algebraic solution.

But the themes are traditional. Nunez, when advertising the capabilities of algebra, feels the need to demonstrate that this wonderful technique is able to resolve both the traditional problems and even more complex problems of the same kind (like No. 12). He presents only one example for each problem type, and thus drops "the four sides." For the last time, however, "the side" appears before the area in No. 10, betraying the Bronze Age descent—and for the last time (before Viète changed the terms in which the problem of homogeneity was discussed) it is explained that what is added to the area is another area, "a root" being the side provided with a "projection 1" (cf. also Nunez' fol. 6^r).

Within a generation, Viète was to show the capability of algebra to elucidate much more complex problems. If algebra was still in need of commercials, much more impressive applications than artificial mensuration geometry were now at hand. After somewhat more than three thousand years, "the area and the four sides," as the totality of mensuration algebra, could leave the world so quietly that nobody noticed its death, and nobody remembered that it had ever existed.

VII. Note added in proof

After having finished the preceding paper I stumbled upon a Greek version of the problems of the four sides and the area. In manuscript S of the pseudo-Heronic *Geometrica* (which is also close to the Near Eastern surveyors' tradition on several other accounts), Chapter 24.3[51] runs as follows:

> A square surface having the area together with the perimeter of 896 feet. To get separated the area and the perimeter. I do like this: In general [i.e., independently of the parameter 896 – JH], place outside ($\varepsilon\kappa\tau\iota\theta\eta\mu\iota$) the 4 units, whose half becomes 2 feet. Putting this on top of itself becomes 4. Putting together just this with the 896 becomes 900, whose squaring side becomes 30 feet. I have taken away underneath ($\nu\phi\alpha\iota\rho\acute{\epsilon}\omega$) the half, 2 feet are left. The remainder becomes 28 feet. So the area is 784 feet, and let the perimeter be 112 feet. Putting together just all this becomes 896 feet. Let the area with the perimeter be that much, 896 feet.

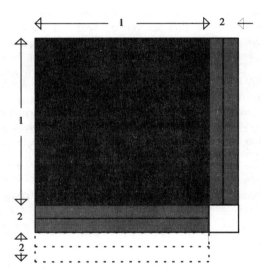

FIGURE 9

The text is thus an almost fully explicit description of the procedure shown in the diagram (whose principle is that of Figure 2, but which is turned around in order to fit the description): whereas the Babylonian text "posits" the "projection," this one "places" the 4 units—which are afterwards told to be *feet*—"outside" the square. One half is "put on top" of the other in the production of the quadratic complement; from the resulting side of the completed square, 2 are identified as that part of the 4 which was left when the half was "taken away underneath," leaving 28 for the side of the original square.

The phrase "in general" tells us that the problem was considered a standard problem where the parameter 896—but not the number of sides—could be varied.

References

*In memory of Niels Arley (1911–1994) who, upon discovering in 1945 that his first published work had been used to dimension the Hanford reactor, turned away from nuclear physics.

1. Otto Neugebauer, ed., *Mathematische Keilschrift-Texte*. 3 vols. (Berlin: Julius Springer, 1935, 1935, 1937), here vol. 3, pp. 1–5. The translation is mine, as are all translations of sources into English in the following.

2. The principles of translation as well as the single operations and terms are discussed in full in pp. 45–69 of my "Algebra and Naive Geometry. An Investigation of Some Basic Aspects of Old Babylonian Mathematical Thought," *Altorientalische Forschungen*, 1990, 17: 27–69, 262–354. In this work I also explain why the detailed investigation of the texts and their terminology invalidates the received interpretation of the Babylonian technique as a numerical algebra, and suggests a reading as "naive" cut-and-paste geometry (to be explained below).

3. Most of our text is written in Akkadian, the spoken language of the Old Babylonian period. Akkadian terms are transcribed in italics. The present text contains only few Sumerian terms (indicated by spaced writing), most of which are genuine loan words, and which go back to the mathematics of the Sumerian epoch (before 2000 B.C.). Other texts (mathematical as well as non-mathematical) may contain many more Sumerian terms, but as a rule these functioned as word signs for Akkadian speech.

4. Imagining lines as provided with an implicit standard width seems indeed to be quite common in field measurement which has not interacted (or not interacted intensely) with Euclidean abstraction. It was the practice of Ancient Egypt—see T. E. Peet, *The Rhind Mathematical Papyrus*, British Museum 10057 and 10058 (London: University Press of Liverpool, 1523), p. 25; Leonardo Fibonacci describes it as the system used when land was bought and sold in thirteenth-century Pisa (*Pratica geometrie*, pp. 1–224 in Leonardo Pisano, *Scritti*. 2 vols., Vol. II: *Practica geometriae et Opusculi*, ed. B. Boncompagni (Roma: Tipografia delle Scienze Matematiche e Fisiche, 1862), p. 3f); Luca Pacioli, finally does the same for fifteenth-century Florence (*Summa de Arithmetica geometria Proportioni: et proportionalita.* (Novamente impressa. Toscolano: Paganinus de Paganino, 1523), part II, fol. 6^V–7^r).

   One may ask whether the Euclidean definition of the line ("a length without width") was originally (and well before Euclid, of course) introduced with the purpose of barring this "misunderstanding" (as a Greek geometer would see the matter).

5. Those who know the terminology of Greek geometry may observe that the dýnamis is a square considered in the same way—cf. my "Dýnamis, the Babylonians, and Theaetetus 147c7–148d7," *Historia Mathematica*, 1990, 17: 201–222.

6. It is forgotten in most general histories of mathematics but should be strongly emphasized that the place value system used in the Babylonian mathematical texts appears to have been used only for intermediate calculations (like a slide-rule, it was a pure floating point system, presupposing that

the reckoner knew the order of magnitude) and in the mathematical school texts. Economical texts (of course) use other number systems where the absolute order of magnitude is fixed.

7. This is what al-Khwārizmī tells in the preface (F. Rosen, (ed., trans.), The *Algebra* of Muhammad ben Musa (London: The Oriental Translation Fund, 1831), p. 3, cf. J. Ruska's corrections, "Zur ältesten arabischen Algebra und Rechenkunst," *Sitzungsberichte der Heidelberger Akademie der Wissenschaften. Philosophisch-historische Klasse,* Jahrgang 1917, 2. Abhandlung, p. 5). Rosen's translation was made from the manuscript Oxford, Bodleian I CMXVIII, *Hunt.* 214, fol. 1–34, as was Rozenfeld's Russian translation and the Arabic edition (al-Khwārizmī , *Kitāb al-muḫtaṣar fi ḥisāb al-jabr wa'l-muqābalah,* ed. A. M. Muśarrafa & M. M. Ahmad (Caïro, 1939)). A close analysis of the text and comparison with Latin translations made in the twelfth century by Robert of Chester and Gherardo of Cremona shows that the text of the Oxford manuscript has been amended by at least three different editors (two of whom must antedate Robert of Chester)—see chapter V in my "'Oxford' and 'Cremona': On the Relations between two Versions of al-Khwārizmī 's *Algebra*," *Filosofi og vidensk-absteori paRoskilde Universitetscenter.* 3. Række: *Preprints og Reprints* 1991 nr. 1. (To appear in *Proceedings of the 3rd Maghrebian Symposium on the History of Mathematics Alger, 1–3 December 1990*). For most purposes, Gherardo's version (now available in a critical edition prepared by B. B. Hughes—"Gerard of Cremona's Translation of al-Khwārizmī's *Al-Jabr*," *Mediaeval Studies,* 1986, 48: 211–263) is to be preferred to the revised Arabic text; unfortunately for historians of mathematics, however, Gherardo omitted the preface as well as the second and third part of the work (the practical geometry and the arithmetic of legacies), for which we have to trust the corrected Arabic text or one of its derivatives.

8. *Verification of the Problems of Algebra through Geometrical Demonstrations,* ed., trans. P. Luckey, "Tābit b. Qurra über den geometrischen Richtigkeitsnachweis der Auflösung der quadratischen Gleichungen," *Sächsischen Akademie der Wissenschaften zu Leipzig. Mathematisch-physische Klasse. Berichte,* 1941, 93: 93–114.

9. That second-degree problems constituted at least the core of *al-jabr* follows from al-Khwārizmī's introduction. Most likely, however, the formulation and solution of second-degree problems by means of "treasures," "roots," and "dirhems" (cf. below) was not only the core of *al-jabr* but also the meaning of the term *stricto sensu.*

10. An analysis of the relevant parts of this treatise, together with arguments for the dating, will be found in my "Al-Khwārizmī, Ibn Turk, and the Liber Mensurationum: on the Origins of Islamic Algebra," *Erdem* (Ankara), 1986, 2: 445–484; cf. also my "'Algèbre d'*al-ǧabr*' et 'algèbre d'arpentage' au neuvième siècle islamique et la question de l'influence babylonienne," pp. 83–110 in Fr. Mawet & Ph. Talon (eds), *D'Imhotep à Copernic. Astronomie et mathématiques des origines orientales au moyen âge* (Leuven: Peeters 1992).

11. Critical edition in H. L. L. Busard, "L'algèbre au moyen âge: Le 'Liber mensurationum' d'AbûBekr," *Journal des Savants,* Avril-Juin 1968, 65–125.

12. The text is corrupt (or possibly intentionally enigmatic, as is indeed No. 50). More or less at the corresponding place in his exposition, Leonardo (*Pratica geometrie* (cit. n. 4), p. 61) discusses the problem $_4 s + \frac{3}{8} Q = 77\frac{1}{2}$.

13. The datum of Nos. 16 and 18 ($d - s = 4$) points back to the crude idea that $s = 10$, $d = 14$ (the result, however, is found correctly as $4 + \sqrt{32}$). In No. 19, the diameter is found as $2 \cdot 7\frac{1}{14}$, and the problem is thus constructed backwards from the value $d = 14\frac{1}{7}$, an approximation to the length of the diagonal in a 10×10 square that is given in the beginning of the chapter. The quasi-identity between Nos. 16 and 18 shows that the tradition has been jumbled at some point (whether during the copying of Abū Bakr's manual or in the sources on which he draws), No. 18 representing the original formulation (traditionally, differences had been told as excesses). For this reason, the "seven and one half" of No. 15 is probably to be understood as a distorted version of the "seven and one half of a seventh" of No. 19.

One may add that No. 12 builds on the same geometrical configuration as No. 6, a 6×10 square, and that cut-and-paste solutions of the two problems will be identical (any solvable problem "area minus four sides" stands in this relation to a problem of the type "area plus four sides"). No. 13, on its part, deals with the same square as No. 12. These two deviations from the norm may thus be understood as a cascade of derived problems. In a similar way, No. 17 may have been constructed as a parallel to No. 16 on the basis of the approximation $\sqrt{2} : 1 \approx 17 : 12$, in which case Nos. 16 and 17 could be related via the algorithm for side-and-diagonal numbers.

14. Cf. the following passages in Hughes' edition of Gherardo's translation (cit. n. 7): IIA: 11f; IIB: 12–14; VI: 18, 45f, 70; VII: 6f, 30–34, 52f, 84, 92, 119f, 121f. "Opposition" occurs in VI: 74 and VII: 19.

15. In the passage from No. 55 where "restoration" meant the elimination of *144 dragmas* from *one treasure and 144 dragmas,* the "opposition" stands for the subtraction of 144 from the other side of the equation.

16. Speaking about "the opposite side" comes naturally when we refer to our own equations, where the sign of equation separates two sides. It comes less easily if equations are formulated in spoken words, as are al-Khwārizmī's and Abū Bakr's "rhetorical" equations. Abū Bakr's use of "opposition" thus suggests that this terminology was formed around some kind of material representation of equations (as we shall see, al-Khwārizmī's usage must be secondary), most probably a sort of scheme.

The use of schemes with opposing sides is indeed known from India—see B. Datta & A. N. Singh, *History of Hindu Mathematics. A Source Book.* (Bombay: Asia Publishing House, 1962. 1st ed. Lahore: Motilal Banarsidass, 1935–38), part II, pp. 28–32. *Al-jabr* can hardly have been borrowed from the "scientific" algebra of "scientific" mathematicians like Āryabhaṭa and Brahmagupta, it is true—cf. my "Formation of 'Islamic Mathematics'. Sources and Conditions," *Science in Context,* 1987, 1: 281–329, here p. 286.

However, a connection to Indian practical mathematics is strongly suggested by the term "root" (Arabic *jadr*), used first about the square root of a number and next (via the square root of the unknown *treasure*) about the unknown of an equation. The term makes no metaphorical sense in the Arabic *al-jabr* tradition, where the root was taken of a number or of an amount of money and had no geometrical connotations (cf. below); in India, however, the square root was understood as the base of a geometrical square, and designated since early times by the term *mūla*, "base" or "root [of a tree]"—see Datta & Singh, *op. cit.,* part I, pp. 169f.

17. The al-Khwārizmīan definition is found in the *Kāfi* (A. Hochheim, *Kâfîfîl Hisâb (Genügendes über Arithmetik) des Abu Bekr Muhammed ben Alhusein Alkarkhi.* 3 vols. (Halle: Louis Nebert, 1878), vol III, p. 10); on the use, see G. A. Saliba, "The Meaning of al-jabr wa'l-muqābalah," *Centaurus,* 1972–73, 17: 189–204, here pp. 199f.

18. In first-degree problems (e.g., in the inheritance algebra treated as part III of al-Khwārizmī 's Algebra), it is customary to label the unknown "a thing"; "a root," as a matter of fact, would make no sense. One text published in Medieval Latin translation in G. Libri's *Histoire des mathématiques en Italie,* vol. I (Paris, 1838) uses "a treasure" (p. 304f), which is of course also a meaningful name for an unknown amount of money.

    While the "root" may point to Indian practical mathematics, weak indications exist that "the thing" is related to Greco-Egyptian practice, either by descent or by common descent. (The evidence is listed but not thoroughly discussed in my "Sub-scientific Mathematics: Undercurrents and Missing Links in the Mathematical Technology of the Hellenistic and Roman World," *Filosofi og videnskabsteori pa Roskilde Universitetscenter.* 3. Række: *Preprints og reprints* 1990 nr. 3, end of chapter IV—to appear in *Aufstieg und Niedergang der römischen Welt,* II vol. 37, 3.) However, since mathematical problems circulated between China and the Mediterranean no later than the first Christian century (cf. note 47), Indian and Greco-Egyptian connections are not mutually exclusive. However, since mathematical problems circulated between China and the Mediterranean no later than the first Christian century, Indian and Greco-Egyptian connections are not mutually exclusive.

19. The fragment was published by Marshall Clagett, *Archimedes in the Middle Ages.* 5 vols. Vol. V: *Quasi-Archimedean Geometry in the Thirteenth Century,* p. 599 (Philadelphia: The American Philosophical Society, 1984); references to the Indian practice will be found in my "'Algèbre d'*al-ğabr*' et 'algèbre d'arpentage',", n. 22.

20. Even Leonardo and Nunez, when they are to explain the geometrical interpretation of *al-jabr,* refer to the "root" as a rectangle with length equal to the side of the square, and with width 1. Cf. below.

21. The conclusions do not hold for all problems: the voluntarily abstruse standard solution of No. 50, for instance, is a mere translation of the tortuous *al-jabr* solution which follows it; other standard solutions appear to be geometrical but do not use cut-and-paste techniques. These exceptions, however, do not concern us here, however relevant they are for a complete analysis of Abū Bakr's eclectic manual.

22. This does not go by itself even within a naive cut-and-paste algebra, as demonstrated by the Old Babylonian algebraic corpus: Old Babylonian lines and surfaces may not only represent pure numbers or prices, which permitted the scribal mathematicians to solve non-geometric problems by means of their naive-geometric technique. A line could also represent an area, which made possible the treatment of biquadratic problems (e.g., BM 13901 No. 12, which is solved as a biquadratic even though a simple quadratic solution is possible).

    In contrast, a technique which restricts itself to manipulating those geometrical entities which enter the problem directly is by necessity prevented from developing into an all-purpose algebra. One might even be tempted to use this as part of a definition of *algebra* as "complex analytical computation, or theory for such computation, where intermediate steps need have a meaning only with relation to a representation but not necessarily with relation to the entities that define the problem"—in which case Old Babylonian algebra is algebra, but mensuration algebra is not.

23. Ed. Neugebauer, MKT I, pp. 108f (cit. n. 1). Analysis in my "Algebra and Naive Geometry," pp. 309ff (cit. n. 2).

24. An edition of the Latin text with German translation was published by Maximilian Curtze, in *Urkunden zur Geschichte der Mathematik im Mittelalter und der Renaissance,* pp. 1–183 (Abhandlungen zur Geschichte der mathematischen Wissenschaften, vol. 12–13. Leipzig: Teubner, 1902). In the footnotes to the edition, Curtze also traced the parallels between Savasorda's text and Leonardo's *Pratica geometrie.*

25. It is thus wholly wrong despite a generally accepted view that the treatise is "the earliest exposition of Arab algebra written in Europe" (Levey, "Abraham bar Ḥiyya ha-Nasi," *Dictionary of Scientific Biography,* vol. I, pp. 22f, quotation p. 22 (New York: Scribner, 1970)).

26. Savasorda's treatment of his §18 might be taken as an argument against his being familiar with traditional cut-and-paste procedures. Here he finds the difference between the sides of a rectangle from the area and the diagonal by means of the rule that $d^2 = 2A + (\ell_1 - \ell_2)^2$, which is stated in §14 and argued there from *Elements* II.7. After that he solves the problem from the area and the difference between the sides. If he had thought of the naive diagram probably underlying his rule, however, it might also have told him that $(\ell_1 + \ell_2)^2 = d^2 + 2A$, which would have simplified the solution (cf. Figure 7). However, an early Old Babylonian problem from Tell Dhiba'i to which we shall return (p. ??? and later) applies precisely the same method as Savasorda. Both authors (and the whole tradition) may thus have used the problem to show the combination of several standard methods.

    It is noteworthy that the proof of *Elements* II.7 builds on the sub-diagram $MGCJ$ of Figure 7 (without diagonals), while that of *Elements* II.4 (from which follows that $(\ell_1 + \ell_2)^2 = d^2 + 2A$, of which Leonardo Fibonacci makes use when solving the corresponding problem) employs the complete diagram (without the lines EJ and KH and without diagonals).

27. We may also mention Leonardo's counterpart of Savasorda's §18 (cf. above, note 26), where Leonardo (like Abū Bakr) finds the sum of the sides, and refers in his proof to *Elements* II.4.

28. The two translations have been made so as to show precisely the extent and character of the agreements/disagreements between the two texts, in vocabulary as well as in the choice of grammatical forms. For the sake of creating one-to-one-correspondences, the translation "expanse" has been used for *embadum*, a term for the area which Leonardo shares with Savasorda/Plato.

29. In the completion of the *al-jabr* procedure, the 4 to be added to 60 is to be found as the square on half the number of roots, not as 2 plus this half. The root (and thus the shorter side), furthermore, is found as $\sqrt{64}$ minus half the number of roots, and the longer side finally as the shorter plus 2 the difference between the sides.

   All this would certainly have been recognized by Leonardo. In all probability, his "and so on" serves to conceal that he does not understand what goes on.

30. There is a vague possibility that Leonardo still had access to the habitual diagrams for a number of complex problems involving the diagonal of a rectangle (e.g., $\ell_1 + \ell_2 + d = 24$, $A = 48$, *Pratica geometrie* p. 68 (cit. n. 4), where he introduces diagrams that generalize the one shown in Figure 7. But he may also have developed these diagrams anew, since they follow without too much difficulty from the procedure.

31. Hughes, "Gerard of Cremona's Translation," p. 233 (cit. n. 7).

32. *Propositiones ad acuendos iuvenes,* problem 52, version II, ed. M. Folkerts, "Die „lteste mathematische Aufgabensammlung in lateinischer Sprache: Die Alkuin zugeschriebenen *Propositiones ad acuendos iuvenes,*" *Österreichische Akademie der Wissenschaften, Mathematisch-Naturwissenschaftliche Klasse. Denkschriften* (Wien, 1978), 116. Band, 6. Abhandlung, here p. 74. Emphasis added.

33. This relation between professional mathematical practice and recreational mathematics is a focal theme in my "Sub-Scientific Mathematics. Observations on a Pre-Modern Phenomenon," *History of Science,* 1990, 28: 63–86.

34. This characteristic has a double explanation: A riddle is always better the more surprising its formulation. Moreover, as long as the parameters of a problem are not noteworthy, they are likely to change when transmitted within a semi-oral culture; once somebody has chosen a remarkable parameter it is likely to be remembered, both because this follows from remarkability *per se,* and because it makes the riddle as a whole better.

   Mathematical riddles are hence liable to be born striking, and to conserve this characteristic when they are transmitted. If by accident they are born without marked parameters, a kind of attraction law guarantees that they will acquire them soon (or that they will be forgotten).

   A particular variant of the quest for the extraordinary was mentioned above: The presence in the Liber mensurationum of deliberately opaque and perplexing problem solutions, which the disciple is asked to look through.

35. The texts were published by Taha Baqir, in "Some More Mathematical Texts from Tell Harmal," *Sumer,* 1951, 7: 28–45, and in "Tell Dhiba'i: New Mathematical Texts," *Sumer,* 1962, 18: 11–14, pl. 1–3, respectively.

36. See my "Algebra and Naive Geometry," p. 326 (cit. n. 2); R. M. Whiting, "More Evidence for Sexagesimal Calculations in the Third Millennium B.C.," *Zeitschrift für Assyriologie und Vorderasiatische Archäologie,* 1984, 74: 59–66, here p. 65f; and J. Friberg, "Mathematik," in *Reallexikon der Assyriologie und Vorderasiatischen Archäologie,* vol. VII, 531–585, here p. 541 (Berlin & New York: de Gruyter, 1990).

37. Since no traces of genuine second-degree algebra are found in the Old Akkadian school texts, we may also surmise that the discovery of the quadratic completion (the "Akkadian method") took place somewhere between the 22nd and the 19th century B.C..

38. BM 13901 Nos. 8 and 9 deal with two squares, about which the sum of the areas and the sum of/difference between the sides are stated. The square sum of the sides sides (20′ and 30′) is no square, and thus the problems cannot be transformed into rectangle-diagonal problems without a change of parameters. Evidently it is not excluded that surveyors' rectangle-diagonal problems have been adopted and transformed, and the parameters then changed. However, reflections of our tradition in classical sources (in particular *Elements* II, cf. below) and the unquestionable presence of two-square problems where $Q_1 - Q_2$ is given speak in favour of the two-square assumption with given sum. A sequence of problems about the same two squares in the late Old Babylonian text TMS V (one of which coincides with BM 13901 No. 8) speaks about the smaller square as located concentrically within the larger one—a configuration that refers to geometrical practice (E. M. Bruins & M. Rutten, eds., *Textes mathématiques de Suse.* Paris: Paul Geuthner, 1961, here pp. 46f). One of the problems (col. III, l. 4, unmentioned and untranslated in the edition) tells the difference between the areas and the difference between the sides.

39. Ed. Neugebauer, MKT III, pp. 14–17 (cit. n. 1).

40. ℓ_1 and ℓ_2; ℓ_1 and d; $\ell_1 + d$ and ℓ_2; $\ell_1 + \ell_2$ and A; $\ell_1 + \ell_2$ and d; $\ell_1 + d$ and ℓ_2; $\ell_1 + d$ and $\ell_2 + d$; $\ell_1 + \ell_2 + d$ and A.

41. This discontinuity can be traced on several levels beyond those already mentioned (Sumerian word signs and problem types): the structure of the terminology; the construction of problems from integral solutions and integral coefficients (evidence that the problems have been borrowed rather directly from the mensuration tradition, without much further systematization or tinkering); and a tendency to construct solutions from sum and difference rather than semi-sum and semi-difference (as had been the Old Babylonian habit, and as Abū Bakr would mostly still do in the old problems).

42. For convenience I translate the propositions into symbols (it should be remembered that such a translation is always somewhat arbitrary—cf. the two different translations of prop. 7):

   1. $\square(a, p+q+\cdots+t) = \square(a, p)+\square(a, q)+\cdots+\square(a, t)$.
   2. $\square(a) = \square(a, p) + \square(a, a - p)$.
   3. $\square(a, a + p) = \square(a) + \square(a, p)$.

4. $\square(a+b) = \square(a) + \square(b) + 2\square(a,b)$.

5. $\square(a,b) + \square(\frac{a-b}{2}) = \square(\frac{a+b}{2})$.

6. $\square(a,a+p) + \square(\frac{p}{2}) = \square(a+\frac{p}{2})$.

7. $\square(a+p) + \square(a) = 2\square(a+p,a) + \square(p)$; or, alternatively, $\square(a) + \square(b) = 2\square(a,b) + \square(a-b)$.

8. $4\square(a,p) + \square(a-p) = \square(a+p)$.

9. $\square(a) + \square(b) = 2\left[\square(\frac{a+b}{2}) + \square(\frac{b-a}{2})\right]$.

10. $\square(a) + \square(a+p) = 2\left[\square(\frac{p}{2}) + \square(a+\frac{p}{2})\right]$.

We observe that proposition 6 coincides with proposition 5 if only $b = a+p$. proposition 5 corresponds, however, to the situation where the sum of the two sides is known (as in proposition 9, a and b result from the splitting of a line in unequal segments), and where they are thus drawn in continuation of each other in the proof; proposition 6, on its part, is adapted to the situation where one exceeds the other by p, and the proof thus draws them in superposition. Precisely the same relation holds between proposition 9 and proposition 10, while proposition 4 and proposition 7 are similarly but not identically correlated.

43. Cf. note 26. It should perhaps be stressed once more that Savasorda's and Leonardo's use of propositions from *Elements* does not mean that they were employed within the tradition of mensuration algebra in the form we (and Leonardo and Savasorda) know them, only that they were still close enough to this tradition to be serviceable.

44. Strictly speaking, proposition 9 is cited, but in what seems to be an interpolated lemma. As pointed out by Ian Mueller, propositions 8 and 10 *might* have been cited in the same way, as justifications of unproved assumptions—*Philosophy of Mathematics and Deductive Structure in Euclid's Elements*, p. 301 (Cambridge, Mass., & London: MIT Press, 1981). It seems as if the kind of knowledge contained in the three propositions was too familiar to require explicit citation once it had been proved.

45. They also solve problems about rectangles where the diagonal and either the sum of or the difference between the sides are known. As argued above (see note 38), at least one of these groups (most likely the two-square problems) will have belonged to the early phase of the mensuration algebra.

46. See my "Dýnamis" (cit. n. 5), where further references to work by earlier authors (not least Wilbur Knorr) on this question are given.

47. Since the second-degree problems which turn up in the first century (CE) Chinese *Nine Chapters on Arithmetic* (*Chiu chang suan shu. Neun Bücher arithmetischer Technik*, ed. trans. Kurt Vogel, pp. 91f (Braunschweig: Friedrich Vieweg & Sohn, 1968)) are related to the "new" Seleucid problems (and the dress of one of them, the leaning reed, an obvious borrowing), conquest can hardly be the only factor involved.

48. Part II, fol. 15^{r} (cit. n. 4).

49. Another suggestive deviation from Leonardo is Pacioli's version of Abū Bakr's No. 38 (above, p. ???): It is more correct than the Gherardo translation, which had been repeated so faithfully by Leonardo. Pacioli, indeed, finds the completing square 4 as "half the number of sides squared" (fol. 19^{r}). Since the Gherardo/Leonardo text is meaningless as it stands, it is highly unlikely that Pacioli could have used this version and just improved it. If he had done so (for example, supported by an *al-jabr* analysis), he could have produced a fully correct solution: instead, his explanation still presupposes tacitly that the excess and half the number of sides coincide.

We may infer that Pacioli's source for the pattern "sides and area" is thus not likely to have been the Gherardo version of the *Liber mensurationum*.

50. P. Nunez, *Libro de Algebra en Arithmetica y Geometria,* fol 277^{V}ff (Anvers: En casa de los herederos d'Arnaldo Birckman, 1567).

51. Ed. J. L. Heiberg [1912: 418] Heronis Definitiones cum variis collectionibus. Heronis quae feruntur *Geometrica*, p. 412 (Heronis Alexandrini Opera quae supersunt omnia, IV. Leipzig: Teubner, 1912).

The Method of Indivisibles in Ancient Geometry

Wilbur R. Knorr
Stanford University

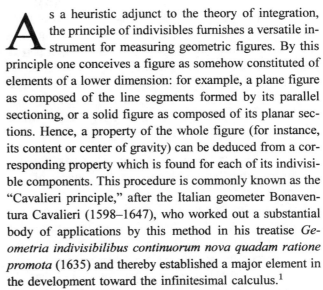

A s a heuristic adjunct to the theory of integration, the principle of indivisibles furnishes a versatile instrument for measuring geometric figures. By this principle one conceives a figure as somehow constituted of elements of a lower dimension: for example, a plane figure as composed of the line segments formed by its parallel sectioning, or a solid figure as composed of its planar sections. Hence, a property of the whole figure (for instance, its content or center of gravity) can be deduced from a corresponding property which is found for each of its indivisible components. This procedure is commonly known as the "Cavalieri principle," after the Italian geometer Bonaventura Cavalieri (1598–1647), who worked out a substantial body of applications by this method in his treatise *Geometria indivisibilibus continuorum nova quadam ratione promota* (1635) and thereby established a major element in the development toward the infinitesimal calculus.[1]

Although Cavalieri's title describes his geometry as "advanced through a certain *new* method," the technique of measurement via indivisibles was in fact far from new at his time. Attention has been drawn in recent studies, for instance, to elaborations of indivisibilist techniques by the Chinese geometers Liu Hui (3rd century A.D.) and Zu Geng (5th century A.D.).[2] But earlier still, Archimedes (3rd century B.C.) had made prominent use of the technique for the determination of volumes and centers of gravity in the *Method.*[3]

It is the ancient background to this technique that I wish to explore here: I shall maintain that the use of indivisibles was not unique to Archimedes, but owed its origins to geometers before him and continued in use after him. Then, taking up the appearance of indivisibilist reasonings among later geometers, I shall consider the question of their possible dependence, directly or indirectly, on ancient precedents.

Section I

In his *Commentary on Ptolemy's Book I,* Theon of Alexandria (ca. A.D. 360) includes a derivation of a formula for the volume of the sphere (namely, as 11/21 the volume of the cube of its diameter).[4] As one would expect, Theon depends expressly on theorems from Archimedes' *Sphere and Cylinder.* To cover one of the steps assumed in the derivation, Theon attaches a lemma, showing that the ratio of a cube to the cylinder inscribed in it equals the ratio of the square to the inscribed circle, their respective bases.[5] The canon A. Rome, editor of Theon's *Commentary,* has described this lemma as a "pseudo-preuve," in light of its

circular reasoning.[6] But if we consider the passage more closely, it becomes possible to salvage the reasoning without imputing either gross insensitivity on Theon's part or major disruption in the transmission of the text (for instance, one need not reject the lemma as a crude interpolation). For the difficulties are concentrated in just two lines (numbered 4 and 6 in the translation I give below) whose removal leaves a basically coherent argument.[7]

1   I say then that as the square of the diameter to the circle, so is the cube to the cylinder of equal height.

2   For let the circle AB be set out on the diameter GD, and let there be drawn the square EZ on the (line) GD, and let there be erected a cube on the square and a cylinder on the circle, of equal height to the cube. I say that as the square EZ to the circle AB, so is the cube to the cylinder.

3   For let there be set about the diameter KL the (circle) $H\theta$, equal to circle AB, and (let there be set) about it ⟨sc. the diameter KL⟩ the square ⟨MN⟩, manifestly equal to the (square) EZ. Since then, as the square EZ is to the (square) MN, so is the circle AB to the (circle) $H\theta$,

4   [and the cube on EZ to the cube on MN, and further, the cylinder on the circle AB to the (cylinder) on the (circle) $H\theta$,]

5   and alternately, as the square EZ is to the circle AB, so is the square MN to the circle $H\theta$

6   [and further, the cube on EZ to the cylinder on AB, and the cube on MN to the cylinder on $H\theta$].

7   But all the (squares) in MN are equal to those in EZ.

8   Therefore, as the square EZ to the circle AB, so is the cube on EZ to the cylinder on AB.

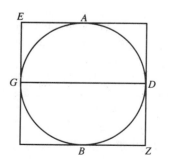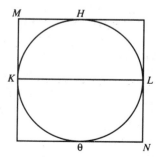

FIGURE 1

The circularity of reasoning which provoked Rome's adverse judgment relates to lines 4 and 6; for these lines simply assert the proportionality of the cubes and cylinders as an immediate inference from the proportionality of the associated squares and circles, which thus assumes what is intended to be proved: that these solids are as their bases. When these lines are bracketed, this logical defect disappears. The lines contribute nothing to the argument anyway, and one would suppose that they were inserted, either by Theon or by a later scribe, as elaborations of lines 3 and 5.[8]

The basic conception of the proof, if not yet entirely clear, can be construed by Theon's phrase "all the squares . . ." in line 7, which indicates a form of summation operation (see also below). Although the circle $H\theta$ and the square MN in line 3 are taken in line 4 to be the bases of a second cylinder and cube, respectively, in the absence of line 4 we naturally take them to be arbitrary sections of the cylinder and cube introduced in line 2. Line 7, as it stands, is unclear: MN has been introduced only as a square and $H\theta$ only as a circle, but the argument requires now that MN be the cube on square MN, and similarly, $H\theta$ must be the cylinder on circle $H\theta$. To straighten out the argument, I emend line 7 thus:

7′ But all the squares, as MN, are equal to the (cube) on EZ, ⟨while all the circles, as $H\theta$, are equal to the cylinder on AB.⟩[9]

Thus restored, the passage offers an acceptable argument within the context of a method of indivisibles. That is, if we conceive the cube as the sum of the squares which result from its planar sectioning, and the cylinder, analogously, as the sum of its circular sections, then the fact that each pair of corresponding planar sections has the same ratio (namely that of square EZ to circle AB) entails that the sums, the cube on EZ and the cylinder on AB, have this same ratio.

The transmission of the word "all" (*panta*) in the extant line 7 is the key to construing the argument in this way, and, I believe, offers us the warrant to do so. For among the standard mathematical authors the same term enters into measurement theorems wherever a process of summation is involved. Archimedes, in particular, employs the term in the *Method* to designate sums of indivisible sections; for example, the circular sections of the paraboloidal segment ("*all* the circles in the segment . . .") are collectively in equilibrium with the corresponding sections of a specified cone ("*all* the circles in the cone . . .").[10] In the formal proofs parallel to the heuristic derivations in the *Method,* Archimedes replaces the indivisibles with finite elements of area or volume, as appropriate, but retains the phraseology with "all" to designate their summation, as in *Conoids and Spheroids,* prop. 21: that "all the cylinders in the cylinder ... are to all the cylinders in the figure inscribed ⟨in the paraboloid⟩ ...".[11] In the summation theorem underlying these measurements, *Conoids and Spheroids,* prop. 1, Archimedes likewise adopts the term "all" to designate the summation (ed. Heiberg, I, 262.3–6, etc.), and this is typical of the summation theorems in his other writings as well.[12] In Archimedean-related discussions by Pappus, the same diction appears (see section II below). A parallel usage can be cited from Euclid, among the Eudoxean measurement theorems of Book XII, namely, in prop. 4, "all the prisms in one pyramid ..." (ed. Heiberg-Stamatis, III, 89.3–4, 10; 91.20–21). More generally, Euclid adopts *panta* to designate sums in theorems on proportion in Book V, as in props. 1 and 12 (*ibid.,* II, 4.5, 19.9–10), and in his theorem on the sum of a geometric progression in Book IX (prop. 25, *ibid.,* II, 223.6, 224.3–4). Indeed, the term *ho sympas arithmos* (literally, "the entire number") designates the summation required in the construction of perfect numbers in prop. 26 of Book IX (*ibid.,* 224.14–15, 20–21).

Theon's treatment of the measurement of the sphere can be compared with an alternative proof extant in the last proposition of the medieval tract *De curvis superficiebus.*[13] Both establish the same rule, that the sphere is 11/21 the cube of its diameter, and both derive it in much the same way. But the medieval proof handles more concisely the step which Theon establishes by the lemma just given:

> But from the multiplication of altitude *OM* into square *GF* there arises the cube, and from *OM* into circle *O* is the cylinder. Therefore, since the ratio of products and multipliers is the same, the ratio of the cube to the cylinder will be as base *GF* to circle *O*.

Thus, the medieval writer conceives of the volumes in a quasi-arithmetical way, the cube as the product of its square base times its altitude, and, similarly, the cylinder as the product of its circular base and its altitude. Since the altitudes are equal, the ratio of the solids will be that of their bases.

The author of *De curvis superficiebus,* named in the manuscripts as Johannes de Tinemue (that is, John of Tynemouth), appears to have been active in Oxford early in the 13th century, and was possibly an associate of Robert Grosseteste.[14] But the technical core of the tract cannot be due to this scholar, since the expertise in advanced aspects of Archimedean geometry, including command of the fine details of his approach in *Sphere and Cylinder,* does not seem to have become available to Latin schol-

ars before 1269, when Willem of Moerbeke translated the Archimedean corpus.[15] One infers that *De curvis superficiebus* must have been based on a Greek prototype, now lost, abridging the exposition of *Sphere and Cylinder,* presumably in accordance with the teachings of Theon of Alexandria.[16]

Ostensibly, the alternative derivation in *De curvis superficiebus* has no connection to the indivisibilist conception that I have discerned with Theon. But the connection becomes evident when one considers how the alternative treatment could justify its replacement of Theon's lemma by the product-rule for uniform solids. An account of precisely this sort can be found in the manual of geometry, the *Metrica,* by Hero of Alexandria (middle of the first century A.D.).[17] In the preface to his survey of solid measurement in Book II, Hero shows that the volume of a given rectangular solid is found by multiplying the numbers that measure its length, width, and depth, inasmuch as the solid can be broken down into that many unit cubes.[18] He then states a rule for measuring any solid:

> in general, any solid figure of whatever thickness, whose height is at right angles to the base, is measured by measuring its base and multiplying that into the height.[19]

To be valid, the rule requires further restriction. For instance, it applies for any uniform solid, namely one for which all its sections by planes parallel to the base are congruent figures. That Hero intends such a restriction is indicated by the example he next cites: when an ellipse is taken as base, and it is moved such that its center moves along a straight line at right angles to the plane of the base while the ellipse itself always remains parallel to its initial orientation, then the resultant "quasi-cylindrical figure" (*schêma hôsperei kylindros*) is measured by the stated rule. Hero allows that the base may be any arbitrarily chosen plane figure and that its height, the line of motion, may be inclined at any angle to the base,[20] as long as "the solid be such that when it is cut by a plane parallel to the base, it makes the sections equal to the base."[21] He then explains how the motion he has described, under which the generating plane figure always remains parallel to its initial position, assures that the parallel sections will be equal, as stipulated in his refined statement of the rule.

Hero does not justify the rule, and his explanation for the rectangular solid, given earlier, could not apply, since in general the solid cannot be subdivided precisely into unit cubes. But the gap could be filled by appeal to the principle of indivisibles, in the manner of Theon's lemma: consider the rectangular solid whose base equals that of the solid in question and whose height equals its height; since each solid is constituted out of the plane figures formed as its parallel sections, and since the sections are equal, taken in pairs, then the volumes must also be equal; since the volume of the rectangular solid is the product of its base and height, it follows that the volume of the given solid is likewise the product of its base and height.

Short of a Eudoxean-style argument, which Hero does not seem to presuppose here, extant ancient sources do not indicate any alternative reasoning by which such a rule can be justified. Why Hero omits a justification of his rule is unclear. Textual corruption is possible, or, more likely, an oversight on Hero's part, perhaps deeming that no further explanation was needed. In any event, the rule is not original with Hero, but rather an established method familiar from his sources. Within this tradition it thus appears that the indivisibilist conception has been taken as intuitively obvious.

Remarkably, Hero does not apply his rule when he measures specific solids. For instance, in chap. 2 to show that the scalene cylinder equals the right cylinder having equal altitude and base, Hero could simply have noted that the two cylinders, when simultaneously cut by a plane parallel to their bases, have equal circles as sections. Introducing the right cylinder should have been superfluous, since his rule establishes directly that the volume of *any* cylinder is found as the product of its base and its altitude. Instead, Hero offers an almost verbatim quotation from Euclid's *Elements* (Book XII, prop. 11): "cones and cylinders of equal altitude are to each other as the bases"—in order to establish the equality of the two cylinders.[22]

Similarly, in chapter 9, to measure the volume of a truncated cone (*kônos kolouros*), Hero reduces it to that of the truncated pyramid (already given in his chapter 7) by showing that the truncated cone and pyramid have to each other the ratio of the circle and square, their respective bases. To justify this latter proportionality, Hero could have appealed to the indivisibilist principle: since any plane simultaneously sectioning the two solids yields a circle and a square in the same ratio. But Hero instead conceives each truncated figure as the difference of two cones or two pyramids, respectively: since the first cone and pyramid have the same ratio as the second cone and pyramid, namely that of the circle and square, their differences will likewise have that same ratio.

It is interesting that, when Hero states the proportionality of these solids, namely that when they have equal altitudes, the pyramid is to the cone as the square is to the circle, he refers to the proportionality of the corresponding parallelepiped (sc. prism) and cylinder, for these too, when they have equal altitudes, will have the same ratio.[23] By

implication, one would suppose that Hero again intends an application of Euclidean results: that every pyramid is one-third the corresponding prism (XII, prop. 7), and every cone is one-third the corresponding cylinder (XII, prop. 10). But Euclid does not establish the proportionality linking prisms to cylinders. This is, of course, precisely the result Theon establishes in his lemma, based on the indivisibilist conception, as proposed above. But how Hero intended to justify it, or whether he assumed it to be evident, the extant text does not say.

To summarize: Hero proposes an arithmetic rule for measuring the volume of any uniform solid as the product of its altitude times the area of its base. He recognizes that a sufficient condition for the validity of this rule is that cutting the solid by planes parallel to the base produces congruent figures as sections; he recognizes also that a solid generated by means of the parallel translation of a given plane figure as base has this property. Hero does not prove these claims, however, and he fails to apply either the geometric principle or the arithmetic rule that follows from it in the measurements of specific solids subsequently in the *Metrica*. The medieval author of *De curvis superficiebus* introduces the same arithmetic rule for solids as a step in his theorem on the sphere (sc. its being 11/21 the volume of the cube of its diameter) and, although he too assumes it without proof, at the precisely analogous place in Theon's version of this theorem on the sphere there is inserted the lemma, that the ratio of the cylinder and cube of equal altitude equals that of their bases, the circle and square, respectively. The extant text of Theon's proof is defective, but we have seen that the argument is basically sound, once we bracket the two superfluous lines 4 and 6, whose presence seems to render the proof logically circular. Accepting the lemma as Theon's own composition,[24] we must assign the inconsistent elements, namely the insertion of lines 4 and 6 and the corruption of line 7, to some later interpolator (presumably, in the Byzantine period, after the ninth century) who did not grasp the sense of Theon's argument. Such incomprehension could not have obtained, of course, for the originating author of the lemma.

Since Theon knew Hero's *Metrica*,[25] it should have been possible for him to adopt the product rule as the basis of his argument, just as the author of *De curvis superficiebus* does. But, instead, he employs the conception of summing the planar sections to constitute the solids. Theon's commentary on Ptolemy, within which this account appears, is directed to students receiving their first systematic introduction to mathematical astronomy. In such a context it would be inappropriate to frame an auxiliary proof like that in the lemma around new or unfamiliar techniques. Thus, it appears that the concept of indivisibles underlying the lemma, far from being original with Theon—or even unusual (he makes no effort either to claim responsibility for the conception, or even to explain it)—had an established place within the tradition of elementary geometry that his students already knew. Presumably, that tradition is typified by writings like the *Metrica* of Hero, whose audience of learners in applied geometry could combine some exposure to the theoretical type of geometry, such as Euclid's *Elements,* with the heuristic and informal procedures characteristic of practical geometry.

While the extant documentation on the latter tradition is fragmentary, we have evidence in the examples cited from Hero and Theon that the conceptions of indivisibles and of motion-generated figures were linked, in the effort to ground the basic rules for the practical, arithmetic measurement of geometric figures.

Section II

The passages from Hero and Theon considered above reveal how an informal tradition of ancient geometry embraced the conception of indivisibles. Although the state of this evidence is scant and incomplete, it can be amply supplemented by earlier testimony, most notably from Archimedes. In particular, the propositions in Archimedes's *Method* depend on indivisibilist conceptions. Examples of Archimedes's method in this work have been widely discussed, especially his treatment of the area of the parabolic segment in prop. 1 and of the volume of the sphere in prop. 2.[26] I shall thus take up a less familiar case to illustrate his procedure.

The first part of the *Method* (props. 1–11) presents a retrospective of Archimedean results whose formal demonstrations had already been published, but which Archimedes now presents in accordance with his heuristic mechanical procedure. In the remaining propositions he presents the heuristic treatment of two new results, on the measurement of two solids formed as segments of cylinders, together with their formal demonstration. The existing manuscript, as inspected by Archimedean editor J. L. Heiberg, preserves most of the heuristic derivation for the first solid, but lacks both its formal sequel and the entire treatment of the second solid.[27] Nevertheless, the basic line of Archimedes's approach is clear. The first of the two solids, as described in the preface of the *Method* and introduced in prop. 12, is formed thus: in a given rectangular prism (specifically, a cube) one constructs the cylinder whose altitude equals that of the prism and whose base is

the circle inscribed in the square face of the prism; if one marks a diameter in the base of the cylinder parallel to a side of the prism and one sections the cylinder by a plane through this diameter and an edge parallel to it of the opposite face of the prism, then the resulting cylindrical section equals one-sixth the prism. Archimedes does not assign a name to this figure, but it is now known as the "cylinder hoof."[28] The second solid is similarly formed, but will not concern us here.[29]

The heuristic measurement of the cylinder hoof is done in two steps. In the first step (prop. 12) the figure is erected between the diameter OR and the opposite edge (elevated over HM), as in Figure 2, and a hemicylinder is erected over the semicircle OPR, with altitude equal to that of the cube; then, if one conceives of $P\theta X$ to be the axis of a balance with fulcrum at θ, it is shown that the hoof when suspended from X will balance the hemicylinder in its given position.

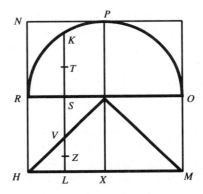

FIGURE 2

In prop. 13 Archimedes introduces a prism erected over the triangular base θHM with altitude equal to that of the cube, and shows that in its given position this prism will be in equilibrium with the hemicylinder, just described, in its given position. It follows that the hoof, if now suspended from P, will balance the prism erected over the opposite side of the balance, in its given position. Since the center of gravity of the prism is known, one can compute the volume of the hoof from the fundamental principle of the balance.[30]

To establish the equilibrium of the prism and hemicylinder in prop. 13, Archimedes considers the sections made in each solid by planes perpendicular to the base and parallel to the axis PX, that is, the rectangles whose altitude equals that of the cube and whose bases are the linear sections made with the base (for example, LV and

SK). Archimedes's proof is not preserved in the existing manuscript, but one can be provided.[31] Since the solids are of the uniform type, such that the planar sections all have the same altitude, it suffices to consider only the linear sections in the plane of the base: that is, to show that the semicircle OPR and the triangle θHM precisely balance each other in their given positions about the fulcrum θ (Figure 2). Consider, then, an arbitrary pair of sections of the circle and the triangle, such as the lines KS and VL, respectively. Let $r = \theta P$, the radius of the semicircle, let $x = VL$, $y = KS$, and let Z, T be the centers of gravity of VL, KS, respectively. Then, since the center of gravity of each line is at its center, $ZS = r - x/2$, $ST = y/2$, so $ZS : ST = 2r - x : y$. Further, KS, being a half-chord perpendicular to the diameter OR, is mean proportional of the segments of the diameter RS, SO; that is, $RS : KS = KS : SO$; since also $RS = HL = VL$ (for VHL is a right isosceles triangle), $x : y = y : 2r - x$. Therefore, $VL : KS = ST : ZS$; that is, the magnitudes VL, KS are inversely proportional to ZS, ST, their respective distances from the axis. This means that VL is in equilibrium with KS, each in its given position relative to the fulcrum θ. Since these lines have resulted from an arbitrary sectioning of the solids, it follows that *all* such lines taken ensemble in the first figure (the prism) will be in equilibrium with *all* such lines taken ensemble in the second (the hemicylinder); that is, the triangle θHM in its position balances the semicircle OPR in its position.

Not only does this establish the striking result, that the triangle and semicircle in this configuration exactly balance each other, but it also provides a determination of the center of gravity of the semicircle: namely, at the point that divides the radius $P\theta$ into segments in the ratio $4 : (3\pi - 4)$, or about $14 : 19$.[32] The *Method*, as we have it, does not state this latter result. But, as noted before, the extant text is incomplete, while the statement of such a corollary would not bear directly on the principal matter, finding the volume of the cylinder hoof, to which the result on the equilibrium of the hemicylinder and the prism in prop. 13 is just a preliminary. Nevertheless, a major project of the *Method* is to determine centers of gravity, such as for the segments of spheres and segments of conics of revolution; Archimedes's regular procedures for such cases would immediately yield the determination for the semicircle in the light of prop. 13. Thus, I see no reason to doubt that Archimedes was aware of this result.

Archimedes's heuristic method thus has two characteristic features, the one mechanical, the other geometric. The *mechanical* aspect, his notion of establishing equilibrium between given figures, is elaborated carefully, by describing explicitly the hypothetical balance in which the various figures are to be weighed. Further, the necessary principles of mechanics as well as the determinations of centers of gravity of specific elementary figures (namely, of plane figures like triangles and parallelograms, and of solid figures like the cylinder and the cone) are stated as "assumptions" (*prolambanomena*) in the prefatory section of the *Method*, since their proofs have already been set out in earlier works, such as *Plane Equilibria* Book I.[33]

By contrast, the essential *geometric* feature of his method, the subdivision of figures into their indivisible elements via parallel sectioning, is neither explained in the *Method* nor employed in any other known Archimedean work. To be sure, Archimedes adopts an explicit terminology to denote the constitutive relation of indivisibles: in prop. 1 he speaks of given plane figures, such as the parabolic segment or the triangle, as being "comprised out of" (*synestêke ek*) all the lines formed as their sections; alternatively, solids of revolution, like the cone, sphere, and conoids, are said to be "filled out" (*symplêrôthenta*) when all the circles formed as their planar sections are taken (props. 2, 3, 4, 5, 6, 14). In each case, the word "all" (*panta*) denotes a summation process: having established that the indivisibles formed by arbitrary sections of the given figures are in equilibrium, he concludes that "all" the indivisibles in those figures, that is, the figures themselves, likewise are in equilibrium.

In the corresponding formal proofs, a related terminology is adopted: in prop. 15, for instance, where the formal demonstration of the measure of the hoof is given, Archimedes speaks of "all" the prisms in a given sectioned prism as having the same ratio to "all" the prisms circumscribed about the corresponding sectioned cylinder (sc. the hoof) that "all" the parallelograms in a given parallelogram have to "all" the parallelograms circumscribed about a specified parabolic segment.[34] The same transfer of terminology is evident in the other places in the Archimedean corpus where Archimedes provides the formal demonstrations of propositions of the *Method* (specifically, *Quadrature of the Parabola*, prop. 14; *Conoids and Spheroids*, props. 19, 21, 25, 26, 27, 29).[35] In all these cases, the indivisible elements in the figures of the *Method* are replaced by finite components, and the phrases with *panta* now denote collections of them, finite in number, that constitute figures inscribed or circumscribed to the figure to be measured. From these figures inequalities are deduced in the context of indirect proofs of the familiar Eudoxean type, as in Euclid's *Elements*, Book XII.[36]

In the judgment of Archimedes himself, his "mechanical method" is not acceptable as a formal demonstration.

Speaking of "the investigation by means of mechanics" (*theôrein dia tôn mêchanikôn*) in the preface to the *Method,* for instance, he writes, "I am persuaded that it is useful, nevertheless, toward the demonstrations (*apodeixeis*) of the theorems themselves."[37] The mechanical analysis makes certain properties "apparent" (*phanenta*), or provides a certain "recognition" (*gnôsis*) of things that can then be secured by demonstration.[38] In a note following his heuristic analysis of the parabolic segment in prop. 1, Archimedes affirms that the theorem is not yet demonstrated by the reasoning just given, but that it provides an "impression" (*emphasis*) of a certain sort which, when we "suppose (*hyponoountes*) it to be true," we can advance to the demonstration.[39]

It is undeniably true that indivisibles, of the sort employed in the *Method,* are unacceptable in a formal demonstration. But the Archimedes scholar E. J. Dijksterhuis, and others following him, have made a stronger claim: that Archimedes considers the use of indivisibles the *only* invalidating feature of his method.[40] By this, Dijksterhuis would like to suppose that Archimedes confers demonstrative validity to the mechanical aspect of his method.[41] But this cannot be right. In Archimedes's account the indivisibles are merely a secondary aspect; for the essence of his method lies in its appeal to mechanical principles. In his preface he twice refers to it by the phrase *dia tôn mêchanikôn,* never by a phrase denoting indivisibles, and he contrasts "demonstrative" (*apodeiktikôs*) with "mechanical" (*mêchanikôs*). In such a context, then, when he says that "investigation via *this* method lacks the force of demonstration,"[42] he can mean only the *mechanical* aspect of the method. What seems especially to stigmatize the indivisibilist aspect, in the view of Dijksterhuis, is that when Archimedes presents the theorem on the parabolic segment in the *Quadrature of the Parabola,* he follows the mechanical mode, much as in prop. 1 of the *Method,* with the principal difference that the indivisibles are replaced by finite elements of area. According to Dijksterhuis, this vindicates the mechanical mode. But in his prefatory remarks in *Quadrature of the Parabola,* just as in the *Method,* Archimedes distinguishes between the two modes (namely, mechanical and geometrical) and, consequently, includes both forms in his exposition: "we have undertaken to write ... first what was *discovered* (*heurethen*) by means of mechanical principles, next what has been *exhibited* (*epideichthen*) by means of geometrical principles."[43] If the mechanical treatment were formally acceptable, there would be no need for a second "geometric" demonstration. By attaching it, Archimedes apparently wishes to forestall possible objections to the assumption of mechanical properties, like weight and equilibrium, in

demonstrations dealing exclusively with geometric properties of figures.

To a modern historian of ancient mathematics, the appearance of indivisibles in Archimedes's heuristic accounts is especially striking, precisely because the documentation surviving from antiquity is so strongly dominated by the formal tradition, within which indivisibles are not allowed. But it is far from clear that the view from the ancients' own perspective is the same. In the propositions of the *Method* Archimedes elaborates with care the mechanical features of his exposition, and in his preface to Eratosthenes he explicitly claims originality for the method. We infer that this manner of applying mechanical principles was novel within the discipline as known to Archimedes's contemporaries. By contrast, Archimedes volunteers *no* explanation of the indivisibles: although they are formed as parallel sections of the given figures, he fails to specify how such elements "in" these figures can be taken to "constitute" them or "fill them out," or what manner of magnitude (whether finite or infinitesimal) they can have in relation to the magnitude of the figures comprised of them. Of course, the provisional character of the analyses in the *Method* makes such an account dispensable; they are a scaffolding, as it were, for the formal demonstrations that will subsequently supersede them. The indivisibles are thus merely a simplifying ploy, serviceable for a heuristic investigation, but readily replaceable by finite magnitudes in a properly executed convergence argument of the Eudoxean type.

But even if Archimedes was not compelled on logical grounds to give an account of the indivisibles, he surely needed *some* account, simply to assure that Eratosthenes understood what he meant. By omitting such clarifications, Archimedes tacitly indicates his ability to assume that the indivisibilist aspects of his technique are familiar.

All the more, Hero and Theon must have assumed from well established geometric teachings the comparable heuristic features that we have encountered in their works, as cited above. Both writers, accomplished synthesizers of technical learning, issue commentaries and manuals that offer little scope for original concepts, methods, or results; and, if they do claim originality in rare cases, neither does so in connection with these passages. Both have access to Archimedean sources: Hero cites the *Method* and other Archimedean works several times in the *Metrica,* while Theon cites the *Sphere and Cylinder* and *Dimension of the Circle* in his Ptolemy commentary.[44] But neither cites Archimedes for the indivisibles or the associated techniques, and, at any rate, the *Method* could not have provided them, since, as we have seen, Archimedes does not give any explanation of them there.

It would be gratuitous to postulate a lost Archimedean work as the source for the later writers' heuristic techniques, although we cannot entirely rule this out. Just as Archimedes treats the method of indivisibles as familiar within the common body of geometric teachings, it seems no less appropriate to suppose the same of Hero and Theon. We can thus turn to the earlier period and consider what the precedents for the use of indivisibles might have been among Archimedes's predecessors.

Section III

The search for pre-Euclidean precedents of Archimedes's technique of indivisibles turns at once to Democritus, the Presocratic whose atomist doctrines sparked controversy among the Greek philosophers of nature.[45] For in the preface to the *Method,* where Archimedes singles out two theorems (that any pyramid is one-third the prism of equal altitude and base, and similarly, that any cone is one-third the cylinder of equal altitude and base), he reserves to Eudoxus the honor of having been first to provide their geometric proofs, but, he says, "one could assign no small credit to Democritus, who was first to make the statement (*apophasis*) about the said figure, having stated it without proof (*apodeixis*)."[46]

While it is plausible that the exponent of atomist doctrines in natural philosophy might have applied these notions to geometry, we have explicit testimony that Democritus was aware of the indivisibilist conception of geometric magnitude. For we learn from the Stoic philosopher Chrysippus (3rd century B.C.), as reported by the later philosopher Plutarch (late first to early 2nd century A.D.), that Democritus proposed "physically (*physikôs*) and successfully (*epitychôs*)" the following paradox relating to the atomist conception of continuous magnitude:

> If a cone were cut by a plane parallel to the base, how must one conceive of the surfaces of the segments: as becoming equal or unequal? For being unequal, they make the cone irregular, taking many step-like indentations and roughnesses. But if they are equal, the segments will be equal and the cone will appear to have the property of the cylinder, being composed of equal, and not unequal, circles; which is most absurd.[47]

The diction here, that the cone is "composed out of" (*synkeimenos ek*) its constituent circles reminds one of the phrase used by Archimedes in the *Method* (prop. 1): that the parabolic segment is "constituted out of" (*synestêke ek*) its linear elements.

The "cone fragment" thus attests to Democritus's awareness of the indivisibilist conception. But it falls short of establishing Democritus as a source for Archimedes. If anything, Democritus here emerges as a *critic* of the conception of indivisibles, not the advocate of a view which might justify their use in a working geometric technique—and, indeed, our reporter Chrysippus (a Stoic, and no friend of atomist doctrines) appears to take Democritus' puzzle this way, in describing it as "successful." One cannot automatically assume, moreover, that a Presocratic thinker, such as Democritus, would have distinguished between physical and mathematical modes of being, thereby to limit the divisibility of the former, but to admit the unlimited divisibility of the latter. Indeed, the later Epicurean atomists, who followed in the tradition of Democritus, tended to dispute the validity of the idealizing conceptions made by geometers.[48]

As for the two theorems mentioned in the preface of the *Method,* Archimedes says nothing about the manner of derivation Democritus set out, or even whether he did anything more than merely "state" them. In the preface to his earlier work, *On the Sphere and Cylinder,* Book I, however, where Archimedes speaks of the same two theorems on the pyramid and the cone, he *omits* any mention of Democritus, but instead asserts unequivocally that "although there were many geometers worthy of note before Eudoxus, it happened that ⟨these theorems⟩ were unknown by all, nor were they recognized by anyone."[49] This statement is incompatible with what he says in the *Method,* as cited above, reserving "no small credit to Democritus" for having stated these same theorems. The discrepancy indicates that in the time between writing the *Sphere and Cylinder* and the *Method* (taken to be a later work) Archimedes became aware of the relevant contribution by Democritus. On any other account of the chronology the inconsistency is inexplicable.[50]

Even before *Sphere and Cylinder,* namely, in *Quadrature of the Parabola,* although Archimedes adopts a purely finitist limiting procedure in a "mechanical" argument for the measurement of the parabolic segment, nevertheless, the technique clearly betrays dependence on an indivisibilist heuristic such as that presented in *Method,* prop. 1. It is thus clear that Archimedes was using the indivisibilist notion before he became aware of Democritus's contribution. Yet we cannot suppose that Archimedes invented the notion of indivisibles independent of any precedent, since we have seen the same conception firmly attested in Democritus's "cone fragment." It must be the case, then, that Archimedes and Democritus both drew from a common pre-Euclidean geometric technique of indivisibles.

Among pre-Euclidean thinkers, in fact, there is an extensive body of speculation about indivisible magnitudes. In natural philosophy, for instance, both the Eleatics (notably, Zeno) and the Atomists (such as Leucippus and Democritus) presented controversial doctrines about the constitution of substance: the Eleatics insist that, if finite magnitudes were constituted out of irreducible parts (for instance, lines out of points), those magnitudes must be either zero (if the parts were of zero magnitude) or infinite (if the parts were of finite magnitude).[51] By contrast, the Atomists introduce the indivisibles precisely in answer to puzzles of this sort. In a series of arguments that Aristotle apparently assigns to Democritus, for instance, the question is raised, if a magnitude is divisible throughout, and one conceives it to have actually been divided throughout, then what remains after this division?—if there remain divisible parts, then the division has not actually been carried through, against the hypothesis; but, if indivisible parts (that is, points) are left, then the whole magnitude will be constituted out of parts of no magnitude, and thus it will itself be no magnitude.[52] Upon further consideration of difficulties entailed by this conception, the atomist argument concludes that there must be indivisible magnitudes; that is, the division process cannot go on indefinitely.

Aristotle, who subscribes to a strict continuist view, not only for geometric entities, but also for physical substances, devotes Book VI of the *Physics* to refuting conceptions of indivisibles in the theory of moving bodies. For instance, if one supposed that a motion traversed an indivisible distance in an indivisible interval of time, then a faster motion would traverse a smaller distance in the same time, or would traverse the same distance in a smaller time; in each case, the hypothetical indivisible will have been divided.[53] In a more purely geometric vein, the Peripatetic author of *De lineis insecabilibus* reviews a whole repertory of arguments refuting what had been purported grounds for supposing indivisible magnitudes. Some of these arguments are of an abstract type, based on notions of the great and the small or on the Ideas in general (chaps. 1–4),[54] or the inability, as claimed by Zeno, for one to count mentally an infinity in finite time (969a27–b7). Others are more concretely geometric: that under the hypothesis of indivisibles, all magnitudes must be commensurable with each other (970a1–4); or that in an isosceles triangle, the base, if supposed indivisible, must nonetheless be bisected by its altitude (970a9–12); or that in a square whose sides are hypothesized to be indivisibles, the diagonal, although greater than a single indivisible (namely, the side), must yet be less than two indivisibles (970a12–14); and so on.

None of these passages suggests that indivisibilist conceptions, like the Democritean notion of slicing the cone into its indivisible circular sections, could be deployed toward sophisticated insights into the measurement of geometric figures, as is done in Archimedes's exposition in the *Method*. It is clear, then, that this philosophical literature did not supply precedents for the technical applications, but, to the contrary, was responding to conceptions assumed in a technical literature. If anything, the philosophical writers took it to be their business not to explicate or defend assumptions of indivisible magnitudes, where geometers may have introduced them, but rather to expose the logical incoherence of such assumptions.

As far as the sources of his geometric technique are concerned, Archimedes provides only one explicit indication: he refers to Eudoxus as having effected the first proper demonstration of the theorems on the pyramid and the cone. It is appropriate to consider, then, whether we can discern implicit in the Eudoxean proofs of these theorems some vestige of the heuristic use of an indivisibilist conception, such as one can indeed discern behind Archimedes's formal demonstrations in *Quadrature of the Parabola* and *Conoids and Spheroids*. These Eudoxean proofs survive now only in the form given by Euclid in Book XII of the *Elements,* and we cannot be certain that Euclid transmits a pure rendition of the older proofs. Indeed, there are signs that Euclid has modified them in certain respects.[55] Despite this, I think it is possible to see how the pyramid theorem in Euclid's treatment (taken as representative of Eudoxus's treatment) had an indivisibilist preliminary.

The key result is set out in *Elements* XII, prop. 5: that pyramids of equal height are as their bases. An indivisibilist argument could proceed along these lines:[56] assuming first two pyramids having triangular bases and lying within the same parallel planes, we can conceive them to be cut by a plane parallel to the base: the sections are triangles which, as can readily be shown, are similar to the bases of their respective pyramids. Since every pair of triangular sections is in the ratio of the bases, it follows when the solids are filled out that all the triangles in one pyramid have this same ratio to all the triangles in the second; that is, the pyramids themselves are in the ratio of their bases (Figure 3).

To formalize this construction one must find a way of replacing the indivisibles with finite elements of volume. In the Archimedean-style method adopted by A. M. Legendre, one forms a system of n prisms of equal height inscribed in one of the pyramids (Figure 4a), in correspondence with the system of $n+1$ prisms of the same height circumscribed about it (Figure 4b).[57] The difference between the two sums

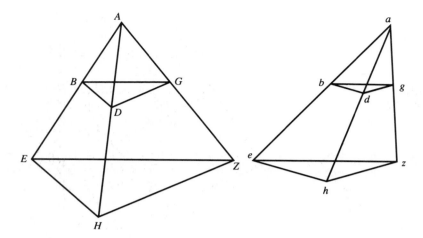

FIGURE 3

is exactly one prism (namely, prism $EMLDGB$ in Figure 4b), having as its base the base of the pyramid, and as its height the height of one of the prismatic elements. Since n can be taken arbitrarily large, this volume difference can be made arbitrarily small. Hence, the pyramid can be approximated as closely as desired by suitable lower- and upper-bounding figures. Since the corresponding prismatic sums inscribed and circumscribed, respectively, in the two pyramids will have the ratio of their bases, one can deduce, by the familiar form of indirect argument, that the pyramids must have the same ratio.

In the Euclidean scheme for convergence, one depends on consecutive bisection for reducing the difference between the inscribed figure and the pyramid.[58] Beginning from the pyramid, one separates it into a prism and remainder by dividing each of its sides in half (Figure 5a); applying the same division process in turn to this remainder, again to the resultant remainder, and so on, one can ultimately reduce the difference to less than a preassigned, arbitrarily small volume. The Euclidean lemma on convergence (*Elements,* X, prop. 1) requires that at each stage the remainder is made less than half of the preceding remainder. Now, at the first stage, the removal of the single prism subtracts only 3/8 of the pyramid's volume, leaving a remainder of 5/8, consisting of a prism and two small pyramids. At the next stage, exactly one-half of the prism will be removed, but, again, only 3/8 of each of the pyramids (Figure 5b). Thereafter, a remainder consisting of smaller prisms and pyramids will result, from which the next reduction will remove half from each of the prisms, but only 3/8 from each of the pyramids. Thus, the reduction at each

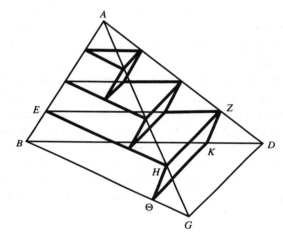

FIGURE 4a

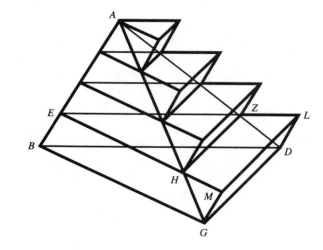

FIGURE 4b

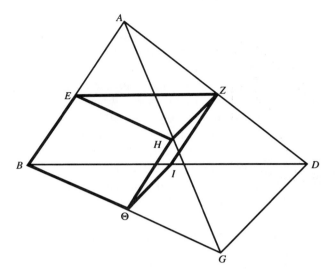

FIGURE 5a

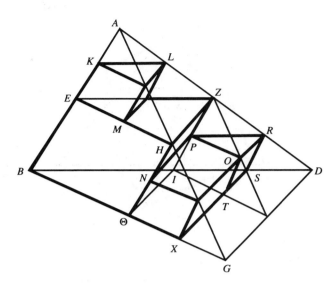

FIGURE 5b

stage is never more than half the previous remainder, although it approaches closer to one-half as the procedure continues.[59]

In this way, the parallel sectioning suggested by the indivisibilist construction results in a convergence slightly too slow to conform to the requirements of the standard Euclidean procedure. The geometer seeking a proof of this sort would realize this impasse, since it follows from the anticipated conclusion, the proportionality of pyramids and bases. But examining these constructions reveals an adaptation: to remove not only one prism, but *both* of the prisms

obtained by dissecting each of the small pyramids. The net reduction then becomes 3/4 of the preceding remainder. For the purposes of the proof, one needs to claim only that each of the two prisms formed in the dissection is greater than each of the pyramids, so that the removal of the prisms reduces the remainder by more than one-half. This is in fact the form of the argument followed in the Euclidean construction in *Elements* XII, prop. 5 (Figure 6).

To conclude the derivation that the pyramid equals one-third of the associated prism, one can follow the dissection procedure used by Euclid (*Elements* XII, prop. 7),

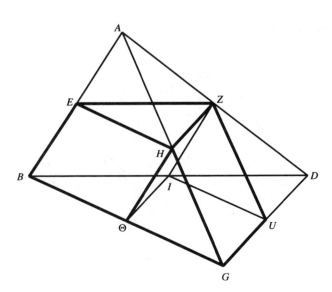

FIGURE 6a

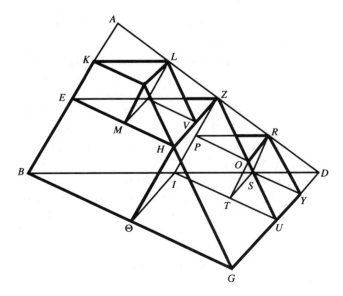

FIGURE 6b

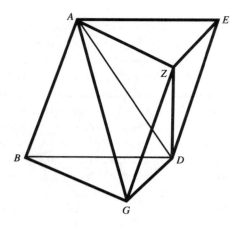

FIGURE 7

where the triangle-based prism is divided into three equal pyramids (Figure 7): pyramids $AZDG$, $AZDE$ are equal, since they share the vertex A and have equal bases, the triangles ZDG, ZDE; similarly, pyramids $DAGZ$, $DAGB$ are equal, since they share vertex D and have the equal triangles AGZ, AGB as bases.[60]

The second Eudoxean theorem, that the cone is one-third the associated cylinder, is demonstrated in *Elements* XII, prop. 10. Here, the circular base of the cone is approximated by the polygonal base of an inscribed pyramid in the context of an indirect proof in the standard Euclidean manner. If we consider the possibility of a preliminary construction with indivisibles, an alternative form can be proposed: taking the cone and cylinder of equal base and equal

height h (Figure 8a), we introduce a pyramid and prism with common base and with height equal to h (Figure 8b). If we now section the cone and cylinder by a plane at distance k below the vertex of the cone, it will produce two circles (HT, HK), while in the prism and the pyramid the corresponding plane, also at distance k, will cut off triangles (hti and hkl, respectively). Since the two circles have the same ratio as the two triangles (namely, $k^2 : h^2$), it follows after summation that all the circles of the cone have to all the circles of the cylinder the ratio that all the triangles of the pyramid have to all the triangles of the prism; that is, the cone and cylinder have the same ratio of the pyramid and the prism, namely 1 : 3.

Formalizing this version of the construction is straightforward under the two-sided "compression" method of convergence commonly applied by Archimedes.[61] But it poses considerable difficulty within the standard Euclidean one-sided "approximative" convergence procedure. Thus, the construction in the *Elements,* as one would expect, adopts a different approach. Nevertheless, one can cite from Pappus a construction reminiscent of the sectioning argument just given. In his account of the measurement of the Archimedean spiral, he sets up a relation between the circular sectors inscribed in the spiral and the associated circle, and the cylindrical sections inscribed in a cone and an associated cylinder: since the sectors have the same ratio as the cylindrical sections (namely, the squares of their radii), the spiral and the circle, obtained through summation of the former elements, will have the same ratio as the cone and the cylinder, obtained by summing the latter, that is, the ratio of 1 : 3. This heuristic argument lends itself

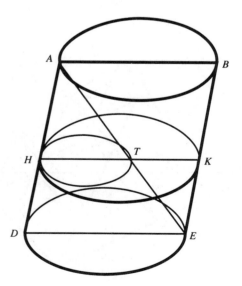

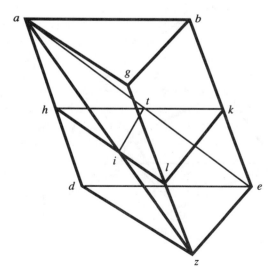

to formalization under the Archimedean scheme of "compression," and Pappus appears to assign it to Archimedes himself.[62] One can thus propose that it was inspired by a similar construction, using parallel sections, for working out the relation of the cone and the cylinder.

These examples thus reveal an aspect that discourages the simple modification of the indivisibilist preliminaries as the basis for regular proofs. The scheme of parallel sectioning of plane or solid figures is not readily convertible to a convergence construction of the one-sided "approximative" type, adopted by Euclid. It does lend itself to the two-sided ("compression") method, however, the manner favored by Archimedes, its apparent inventor. Indeed, we can derive from indivisibilist precedents, like the sectioning of the cone and pyramid, a possible incentive for Archimedes's development of the alternative convergence technique.[63]

My aim in discussing these examples has been to show how the indivisibilist technique in geometry was not an isolated and idiosyncratic whim of Archimedes, published in only a single work, the *Method*; but rather that it was a technique familiar to Archimedes from his own technical sources, whose tradition was still drawn upon by later commentators like Hero and Theon. Considerations of rigor do, of course, exclude this technique from the formal tradition, as represented by the treatises of Euclid and Archimedes (apart from the *Method*). Even Archimedes, as we perceive from the prefaces to the *Method* and *Quadrature of the Parabola,* feels on the defensive, and has to argue the value of his heuristic method in the face of formalist objections. This attitude, doubtless shared by others in his tradition, would discourage the exposition of heuristic methods, like that of indivisibles, in their published writings.

Theon's text of the cylinder lemma reveals how the method of indivisibles has become unfamiliar within the later geometric tradition, when training in the formal manner has discouraged exposure to heuristic methods. In contrast to the priorities of researchers, who would be interested in expanding the domain of known technical results, the commentators' agenda seems insensitive to the aims of discovery, but appears to reflect instead the exegetical and analytical aims of the philosophical curriculum that dominated the schools of later antiquity. The same insensitivity, however, need not have marked the teaching of geometry in the early period, around the time of Archimedes.

Section IV

Two recent studies by Lam Lay Yong and Shen Kangsheng [1985] and D. Wagner [1978] have noted the adoption of

the Cavalieri principle in early Chinese efforts.[64] In particular, the discussion of the measurement of the sphere by Liu Hui (3rd century A.D.) and his commentator Zu Geng (5th century) turns on manipulations of indivisibles. Liu reduces the determination of the volume of the sphere to that of the "umbrella" (or "box lid") figure formed as the common space defined by two cylinders intersecting at right angles within a cube (Figure 9a): since the planar sections of this figure are the squares circumscribed about the circular sections of the sphere (Figure 9b), it is deduced that the solid and the sphere have the ratio of the square and the circle. But there Liu stops, unable to evaluate the volume of the derived figure. Zu completes the analysis of the "umbrella" figure via another application of indivisibles, whence the sphere is found.

FIGURE 9a

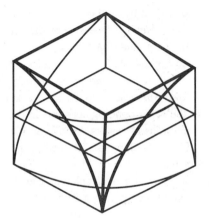

FIGURE 9b

Liu's argument for the ratio of the "umbrella" and the sphere recalls so strikingly Theon's lemma on the ratio of the cube and the cylinder, that one is led to inquire whether some objective connection exists between the Chinese and Greek traditions. Such a link, proposed by some,[65] has been doubted by Lam and Shen, however, whose position, I think, is correct. In the *Method*, Archimedes adopts a related technique of indivisibles to work out the measurements of both the sphere and the cylindrical section (the latter being the same figure as Liu's "umbrella"), but by independent constructions without any apparent connection to each other. Any writer conversant with this tradition, whether through the *Method* or through paraphrases, as in Hero's *Metrica*, would thus know what the volumes are, but have no precedent for seeing, as Liu does, that their ratio is that of the circle and the square. Moreover, the Archimedean introduction of mechanical assumptions finds no parallel in the Chinese examples. Thus, if the Chinese geometers owed their use of indivisibles to a Greek tradition, it would have to be in an alternative form, lacking the subtle refinements made by Archimedes.

This does not exclude an input from the Greek into the Chinese. But it would be an input of a sort no longer represented by extant Greek sources. In view of the ancient precedents for the indivisibilist technique, at any rate, we cannot assign absolute priority to the Chinese geometers, although they may well share credit as independent inventors.

With Johannes Kepler and Bonaventura Cavalieri, however, I think that even such a restricted form of attribution is doubtful. To be sure, the ancient works that have provided our examples of the indivisibilist analysis were not available in the seventeenth century. Although the *Method* was still an important source for Hero in the first century A.D., only a single manuscript of the work is known from after the tenth century, and that remained lost until it was identified by Heiberg and published in 1906.[66] Similarly, Hero's *Metrica*, preserving the related conception of motion-generated figures, was unavailable before H. Schöne's edition of 1903.[67] Nevertheless, Archimedes's efforts expounded in the *Quadrature of the Parabola* and the *Conoids and Spheroids*, which, as we know by comparison with the corresponding accounts in the *Method*, are the formal renditions of results discovered via indivisibles, were studied with zeal by the sixteenth and seventeenth century geometers.

Thus, the invention by Kepler or Cavalieri of an indivisibilist method amounts to a *reinvention*, a retrieval of the underlying heuristic presupposed by the formal treatments extant from Archimedes.[68] The later geometers' aim here, as in other areas of geometry, was to find alternative procedures making the formidable, tightly structured proofs in the ancient treatises more readily comprehensible. Thus, where a comparable heuristic already existed among the ancients, the modern accounts inadvertently restored what was implicit in the extant ancient source.

In the course of his measurements of solids, Kepler often cites Archimedean theorems explicitly.[69] Similarly, in his own exposition of the method of indivisibles, Cavalieri first goes over the familiar terrain of the Archimedean measurements of the parabola, the spiral and the conoids, while, as he does so, retaining the Archimedean terminology, such as "spheroid" for ellipsoids of revolution.[70] A clear sign of the Archimedean connection can be detected in Cavalieri's adoption of the "*omnes*" terminology (for instance, speaking of "*all* the lines" which constitute a plane figure). This use of *omnes* parallels Archimedes's term *panta* to denote the same kind of summations (applied for the constituents of a figure, whether indivisible or finite); indeed, in these very contexts the term *panta* is translated as *omnes* in the Latin versions of the Archimedean corpus, most notably, the version by Federico Commandino [1558]. This is not to say that Cavalieri meant by his "*omnes*-concept" (in the expression adopted by K. Andersen [1985]) nothing other than what Archimedes denotes in his use of the terms. As E. Giusti and K. Andersen have recently argued, Cavalieri appears to have intended a modification of the notion of magnitude that could place the indivisibilist technique on an acceptable logical footing. By contrast, for Archimedes the *panta* locution denotes summation, whether of finite magnitudes in a regular proof or of indivisibles in an argument of the heuristic type. Archimedes can leave imprecise the notion of the indivisibles and how they actually constitute a finite magnitude, since for him they are only provisional.

Defenders and critics alike of the method of indivisibles in the seventeenth century considered it a simplified equivalent of the formal "exhaustion" procedures.[71] Andersen argues that theirs was a misconception, for, in her view, Cavalieri intended by indivisibles, specifically in the "*omnes*-concept," a notion subtler than either his contemporaries or modern scholars commonly realize. The entity Cavalieri denotes by "all the lines" in a given figure, for instance, is not a naive summation of indivisible components somehow resulting in the continuous magnitude; it is a new type of entity, a set, in a specific relation to the magnitude itself. If Andersen's is the correct interpretation of Cavalieri's meaning, one can nevertheless understand the cause of the misconception. For Cavalieri's readers were certain to pick up the Archimedean allusion in the "*omnes*" terminology, and to suppose (quite rightly) that the ancient conception denoted simply a summation of the constituents in a given magnitude. Cavalieri's follower Evangelista Torricelli simply identified the collection of indivisibles with the figure containing them and thus, according to Andersen's account, misled others about the subtle logical refinements intended by Cavalieri.

Having access to the *Method,* we can perceive, as the seventeenth century geometers could not, that Archimedes had extrapolated the same summation concept for the case of indivisibles, much in the naive manner that Torricelli espoused and that drew criticism on Cavalieri.

Doubtless, the foundational issues which must be resolved in order to convert the indivisibilist technique into an acceptable proof procedure were more vexing than either the ancient or the early modern geometers supposed. But I think the view of their ultimately simple correspondence may well capture the insight a working geometer would maintain about the method. I have proposed that an indivisibilist heuristic was included within the pre-Eudoxean geometric tradition as a background for the rigorous limiting theorems of Eudoxus and Archimedes, and, conversely, that the early modern methods of indivisibles are at base a reconstruction of the lost heuristic by geometers who were impatient over the demands of formal demonstration. In both the ancient and modern cases, the active interest in heuristic methods, like the method of indivisibles, owed its appearance to the geometers' willingness to suspend the reservations of the formalist sort and to appreciate the value of informal techniques.

NOTES

1. For accounts of Cavalieri's methods, see E. Giusti, *Bonaventura Cavalieri and the Theory of Indivisibles* (Bologna: Edizioni Cremonese, 1980) and K. Andersen, "Cavalieri's Method of Indivisibles," *Archive for History of Exact Sciences* 1985, 31: 291–367. Specimens of the method are presented by M. Baron, *The Origins of the Infinitesimal Calculus* (Oxford: Pergamon, 1969), pp. 122–135.

2. See D. Wagner, "Liu Hui and Tsu Keng-chih on the Volume of a Sphere," *Chinese Science* 1978, 3: 59–79; Lam Lay-Yong and Shen Kangsheng, "The Chinese Concept of Cavalieri's Principle and its Applications," *Historia Mathematica* 1985, 12: 219–228. These are considered further in section IV below.

3. The critical Greek edition of Archimedes's text is in *Archimedis Opera* (ed. Heiberg, 2nd ed., Leipzig: Teubner, 1910-15, II, pp. 426–507), which supersedes the earlier accounts in J. L. Heiberg and H. G. Zeuthen, "Eine neue Schrift des Archimedes," *Bibliotheca Mathematica* 1906–07, 7 (ser. 3): 321–363; and Heiberg, "Eine neue Archimedeshandschrift," *Hermes* 1907, 42: 234–303. The

Method appears in all the major translations and commentaries on Archimedes's work, e.g. T. L. Heath, Supplement to *The Works of Archimedes* (Cambridge: Cambridge University Press, 1912; reprint New York: Dover, n.d.); P. Ver Eecke, *Archimède: Les oeuvres complètes* (first ed., Paris & Brussels: Desclée, de Brouwer, 1921), pp. 475–519; C. Mugler, *Les oeuvres d'Archimède* (Paris: Association Budé & Société d'éditions 'Les belles lettres,' 4 vols., 1970–72); E. S. Stamatis, *Archimêdous Hapanta* (Athens: Technikou Epimelêtêriou tês Hellados, 3 vols. in 4, 1970–74), II, pp. 383–465; E. J. Dijksterhuis, *Archimedes* (Copenhagen: Munksgaard, 1956; reprint Princeton: Princeton University Press, 1987), pp. 313–336. For references to studies of this work, see I. Schneider, *Archimedes* (Darmstadt: Wissenschaftliche Buchgesellschaft, 1979), pp. 109–121; and W. R. Knorr, "Archimedes after Dijksterhuis," in Dijksterhuis, *Archimedes* (*op. cit.,* Princeton, 1987), pp. 436–437.

4. Theon expresses the constant (equivalent to $\pi/6$) in unreduced form, as the ratio $7\frac{1}{3}$: 14; for the derivation see Theon's *Commentary on Ptolemy's Almagest,* Book I (ed. A. Rome, Vatican City: *Studi e Testi* 72, 1936), pp. 394–399. It is discussed in W. R. Knorr, "On Two Archimedean Rules for the Circle and the Sphere," *Bollettino di Storia delle Scienze Matematiche* 1986, 6: 145–158; and *Textual Studies in Ancient and Medieval Geometry* (Boston/ Basel/ Berlin: Birkhäuser, 1989), pp. 604–606. For a summary of Theon's work see G. J. Toomer, "Theon of Alexandria," in *Dictionary of Scientific Biography* (New York: Scribner's Sons) 1976, 13: 321–325.

5. Theon (ed. Rome), pp. 398–399.

6. Theon (ed. Rome), p. 398n.

7. By square brackets "[...]" I indicate lines that I recommend for exclusion; by angled brackets "⟨...⟩" lines that I suggest be added; and by parentheses "(...)" phrases that are tacitly understood in the elliptical idiom of the Greek.

8. In both lines the phrase "and further" (*kai eti*) suggests the hand of Theon, since phrases with *eti* often mark additions he makes in his commentaries; on the appearance of this term in texts edited by Theon, see Knorr, *Textual Studies* (cit. n. 4), pp. 759, 772.

9. The change is actually slight: in line 7 "in MN" (*en tôi MN*) and "to those in EZ" (*tois en tôi EZ*) correspond to the emended phrases in line 7′ "as MN" (*hôs MN*) and "to the (cube) on EZ" (*tôi apo tou EZ*), respectively. Any account of the lemma must assume the omission of some counterpart to the line in angled brackets, since otherwise there is no accommodation for the case of the cylinder.

10. *Method,* prop. 5 (ed. Heiberg, II, p. 462.22–24); for citations of other instances, see section II below.

11. *Conoids and Spheroids,* prop. 21 (ed. Heiberg, I, p. 350.13–20); for other examples, see section II.

12. For instance, *Spiral Lines,* prop. 10 (ed. Heiberg, II, p. 32.27–31), with both *panta* and *sympanta; Quadrature of the Parabola,* prop. 23 (*ibid.,* p. 310, lines 7, 18–19 and 22).

13. For the text and translation, see M. Clagett, *Archimedes in the Middle Ages* (Madison: University of Wisconsin Press, 1964), I, pp. 504–505; for discussion, see Knorr, "Two

Archimedean Rules" (cit. n. 4), pp. 152–154; and *Textual Studies* (cit. n. 4), pp. 604–606.

14. On the work of John of Tynemouth, see Knorr, "John of Tynemouth *alias* John of London," *British Journal for the History of Science* 1990, 23: 293–330. (Note, however, that some of my claims there, particularly on the proposed identification of John of Tynemouth with a certain John of London, are currently being reconsidered.) On the connections with Grosseteste, see my "*Falsigraphus* vs. *adversarius*: Robert Grosseteste, John of Tynemouth, and Geometry in the 13th Century" (forthcoming).

15. Moerbeke's translation has been edited by Clagett, *Archimedes in the Middle Ages* (Philadelphia: American Philosophical Society, 1976), vol. II. For analysis of *De curvis superficiebus* in comparison with other treatments of Archimedes' sphere measurement, see my "The Medieval Tradition of Archimedes' *Sphere and Cylinder*," in *Mathematics and its Applications to Science and Natural Philosophy in the Middle Ages: Essays in Honor of Marshall Clagett,* ed. E. Grant and J. E. Murdoch (Cambridge: Cambridge University Press, 1987), pp. 3–42; and my *Textual Studies* (cit. n. 4), pp. 595–615, 618–624.

16. The connections to Theon are presented in Knorr, "Two Archimedean Rules" (cit. n. 4); "Medieval Tradition" (cit. n. 15); and *Textual Studies* (cit. n. 4), pp. 604–606.

17. On Hero, see A. G. Drachmann, "Hero of Alexandria," in *Dictionary of Scientific Biography* 1972, 6: 310–314; and M. S. Mahoney, "Hero of Alexandria: Mathematics," *ibid.,* pp. 314–315.

18. Hero (ed. Schöne), pp. 92.14–94.7.

19. *Ibid.,* p. 94.7–11.

20. Hero distinguishes between the "height" (*hypsos*) of the solid and the altitude, or "perpendicular" (*kathetos*), that one uses for measuring it; compare these passages: "even if the *hypsos* is not at right angles to the base, but is inclined" (*ibid.,* p. 94.23–25) and "one must take the area of the base and multiply it by the said *kathetos*" (p. 94.28–30).

21. *Ibid.,* p. 94.25–26.

22. *Ibid.,* p. 98.7-8.

23. *Ibid.,* pp. 116.28-118.8.

24. The authenticity of Theon's lemma as a whole need not be doubted, since the ploy of attaching a lemma of this sort, as here, to supplement a mathematical exposition is consistent with his style as a commentator. Moreover, the manuscripts give no evidence of significant corruption or intervention by interpolators in this or adjacent sections of Theon's *Commentary on Ptolemy.*

25. An excerpt from Theon's commentary on Hero's *Metrica* is quoted by the anonymous compiler of the so-called *Introduction to Ptolemy's Almagest*; this work has not been edited. For discussion, including citation and critique of the study by J. Mogenet (1956), see Knorr, *Textual Studies* (cit. n. 4), pp. 155–211.

26. See, for instance, S. Y. Luria, *Archimedes* (Vienna: 'Neues Oesterreich', 1948; translation of the Russian edition of 1945), pp. 61–63; Baron, *Origins* (cit. n. 1), pp. 46–50; Schneider, *Archimedes* (cit. n. 3), pp. 109–121; also the translation of the *Method* by Heath (cit. n. 3), pp. 15–21; the translation by Ver Eecke (cit. n. 3), pp. 481–488; and

the commentary by Dijksterhuis, *Archimedes* (cit. n. 3), pp. 316–324; for additional references, see Knorr, "Archimedes after Dijksterhuis" (cit. n. 3), pp. 436–437.

27. The *Method* is extant only in the Istanbul (Constantinople) manuscript identified by Heiberg and collated in his second edition of Archimedes (Leipzig, 1910–15); for a description see Heath's Supplement to the *Works of Archimedes* (cit. n. 3), pp. 5–7; and Dijksterhuis, *Archimedes* (cit. n. 3), pp. 44–45. The manuscript has been inaccessible for study since its unauthorized removal from Istanbul in the 1920s.

28. Dijksterhuis (*Archimedes,* cit. n. 3, p. 331) calls this figure the "cylinder hoof," corresponding to the German term *Zylinderhuf.* The same figure is measured by J. Kepler in *Nova stereometria doliorum vinariorum* (Linz, 1615). In his commentary on Kepler's stereometry, F. Hammer employs the term *Zylinderhuf;* cf. *Keplers Gesammelte Werke,* vol. 9 (Munich, 1960), p. 438, relative to Kepler's prop. 17. Kepler, however, calls the figure by the generic name *segmentum cylindri* (*ibid.,* p. 33) or *Walger-schnitz* (*-schnitzlein, -schnittlin*) (*ibid.,* pp. 183–184, 190). The same figure is included among the solid measurements in Hero's *Metrica* (Book II, chap. 14), but Hero, citing Archimedes's theorem from the *Method,* calls it merely the "segment of a cylinder" (ed. Schöne, p. 130.12–24).

29. The second solid is constructed thus: if in a given cube two equal cylinders are inscribed, each having diameter and height equal to the side of the cube and its axis perpendicular to the other's, then their common section equals two-thirds the cube (see Fig. 9). Later Chinese geometers give this figure the name "box lid" (see section IV below). It can be formed out of eight figures of the type of Archimedes's first cylindrical section, as Juel has shown; cf. Heath, Supplement (cit. n. 3), p. 51.

30. That is, the magnitudes of the figures are inversely proportional to the distances of their respective centers of gravity from the fulcrum. Since the center of gravity of the triangular base of the prism lies on θX, dividing it into segments in the ratio 2 : 1, while the suspension distance of the hoof is $P\theta$ (= θX), the prism will be three-halves the volume of the hoof. Since the prism is one-fourth the cube, it follows that the hoof is one-sixth the cube.

31. For reconstructions, see Heiberg's supplement to the proof in his edition (cit. n. 3), vol. II, pp. 494–495; and the commentaries by Heath, *Supplement* (cit. n. 3), pp. 39–40; and Dijksterhuis, *Archimedes* (cit. n. 3), pp. 331–332.

32. By symmetry, the center of gravity of the semicircle lies on the axis θP; let its distance from the fulcrum be z. The areas of the figures are known (the triangle equals r^2, the semicircle $cr/4$, for c the circumference of the circle), and the center of gravity of the triangle is known to lie at distance $2r/3$ from the fulcrum. Thus, $z : 2r/3 = r^2 : rc/4$, that is, $z : r = 8r : 3c = 4 : 3\pi$, or approximately 14 : 33 (for π approximately 22/7). It follows that the center of gravity divides the radius into segments approximately in the ratio 14 : 19. Cf. also Heath, *Supplement* (cit. n. 3), p. 40.

33. "Lemmas" to the *Method* (ed. Heiberg, II, pp. 432–434). The stated results on the centers of gravity of solids appear to have been demonstrated in a lost work, called the *Equilibria* in a citation in *Floating Bodies,* Book II, prop. 2; cf.

Knorr, "Archimedes' Lost Treatise on Centers of Gravity of Solids," *Mathematical Intelligencer* 1978, 1: 102–108.

34. Archimedes (ed. Heiberg), II, p. 506.14–21.

35. Cf. also *Spiral Lines,* prop. 21; *Plane Equilibria,* Book I, props. 7, 9, 13; other passages where the same terminology for summations appears are listed in Heiberg's index in the entries for *pas* and *sympas* (*Archimedis Opera,* III, pp. 381, 391). In his treatments of the spirals, apparently based on an Archimedean exposition, Pappus also employs this terminology; cf. Knorr, "Archimedes and the Spirals," *Historia Mathematica* 1978, 5: 43–75.

36. The Eudoxean method is described in most accounts of ancient geometry; see, for instance, Knorr, *Ancient Tradition of Geometric Problems* (Boston/ Basel/ Stuttgart: Birkhäuser, 1986; repr. New York: Dover, 1993), pp. 78–79.

37. Archimedes (ed. Heiberg), II, p. 428.24–26.

38. *Ibid.,* p. 428.30

39. *Ibid.,* p. 438.16–21.

40. Cf. Dijksterhuis, *Archimedes* (cit. n. 3), p. 319: "the mathematical deficiency is exclusively a consequence of the use of indivisibles." For other references, see Knorr, "Archimedes after Dijksterhuis" (cit. n. 3), p. 437.

41. Dijksterhuis maintains that "there is not the least objection from the mathematical point of view against properly founded barycentric considerations" (*ibid.,* p. 319).

42. Archimedes (ed. Heiberg), II, p. 428.28–29.

43. *Ibid.,* II, p. 262.8–13; *epideichthen* could be a scribal slip for *apodeichthen* (Heiberg translates "demonstratum," as if the latter); cf. *ibid.,* pp. 264.26–266.2: "I have written down the proofs (*apodeixeis*), first as they have been investigated (*theôrein*) by means of mechanics (*mêchanika*), then afterward as they are proved (*apodeiknysthai*) by means of geometry (*geômetroumena*)."

44. Indeed, the property of the cylinder that Theon proves in the lemma in this commentary is elsewhere assigned to Archimedes (whether rightly or not) by an anonymous scholium to a pseudo-Diophantine collection; cf. Knorr, "Two Archimedean Rules" (cit. n. 4). Theon's lemma is auxiliary to his derivation of the arithmetic rule for the volume of the sphere, for which he explicitly cites Archimedes's *Sphere and Cylinder* for the necessary geometric theorems. That he fails to cite any source, Archimedes or another, for the arithmetic rule itself and its dependent lemma may indicate that Theon is here improvising beyond the specific contents of his sources.

45. For accounts of the ancient atomist theories, see W. K. C. Guthrie, *A History of Greek Philosophy,* vol. II (Cambridge: Cambridge University Press, 1965), chap. VIII; G. S. Kirk, J. E. Raven, and M. Schofield, *The Presocratic Philosophers,* 2nd ed. (Cambridge: Cambridge University Press, 1983), chap. XV; D. Furley, *The Greek Cosmologists I* (Cambridge: Cambridge University Press, 1987), chaps. 9–11.

46. Archimedes (ed. Heiberg), II, p. 430.6–9.

47. Democritus, fragment 155 (DK 68 B155); text as given by H. Diels and W. Kranz, *Die Fragmente der Vorsokratiker,* 6th ed., 3 vols., (repr. Dublin & Zurich: Weidmann, 1966), II, pp. 173–174. The text is reproduced with translation in I. Thomas, *Selections Illustrating the History of Greek Mathematics,* 2 vols. (Loeb Classical Library, 1939–41; repr. Cambridge, Mass.: Harvard University Press & London: Heinemann, 1957), I, pp. 228–231. Its meaning and historical significance have been widely debated; see, for instance, Heath, *A History of Greek Mathematics,* 2 vols. (Oxford: at the Clarendon Press, 1921), I, pp. 179–180; Luria, "Die Infinitesimaltheorie der antiken Atomisten," *Quellen und Studien zur Geschichte der Mathematik, Astronomie und Physik,* 1932–33, 2 (Abt. B): 106–185; J. Mau, *Zum Problem des Infinitesimalen bei den antiken Atomisten* (Berlin, 1954); Guthrie, *History* (cit. n. 45), pp. 487–488; H. J. Waschkiess, *Von Eudoxos zu Aristoteles: Das Fortwirken der Eudoxischen Proportionentheorie in der Aristotelischen Lehre vom Kontinuum* (Amsterdam: Grüner, 1977), pp. 279–284; and Furley, *Greek Cosmologists* (cit. n. 45), pp. 130–131.

48. On the mathematical doctrines of the later atomists, specifically in the following of Epicurus, see Furley, *Two Studies in the Greek Atomists* (Princeton: Princeton University Press, 1967); and D. Sedley, "Epicurus and the Mathematicians of Cyzicus," *Cronache Ercolanesi* 1976, 6: 23–54.

49. Archimedes (ed. Heiberg), I, p. 4.11–13.

50. This observation on the chronology was first made by F. Arendt, "Zu Archimedes," *Bibliotheca Mathematica* 1913–14, 14 (ser. 3): 289–311; see also E. Neuenschwander, "Die stereometrischen Bücher der Elemente Euklids," *Archive for History of Exact Sciences* 1974 (14: 91–125), p. 121n; Knorr, "Archimedes and the *Elements*: Proposal for a Revised Chronological Ordering of the Archimedean Corpus," *Archive for History of Exact Sciences* 1978 (19: 211–290), p. 262; Schneider, *Archimedes* (cit. n. 3), pp. 31, 38. The older view, setting the *Method* early in the sequence of Archimedes's writing, was proposed by Zeuthen and Heiberg (in "Eine neue Schrift des Archimedes," cit. n. 3), and accepted by Heath (in *History,* cit. n. 47, II, p. 22) and most other commentators. Dijksterhuis, for instance, notes that the two passages from *Sphere and Cylinder* and *Method* are "at variance," but, committed to Zeuthen and Heiberg's hypothesis, he simply manifests puzzlement as to how to account for it (*Archimedes,* cit. n. 3, p. 143n).

51. For citations and analysis of the basic texts: on the Atomists, see note 45; on Zeno, fragments 1-2 see Guthrie, *History* (cit. n. 45), pp. 88–91; Kirk, Raven, and Schofield, *Presocratics* (cit. n. 45), pp. 265–269; Furley, *Greek Cosmologists* (cit. n. 45), pp. 106–110.

52. *Generation and Corruption,* I, chap. 2, 316a15–b17; for discussion, see Furley, *Greek Cosmologists* (cit. n. 45), pp. 124–126.

53. Cf. Physics VI, chap. 2, 233b16–32; chap. 3, 234a24–32.

54. Cf. Plato's hypothesis on "indivisible lines," as reported by Aristotle, *Metaphysics* I, chap. 9, 992a10–24; for discussion, see Heath, *Mathematics in Aristotle* (Oxford: at the Clarendon Press, 1949), pp. 199–200. The early Peripatetic tradition of criticizing the doctrine of indivisibles is surveyed by Waschkiess, *Von Eudoxus zu Aristoteles* (cit. n. 47).

55. For instance, it appears that Euclid eliminated the use of circumscribed bounding figures employed in his Eudoxean source; see Knorr, *Textual Studies* (cit. n. 4), p. 407.

56. An argument of this type is described by Heath, *History* (cit. n. 47), I, p. 180; *The Thirteen Books of Euclid's Elements*, 2nd ed. (3 vols., Cambridge: Cambridge University Press, 1926; reprint New York: Dover, 1956), III, p. 366; and Baron, *Origins* (cit. n. 1), pp. 20–22.

57. The argument is summarized by Heath, *Euclid's Elements* (cit. n. 56), III, pp. 390–391.

58. In principle one could do the same separately for the circumscribed figure.

59. The first remainder r_1 is $3/8 + 1/4$; the second r_2 is $3/16 + 3/32 + 1/16$ (or $r_1/2 + 1/32$); the third is $3/32 + 3/64 + 3/128 + 1/64$ (or $r_2/2 + 1/128$); etc. Here, the last term $1/4^n$ represents the portion consisting of small pyramids.

60. Note that the $1 : 3$ ratio could be derived simply as an arithmetic corollary of the dissection of the pyramid in Figure 6a. The existence of *some* constant ratio between pyramid and prism of equal height and base follows from the proportionality of volumes and bases just established. Denote this ratio k and assign unit volume to the prism whose height and base are equal to those of pyramid $ABGD$; then the pyramid has volume k. Each prism $EZH\theta IB$, $ZH\theta IUG$ has volume $1/8$, so that their associated pyramids $AEZH$, $ZIUD$, are each of volume $k/8$. Thus, $k = 1/4 + k/4$, or $k = 1/3$.

61. One would simultaneously inscribe and circumscribe figures consisting of n-many cylindrical parts of equal height; since the difference between the two figures would precisely equal the bottom-most cylinder in the series, the difference can be made arbitrarily small, by taking the height of the cylinders suitably small. Since the bounding figures thus converge to each other, they each converge *a fortiori* to the intermediate cone.

62. See Knorr, "Archimedes and the Spirals" (cit. n. 35); *Ancient Tradition* (cit. n. 36), pp. 161–163.

63. The analyses of the parabolic segment (as in *Quadrature of the Parabola*, prop. 17) and the spiral (in the version of Pappus) depend on the "compression by difference" form, and would follow readily from the dissection of the figures by parallel sectioning. In *Sphere and Cylinder*, Archimedes adapts the "compression by ratio" technique for circular figures (e.g., circles, cones, spheres), where the reduction of the difference between circumscribed and inscribed figures is accomplished first through their being brought into a sufficiently close ratio. This alternative "ratio" form could be explained as a modification of the "difference" form of convergence, in the effort to replace the older "approximative" form.

64. See Wagner, "Liu Hui" (cit. n. 2); Lam and Shen, "Chinese Concept" (cit. n. 2). Wagner's account is summarized by B. L. van der Waerden, *Geometry and Algebra in Ancient Civilizations* (Berlin/Heidelberg/New York/Tokyo: Springer, 1983), pp. 204–207.

65. Van der Waerden (*ibid.*, p. 205) reports this to be the opinion of A. P. Youschkevitch, for instance.

66. On this manuscript, see note 27 above.

67. According to Heath (*History* [cit. n. 47], II, p. 317), the Constantinople codex containing the *Metrica* was first identified as such in 1896 by R. Schöne (whose son, H. Schöne edited the work in *Heronis Opera*, vol. 3, 1903). Testimonia and fragments of the *Metrica* were known before from other sources and discussed by F. Hultsch, P. Tannery and others.

68. For examples of treatments from Kepler and Cavalieri, see Baron, *Origins* (cit. n. 1), pp. 108–116, 122–135. On Cavalieri's methods, see also Giusti, *Cavalieri* (cit. n. 1) and Andersen, "Cavalieri's Method" (cit. n. 1).

69. The first part of Kepler's *Stereometria doliorum* is presented as an exposition of results "ab Archimede ... investigata" (ed. Hammer, [cit. n. 28], p. 13), which are explicitly cited from *Dimension of the Circle, Quadrature of the Parabola, Sphaeroids*, etc. This is followed by a *Supplementum ad Archimedem* which presents results on the volumes of solids of revolution beyond what Kepler could read in any Archimedean source then available.

70. See Giusti, *Cavalieri* (cit. n. 1), chap. 2; and Andersen, "Cavalieri's Method" (cit. n. 1), sect. VII.

71. The reaction to Cavalieri's methods is described by Giusti, *Cavalieri* (cit. n. 1), chap. 3; and Andersen, "Cavalieri's Method" (cit. n. 1), sect. X. More generally, on 17th century views of the status of indivisibilist methods vis-à-vis the formal "exhaustion" methods, see D. T. Whiteside, "Patterns of Mathematical Thought in the Later 17th Century," *Archive for History of Exact Sciences* 1961, 1: 179–388.

BIBLIOGRAPHY

EDITIONS OF ANCIENT WORKS

Archimedes, *Opera*, ed. J. L. Heiberg, 2. ed., 3 vol., Leipzig: Teubner 1910–1915.

Euclid, *Elements*, ed. J. L. Heiberg, 5 vol., Leipzig: Teubner, 1883–1888, (2nd ed., ed. E. S. Stamatis, 5 vol. in 6, Leipzig: Teubner 1969–77).

Hero of Alexandria, *Metrica*, ed. H. Schöne (in *Opera*, vol. 3), Leipzig: Teubner, 1903.

Pappus of Alexandria, *Collectionis quae exstant*, ed. F. Hultsch, 3 vol., Berlin: Weidmann, 1876–78.

Presocratics: see Diels and Kranz.

Theon of Alexandria, *Commentaire sur les livres 1 et 2 de l'Almageste*, ed. A. Rome (*Studi e Testi*, 72), Vatican City, 1936.

MODERN STUDIES

Andersen, K. 1985. Cavalieri's method of indivisibles, *Archive for History of Exact Sciences*, 31, 291–367.

Arendt, F. 1913-14. Zu Archimedes, *Bibliotheca Mathematica*, 14 (series 3), 289–311.

Baron, M. E. 1969. *The origins of the infinitesimal calculus*, Oxford: Pergamon.

Cavalieri, B. 1635. *Geometria indivisibilibus continuorum nova quadam ratione promota*, Bologna (2. ed., 1653).

Clagett, M. 1964. *Archimedes in the Middle Ages*, I: The arabo-latin tradition, Madison: University of Wisconsin Press.

——. 1976. *Archimedes in the Middle Ages*, II: The translations from the Greek by William of Moerbeke, Philadelphia: American Philosophical Society.

Commandino, F. 1558. *Archimedis opera non nulla ... nuper in Latinam conversa, et commentariis illustrata*, Venice.

Diels, H. and W. Kranz. 1966. *Die Fragmente der Vorsokratiker,* 3 vols., Dublin/ Zurich: Weidmann (reprint of the 6th edition of 1951).

Dijksterhuis, E. J. 1956/1987. *Archimedes,* Copenhagen: Munkesgaard (first edition 1956; reprinted Princeton: Princeton University Press, 1987, with bibliographic supplement by W. R. Knorr).

Drachmann, A. G. 1972. Hero of Alexandria, in *Dictionary of Scientific Biography,* ed. C. C. Gillispie (New York: Scribner's Sons), vol. 6, pp. 310–314.

Furley, D. 1967. *Two studies in the Greek atomists,* Princeton: Princeton University Press.

———. 1987. *The Greek cosmologists,* I: The formation of the atomic theory and its earliest critics, Cambridge: Cambridge University Press.

Giusti, E. 1980. *Bonaventura Cavalieri and the Theory of Indivisibles,* Bologna: Edizione Cremonese.

Guthrie, W. K. C. 1965. *A history of Greek philosophy,* II: The Presocratic tradition from Parmenides to Democritus, Cambridge: Cambridge University Press.

Heath, T. L. 1897/1912. *The Works of Archimedes,* Cambridge: Cambridge University Press (first edition 1897; reprinted with the *Supplement of 1912: The Method of Archimedes,* New York: Dover, n.d.).

———. 1921. *A History of Greek mathematics,* 2 vols., Oxford: at the Clarendon Press (reprinted New York: Dover, 1981).

———. 1926. *The thirteen books of Euclid's Elements,* 2. ed., 3 vol., Cambridge: Cambridge University Press (reprinted New York: Dover, 1956).

———. 1949. *Mathematics in Aristotle,* Oxford: at the Clarendon Press.

Heiberg, J. L. 1907. Eine neue Archimedeshandschrift, *Hermes,* 42, 234–303.

Heiberg, J. L. and H. G. Zeuthen. 1906–07. Eine neue Schrift des Archimedes, *Bibliotheca Mathematica,* 3. ser., 7, 321–363.

Kepler, J. 1615. *Nova stereometria doliorum vinariorum,* Linz (in *Gesammelte Werke,* vol. 9, ed. F. Hammer, Munich, 1960)

Kirk, G. S., J. E. Raven and M. Schofield. 1983. *The Presocratic philosophers,* 2nd ed., Cambridge: Cambridge University Press.

Knorr, W. R. 1975. *The evolution of the Euclidean Elements,* Dordrecht & Boston: Reidel.

———. 1978a. Archimedes and the *Elements*: Proposal for a revised chronological ordering of the Archimedean corpus, *Archive for History of Exact Sciences,* 19, 211–290.

———. 1978b. Archimedes and the pre-Euclidean proportion theory, *Archives internationales d'histoire des sciences,* 28, 183–244.

———. 1978c. Archimedes and the spirals: The heuristic background, *Historia Mathematica,* 5, 43–75.

———. 1978d. Archimedes' lost treatise on centers of gravity of solids, *Mathematical Intelligencer,* 1, 102–108.

———. 1982. Infinity and continuity: The interaction of mathematics and philosophy in antiquity, in *Infinity and continuity in ancient and medieval thought,* ed. N. Kretzmann, Ithaca, N. Y.: Cornell University Press, pp. 112–145.

———. 1985. Archimedes' *Dimension of the circle*: A view of the genesis of the extant text, *Archive for History of Exact Sciences,* 35, 281–324.

———. 1986a. *The ancient tradition of geometric problems,* Boston/Basel/Stuttgart: Birkhäuser (reprint New York: Dover, 1993).

———. 1986b. On two Archimedean rules for the circle and the sphere, *Bollettino di Storia delle Scienze Matematiche,* 6, 145–158.

———. 1987a. The medieval tradition of Archimedes' *Sphere and Cylinder,* in *Mathematics and its applications to science and natural philosophy in the Middle Ages: Essays in honor of Marshall Clagett,* ed. E. Grant and J. E. Murdoch, Cambridge: Cambridge University Press, pp. 3–42

———. 1987b. Archimedes after Dijksterhuis: A guide to recent studies [bibliographic supplement to E. J. Dijksterhuis, *Archimedes,* Princeton: Princeton University Press, 1987, pp. 419–451].

———. 1989. *Textual studies in ancient and medieval geometry.* Boston/Basel/Berlin: Birkhäuser

———. 1990. John of Tynemouth *alias* John of London: Emerging portrait of a singular medieval mathematician, *British Journal for the History of Science,* 23, 293–330.

———. forthcoming. *Falsigraphus* vs. *adversarius*: Robert Grosseteste, John of Tynemouth, and geometry in the 13th century, (to appear in *Mathematische Probleme im Mittelalter,* ed. O. Schönberg, Wolfenbüttel: Herzog August Bibliothek).

Lam Lay-Yong and Shen Kangsheng. 1985. The Chinese concept of Cavalieri's principle and its applications, *Historia Mathematica,* 12, 219–228.

Luria, S. Y. 1932–33. Die Infinitesimaltheorie der antiken Atomisten, *Quellen und Studien zur Geschichte der Mathematik, Astronomie und Physik,* Abt. B, 2, 106–185.

——— (= S. J. Lurje). 1948. *Archimedes,* Vienna: 'Neues Oesterreich' (German translation from the Russian edition of 1945).

Mahoney, M. S. 1972. Hero of Alexandria: Mathematics, in *Dictionary of Scientific Biography,* ed. C. C. Gillispie (New York: Scribner's Sons), vol. 6, pp. 314–315

Mau, J. 1954. *Zum Problem des Infinitesimalen bei den antiken Atomisten,* Berlin.

Mugler, C. 1970–72. *Les oeuvres d'Archimède,* 4 vols. (series Association Budé), Paris: Société d'édition 'Les belles lettres'.

Neuenschwander, E. 1974. Die stereometrischen Bücher der Elemente Euklids, *Archive for History of Exact Sciences,* 14, 91–125.

Schneider, I. 1979. *Archimedes: Ingenieur, Naturwissenschaftler und Mathematiker,* Darmstadt: Wissenschaftliche Buchgesellschaft.

Sedley, D. 1976. Epicurus and the mathematicians of Cyzicus, *Cronache Ercolanesi,* 6, 23–54.

Stamatis, E. S. 1970–74. *Archimêdous Hapanta,* 3 vols. in 4 (classical Greek text from Heiberg's 2nd edition, with translation into modern Greek), Athens: Technikou Epimelêtêriou tês Hellados.

Thomas, I. 1939–41. *Selections illustrating the history of Greek mathematics,* 2 vols. (Loeb Classical Library; repr. 1957, etc.), Cambridge, Mass.: Harvard University Press & London: Heinemann.

Toomer, G. J. 1976. Theon of Alexandria, in *Dictionary of Scientific Biography,* ed. C. C. Gillispie (New York: Scribner's Sons), vol. 13, pp. 321–325.

Ver Eecke, P. 1921. *Archimède: les oeuvres complètes,* first ed. (French translation from the 2nd edition of Heiberg), Paris/Brussels: Desclée, de Brouwer.

van der Waerden, B. L. 1983. *Geometry and algebra in ancient civilizations,* Berlin/Heidelberg/New York/Tokyo: Springer.

Wagner, D. 1978. Liu Hui and Tsu Keng-chih on the volume of a sphere, *Chinese Science,* 3, 59–79.

Waschkiess, H. J. 1977. *Von Eudoxus zu Aristoteles: Das Fortwirken der Eudoxischen Proportionentheorie in der Aristotelischen Lehre vom Kontinuum,* Amsterdam: Grüner.

Whiteside, D. T. 1961. Patterns of mathematical thought in the later 17th century, *Archive for History of Exact Sciences,* 1, 179–388.

Zeuthen, H. G.: see Heiberg and Zeuthen.

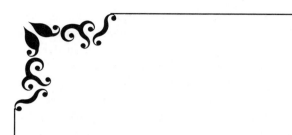

Enigmas of Chinese Mathematics

Frank Swetz
Penn State University

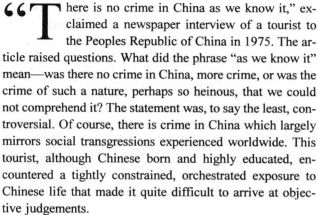

"There is no crime in China as we know it," exclaimed a newspaper interview of a tourist to the Peoples Republic of China in 1975. The article raised questions. What did the phrase "as we know it" mean—was there no crime in China, more crime, or was the crime of such a nature, perhaps so heinous, that we could not comprehend it? The statement was, to say the least, controversial. Of course, there is crime in China which largely mirrors social transgressions experienced worldwide. This tourist, although Chinese born and highly educated, encountered a tightly constrained, orchestrated exposure to Chinese life that made it quite difficult to arrive at objective judgements.

Western observers of the Chinese scientific scene have long been confronted and confounded by a similar situation. Missionaries, teachers, and scientific translators who flocked to China in the early nineteenth century reported a prevailing lack of scientific method and "spirit of inaccuracy" in mathematics and computation.[1] Several conveyed the impression to European audiences that China had no indigenous science or mathematics, "as we know it." Alexander Wylie, a British scientific translator employed by the Manchu court, attempted to alter this impression, that is, to "correct erroneous statements" by publishing a series of papers on traditional Chinese mathematics.[2] Wylie discovered an ancient mathematics literature and notable mathematical accomplishments in algebra which exceeded those of Europe before the sixteenth century. The Chinese, Wylie noted, claimed that European mathematics had originated in China! This information together with reports by E. Biot and L. Matthiessen formed the substance for Moritz Cantor's *Vorlesungen über die Geschichte der Mathematik* [1880–1908] entry on China.[3] Despite these scholarly efforts, the controversy had been established. In 1868, Louis Pierre Sédillot, a noted French orientalist, denounced claims of originality in Chinese mathematics and attributed its results to Greek influences.[4]

Even more carefully documented research has failed to resolve this debate on the twentieth-century nature and origin of Chinese mathematics. In 1913, Japanese researcher Yoshio Mikami published his comprehensive *The Development of Mathematics in China and Japan,* which affirmed the existence of strong mathematical traditions in Imperial China, traditions that extended from at least the Han (206 B.C.–A.D. 200) through the Song dynasties (A.D. 960–1125).[5] At this time (1911–1916), Belgian sinologist/mathematician, Louis Van Hée, published a series of articles on Chinese mathematical classics. Van Hée seemed to affirm Mikami's findings, and yet he noted a general lack of talent for mathematics among the Chinese.

He too attributed traditional Chinese mathematics to foreign influences.[6] In a series of articles published from 1921 to 1929, Italian historian of science, Gino Loria, attributed Chinese mathematical accomplishments to Greek and Babylonian sources.[7] Even some Chinese scholars then assumed that traditional China had no mathematics or science.[8] American mathematical historian, David E. Smith, sympathized with the claims of Chinese mathematical originality and urged further critical research in attempting to resolve the "Unsettled Questions Concerning the Mathematics of China".[9]

Joseph Needham's *opus magnum Science and Civilization in China* in 1953, seemed to satisfy Smith's request. Needham's detailed study affirms that the Chinese had a systematically devised mathematics, in which some developments historically surpassed those of the West. Still, doubts as to the worth and originality of Chinese mathematics have persisted. In 1968, Marco Adamo of the Institute of Mathematics, University of Cagliari, declared that Chinese mathematics was completely copied from Greek and Hindu sources and that its meaning and methodology is incomprehensible;[10] more recently, B.L. van der Waerden has offered the theory that Chinese mathematics originated with a pre-Christian people who inhabited the British Isles.[11]

But Adamo and B. L. van der Waerden notwithstanding, after Mikami, Needham, and at least a score of other modern researchers, the validity of a Chinese mathematical history has been conceded.[12] Where controversies linger, concerns over the nature of that history evolve from the fact that Chinese mathematics differs in many respects from Western mathematics; that is, mathematics *as we know it.* Unfortunately, this situation often results in ethnocentric-based comparisons and judgments of unrealistic expectations. If it is accepted that Chinese mathematics is different and an understanding of that difference is sought, the enigmas of Chinese mathematics can mostly be unraveled.

An Overview of Chinese Mathematical Activity

It appears that from the earliest times in China, mathematics was an activity of societal importance. Legendary founders and mythological sage kings were associated with mathematical activities—measuring the heavens and surveying to calm flood waters, tasks that insured the well-being of the empire.[13] A curriculum for a pre-Confucian gentleman set in the Chou dynasty (1027–256 B.C.) listed mathematics as one of six arts to be studied and noted that knowledge of mathematics concerned the nine operations[14]. Inveter-

ate record keepers, the Chinese compiled dynastic histories and bibliographies from which information on mathematical activity can be gleaned: records of the Han dynasty (206 B.C.–A.D. 220) list 21 titles on astronomy and 18 on calendrical reckoning, of which two are specifically mathematical; by the Sui dynasty (581–618 A.D.), one finds 27 titles on mathematics; the Tang dynasty (618–907 A.D.), 35 titles; and the Song dynasty (960–1279 A.D.) contributions have been estimated at 70 mathematical works.[15] In 1799, scholar official Ruan Yuan (1764–1849) published a compilation, *Biographies of Mathematicians and Astronomers* [Chouren zhuan] listing notable mathematicians and astronomers and their works throughout Chinese history. An updated version of the *Biographies,* which spanned the Qing dynasty (1644–1911 A.D.) contributions, listed 203 mathematicians.[16]

While from ancient times China had a history of mathematical activities, the level of activity within this history was sporadic and uneven. The periods of significant mathematical accomplishments are the Han, late Song, and early Yuan (960–1368 A.D.). Despotic Emperor Qin Shihuangde, unifier of China, ordered a "burning of books" in 213 B.C. How thorough this destruction of knowledge was is open to debate; however, the first two extant references on Chinese mathematics—*The Arithmetical Classic of the Gnomon and the Circular Paths of Heaven* [Zhoubi suanjing] and *The Nine Chapters of the Mathematical Art* [Jiuzhang suanshu]—appear after this period, that is, ca. 100 B.C.. Both works are compilations of previous information. Their contents reveal that the Chinese had a highly developed decimal system of numeration, were familiar with the Pythagorean theorem and its applications, could extract square and cube roots with accuracy, correctly computed volumes for complex geometric solids, and could solve linear equations employing matrix methods. Further, in their computations Chinese mathematicians worked with negative numbers, secured positive roots for quadratic equations and were familiar with indeterminate equations. The *Nine Chapters* was so complete and comprehensive a reference on mathematical procedures and applications that it remained a classic which was studied and edited for the next thousand years. One of the most famous of the *Nine Chapters'* commentators was Liu Hui, an official in the Kingdom of Wei, who lived in the third century.[17] Liu provided mathematical justification for the procedures given in the *Nine Chapters,* and extended its ninth chapter on the right triangle to include problems involving a mathematical technique called *chong-cha* which established protrigonometry in China.[18] He also improved the accepted value of π, which was given as 3.

Using a circle dissection technique involving a 192-sided polygon, Liu obtained a value of π equivalent to 3.141024. A successor, Zu Chongzhi (429–500 A.D.) refined Liu's method and determined π as 355/113 or 3.1415929.[19] Finding a "scar on mathematics," a mistake in the *Nine Chapters,* where the volume of a sphere with radius R was given as $(\pi)^2 R^3/2$, Zu elegantly derived the desired volume as $4\pi R^3/3$.[20]

The first known mathematical problem in China involving an indeterminant situation appears in the eighth chapter of the *Nine Chapters* where four equations in five unknowns evolve. By reducing the system of equations to a ratio between the unknowns, its author obtained a correct solution. In essence, he reduced the system to parametric form and found one particular solution. A more interesting indeterminant situation appeared as the 26th problem in the fourth century collection, *Master Sun's Mathematical Manual* [Sunzi suanjing]:

> There is an unknown number of things. When we count by threes, they leave a remainder of two; when counted by fives, they leave a remainder of three and when counted by sevens, they leave a remainder of two. Find the number of things.

In modern notation this problem gives rise to the set of equations: $N = 3x + 2$, $N = 5y + 3$, $N = 7Z + 2$ or in number congruence form:

$$N \equiv 2 \ (\mathrm{mod}\ 3) \equiv 3 \ (\mathrm{mod}\ 5) \equiv 2 \ (\mathrm{mod}\ 7).$$

The *Manual*'s solution procedure supplies an answer which is equivalent to

$$N = (70 \times 2) + (21 \times 3) + (15 \times 2) - (105 \times 2) = 23.^{[21]}$$

This problem and its solution scheme were transmitted to the West by Wylie in 1852.[22] It eventually supplied the basis for what today is known as the "Chinese Remainder Theorem," that is:

> If $m_1, \ldots, m_k$ are relatively prime in pairs, there exist integers x for which simultaneously $x \equiv a_1 \ (\mathrm{mod}\ m_1), \ldots, x \equiv a_k \ (\mathrm{mod}\ m_k)$. All such integers x are congruent modulo $m = m_1 m_2 \cdots m_k$.

In 656 A.D., the Tang government instituted a department of mathematics at its Royal Academy and set an official mathematics syllabus for training of civil servants. This syllabus, the *Ten Mathematical Manuals* [Suanjing shi shu], was a compilation of ten mathematical works. By this sanctioning action, mathematics was systematized and essentially limited to contents of these ten texts.[23] Embedded in the traditional civil service examination, this syllabus set the standards for mathematical knowledge until 1888 when the contents of the examinations were revised to include western scientific subjects, a reform intended to speed modernization and insure national survival.

The Tang dynasty (618–907 A.D.) marked a period of political and religious syncretism and administrative reform, in which literature and poetry flourished, but little effort was directed towards mathematics. Significant mathematical activity next occurred during the late Song and early Yuan dynasties (960–1368 A.D.). Seeking to strengthen the civil service examination, Song officials published a printed edition of the *Ten Mathematical Manuals* in 1084 thus demonstrating renewed official interest in the study of mathematics. However, the great mathematicians of this period were not associated with the imperial academies: Qin Jiushao (ca 1202–1261) held minor government posts but was not officially recognized as a mathematician; Li Ye (1192–1279) was a reclusive scholar; Yang Hui (fl. 1261–1275) a minor civil servant; and Zhu Shijie (fl. 1280–1303) a wandering teacher.

Qin's major work, *Mathematical Treatise in Nine Sections* [Shushu jiuzhang] appeared in 1247. It focuses on applied mathematics and computational techniques involved with such tasks as securing corvée labor, determining land boundaries, military deployment, and architectural design. The mathematical problem situations described and solved were, however, far more complex than those in bureaucratic manuals. Qin devised a method for solving systems of simultaneous linear congruences, generalizing previous work on this topic by allowing for solution situations where the moduli did not have to be pairwise prime, and extended the traditional methods of root extraction to obtain solutions for cubic and higher degree numerical equations.[24]

Li Ye's principal work was the *Sea Mirror of Circle Measurements* [Ceyuan haying] (1248), a collection of 170 problems on relationships between a right triangle and its inscribed circle. Many of these relationships led to numerical equations of higher degree for which solutions were sought. Li revised a traditional algebraic process known as the "method of the celestial element", *tianyuanshu*, by which he set up a higher-order numerical equation on the traditional counting board. His writing also established a stable algebraic terminology for use in solving higher-order equations.[25]

Yang Hui revised the classical *Nine Chapters* and published his results in *A Detailed Analysis of the Mathematical Methods in the Nine Chapters* [Xiangjie jiuzhang suan] (1261). Yang computed freely with decimal fractions, strengthened the solution of quadratic equations by allowing negative coefficients, and obtained sums for var-

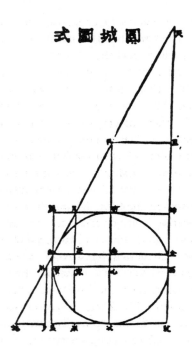

FIGURE 1

ious finite series. In his work, Yang employed a numerical configuration later known outside China as the "Pascal triangle."[26]

The high point of traditional Chinese algebra occurred in Zhu Shijie's *Jade Mirror of the Four Elements* [Siyuan yujian] (1303), in which he explained how to obtain a solution for a system of polynomial equations in four unknowns. In the *Jade Mirror* Zhu also considered methods of employing finite differences in summing numerical series.[27] For example, he cites the problem where a number of soldiers are recruited: 3^3, the first day, $(3+1)^3$ the second day, ... $(3+12)^3$, the thirteenth day; each soldier is paid 250 copper coins daily. It is required to find the number of soldiers recruited and their total salary. Zu solves this problem by using a method of finite differences. His explanation for finding $f(k)$, the total number of soldiers recruited in k days is equivalent to:

$$f(k) = k\Delta + \frac{1}{2!}k(k-1)\Delta^2 + \frac{1}{3!}k(k-1)(k-2)\Delta^3$$
$$+ \frac{1}{4!}k(k-1)(k-2)(k-3)\Delta^4.$$

Throughout the history of the Celestial Empire, its mathematical activity focused on astronomy and calendrical reckoning. From antiquity, there was official interest in spherical measurement and calculation. The *Nine Chapters* contains a problem involving a circle's chord, intercepted arc, and saggita. In the eleventh century, mathematician Shen Kuo (1031–1095) developed a computational relationship between an arc, its intersecting chord, and its sagitta. Yuan astronomer Guo Shoujing (1231–1316 A.D.), utilized Shen's technique to establish measures of longitude and latitude for heavenly bodies. Guo found a variety of relationships between a circle's arc, chord, and sagitta and in doing so devised a spherical prototrigonometry for Chinese astronomy.

Soon after the arrival of European Jesuits in China in the early seventeenth century, the era of chiefly indigenous mathematical development ended. Mathematical activities were now modified by European influences.

Chinese Mathematical Constraints: Philosophical, Social, and Methodological

In examining the history of Chinese mathematics, an observer schooled in western mathematics can easily become beset with questions and doubts—Why was Chinese mathematical activity almost solely algebraic in nature? Was there a deductive method and logic behind Chinese mathematics? If by the thirteenth century, the Chinese were so advanced in mathematics, why did they not continue in their progress? In the past, the easiest way to dismiss these issues was to attribute Chinese accomplishments to other peoples: Greek, Babylonian, Egyptian, and Indian. This view holds that the Chinese did not invent or understand mathematics, they simply copied it. More satisfying answers to these questions can be found in an understanding of the cultural milieu that shaped the mathematics of China.

Philosophical Constraints. Six centuries before the Christian Era, three contending schools of philosophic thought came together to shape the future of China: Confucianism, Taoism, and Mohism. Confucius (551–479 B.C.) believed that correct social order was based on a class division: ruler and ruled. Rulers had to earn their authority and privileges by obtaining righteousness and becoming harmonious beings. These qualities were gained through studies of literature, music, poetry, and calligraphy—humanistic subjects; knowledge of these was verified by the traditional civil service examination. Those who excelled in the examination were admitted into the Imperial bureaucracy and thus assured of social standing and material success for themselves and their family.

Confucianism, which stresses human relationships and ethics, in general, diverted attention from the physical world. Taoism, a more mystical philosophy, contended that man lived in a delicate balance with nature in a vast organistic cosmos. Every being and inanimate object possessed a consciousness that functioned in accommodation with the consciousness of all other objects. This doctrine of balance was enunciated in the classical text known as *The Book of Change* or *I-Ching* [Yi jing]. In contrast, Mohism stressed utilitarianism, advocated asceticism, and urged a world-understanding based on observation and logical argumentation. For the Mohists, reason was based on logic.[28]

Of these three philosophies, Mohism was the most conducive to a scientific spirit and the development of a deductive-based mathematics. It appears that this was one task the Mohists set for themselves, because the Mohist Canon, compiled about 330 B.C., contains a series of geometric definitions which are strikingly similar to those in Euclid's *Elements*.

> The definition of a 'point' is as follows: The line is separated into two parts, and that part which has no remaining part (that is, cannot be divided into smaller parts) and [thus] forms the extreme end [of this line] is a point.[29]

However, the Mohist school of thought did not prevail; Confucian and Taoist doctrines dominated in shaping the Chinese world view. In about the third century, Buddhist beliefs exerted modifying influence.[30] Together these three socio-ethical philosophies combined to form a social and intellectual climate not conducive to a development of mathematics and science. They urged man to look inward rather than to question the world around him. By adhering to a Taoist principle of accommodation, no theory of "cause and effect" was recognized. Further, the Chinese held no beliefs in a supreme divinity whose laws of the universe were beyond human comprehension and thus did not experience the intellectual and scientific liberation felt in the West when men first realized they could comprehend nature through processes of observation, experimentation, and reason. By the seventeenth century, Europeans had formulated a theory of a mechanical universe driven by cause and effect. In western mathematics, this realization of cause-and-effect relationship dependence, eventually evolved into a concept of mathematical function, a building block of modern mathematics. The Chinese never developed an indigenous concept of mathematical function; in 1859, Alexander Wylie, working with the traditional scholar, Li Shanlan translated the word function into Chinese by employing the characters that meant "quantity that contains" accompanied by the explanation, "if the variable quantity contains another variable quantity, then the former is a function."[31]

Societal Constraints. Confucian belief that social harmony rests in the ability of a humanistically-trained literati negated the importance of theoretical mathematics. For the Confucians mathematics was merely a technology, a collection of procedures and techniques, used to solve specific government related problems. Mathematicians were clerks employed in Imperial bureaus; their professional position afforded little social prestige or recognition. The contents and organization of the *Nine Chapters* affirm the outward nature of Chinese mathematics; the book is a bureaucratic manual containing 246 specific problems related to Imperial administration. Most succeeding traditional mathematical texts are of a similar nature; they are not theoretical tracts but rather "how to" manuals written in a concise style. Accordingly, problem-solving situations experienced in administering a government are limited. If sufficient mathematical knowledge exists to accomplish these tasks, 'why seek to extend that knowledge' was a dominant retarding theme in the history of mathematics in China.

The Confucian examination system, the ladder of success in traditional China, also retarded mathematical advance. By regulating the number of civil service examinations set for mathematics positions, the Imperial government controlled the level of mathematical activities within the Empire. Mathematical examinations were held sporadically, and there were periods of time, for instance during the Southern Song dynasty (1127–1280 A.D.), when no mathematicians were solicited. Confucian reverence for, and examination emphasis on, classics elevated the contents of these texts to a central focus of intellectual and academic life. Scholars copied and performed textual annotation and exegesis on sanctioned works. This strategy for scholarship was extended to all official literature, even that involving mathematics. Thus, the *Nine Chapters* became a classic to be annotated, copied, and edited for centuries. The nineteenth century still found Chinese scholars commenting upon this work.[32] Problem situations also became canonical; for example, a *Nine Chapters'* problem involving a circle inscribed in a right triangle became the model for Li Ye's *Sea Mirror* in the thirteenth century. Reverence for classical results promoted inflexible thinking; there was no room for deviating from orthodox dogma or asking "What if?" questions which constitute the essence of scientific thinking.

Confucian scholarship and textual exegesis provided another retarding factor—censorship. Since only approved texts entered the process of preservation, non-approved writings frequently disappeared. Theoretical tracts including those with mathematical proofs were frequently ignored in the preservation process. Proofs were thus lost. As a result, most Chinese mathematical accomplishments remain anonymous. This policy hindered the attainment of a cumulative corpus of mathematical knowledge.

Within Confucian society, social mores established barriers between scholars, merchants, and artisans—scholars did not engage in trade or work with their hands. Theory and practice were separated. China did not have a Thales, Fibonacci, Tartaglia, Stevin, or Galileo to challenge its classical world outlook. Reckoning masters, a professional group in medieval Europe whose activities stimulated interest in applied mathematics, never existed in China.

Methodological Constraints and the Characteristics of Chinese Mathematics. Despite this adverse intellectual, social, and institutional climate, mathematics did develop in the Celestial Empire. Chinese mathematics was empirical and computational in nature, bound to algorithmic manipulation, and frequently conceptualized upon visual patterns. From earliest times, a decimal system of numeration was employed. Numbers were encoded by a series of tally strokes.[33] These symbols in turn gave rise to the construction of a set of operational computing rods used in conjunction with a counting board. This board was partitioned into a rectangular grid allowing for a two-dimensional tabulation and manipulation of numerical data as expressed in rod configurations. Each board position employed during a mathematical procedure had a particular significance and designation, and, in a sense, this orientation of rods on the board was a form of algebraic notation encouraging visual perception and through it a recognition of under-lying mathematical structure. As solution procedures were refined, preconceived patterns were sought in the rod configurations.

How was this rod algebra used to solve a system of linear equations? Consider a counting-board rod arrangement and solution procedure for solving the first problem given in the *fang cheng* [rectangular array] section of the *Nine Chapters*.

Three bundles of high quality rice, two bundles of medium quality and one bundle of low quality rice yield 39 *dou*; two bundles of high quality, three bundles of medium and one bundle of low quality yield 34 *dou*; one bundle of high quality, two bundles of medium quality and three bundles of low quality rice make 26 *dou*. How many *dou* are there in a bundle of high quality, medium and low quality rice respectively?[34]

Using modern notation to express the information of this problem, a system of three linear equations in three unknowns emerges;

$$3x + 2y + z = 39$$
$$2x + 3y + z = 34$$
$$x + 2y + 3z = 26$$

Numerical data is entered onto the board by use of rods working from right to left and from top to bottom. Resulting rod configurations for the set of equations are shown in Figure 3(a). Following rod algorithm, elementary column operations reduce a column to two entries: a variable's coefficient and an absolute term. See first column in Figure 3(b). In this reduced matrix form, a solution for one variable is obtained ($36z = 99$) and back substitution supplies the remaining required values. For the given problem, $z = 2\frac{3}{4}$ *dou*, $y = 4\frac{1}{4}$ *dou* and $x = 9\frac{1}{4}$ *dou*.

FIGURE 2

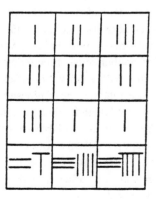

FIGURE 3a

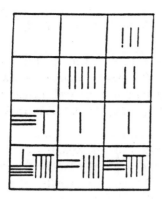

FIGURE 3b

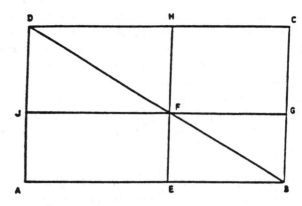

FIGURE 4

In carrying out the above algorithm, several column subtractions result in negative coefficients, that is board entries, upon which further mathematical operations are then undertaken. Mathematicians of this time fully acknowledged negative entries and worked with negative numbers.[35]

The organizational format of the *fang cheng* section demonstrates how mathematicians generalized and extended their results. Of the eighteen problems in this section: eight can be interpreted as representing two equations in two unknowns; six, three equations in three unknowns; two, four equations in four unknowns; one, five equations in five unknowns; and one problem involving five equations in six unknowns. In the last situation, the answer was a series of ratios, indicating that the author appreciated the indeterminate nature of the problem.[36]

Traditional Chinese mathematics is easily criticized as lacking a deductive basis. It is seen as just a collection of problem-solving techniques. However, evidence of mathematical proof and proof theory appear early in China's mathematical history. Zhao Suang (ca. A.D. 200) wrote a *Commentary on the Zhoubi Suanjing* [Zhoubi suanjing zha], which substantiated Pythagorean Theorem statements of the *Zhoubi*. Further, in this work, Zhao proved twenty-one propositions evolving from right-triangle relationships.[37] Liu Hui and Yang Hui provided detailed justifications, that is proofs, for solution procedures found in the *Nine Chapters*. All these proofs are based on a form of geometric-algebraic intuition and encompass a theory called the "out-in complementary principle."[38] This principle recognizes that the complements, or dejoint regions, of a rectangle formed about the diagonal of a given rectangle are equal in area. From this relationship, a series of proportions is obtained. In Figure 4, area of rectangle $AEFJ$ equals area of rectangle $FGCH$ and $AE \times EF =$

$FG \times GC$ from which follows $JF/JD = FG/GB$. By visually or physically rearranging such geometrical regions, Chinese mathematicians arrived at various mathematical relationships. Consider how the "Pythagorean Theorem" as described in the *Zhoubi* can be verified using the "out-in complementary principle." In Figure 5, consider right triangle ABC: the region $BCED$ is a square on leg BC of the triangle; similarly region $EHGF$ forms a square of leg CA of the triangle (note, $CA = GH$); now, by moving triangle IBD to coincide with triangle ABC and triangle GIH to coincide with triangle GAF, a square on side AB, the hypotenuse square has been formed. Since area has been preserved, it is demonstrated that for a right triangle the sum of the squares of the legs equals the square of the hypotenuse. Moreover Liu Hui showed that the diameter D of an inscribed circle within a right triangle with legs, A, B, and hypotenuse C, is given by the expression $D = \dfrac{2AB}{A+B+C}$.

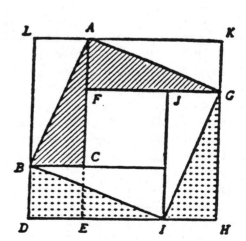

FIGURE 5

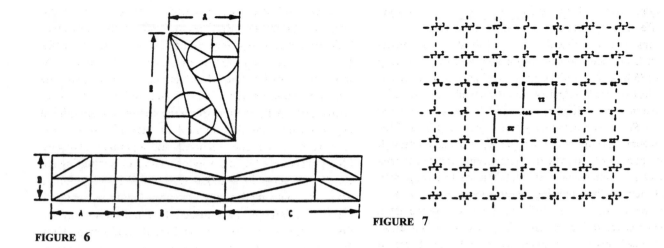

FIGURE 6

FIGURE 7

Such dissection proofs similarly derived volumes for complex solids. Geometric intuition provided a basis for square and cube root extraction, and root-obtaining algorithms on the counting board were extended to find solutions for higher-degree polynomial equations and finally to solve a system of four polynomial equations in four unknowns.[39] In the use of these counting board algorithms Chinese algebra reached its apex; in their use algebraic innovation also reached its limits.

A polynomial was recorded on the counting board as a column of numerical coefficients whose relative position determine the degree of their factors. This method of notation was extended to a two-dimensional array of coefficients to accommodate polynominals in two unknowns, that is, $P(xy)$. In this notation, the absolute term marked a reference point by which columns to its left designated ascending powers of y and rows below it marked ascending powers of x. Zhu Shijie devised a rod layout scheme on a counting board that could accommodate a limited set of up to four polynomial equations in four unknowns. Vertical and horizontal lines on the counting board are associated with coefficients for four unknowns: heaven, earth, man, and matter; the intersection of these lines locate the coefficients for the products of two unknowns. Coefficients for products of non-adjacent unknowns are recorded in corresponding cells according to a diagonal selection process which allows for the input of coefficients for certain factors of unknowns.

Zhu pressed the recording capacity of the counting board to its fullest. Within the two-dimensional matrix, possibilities for rod positioning and manipulation had been exhausted. In turn, resulting intricacies of rod manipulation became very complex and required highly specialized skill.

Popular pressure existed for a simpler, computing device; during the latter half of the Ming dynasty (1368–1644) it appeared as the string abacus, and the use of rod computations diminished.

Foreign Influences and Chinese Claims of Mathematical Preeminence

Questions involving Chinese priority versus foreign influences in the development of mathematical concepts and techniques require further research, but some perspective can be established. There are only speculative suggestions based on isolated similarities that associate Babylonian, Egyptian, or Greek influences to Chinese mathematics of the pre-Christian era; no documented evidence for such claims exist. Starting during the Northern and Southern dynasty period and extending through the Tang dynasty, that is, (420–A.D. 907), Buddhist intellectual and cultural exchanges took place between India and China. Several Indian texts on astronomy were translated for use in China.[40] Indian astronomers served the Tang Court and introduced several mathematical innovations: division of a circle into 360 parts; use of tables of half-chords (sines) and Hindu numerals together with their written algorithms. These innovations were not, however, adopted by the Chinese as a result of this influence. Buddhist concepts for very large and small numbers, for example 10^{120} as an *unimaginable quantity*, were absorbed by the Chinese. Looking westward, clear evidence of Chinese influence can be found in the works of the Indian mathematician Mahavira (ca. 850) and his countryman, Bhaskara (1114–ca. 1185); both duplicated problems found in the *Nine Chapters*. Islamic astronomers also labored in Imperial offices and, in the thirteenth century, may have introduced concepts of Euclidean geometry and trigonometry to the Chinese. No firm evidence has yet been offered to substantiate this belief, however.

Chinese claims that the West received its mathematics from China seem to have originated with Mei Gucheng (1681–1763), a chauvinistic mathematician, who, in reacting against *arrebala* [algebra] as introduced into China by seventeenth century Jesuits, noted in his *Pearls Recovered from the Red River* [Chishui yizhen] that traditional Chinese algebra was superior to that of the Europeans. Through a play upon words, it was determined that the Europeans themselves admitted that their mathematics had originated in the Orient. In the late Ming era of isolationism and chauvinism that resulted after 1707, this theme was perpetuated and was resurrected in the later Qing dynasty when traditional Chinese institutions again felt under threat of foreign influences. Prince Kung, a dynamic Qing reformer, in requesting mathematical education reforms from the Throne in 1867, stated:

> Western science borrowed their roots from ancient Chinese mathematics. Westerners still regard their mathematics as coming from the Orient.[41]

Alexander Wylie and other nineteenth-century European observers of Chinese science encountered this view.

Conclusion

Like the famous Bonsai trees of Asian horticulture, the growth of mathematics in China was constrained and stunted. Certainly China sporadically possessed mathematicians of talent and ingenuity. Mathematics had potential to blossom and flourish; however, this potential was mostly not realized. The development of mathematics requires more than individual genius; it requires a climate of intellectual, institutional and societal support. In the situation of traditional China, this support complex was lacking. Mathematicians, for the most part, were isolated within the system of bureaucratic regimentation which restricted mathematical activities. Often their talents were frustrated by the tasks of annotating ancient texts and being forced to unravel and rediscover forgotten mathematical results.

Even with this adverse climate, Chinese computational and algebraic accomplishments achieved up through the time of the Song dynasty were significant. Mathematics had evolved into more than a mere collection of problem-solving techniques: procedure was generalized beyond the realm of application; proofs were provided for some results but few theories *per se* evolved. For example, Chinese results in solving higher-order equations were impressive but they never developed a theory of equations. A search

for cause and effect was avoided—"just solve the equations; don't question them," seemed the *modus operandi* of Chinese mathematical practitioners. In contrast, the mathematics of the Latin West from the thirteenth century onwards was "cause and effect" driven. From at least the sixteenth century a series of economic and political movements, principally the rise of mercantile capitalism, global expansionism, and the stirrings of industrialization, exerted pressure for computational and algebraic advances. China experienced no such series of science-supporting societal movements. Instead, just when the Celestial Empire's tempo of mathematical and scientific activity increased, Neo-Confucian reformers pushed the Empire back into periods of xenophobic isolation and intellectual retreat. Only the jarring intrusions of western gunboats in the early nineteenth century would reverse China's societal priorities.

References

1. E. C. Bridgman who commented on this situation in his *Chinese Chrestomathy* (Macao: Mission Press, 1841). Later it was discussed in further detail by L. G. Morgan, *The Teaching of Science to the Chinese* (Hong Kong: Kelly and Walsh, 1933).

2. Alexander Wylie, "Jottings on the Science of the Chinese: Arithmetic" *North China Herald* (August– November, 1852).

3. Edouard Biot, a French sinologist wrote extensively on Chinese mathematics and astronomy in the years 1835–1850; the German Matthiessen wrote a series of papers on Chinese algebra in the period 1876–1881. For specific listing of these works see Joseph Needham, *Science and Civilization in China* (Cambridge: Cambridge University Press, 1954–) vol. III, Bibliography C, pp. 744–802. Moritz Cantor's work was published in four volumes (1880–1908) in Leipzig.

4. L. P. Sédillot, "De l'Astronomie et des Mathématiques chez les Chinois," *Bollettino di Bibliografia a di Storia delle Scienze Matematiche e fisiche* (1868) p. 161.

5. Y. Mikami, *The Development of Mathematics in China and Japan* (Leipzig: Teubner, 1913); Mikami had access to many materials previously unexamined by western scholars.

6. "The Chinese display no special taste for and no real aptitude in the study of pure mathematics" L. van Hée, "The Ch'ou-Jen Chuan of Yüan Yüan," *Isis* (1926) 8:103–118; for a listing of van Hée's works see Frank Swetz and Ang Tian Se,"A Brief Chronological and Bibliographic Guide to the History of Chinese Mathematics," *Historia Mathematica* (1984)11:39–56.

7. Gino Loria, "Documenti relativi all'antica matematica dei Cinesi," *Archeion* (1922) 3:141.

8. Perhaps the most striking example of the phenomenon was Yu-lan Fung's "Why China has no Science—an Interpretation of the History and Consequences of Chinese Philosophy." *International Journal of Ethics* (April, 1922) 32:237–263.

9. D.E. Smith, "Unsettled Questions Concerning the History of Mathematics in China," *Scientific Monthly* (1931) 33:244–50.

10. M. Adamo, "La Matematica nell'antica Cina," *Osiris* (1968) 15:175–95.

11. Van der Waerden put forth this theory in his *Geometry and Algebra in Ancient Civilizations* (New York: Springer-Verlag, 1983).

12. For example: E. I. Berezkina, K. Chemla, J. Dhombres, P. Y. Ho, J. Hoe, L. Y. Lam, U. Lebbrecht, J. C. Martzloff, D. B. Wagner, etc.

13. Legendary Emperor Yu (ca. 21st century B.C.) was credited with using mathematics to restore order in the Empire.

14. See Frank Swetz, *Mathematics Education in China: Its Growth and Development* (Cambridge, MA.: MIT Press, 1974). Scholars are in disagreement as to the exact context of the "nine operations"; some believe the phrase designates the multiplication tables, other feel it refers to topics in applied mathematics which would later lend themselves as chapter headings for *Jiuzhang suanshu*.

15. Information provided by Jock Hoe, "The Jade Mirror of the Four Unknowns—Some Reflections," Mathematical Chronicle (1978) 7:125–156.

16. See discussion by van Hée, "The Ch'ou-Jen Chuan of Yüan Yüan," op cit.

17. In the beginning of the third century A.D., the Han dynasty was in a state of decline. Three generals seized power and in A.D. 211 divided the Empire among themselves. Eventually, the three geographic regions claimed by these generals became separate kingdoms—the Three Kingdoms: Wei, Wu and Shu Han. The capital of Wei was Loyang.

18. Liu's work with *chong cha* is discussed in detail in Frank Swetz, *The Sea Island Mathematical Manual: Surveying and Mathematics in Ancient China* (University Park, Pa.: Pennsylvania State University Press, 1922).

19. The evolution of π in China is reviewed in Lee Kiong-Pong, "Development of π in China," *Bulletin of the Malaysian Mathematical Society* (1975) 6:40–47.

20. See D. B. Wagner, "Liu Hui and Tsu Keng-chih on the Volume of a Sphere," *Chinese Science* (1978) 3:59–79 or T. Kiang, "An Old Chinese Way of Finding the Volume of a Sphere," *The Journal of the Mathematical Association* (May, 1972) 56:88–91.

21. See Lam Lay Yong and Ang Tian Se, *Fleeting Footsteps: Tracing the Conception of Arithmetic and Algebra in Ancient China* (Singapore: World Scientific, 1992). This work contains a complete translation of *Master Sun's Manual.*

22. Biot was actually the first European to transmit this problem to the West in his 1839 paper, "Table générale d'un ouvrage chinois intitulé . . . Souan-Fa Tong-Tsong, on Traité Complet de l'Art de Compter," *Journal asiatique* (1839) 7:193–217, but it was ignored until Wylie's later discussion.

23. For a listing and discussion of the particular works included in the *Ten Mathematical Manuals* see Li Yan and Du Sheran, *Chinese Mathematics: A Concise History* (tr. John Crossley and Anthony Lun) (New York: Oxford University Press, 1987) pp. 92–104.

24. This work has been thoroughly studied by Ulrich Libbrecht, *Chinese Mathematics in the Thirteenth Century: The Shu shu Chiu-Chang of Ch'in Chiu-shao* (Cambridge, MA: The MIT Press, 1973).

25. Karine Chemla reviews this work in "Reflets des Mesures du Cercle sur la Mer" Ph.D. dissertation, University of Paris, (1982).

26. Yang Hui's work is discussed by Lam Lay-yong: *A Critical Study of the Yang Hui Suan Fa, a Thirteenth-Century Chinese Mathematical Treatise* (Singapore: Singapore University Press, 1977); "The Chinese Connection between the Pascal Triangle and the Solution of Numerical Equations of Any Degree" *Historia Mathematica* (1980) 7:407–424.

27. See Jock Hoe, *Les Systèmes d'Equations Polynômes dans le Siyuan Yujian (1303)* Collège de France, Institut des Hautes Études Chinoises (1977); Li and Du, op. cit. pp. 156–161.

28. A fuller discussion of these philosophical systems and their impact is given in Derk Bodde, "The Attitude Toward Science and Scientific Method in Ancient China," *T'ien Hsiu Monthly* (February, 1936) 2:139–161. There is a fourth school of major philosophical thought in China, Legalism founded by Han Fei in about 200 B.C.; however, it had little effect on the development of mathematics and is excluded from our consideration.

29. See definitions supplied in Needham, op cit pp. 91–94.

30. Buddhism blended with Taoism stressing the seeking of an "inner" personal sense of harmony and well-being.

31. See Man-Keung Siu, "Concept of Function—Its History and Teaching." *Learn from the Masters!* Proceedings of the Kristiansand Conference on the History of Mathematics and Its Place in Teaching, Kristiansand, Norway (Washington, DC: Mathematical Association of America, 1995) pp. 105–122. Joseph Needham holds strongly to this "cause and effect" intellectual stimulus. See his "Mathematics and Science in China and the West," *Science and Society* (1956) 20:320–343.

32. For example, Li Huang (d 1811) wrote a *Careful Explanation of the 'Nine Chapters on the Mathematical Art' with Diagrams.*

33. The numerals and then computing rod configurations are:

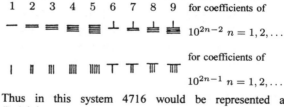

Thus in this system 4716 would be represented as ⫼⊥|⊥ (Occasionally the symbol x was used as an alternative to ≣).

34. This is the fifth problem of chapter eight in the *Nine Chapters*. For a further discussion of Chinese matrix solution techniques, see Lam Lay-Yong and Shen Kangshen, "Methods of Solving Linear Equations in Traditional China" *Historia Mathematica* (1989) 16:107–122.

35. Computing rods differed in color, black or red, or cross-sectional shape, round or oval, depending on whether they represented positive or negative numbers respectively. For further information on Chinese work with negative numbers, see Lam Lay-Yong and ang Tian-Se, "The Earliest Negative numbers: How They Emerged from a Solution of Simultane-

ous Linear Equations" *Archives Internationales d'Histoire des Sciences* (1987) 37:222–262.

36. In modern notation, this problem concerns the system:

$$2x + y = w$$

$$3y + z = w$$

$$4z + v = w$$

$$5v + u = w$$

$$6u + x = w$$

37. See B. S. Gillon, "Introduction, Translation, and Discussion of Chao Chun-Ch'ing's 'Notes to the Diagrams of Short Legs and Long Legs of Squares and Circles'." *Historia Mathematica* (1976) 4:253–93.

38. This principle is discussed more fully in Wu Wenchun, "The Out-In Complementary Principle," *Ancient China's Technology and Science* (Beijing: Institute of the History of Natural Sciences, Chinese Academy of Science, 1983) pp. 66–89; see also Swetz, *Sea Island Manual* op cit, pp 37–41.

39. For further information on Chinese root-extraction procedures see Frank Swetz, "The Evolution of Mathematics in Ancient China," *Mathematics Magazine* (January, 1979) pp. 10–19; the Chinese work with polynominal equations is examined by Lam Lay-Yong, "Chinese Polynomial Equations in the Thirteenth Century," *Explorations in the History of Science and Technology in China* (Shanghai: Chinese Classics Publishing House, 1982) pp. 231–272.

40. For example in the Sui dynasty (581–618 A.D.), *The Method of Calculation of the Brahma School* was translated into Chinese as *Poluomen suan fa* and *Calculation of the Calendar, Brahma School* appeared as *Poluomen yin yang suan li*; in 718 A.D., the Hindu *Catching Nines Calendar* which contained a table of sines, appeared in Chinese as *Jiu zhi li*.

41. The second half of the nineteenth century witnessed an era of Chinese social and intellectual reform. For details on this period see, Frank Swetz, "The Introduction of Mathematics in Higher Education in China, 1865–1887," *Historia Mathematica* (1974) 1:167–179; Cyrus Peake, "Some Aspects of the Introduction of Modern Science into China," *Isis* (1934) 22:173–219; Wann-Sheng Horng, "Li Shanlan, the Impact of Western Mathematics in China During the Late 19th Century", PhD dissertation, The City University of New York, March, 1991.

Combinatorics and Induction in Medieval Hebrew and Islamic Mathematics

Victor J. Katz
University of the District of Columbia

How many ways are there of arranging six people in a row? How many ways are there of picking a committee of three members out of a group of twenty people? What are the odds on winning the lottery, given that you need to pick six correct numbers out of forty-four? The concept of counting what are today called permutations and combinations is an important one, useful in many areas of pure and applied mathematics and often taught, at least initially, in secondary schools. These ideas have wide ranging applications, and the methods used in developing the formulas apply to other counting problems as well—problems which are at the heart of probability computations, for example. This article will explore the early history of these counting methods, a history which even today is not entirely known. What is known, however, is that these concepts, unlike many other aspects of modern mathematics, do not stem from ancient Greece. The ancient Greeks evidently did not consider such problems. The earliest mention of these counting problems is in Indian mathematics, while the detailed development of formulas—and their proofs—was accomplished in the Islamic and Jewish worlds from the twelfth through the fourteenth centuries. As will be clear, many original methods for solving such counting problems can easily be adapted for use in today's classroom.

Although combination and permutation problems appear in ancient Indian mathematics, no proofs or justifications of procedures occur in the extant written material. For example, the medical treatise of Susruta, perhaps written in the sixth century BCE, states that 63 combinations can be made out of six different tastes—bitter, sour, salty, astringent, sweet, hot—by taking them one at a time, two at a time, three at a time, ...[1] In other words, there are six single tastes, 15 combinations of two, 20 combinations of three, and so on. Works attributed to the Jains from about the third century BCE, include similar calculations dealing with such topics as philosophical categories and senses. In all these examples, however, the numbers are small enough that simple enumeration is sufficient to produce the answers. We do not know whether relevant formulas had been developed.

On the other hand, a sixth-century work by Varahamihira deals with a larger value. It plainly states that "if a quantity of 16 substances is varied in four different ways, the result will be 1820."[2] In other words, since Varahamihira was trying to create perfumes using four ingredients out of a total of 16, he had calculated that there were precisely 1,820 $(= C_4^{16})$ different ways of choosing the ingredients. It is unlikely that the author actually enumerated these 1,820 combinations, and so we assume that he knew

a formula to calculate that number. Nevertheless, no such formula exists in the Indian literature of the time.

Bare statements of combinatorial results also appear in an early work in Hebrew, the mystical *Sefer Yetzirah,* which is estimated to have been written in Palestine before the third century CE. The text, whose translated English title is the *Book of Formations,* deals with the letters of the Hebrew alphabet. Consider two sections:

> He fixed the twenty-two letters on the sphere like a wall with two hundred and thirty-one gates ... But how was it done? He combined ... the א (aleph) with all the other letters in succession, and all the others again with א; ב (bet) with all, and all again with ב, and so the whole series of letters. Hence it follows that there are two hundred and thirty-one formations.[3]

> Two stones build two houses, three stones build six houses, four build twenty-four houses, five build one hundred and twenty houses, six build seven hundred and twenty houses, and seven build five thousand and forty houses. From thence further go and reckon what the mouth cannot express and the ear cannot hear.[4]

In the first instance, the author found that there are 231 combinations of two elements taken from a set of 22 ($22 \times 21/2$). The tenth-century commentator Saadia Gaon (892–942), who was born in Egypt but spent much of his life in Babylonia, noted that children in Palestine learn spelling and pronunciation by considering all $22^2 = 484$ possible ordered pairs of letters. He noted further that the author of the *Sefer Yetzirah* dropped all 22 repeated pairs from consideration (e.g., אא) and then halved the remaining number to get his result.

For permutations, the *Sefer Yetzirah* has calculated that the number of possible words which can be formed from 2, 3, 4, 5, 6, and 7 letters respectively are 2!, 3!, 4!, 5!, 6!, and 7!. Saadia Gaon extended this rule as follows:

> If somebody wants to know how many words may be built from a larger number than that, as for instance 8, 9, 10, ... the rule is that one should multiply the result of the first product by the following number and what one thus obtains is the sum total. And its explanation is as follows: The permutations of two letters give 2 words, if you multiply 2 by 3, you get 6, and that is the number of the permutations of the three letters. ... If you want to know the number of the permutations of 8 letters, multiply the 5040 that you got from 7 by 8 and you will get 40,320

words; and if you search for the number of permutations of 9 letters, multiply 40,320 by 9 and you will get 362,880.[5]

Because the longest word in Scripture has eleven letters, Saadia even calculates the number of permutations of eleven letters: 39,916,800.

Shabbetai Donnolo (913–970), who lived in southern Italy, derived this factorial rule more explicitly in another commentary on the *Sefer Yetzirah:*

> The first letter of a two-letter word can be interchanged twice, and for each initial letter of a three-letter word the other letters can be interchanged to form two two-letter words—for each of three times. And all the arrangements there are of three-letter words correspond to each one of the four letters that can be placed first in a four-letter word: a three-letter word can be formed in six ways, and so for every initial letter of a four letter word there are six ways—altogether making twenty-four words, and so on.[6]

He even notes that, if one used all 22 letters and rearranged them in various ways, one would obtain all the words of all the languages on earth. "But the number is too great for flesh and blood to calculate."[7]

Interestingly, one of the earliest Islamic authors considering permutations also saw them in regard to letters of the alphabet, in his case the Arabic alphabet. He was al-Khalil ibn Ahmad (717–791), who calculated the actual number of possible words one could get by taking 2, 3, 4, or 5 letters out of the Arabic alphabet of 28 letters. Other early Islamic authors, including Thabit ibn Qurra (830–890) and Abu Kamil ibn Aslam (850–930), gave answers involving permutations and combinations for other small sets of objects.

Just as their Indian predecessors did, the Islamic authors of this time period only provided bare answers to combinatorial questions. In the ninth century, however, the Indian work *Ganitasarasangraha* by Mahavira finally gave an explicit algorithm for calculating the number of combinations:

> The rule regarding the possible varieties of combinations among given things: Beginning with one and increasing by one, let the numbers going up to the given number of things be written down in regular order and in the inverse order (respectively) in an upper and a lower horizontal row. If the product of one, two, three, or more of the numbers in the upper row taken from right to left be divided by the corresponding product

of one, two, three, or more of the numbers in the lower row, also taken from right to left, the quantity required in each such case of combination is obtained as the result.[8]

Mahavira does not, however, give any proof of this algorithm, which can be translated into the modern formula

$$C_r^n = n(n-1)(n-2)\cdots(n-r+1)/r!$$

He simply applies the rule to two problems, one about combinations of the tastes—as his predecessor did—and another about combinations of jewels on a necklace, where these may be diamonds, sapphires, emeralds, corals, or pearls.

In ensuing centuries other similar computations appear in both Indian and Islamic literature. For example, al-Samaw'al ben Yahya ben Abbas al-Maghribi (d. 1175), a Jew from Baghdad who converted to Islam, in discussing his methods for solving large systems of equations actually wrote down in a systematic fashion all 210 combinations of 10 unknowns taken six at a time in his *Al-Bahir*.[9] He did not, however, indicate how to calculate the number for other cases. Also around the middle of the twelfth century, Bhaskara (1114–1185) in India similarly gave rules for solving various combinatorial problems, including essentially the rule given by Mahavira:

> Let the figures from one upwards, differing by one, put in the inverse order, be divided by the same arithmeticals in the direct order; and let the subsequent be multiplied by the preceding, and the next following by the foregoing result. The several results are the changes, ones, twos, threes, etc. This is termed a general rule. It serves in prosody, for those versed therein, to find the variations of metre; in the arts as in architecture to compute the changes upon apertures of a building, and in music the scheme of musical permutations, in medicine, the combinations of different savours.[10]

Bhaskara gives an example in prosody:

> In the permutations of the gayatri metre, say quickly, friend, how many are the possible changes of the verse and tell severally, how many are the combinations with one, two, three, etc. long syllables?[11]

He then writes down the numbers:

$$6 \quad 5 \quad 4 \quad 3 \quad 2 \quad 1$$
$$1 \quad 2 \quad 3 \quad 4 \quad 5 \quad 6$$

and records that the combinations with one long syllable are 6; with two, 15; with three, 20; with four, 15; with five, 6; and with six, 1. He further finds that there is 1 way with all syllables short. It follows that the entire number of combinations is 64.

Bhaskara further notes that the number of permutations of a set of n objects is $n!$. He then asks,

> how many are the variations of form of the god Sambhu by the exchange of his ten attributes held reciprocally in his several hands: namely the rope, the elephant's hook, the serpent, the tabor, the skull, the trident, the bedstead, the dagger, the arrow, and the bow?[12]

He calculates the answer as $10! = 3,628,800$. Bhaskara further shows how to compute the number of permutations in a set of digits where some digits occur more than once. For example, he shows that there are 6 permutations of the digits 2, 2, 1, 1 and 20 permutations of the digits 4, 8, 5, 5, 5.

Again, however, although he gives detailed rules, Bhaskara, like Mahavira, provides no proof or even a derivation of his results. A contemporary of his, Abraham ibn Ezra (1090–1167), a rabbi who lived in southern France, did give a detailed derivation of a rule for determining the number of combinations. Rabbi ibn Ezra did this as part of an astrological discussion in which he determined the number of possible conjunctions of the seven planets—the sun, the moon, Mercury, Venus, Mars, Jupiter, and Saturn. In essence he showed how to calculate C_k^7 for each integer k from 2 to 7 and then noted that the total was 120.

Beginning with the simplest case, Rabbi ibn Ezra found that the number of binary conjunctions was 21. Because there were six planets which could have a conjunction with Jupiter, say, five (other than Jupiter) with Saturn, and so on, this number equals the sum of the integers from 1 to 6. Ibn Ezra noted that the sum of the numbers from 1 to any desired number can be found by multiplying the number by half of itself and by half of unity, that is, in modern notation,

$$\sum_1^n = n\left(\frac{n}{2}\right) + n\left(\frac{1}{2}\right) = \frac{n(n+1)}{2}.$$

In his particular case it followed that $C_2^7 = 6(7)/2 = 21$. In general, $C_2^n = n(n-1)/2$.

To calculate ternary combinations, ibn Ezra explains, "We begin by putting Saturn with Jupiter and with them one of the others. The number of the others is five ... The result is 15. And these are the conjunctions of Jupiter."[13] Namely, there are five ternary combinations involving Jupiter and

Saturn, four involving Jupiter and Mars, but not Saturn, and so on. Hence there are $C_2^6 = 15$ ternary conjunctions involving Jupiter. Similarly, to find the ternary conjunctions involving Saturn but not Jupiter, ibn Ezra needs to calculate the number of choices of two planets from the remaining five: $C_2^5 = 10$. He then finds the ternary conjunctions involving Mars, but neither Jupiter nor Saturn, and finally concludes with the result:

$$C_3^7 = C_2^6 + C_2^5 + C_2^4 + C_2^3 + C_2^2$$
$$= 15 + 10 + 6 + 3 + 1$$
$$= 35.$$

Ibn Ezra next calculates the quaternary conjunctions by analogous methods. The conjunctions involving Jupiter require choosing three planets from the remaining six. Those with Saturn but not Jupiter require choosing three from five. So finally,

$$C_4^7 = C_3^6 + C_3^5 + C_3^4 + C_3^3$$
$$= 20 + 10 + 4 + 1$$
$$= 35.$$

Ibn Ezra then states the results for conjunctions involving five, six, and seven planets. Essentially, he has presented an argument for the case $n = 7$ which is easily generalizable to the general combinatorial rule

$$C_k^n = C_{k-1}^{n-1} + C_{k-1}^{n-2} + \cdots + C_{k-1}^{k-1}.$$

Rabbi ibn Ezra's astrological work which included this material on combinations appeared in Latin translation in 1281. The Latin version seems to indicate that the translation was made from the Arabic, so it must sometime earlier have been translated into that language from the Hebrew original, although no such Arabic manuscript has yet been found. Interestingly, however, early in the thirteenth century Ahmad al-Ab'dari ibn Mun'im discussed the same rule that ibn Ezra did, although in more detail. Little is known about ibn Mun'im, but he probably lived at the Almohade court in Marrakesh (now in Morocco) during the reign of Mohammed ibn Ya'kub al-Nasir (1199–1213). Although the Almohade dynasty originally ruled over a large empire including much of North Africa and Spain, al-Nasir was defeated by a coalition of Christian Kings at the battle of Las Navas de Tolosa in Spain in 1212 and lost many of his Spanish domains.

Ibn Mun'im was basically examining the old question of the number of possible words which could be formed out of the letters of the Arabic alphabet. But before dealing with that question, he considered a different problem: how many different bundles of colors can one make out of ten different colors of silk. He calculated these carefully. First of all, he notes that with only one color, there are 10 possibilities, i.e. $C_1^{10} = 10$. To calculate the possibilities for two colors, ibn Mun'im lists the pairs in order:

$$(c_2, c_1); (c_3, c_1), (c_3, c_2); \ldots (c_{10}, c_1), (c_{10}, c_2),$$
$$\ldots, (c_{10}, c_9)$$

and then notes that

$$C_2^{10} = 1 + 2 + \cdots + 9$$
$$= C_1^1 + C_1^2 + \ldots + C_1^9$$
$$= 45.$$

C_2^k can be calculated similarly for values of k less than 10. To calculate C_3^{10}, ibn Mun'im proceeds analogously.

To determine the number of combinations of three colors, one first combines the third color with the first and the second, then combines the fourth color with each pair of colors among the three colors which precede it—the first, the second and the third—then combines the fifth color with each pair of colors among the four preceding colors, ... up to the combination of the tenth color with each pair of colors among the nine preceding colors. But each pair of colors is a combination from the second line. For this reason, ... we write in the case of the fourth color, that the number of combinations obtained by the combination of the fourth color with each pair among the preceding colors is equal to the number of bunches of two colors composed of colors preceding the fourth color, and this is also equal to the sum of the two first numbers of the second line, namely 3. We therefore write 3 in the second case in the third line.[14]

In other words, for each c_k for $k = 3, 4, \ldots, 10$, ibn Mun'im considers the pairs from the previous calculation which have all indices less than k, for example, $(c_3, (c_2, c_1)); (c_4, (c_2, c_1)), (c_4, (c_3, c_1)), (c_4, (c_3, c_2)); (c_5, (c_2, c_1)), \ldots$ Thus C_3^{10} is the sum $1 + 3 + 6 + \cdots + 36 = C_2^2 + C_2^3 + C_2^4 + \cdots + C_2^9$, each of which numbers can be easily calculated from the calculation of the previous line. The word "line" refers to the fact that ibn Mun'im presents these results in a table, the first line of which lists the numbers $1, 2, \ldots 10$ ($= C_1^1, C_1^2, \ldots C_1^{10}$), the second line the numbers $1, 3, 6, \ldots 36$ ($= C_2^2, C_2^3, C_2^4, \ldots C_2^9$), $\ldots$. Ibn Mun'im constantly refers to this table, in effect, the Pascal triangle, to show how the various calculations can be easily made. (See illustration.)

وهكذا تخطيط المثال في الجدول	لون اول	لون ثان	لون ثالث	لون رابع	لون خامس	لون سادس	لون سابع	لون ثامن	لون تاسع	لون عاشر	جدول جمع الجدول أى / جمع الألوان
من عنصرة الألوان										1	1
جدول الشراريب التي من تسعة ألوان تسعة ألوان									1	9	10
جدول الشراريب التي من ثمانية ألوان ثمانية ألوان								1	8	36	45
جدول الشراريب التي من سبعة ألوان سبعة ألوان							1	7	28	84	120
جدول الشراريب التي من ستة ألوان ستة ألوان						1	6	21	56	126	210
من خمسة ألوان خمسة ألوان					1	5	15	35	70	126	252
من اربعة ألوان اربعة ألوان				1	4	10	20	35	56	84	210
من ثلاثة ألوان ثلاثة ألوان			1	3	6	10	15	21	28	36	120
من لونين لونين		1	2	3	4	5	6	7	8	9	45
من لون لون	1	1	1	1	1	1	1	1	1	1	10

Although he concentrated on the value $n = 10$, ibn Mun'im did discuss some important properties of the numbers C_k^n in the process of this construction. His manuscript deals with the question of permutations in a way slightly different from that of Rabbi Donnolo:

The problem is: We want to determine a canonical procedure to determine the number of permutations of the letters of a word of which the number of letters is known and which does not repeat any letter. If the word has two letters, it is clear that there will be two permutations, since the first letter may be made the second and the second the first. If we augment this by one letter and consider a three letter word, it is clear that, in each of the permutations of two letters of a two letter word, the third letter may be before the two letters, between the two letters, or in the final position. The letters of a three letter word therefore have six permutations. If the word is now augmented by another letter to make a four letter word, the fourth letter will be in each of the six permutations [in one of four positions]. The four letter word will thus have twenty four permutations.[15]

Ibn Mun'im concludes that, no matter how long the word, the number of permutations of the letters is found by multiplying one by two by three by four, by five, etc., up to the number of letters of the word. "The result is equal to the number of permutations of the letters of the word, and this is what we wanted to demonstrate."[16]

Ibn Mun'im pursues several other problems, including permutations with repetitions, before dealing with matters of pronunciations and vowel signs. His aim is to determine the number of possible Arabic words and, after some discussion of exactly what this means, he uses some of the above ideas to calculate explicitly the number of words of nine letters, each word having two non-repeated letters, two letters repeated twice, and one letter repeated three times. The number turns out to have 16 decimal digits.

Later in the thirteenth century Abu-l-'Abbas Ahmad al-Marrakushi ibn al-Banna (1256–1321), also of Marrakech, continued the work of ibn Mun'im.[17] Ibn al-Banna gave effectively the same proof of the result on the number of permutations of a set of n elements, but also discussed the values C_k^n, which enabled him to develop the Indian formula to compute them. He first showed that $C_2^n = n(n-1)/2$ by a simple counting argument: a_1 is associated with each of $n-1$ elements, a_2 is associated with each of $n-2$ elements, etc., so C_2^n is the sum of $n-1, n-2, n-3, \ldots, 2, 1$. Ibn al-Banna then showed, in modern notation, that

$$C_k^n = \frac{n - (k-1)}{k} C_{k-1}^n.$$

How did he do this? First consider C_3^n. To each set of 2 elements from the n elements, one associates one of the $n-2$ remaining elements. One obtains then $(n-2)C_2^n$ different sets. But since $C_2^3 = 3$, each of these sets is repeated three times. For example, $\{a, b, c\}$ occurs as $\{\{a, b\}, c\}$, $\{\{a, c\}, b\}$, and as $\{\{b, c\}, a\}$. Therefore,

$$C_3^n = \frac{n-2}{3} C_2^n$$

as claimed. For the next step, we know that $C_3^4 = 4$. It follows that, if we associate to each set of three elements one of the $n-3$ remaining elements, the total $(n-3)C_3^n$ is four times larger than C_4^n. In other words,

$$C_4^n = \frac{n-3}{4} C_3^n$$

as desired. A similar argument holds for other values of k. And upon putting together these results, it becomes clear that

$$C_k^n = \frac{n(n-1)(n-2)\cdots(n-(k-1))}{1 \cdot 2 \cdot 3 \cdots k},$$

the standard formula for the number of ways to pick k elements out of a set of n. Using this result and the result on permutations, ibn al-Banna easily showed by multiplication that the number P_k^n of permutations of k objects out of a set of n is

$$P_k^n = n(n-1)(n-2)\cdots(n-(k-1)).$$

In ibn al-Banna's proofs of the basic combinatorial results one can see the idea of proof by induction. In each proposition, he begins with a known result for a small value

and uses it to build up step by step to higher values. But ibn al-Banna does not state this induction result as a principle of proof. His younger contemporary, Levi ben Gerson (1288–1344), who lived around Orange, in the south of France, first did that. Whether Levi knew of ibn al-Banna's or ibn Mun'im's work is not known, although Levi had undoubtedly read the work of some Islamic mathematicians which had recently been translated into Hebrew.[18] In any case, Levi presented the same basic results as ibn al-Banna in his *Maasei Hoshev* of 1321, but rewrote the proofs into the form of a proof by induction. Namely, in each of his basic results, he stated both the inductive step (which gets from k to $k+1$) and that the result is true for a small number, and then noted that these together imply that the result is true in general. We will therefore consider Levi's work on the combinatorial ideas.

Levi, like ibn al-Banna, states his results abstractly, not in terms of any concrete objects such as planets or letters of the alphabet. In other words, his results are no longer part of other fields but have become part of mathematics. Levi begins his series of theorems with the proposition that the number of permutations of a given number n of elements is $n!$. His statement of the inductive step of this proof is in words— after all, Levi, like his predecessors, used no symbols:

> If the number of permutations of a given number $[n]$ of different elements is equal to a given number $[P_n]$, then the number of permutations of a set of different elements containing one more number equals the product of the former number of permutations and the given next number $[P_{n+1} = (n+1)P_n]$.[19]

Levi's discussion is similar to that of Rabbi Donnolo. Given a permutation, say $abcde$, of the original n elements (here taken as 5) and a new element f, he notes that $fabcde$ is a permutation of the new set. Because there are P_n permutations of the original set, there are the same number of permutations of the new set beginning with f. Also, if one of the original elements, for example e, is replaced by the new element f, there are also P_n permutations of the set a, b, c, d, f and therefore also the same number of permutations of the new set with e in the first place. Because any of the n elements of the original set, as well as the new element, can be put in the first place, it follows that the number of permutations of the new set is $(n+1)P_n$. Levy concludes this proof by showing that all of these permutations are in fact different. He then concludes,

> Thus it is proved that the number of permutations of a given set of elements is equal to that number

formed by multiplying together the natural numbers from 1 up to the given number of elements. For the number of permutations of 2 elements is 2, and that is equal to 1×2, the number of permutations of 3 elements is equal to the product 3×2, which is equal to $1 \times 2 \times 3$, and so one shows this result further without end.[20]

Levi thus gives the beginning step and then notes that, with the inductive step proved, the complete proposition is also proved.

After next proving by using a counting argument that the number P_2^n of permutations of 2 elements in a set of n is $n(n-1)$, Levi proves by induction on k that the number P_k^n of permutations of k elements in a set of n is $n(n-1)(n-2)\ldots(n-(k-1))$. As before, he states the inductive step as a theorem:

> If a certain number of elements is given and the number of permutations of order a number different from and less than the given number of elements is a third number, then the number of permutations of order one more in this given set of elements is equal to the number which is the product of the third number and the difference between the first and second numbers.[21]

Modern symbolism replaces Levi's convoluted wording with a brief phrase: $P_{j+1}^n = (n-j)P_j^n$. Levi's proof is similar to that of the previous proposition. He then states his desired result:

> It has thus been proved that the permutations of a given order in a given number of elements are equal to that number formed by multiplying together the number of integers in their natural sequence equal to the given order and ending with the number of elements in the set.[22]

To clarify this statement, Levi first gives the initial step of the induction by quoting his previous result in the case $n = 7$: $P_2^7 = 6 \times 7$. Then, the number of permutations of order 3 in this set is $5 \times 6 \times 7$ (since $5 = 7 - 2$). Similarly, the number of permutations of order 4 equals $4 \times 5 \times 6 \times 7$ "and so one proves this for any number."[23]

In the next three propositions Levi completes his development of formulas for permutations and combinations. He shows by considering the permutations of each set of k elements that $P_k^n = k!C_k^n$ or that $C_k^n = P_k^n/k!$. But, since he has already given a formula for P_k^n in the previous theorem, he now has the general formula for C_k^n, which ibn al-Banna also gave. Levi concludes his treatment of permutations by noting the symmetry of the combinatorial

values, that is, that $C_k^n = C_{n-k}^n$, a result not mentioned by ibn al-Banna.

Although their methods of proof were slightly different, both Levi ben Gerson and ibn al-Banna clearly were able to establish in reasonable ways the same important results, results which did not appear in mathematics in the Latin West until the mid-seventeenth century in the work of Mersenne and Pascal, among others. Whether or not Levi was influenced by ibn al-Banna or some of his contemporaries, it nevertheless seems that, given the close proximity of Jew and Moslem on the Mediterranean coast in the eleventh through the fourteenth centuries, each civilization influenced the other in some way. More archival research will be necessary to ascertain exactly the direction of this influence in the discovery of the basic rules of counting permutations and combinations.

References

1. Gurugovinda Chakravarti, "Growth and Development of Permutations and Combinations in India," *Bulletin of Calcutta Mathematical Society* 24 (1932), 79–88. More information can be found in N. L. Biggs, "The roots of Combinatorics," *Historia Mathematica* 6 (1979), 109–136.
2. Ibid. The quotation is from the *Brhat Samhita,* chapter 77, rule 20 as translated in J. K. H. Kern, "The Brhatsamhita of Varahamihira," *Journal of Royal Asiatic Society* (1875), 81–134.
3. Isidor Kalisch, ed. & trans., *The Sepher Yezirah (The Book of Formation),* (Gillette, NJ: Heptangle Books, 1987), p. 11.
4. Ibid., p. 23.
5. Quoted in Solomon Gandz, "Saadia Gaon as a Mathematician," *Saadia Anniversary Volume* (New York: American Academy for Jewish Research, 1943); reprinted in Solomon Gandz, *Studies in Hebrew Astronomy and Mathematics,* Shlomo Sternberg, ed., (New York: Ktav, 1970).
6. Quoted in Nachum L. Rabinovitch, *Probability and Statistical Inference in Ancient and Medieval Jewish Literature,* (Toronto: University of Toronto Press, 1973), p. 144.
7. Ibid., p. 145.
8. Mahaviracarya, *Ganita-sara-sangraha,* translated by M. Rangacarya (Madras: Government Press, 1912), p. 150.
9. Al-Samaw'al, *Al-Bahir en Algèbre d'As-Samaw'al,* edited, with notes and introduction, by Salah Ahmad and Roshdi Rashed (Damascus: Ministère de l'Enseignement Supérieur R.A.S., 1972).
10. H. T. Colebrook, *Algebra with Arithmetic and Mensuration from the Sanskrit of Brahmagupta and Bhascara* (London: John Murray, 1817), p. 49.
11. Ibid.
12. Ibid., p. 124.
13. Jekuthiel Ginsburg, "Rabbi Ben Ezra on Permutations and Combinations," *The Mathematics Teacher* 15 (1922), 347–356.
14. Ahmed Djebbar, *L'Analyse Combinatorie au Maghreb: L'Exemple d'Ibn Mun'im (XIIᵉ–XIIIᵉ s.),* (Orsay: Université

de Paris-Sud, Publications Mathématiques D'Orsay, 1985), pp. 51–52.

15. Ibid., pp. 55–56.
16. Ibid.
17. For more on ibn al-Banna, see Ahmed Djebbar, *Enseignement et Recherche Mathématique dans le Maghreb des XIIIe - XIVe siecles* (Orsay: Université de Paris-Sud, Publications Mathématiques d'Orsay, 1981).
18. Ibid., p. 74.
19. Levi ben Gerson, *Sefer Maassei Choscheb,* edited and translated into German by Gerson Lange (Frankfurt am Main: Louis Golde, 1909), pp. 47–48.
20. Ibid., p. 49.
21. Ibid., p. 50.
22. Ibid., p. 51.
23. Ibid.

The Earliest Correct Algebraic Solutions of Cubic Equations

Barnabas Hughes
California State University, Northridge

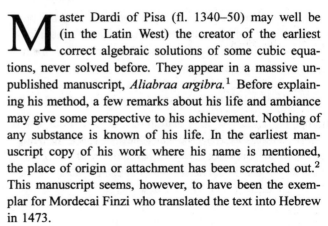

aster Dardi of Pisa (fl. 1340–50) may well be (in the Latin West) the creator of the earliest correct algebraic solutions of some cubic equations, never solved before. They appear in a massive unpublished manuscript, *Aliabraa argibra.*[1] Before explaining his method, a few remarks about his life and ambiance may give some perspective to his achievement. Nothing of any substance is known of his life. In the earliest manuscript copy of his work where his name is mentioned, the place of origin or attachment has been scratched out.[2] This manuscript seems, however, to have been the exemplar for Mordecai Finzi who translated the text into Hebrew in 1473.

Little is known of the translator except that he lived from time to time in Bologna and Mantua, and that he translated several mathematical works into Hebrew.[3] In his translation, Finzi supplied the items obliterated in the Tempe manuscript, particularly that the author was "Maestro Dardi da Pisa"[4] who completed the text on 6 November 1344 (f. 8r).[5] The only known Dardi of this period is one Ziio Dardi (fl. 1346), who apparently taught abacist mathematics in Venice.[6] At the moment these two abacists have not been identified as one and the same person.

Notably, this text departs radically from the common mathematical texts of the time. They were abacist notebooks written by and for authors who conducted their own schools. A typical abacist text might contain separate chapters on

- representation of numbers by symbols and fingers
- multiplication and division of whole numbers
- addition, subtraction, multiplication and division of fractions and mixed numbers
- computation of square and cube roots
- basic geometric concepts
- the Rule of Three
- the Florentine monetary system
- algebra

All of the above would be exemplified in many problems, both theoretical and practical.[7] Dardi's text, on the contrary, devotes most of its space to algebra, approximately 136 folios (*recto* and *verso*) of 162 in *quarto*.[8] Herein the reader finds a catalog of 198 different types of equations,[9] general rules for solving them, specific examples for each type, rules for operating on radicals and finding square and cube roots, a synopsis of al-Khwarizmi's *Algebra,* and finally various problems in the abacist genre. There are a large number of types of equations because,

in the early fourteenth century, the science of algebra had not developed conceptually or symbolically the generalization probably achieved by Thomas Harriot[10] wherein the equation is set equal to zero,

$$a_0 x^n + a_1 x^{n-1} + a_2 x^{n-2} + \cdots + a_{n-1} x + a_n = 0.$$

Rather, each possible combination of known and unknown quantity to the fourth degree was stated in words, all being set equal to some positive number other than zero. For example, "li censi e numero e radici chuba de numero sono vguale ale cose" (f. 115r); that is,

$$ax^2 + n + \sqrt[3]{m} = bx,$$

as well as verbal equivalents of

$$ax^4 = n \quad n = ax^3 + \sqrt[3]{bx3} \quad bx^2 = ax^4 + n.$$

The thinking here was geometric despite the presence of the word algebra. Al-Khwarizmi had used geometric figures and arguments to establish his algebraic algorithms.[11] Fibonacci continued the tradition.[12] Three hundred years later, Cardano continued the comparison in his rules for algebra; for instance,

> For a *positio* [the first power] refers to a line, *quadratum* [the square] to a surface, and *cubum* [the cube] to a solid body, it would be very foolish for us to go beyond this point. Nature does not permit it.[13]

Hence, numbers were set equal only to a number, never to nothing (although I think that Dardi was in the neighborhood of the right idea; see Appendix I).

Dardi's text is an entirely different genre from the abacist rule and exercise. His was an advanced algebra that begins with a clear explanation of the tools (listed above) necessary for understanding and applying the rules he gives. The primary tool is al-Khwarizmi's *Algebra*. Dardi's synopsis is nearly 10 pages long, folios 23r to 28v. From a study of it, the reader must learn to solve each type of equation by manipulating and computing the numbers given in the problem. For instance, while it is convenient for modern students to solve a quadratic equation by the quadratic formula, the late medieval student faced a sequence of statements that told the student what to do. (My discussion below of the actual contributions offers examples.) Modern students customarily think in terms of *x-equals-some-formula* that will produce a number to satisfy an equation. The student of Dardi's day had to learn to think in terms of a procedure operating on given numbers that would produce the answer to a problem. While I discuss Dardi's contributions in modern terminology, the reader must see through

my anachronistic wording to the mental framework of the fourteenth-century student.

Techniques for solving such equations as $ax^3 = n + \sqrt{m}$, $ax^3 + bx^2 = cx$, and other equations easily reducible to quadratics were in the tool kits of mathematical practitioners. Maestro Dardi of Pisa created procedures for solving more kinds of cubic equations. He added techniques for solving

$$ax^3 + \sqrt{bx^3} = c \qquad \text{(A)}$$

and six other equations like it, together with a specific case of

$$ax^3 + bx^2 + cx = d \qquad \text{(B)}$$

To begin with, both types of equations, (A) and (B), must be modified by dividing by the coefficient of the highest power of the unknown. General formulas for solving these equations, however, can be adjusted so that the coefficient, a , does appear in the formula. Equations of the first type are solved as a quadratic by substituting $y^{2/3}$ for x. Equations of the second type are solved under a new conditiion and by a unique procedure. The condition requires that the coefficient of the first-degree term equal one-third the cube of the coefficient of the second-degree term. The procedure substitutes $y - \frac{b}{3a}$ for x. The condition and procedure remove the first and second-degree terms. The procedure is easily generalizable for removing the second term in any equation of the nth degree arranged in descending powers of the unknown. To the best of my knowledge, Dardi was the first algebraist who transposed equations by this reduction process.

While the foregoing explanation briefly details what Dardi accomplished, it lacks the flavor and mystery of his own words. What follows is my translation of Chapter 41 of Dardi's text which focuses on equation (A), $ax^3 + \sqrt{bx^3} = c$. The letter C in the example is Dardi's abbreviation for *cosa* or first degree unknown. My commentary follows the translation.[14]

> When the numbers are equal to the cube and the root of the cube, you want to partition (divide trans.) the numbers by the quantity of the unknown cube or root, and set aside the outcome.
>
> Then, multiply the quantity of the unknown cube or root into itself and divide the known quantity of the cube or root into that multiplication, and connect the fourth of what results to the (outcome of the) partition set aside,
>
> and the root of that sum less the root of the fourth which you connected, namely the root of the fourth which came by partitioning the known

(quantity of the) cube or root in the multiplication of the unknown cube or root, becomes the root of the cube. And multiply this root into itself to become a cube. The cube root of this multiplication becomes the cosa.

For example: find me two numbers in the ratio of 2 to 3, such that the product of the square of the first multiplied by the second together with the root of this product equals 342.

This is the rule. Let the first number be $2C$ and the second $3C$ to create 12 cube and the root of 12 cube to be 342. Now follow the rule. Partition the 342 by 12 to get $28\frac{1}{2}$ and save this. Then, multiply the 12 of the unknown cube itself to get 144 and divide it into the 12 of the root to get $\frac{1}{12}$. Take $\frac{1}{4}$ of the $\frac{1}{12}$ to get $\frac{1}{48}$ which you add to the $28\frac{1}{2}$ which you saved to reach $28\frac{25}{48}$. Now the root of $28\frac{25}{48}$ less the root of the $\frac{1}{4}$, that is less the root of $\frac{1}{48}$, becomes the root of the cube. Next multiply this root into itself to find the cube, 27. Then take the cube root of it to reach the *cosa* which is 3. Multiply the *cosa* by 2 to get the first number or 6, by 3 to get the second number or 9.

Lines 1 through 4 describe the general equation and the initial operation. Dardi had no word for *coefficient*; yet he makes himself clear by the phrase "the quantity of the unknown cube or root." In so many words he states the equivalent of the following:

$$c = ax^3 + \sqrt{bx^3} \quad (1)$$

$$\frac{c}{a} = x^3 + \sqrt{\frac{b}{a^2}x^3} \quad (2)$$

or more easily

$$r = x^3 + \sqrt{\frac{x^3}{s}}. \quad (3)$$

From lines 5 to 14 Dardi computes with *known* numbers. He does not discuss variables for the good reason that he never writes of them nor was the terminology available. Further, after reading lines 5 to 8 we realize that Dardi is completing the square on (3) above; that is

$$r + \frac{1}{4s} = x^3\sqrt{\frac{x^3}{s}} + \frac{1}{4s} \quad (4)$$

Line 9 then finds the square root of (4) from which it subtracts $\sqrt{1/4s}$. This, Dardi says in lines 12 and 13, is "the root of the cube."

Only it is not a cube root! The reason follows. The continuing sentence says to "multiply this root into itself to

become a cube." Certainly he did not mean that to multiply a number by itself is to cube a number! Rather, I believe that Dardi was deliberately obfuscating the hidden operation; being able to solve more complex equations gave him an advantage in commerce and teaching. What he actually did (equivalently) in order to make line 15 valid was to substitute

$$y^{2/3} \quad \text{for } x \quad (5)$$

between lines 4 and 5. This effectively reduces the cubic equation to the quadratic

$$r = y^2 + y\sqrt{\frac{1}{s}} \quad (6)$$

whose square can be completed:

$$r + \frac{1}{4s} = y^2 + y\sqrt{\frac{1}{s}} + \frac{1}{4s}. \quad (7)$$

This becomes

$$r + \frac{1}{4s} = \left(y + \sqrt{\frac{1}{4s}}\right)^2 \quad (8)$$

Hence it follows

$$\sqrt{r + \frac{1}{4s}} - \sqrt{\frac{1}{4s}} = y \quad (9)$$

By squaring (9) and taking its cube root, the reduction substitution (5) is undone and the *cosa* found. The example in lines 15 to 29 reflects exactly the foregoing. The initial problem is then solved, all done without the benefit of modern symbols.

The six equations, identifiable by marginal numbers in the manuscript and with solution processes similar to that of (A) are

115. $ax^3 + \sqrt{bx^3} = \sqrt{c}$ 127. $ax^3 + \sqrt{bx^3} = \sqrt[3]{c}$

116. $ax^3 + \sqrt{c} = \sqrt{bx^3}$ 128. $ax^3 + \sqrt[3]{c} = \sqrt{bx^3}$

117. $\sqrt{bx^3} + \sqrt{c} = ax^3$ 129. $\sqrt{bx^3} + \sqrt[3]{c} = ax^3$

The statements of the equations differ from (A) only in some root being taken of the "number" or constant term. Note also the pairing of equations between sets of equations. Each equation is solved by the same or equivalent substitution methods to those employed in (A). That is, reduce the cubic equation to a quadratic by replacing x with $y^{2/3}$. After the quadratic is solved, square the value of y and finally take the cube root of the square. Thus each of the six rules concludes with the same step as does (A): "this produces the root of the cube. Its multiplication (on itself) produces the cube, and the root of the cube from that multiplication produces the *cosa*."[15]

Symbolically these formulae represent the processes:

115. $x = \sqrt[3]{\left(\sqrt{\dfrac{b}{4a^2}} + \sqrt{\dfrac{c}{a^2}} - \sqrt{\dfrac{b}{4a^2}}\right)^2}$

127. $x = \sqrt[3]{\left(\sqrt{\dfrac{b}{4a^2}} + \sqrt[3]{\dfrac{c}{a^3}} - \sqrt{\dfrac{b}{4a^2}}\right)^2}$

116. $x = \sqrt[3]{\left(\sqrt{\dfrac{b}{4a^2}} \pm \sqrt{\dfrac{b}{4a^2} - \sqrt{\dfrac{c}{a^2}}}\right)^2}$

128. $x = \sqrt[3]{\left(\sqrt{\dfrac{b}{4a^2}} \pm \sqrt{\dfrac{b}{4a^2} - \sqrt[3]{\dfrac{c}{4a^3}}}\right)^2}$

117. $x = \sqrt[3]{\left(\sqrt{\dfrac{b}{4a^2}} + \sqrt{\dfrac{c}{a^2}} + \sqrt{\dfrac{b}{4a^2}}\right)^2}$

129. $x = \sqrt[3]{\left(\sqrt{\dfrac{b}{4a^2}} + \sqrt[3]{\dfrac{c}{a^3}} + \sqrt{\dfrac{b}{4a^2}}\right)^2}$

Note that the pair of solutions, 116 and 128, alone produce double roots, both of which are positive. At that time in history solution processes did not produce negative roots, despite the contradictory "debit roots" (see Appendix I). Finally, the algebraic process shown above is quite different from the radically geometric technique used by Girolamo Cardano in his seminal text, *The Great Art* or *The Rules of Algebra*.[16]

The second and different cubic equation which Dardi solves is

$$x^3 + ax^2 + bx = c.$$

It appears as the second of four extraordinary problems in an appendix. They are called "extraordinary" because, as he remarks in a prefatory paragraph, he does not have general rules for solving them. Indeed, the rules "came upon [him] accidently" (f. 121r) and are valid for the specific cases only. The problem discussed here appears as the first of three cases in Cardano's treatment of this type of cubic equation.[17]

Dardi's method for solving the equation is stated briefly:

When the *cosa* and the *censo* and the *chubo* are equal to the number, then you must divide the entire adequation by the quantity of the *chubo*. Then divide the quantity of the *cosa* by the quantity of the *censo*. Then having cubed this quotient, add to it the number of the adequation. Take the cube root of this sum less the quotient you obtained by dividing the quantity of the *cosa* by the quantity of the *censo*. The result is the value of the *cosa*.[18]

Once again Dardi computes only with known numbers. What he says to do after dividing through by the coefficient of x^3, is represented by:

$$x = \sqrt[3]{\left(\frac{b}{a}\right)^3 + c} - \frac{b}{a}.$$

His method for reducing the problem so that its solution was merely a matter of finding a cube root suggests his ingenuity, although he wrote that the technique came upon him "by accident."

There are two parts to the method. First substitute $y - \frac{a}{3}$ for x which produces

$$y^3 + \left(b - \frac{a^2}{3}\right)y + \frac{2a^3}{27} - \frac{ab}{3} = c.$$

The substitution removes the second degree term. Next, the first-degree term disappears provided the necessary condition is fulfilled; namely, $b = \frac{a^2}{3}$. Only the third-degree term remains, and the solution follows by implementing the final directions stated above.

The example attached to this problem considers a man who, having loaned another man 100 lire, received back 150 lire after three years. Dardi wants to know how much interest the loan made each month, x, to produce the total interest of 50 lire. The equation is

$$\frac{x^3}{80} + \frac{3x^2}{4} + 15x = 50.$$

Once the initial substitution has been made, the resulting equation is easily solved.

Without the benefit of algebraic symbols, Master Dardi of Pisa made significant advances in the analysis and solution of cubic equations. Although as many as six manuscript copies of his text exist, there is no evidence that he influenced other mathematicians. Nonetheless, other abacists were making progress toward solving more kinds of cubic equations. By the end of the fourteenth century, an anonymous abacist included in his book important techniques toward finding the solution of "cubes and squares equal to numbers."[19]

Appendix I
An equation set equal to nothing

At the conclusion of the tract is this note:[20]

> Note that there may be *cose* and numbers on one side equal to nothing, as can be seen in some ad-equations mostly characterized, as we have been saying, by differences or other things. You must take the numbers from each side and you will have *cose* equal to nothing minus the number. Then you divide the number that is minus by the quantity of the *cose*. However much results that much you debit to the *cosa*.

Dardi's explanation follows the pattern used throughout his text: first the general equation, $ax + b = 0$; then, the rule for solving the general equation, *subtract the number from both sides and divide by the coefficient of the unknown*; finally, an example completely solved, $5x + 15 = 0$. The "minus the number" in line five should be interpreted as a subtraction rather than as a negative number. At this point in time, as far as I have been able to discover, our negative numbers were considered *debts* to be subtracted from whatever a person had, even if it were nothing. This idea was at least discussed by Fibonacci.[21] Nonetheless, Dardi certainly took a step toward setting equations equal to zero and considering negative roots.

References

1. My principal source is the 1429 copy of Dardi's text in the Hayden Library of Arizona State University, Tempe, from which I have made all quotations cited here. I gratefully acknowledge the assistance of Warren Van Egmond who has done more than anyone else in the study of abacist texts; he supplied me with copies of several sections from his forthcoming critical edition of Dardi's *Aliabraa argibra*. His *Practical Mathematics in the Italian Renaissance: A Catalog of Italian Abbacus Manuscripts and Printed Books to 1600.* (Firenze: Annali dell'Istituto e Museo di Storia della Scienza, mongrafia n. 4, 1980) is a *tour de force*. For whoever wishes to pursue further study in this field, the catalog together with other publications of his noted below is absolutely necessary . Copies of Dardi's text, except for the ASU copy which was not available to Van Egmond in 1980, are noted on pp. 188–189. For a descriptive analysis of the contents of the text based on the Sienna MS, see Raffaella Franci and Laura Toti Rigatelli, "Toward a history of algebra from Leonardo of Pisa to Luca Pacioli," Janus, 1985,72 (n. 1–3): 36–39. See also Raffaella Franci, "Contributi alla risoluzione dell'equazione di 3o grado nel xiv secolo," in Menso Folkerts and Uta Lindgren, eds., *Mathemata Festschrift für Helmuth Gericke* (Stuttgart: Franz Steiner, Reihe "Boethius," Bd. 12, 1985), pp. 221–228.

2. Barnabas Hughes, "An early 15th-century algebra codex: a description," *Historia Mathematica*, 1987, 14: 167–172.

3. Moritz Steinschneider, *Mathematik bei den Juden* (Hildesheim: Georg Olms 1964-rpt), pp. 193–194.

4. Warren Van Egmond, "The algebra of Master Dardi of Pisa," *Historia Mathematica*, 1983, 10: 420.

5. f. introduces the numerous references in my text to folio numbers in the Tempe manuscript.

6. Warren Van Egmond, *The Commercial Revolution and the Beginnings of Western Mathematics in Renaissance Florence, 1300–1500* (PhD dissertation, Indiana University, 1976), p. 411.

7. Barnabas Hughes, "Learning algebra in 14th century Italy," in Nancy Van Deusen and Alvin Ford, eds., *Paradigms in Medieval Thought—Applications in Medieval Disciplines* (Lewiston, NY: Edwin Mellen Press [Medieval Studies Volume 3], 1990), pp. 6–7.

8. Hughes, "Algebra Codex," pp. 168–169.

9. Warren Van Egmond, "The algebra of Master Dardi of Pisa," *Historia Mathematica*, 1983, 10: 402–417.

10. John Wallis, *A treatise of algebra, both historical and practical* (London 1685), p. 198. For information about Harriot, see E.G.R. Taylor, *The Mathematical Practioners of Tudor & Stuart England* (Cambridge: Cambridge University Press, 1968), pp. 182–183.

11. Barnabas Hughes, ed., "Gerard of Cremona's Translation of al-Khwarizmi's *al-Jabr*: A Critical Edition," *Mediaeval Studies*, 1986, 48: 236–241.

12. Baldassarre Boncompagni, ed., *Scritti di Leonardo Pisano*, 2 vols., Vol. 1: *Il Liber Abbaci di Leonardo Pisano* (Rome, 1857), pp. 408–409.

13. T. Richard Witmer, ed., *The Great Art or The Rules of Algebra by Girolamo Cardano* (Cambridge MA: M.I.T Press, 1968), p. 9.

14. *The original text of the rule alone*: "Qvando li numeri sono vguali ali chu e a R de chu, else vole partire li numeri in laquantitade deli chu non mentoadi auer R e quello chene viene salva. Poy moltiplicha laquantitade non mentoadi auer R in si medema e parti laquantitade deli chu mentoadi auer R in quella moltiplichation, elo $\frac{1}{4}$ dequello chene viene çonçi sopra lo partimento chefo salvado, ela R de quella somma minus la R dequello $\frac{1}{4}$ chetu çonçessu, çoe la R del $\frac{1}{4}$ chene viene partando li chu mentoadi aver R in la moltiplichatione deli chu non mentoadi auer R, viene a essere la R del chubo. E moltiplicha questa R in si medema viene a essere lo chubo, ela R chu de questa moltiplichation viene a essere la C." (f. 48v; note: R = radice; C = cosa).

15. "e elne viene a esse la R del chu. La soa moltiplichation viene a essere lo chubo ela R chubo de quella moltiplication viene a essere la C." (ff. 48v, 84r, 84v, 85r [has *valer* in place of *essere*], 89v, 90r, 90v).

16. Witmer, *The Great Art*, passim.

17. *Ibid.*, pp. 121–124, n. 6.

18. The original text of the rule alone: "Qvando le C eli Ç eli chu sono vguali al numero else vole partire tuta la adequatione perla quantitade deli chu e poy partire la quantitade dele C in la quantitade deli Ç e quello chene viene redu a R chu e quella moltiplichatione çonçi sopra lo numero chete viene inla adequatione ela R chu de quella somma minus

lo partimento chete vene partando laquantitade dele C in la quantitade deli Ç. viene a valer la C." (f. 121v; note: Ç = censo)

19. For the text, see Raffaella Franci and Marisa Pancanti, eds., *Il Trattato d'Algibra* (Siena: Universita degli Studi 1988, Quaderni del Centro Studi della Matematica Medioevale, n. 18), pp. 99–105. For discussion, see Raffaella Franci and Laura Toti Rigatelli, "Fourteenth-century Italian algebra," in Cynthia Hay, ed., *Mathematics from Manuscript to Print 1300–1600* (Oxford: Clarendon Press, 1988), pp. 21–22.

20. E nota che selte auenisse C e numeri da vna parte vgual a niente como po essere alchuna adequatione per molti traxi che auene digando che in molti sia deferentia ouer oltre cose debi trar li numeri di caschauna dele parte et aueray C vgual a niente minus numero che tu debi partire lo numero che minus per la quantitade dele C. E quello chene viene tanto sara debita la C ponamo chelte venisse $\frac{5}{C}$ e $\frac{15}{N}$ vgual a niente, parti 15 per 5 elne viene 3 e tanto sara debita la cosa. (f. 121r. The apparent fractions in the next to last line show Dardi's way of writing $5x + 15$.)

21. In his text *Flos super solutionibus quarundam questionum*, Fibonacci interpreted a negative root in a financial loss as a debt, "Hanc quidem quaestionem insolubilem esse monstrabo, nisi concedatur, primum hominem habere debitum," Boncompagni, *Scritti,* II, p. 238, quoted in David Eugene Smith, *History of Mathematics,* 2 vols., II (1925; rpt. New York: Dover, 1958), p. 258 n.4.

II

Historical Studies

From the Scientific Revolution to the Present

The Early Geometrical Works of Marin Getaldić

Žarko Dadić
University of Zagreb

The Croatian mathematician and physicist Marin Getaldić (1568–1626) was born in Dubrovnik where he spent most of his life. At the end of the sixteenth century he traveled around Europe, spending some time in England, France, and what became Belgium and Italy. Five years later, at the beginning of the seventeenth century, he returned to Dubrovnik, where he died in 1626. From 1603 to 1613 he published five works on mathematics and one on physics, dealing with specific weights. At he same time he conceived his comprehensive mathematical work, *De resolutione et compositione mathematica,* which he finished shortly before he died. It was published posthumously in Rome in 1630.[1]

In Paris, Getaldić met François Viète, who greatly influenced his work in mathematics. This influence was particularly reflected in Getaldić's consistent application of Viète's algebraic analysis in his *De resolutione et compositione mathematica,* as well as in his decision to solve certain mathematical problems from his earlier works again. Indeed, the chief difference between Getaldić's *De resolutione et compositione mathematica* and his earlier mathematical treatises is that he used Viète's algebraic analysis only in the former. Since in this respect his earlier works can be regarded as a whole, we shall present them here. They are as follows: *Nonnullae propositiones de parabola* (1603), *Variorum problematum collectio* (1607), *Supplementum Apollonii Galli seu exsuscitata Apollonii Pergaei tactionum geometriae pars reliqua* (1607), *Apollonius redivivus seu restituta Apollonii Pergaei inclinationum geometria* (1607) and *Apollonius redivivus seu restitutae Apollonii Pergaei De inclinationibus geometriae, liber secundus* (1613).

Getaldić's *Nonnullae propositiones de parabola,* (1603)[2] resulted from his great interest in the construction of parabolic mirrors which could be used for burning, but is basically a work about mathematical problems. His study of conic sections for the construction of such mirrors, led him to conclude that all parabolas obtained by cutting the right-angled, the acute-angled, the obtuse-angled, and the oblique cones are congruent to one another.

To prove his proposition, Getaldić used theorems from the first four books of *Conics* by Apollonius of Perga, the only one of these books known to him. But since they did not suffice, he proved certain propositions independently and demonstrated some theorems which were in other books of Apollonius that were unknown to him. His proofs are more exhaustive than those of Apollonius of Perga, because he proved the theorem for any diameter and conjugate chords of a parabola, and not for only a specific case of an axis as did Apollonius. The new arrangement

115

of his explanations helped him conclude that all parabolas obtained by cutting a cone are congruent to one another.

In the fourth theorem Getaldić tried to prove that a parabola obtained as a section of any cone was congruent with a parabola obtained as a section of a right angled cone. This proof has a limited value being somewhat incomplete. That is, Getaldić first should have assumed that there are two kinds of parabolas, the one obtained as a section of a right cone and the other obtained as a section of an oblique cone, and then prove their congruence. So his proof refers only to parabolas obtained as sections of right cones and not as sections of oblique cones perpendicular to the plane of symmetry.[3]

Convinced that he had proved the congruence of all parabolas, Getaldić formulated the first part of his main conclusion in optics in his second corollary. He asserted that, since the parabolas obtained from any cone are congruent, and since the parabolas obtained as sections of a right-angled cone are suitable for the construction of mirrors for burning, all parabolas are suitable for this purpose, that is, those obtained by sections of acute-angled, right-angled, and the obtuse-angled cones and even oblique ones. The second part of his conclusion ensued from the fifth theorem. Getaldić attempted to prove that all sun rays falling on the mirror parallel to the axis are reflected in one point which is removed from the apex of the parabola by one-fourth of the parameter.

During the sixteenth and seventeenth centuries interest in ancient works on mathematics increased and several mathematicians tried to restore and reconstruct works that had been lost, drawing upon the quotations and references in other works of ancient mathematicians.[4] The works of Apollonius of Perga were partly unknown or lost. The preface to the seventh book of *Mathematical Collection* by Pappus outlined the contents of Apollonius's treatises *On tangencies* and *On inclinations*. Here mention is made of problems Apollonius discussed and tried to solve. They were set forth and formulated in such a complicated way, however, that various interpretations were possible.

Viète restored *On tangencies* in his *Apollonius Gallus seu exsuscitata Apollonii Pergaei Περίεπαφων geometria*.[5] Viète reconstructed ten problems of Apollonius, which today are formulated as follows: construct a circle which passes through m points, touches n straight lines and p circles such that $m + n + p = 3$. If zero and whole positive numbers are used for m, n, and p, we obtain ten problems that Viète tried to solve.

When Viète published this work in Paris, Marin Getaldić was there, and the two men could have worked together. Getaldić was well acquainted with Viète's writ-

ings. His restoration of Apollonius's treatise *On tangencies* prompted Getaldić to write *Supplementum Apollonii Galli*,[6] which was published in Venice in 1607. The preface to the seventh book of the *Mathematical Collection* by Pappus led Getaldić to conclude that Apollonius's treatise *On tangencies* contained more than ten problems, actually six more, which today are formulated as follows: construct a circle of a given radius which passes through m points, touches n straight lines, and p circles so that $m + n + p = 2$. If we use zero and whole positive numbers for m, n, and p, we obtain six problems which Getaldić tried to solve. Getaldić showed that these problems had been discussed by Apollonius by quoting a passage about Apollonius's work *On tangencies* from Pappus's text translated by Frederico Comandino. He concluded that with his addition, the whole treatise *On tangencies* would be restored. Dissatisfied with particulars of Viète's solutions, Getaldić also gave his solution of the eight theorems from Viète's book.

Getaldić's amendment to Viète's restoration must have encouraged him to reconstruct another work by Apollonius, *On inclinations*, which he entitled *Apollonius redivivus seu restituta Apollonii Pergaei inclinationum geometria*.[7] He published it in Venice in 1607 when his first restoration was published.

As Getaldić understood Pappus's text, Apollonius's two books on inclinations included five problems: Four in one book and one in the other. Getaldić seemingly wanted to analyse all five problems in one volume, but his work for the Dubrovnik Republic and the journey to Constantinople distracted him from finishing the last problem (though he had already solved it in principle) and publishing it together with the other four. He published the fifth problem separately as the second book and entitled it *Apollonius redivivus seu restitutae Apollonii Pergaei De Inclinationibus geometriae, liber secundus*.[8] This book was published in Venice in 1613.

Though Getaldić called his work *Apollonius revived*, all solutions in it were his own. Since Pappus's formulations were distorted, Getaldić had to start from the beginning. He had to formulate the problems and solve them. Though Getaldić put great effort into this work, it was not suggested in the title. Undoubtedly Getaldić was the first to formulate Apollonius's problems on inclinations, and his formulations are not just an encouragement but also a basis for later restorations as we shall see below.

Getaldić's restorations of Apollonius's works are important, and so are his other writings. He used the synthetic procedure, that is construction, to solve problems in restoring Apollonius.

In ancient Greece, all mathematicians solved geometrical problems by construction, starting from the given values and working towards the desired ones, and then the construction had to be proved. This is the synthetic procedure, and the construction itself is called the synthesis. In a simple case, such a construction can easily be found, but in more complicated cases, this is not an easy matter. However, the construction procedure can be conducted more easily if first the relations between the given and desired values are considered, and the ancient Greeks seem to have approached the problems in this way. They never described the steps that led them to the construction, and the process of finding a solution was never revealed in their writings.

The hidden process of finding the construction was, in fact, an analysis of the problem and a study of the relations between the given and the desired magnitudes. Probably under the influence of Aristotle, Greek mathematicians gradually started to mention this procedure in their texts. In his *Elements,* Euclid put forward only the construction of the problem, or the synthetic solution, without any reference to an analysis, but in his *Data,* he mentioned the analysis of the problem.

Later on, other mathematicians, including Apollonius of Perga, placed greater emphasis on the analysis of a problem. Some of Apollonius's works presented the analysis, while others included only the synthesis. At the turn of the sixteenth century Viète introduced algebraic analysis, a more efficient method of solving problems by construction. Getaldić obviously became familiar with algebraic analysis since he used it extensively in his *De resolutione et compositione mathematica* but not in his earlier works.

While restoring the two works by Apollonius, Getaldić had to decide on the method. He seems to have wanted to restore them in the manner that corresponded to the original. Adding the algebraic analysis to the construction of Apollonius's problems was tempting but that was out of the question because it was a new achievement in mathematics. However, Getaldić could have used the geometrical analysis, indicated it in the text, and on that basis developed the synthesis or the construction of a problem. His indication of steps to the construction, or the geometrical analysis and the construction, would have been justified, because Apollonius of Perga might have done either or both. Though Getaldić could not have been certain of that, he chose to present only the synthesis. He did his best to make his three books containing the two restored works of Apollonius well balanced and to present only the construction, which does not mean that he did not use algebraic analysis while looking for a solution.

Getaldić worked on restoring Apollonius and wrote his *De resolutione et compositione mathematica* at the same time. The problems he solved regarding Apollonius were included in his main work, although the solutions in them differed. In the restoration of the Apollonius, Getaldić brought forward only the construction, while in *De resolutione et compositione mathematica* he included the algebraic analysis as well. Thus his *De resolutione et compositione mathematica* shows how he reached the solution. During his work on Apollonius, Getaldić also studied a collection of problems entitled *Variorum problematum collectio,* which does not contain algebraic analysis. Many of these problems are included in his *De resolutione et compositione mathematica.*

Getaldić certainly had difficulties in solving Apollonius's problems, first of all in finding all the cases of the solutions. Fortunately, the first version of the solutions to the first two problems from *Apollonius redivivus* (1607) has been preserved. Getaldić sent these two problems and their solutions without the geometrical analysis to Galileo Galilei to review and give his opinion, so that today they can be found in Galilei's legacy in Biblioteca nazionale centrale in Florence (Gal. 110, pp. 6r–8r).[9] This version makes clear that it was with great difficulty that Getaldić reached all cases of the solutions. In comparing the solutions of the first problem in the first version of the manuscript, *Apollonius redivivus* (1607) and *De resolutione et compositione mathematica* (1630), one can see that they are almost identical as far as the construction is concerned. The solution in the 1630 work however presented the algebraic analysis first from which a porism was deduced. The synthesis of the problem was deduced with the help of a porism. In the synthetic solution of 1607, Getaldić did not refer to the porism in the construction, but the construction was deduced as if the porism had been included. This is not a lack in the construction, because the porism was just a guide in the construction and the proof of the construction appears separately. If Getaldić had not used algebraic analysis in the construction in the 1607 work, this construction would hardly have been identical to the one presented in the 1630 work.

The second problem that Getaldić sent to Galileo was considerably changed both in approach to the solution and in the addition of new cases in the published version. The text sent to Galileo presented only three cases of the solution, while the 1607 text quoted five. In *De resolutione et compositione mathematica* Getaldić presented six cases of the solution to the same problem. This shows that Getaldić used algebraic analysis more effectively in finding new cases of a solution.

It follows that Getaldić must certainly have employed algebraic analysis to solve the problems in the restored works of Apollonius. Paul Guldin's letter of February 24, 1617, written to Getaldić and preserved in Vatican Library (Vat. 6921, pp. 103r–104r) also points to that.[10] Getaldić wrote to Guldin that he found difficult synthetic solutions by means of the algebraic analysis, namely that he had found the conditions of the problem and different cases by the algebraic method. In his reply Guldin encouraged Getaldić, referring to his *De resolutione et compositione mathematica*, "to open the sources whence so many proofs, true cases, determinations and deductions have flowed, which we have admired so much in the second book of Apollonius revived." Guldin also wrote that Getaldić had promised him "to describe in detail the technique, the method and the way to do it in the book on mathematical analysis and synthesis."[11] One may conclude that Getaldić deliberately presented the restorations of Apollonius's works in the synthetic form regardless of whether he used the geometric or algebraic analysis while solving problems.

Getaldić had understood and formulated the fifth problem of Apollonius when he prepared the first four problems for publishing, and so he published the formulation of the fifth one at the end of his work *Apollonius redivivus*. This formulation prompted Alexander Anderson to solve it himself. In Paris in 1612, Anderson published the treatise *Supplementum Apollonii redivivi*[12] in which he solved this fifth problem according to the Getaldić formulation. In the meantime Getaldić solved the problem, and was going to publish it when George Strachan came to Dubrovnik bringing Anderson's treatise. Getaldić discussed it in the preface to his work *Apollonius redivivus seu restitutae Apollonii Pergaei De inclinationibus geometriae, liber secundus* (1613) dealing with Apollonius's fifth problem on inclinations. A free formulation of this problem is:

> Two semicircles which have their basis (diameter) on the straight line are given. The given length which lies on the straight line passing through the intersection of one semicircle with the basis must be put between these semicircles.

This fifth problem is far more complicated than the other four problems in Apollonius's work on inclinations. The problem allows many different solutions and positions of the semicircles in relation to each other. Getaldić solved it synthetically and did not mention analysis. Anderson started differently with an analysis of the problem, regarding the desired magnitude as the given one, and finding relations between particular magnitudes. Samuel Horsley in the work *Apollonii Pergaei Inclinationum libri duo* pub-

lished in Oxford in 1770 considered this analysis algebraic though given in the geometrical form of ancient mathematicians. Anderson's analysis can indeed be presented in algebraic form with algebraic symbols, and the result can be deduced by means of an algebraic equation. However, this was not Anderson's approach. He formulated several ratios between the lengths viewing them purely geometrically in the way ancient Greek geometers understood the ratio between lengths. Thus he obtained a geometrical ratio with the unknown length a ratio which could also be written in algebraic form, but Anderson did not see it as such. Proceeding from a geometrical ratio, Anderson gave a geometric solution.

Getaldić showed great interest in Anderson's solution, as he said in the preface to his second book of restorations, entitled *Apollonius redivivus*. He reproached Anderson for having chosen too uncertain a way to solve the equation, or the ratio, but he did not explain why he considered this procedure uncertain.

After Getaldić published his restoration of Apollonius's fifth problem on inclinations in 1613 and Anderson published his solution to the same problem in 1612, Anderson published another treatise which dealt with the same problem. His treatise Αἰτιολογία *Pro zetetico Apolloniani problematis a se iam pridem edito in supplemento Apollonii redivivi,* published in Paris in 1615, was dedicated to Marin Getaldić.[13] The author directly addressed Marin Getaldić's work. Among other things, Anderson mentioned how his friend Strachan was entertained by Getaldić, in Dubrovnik when Strachan brought him Anderson's treatise.

Because of Getaldić's remarks or for some other reason, Anderson tried to solve the same problem again, this time he employed the algebraic approach within Viète's algebraic method. He pointed out that this new solution to the problem favoured the use of the algebraic method and that this method more clearly revealed the solution to the problem. This shows that Anderson did not care to present the solution to Apollonius's problem in the way Apollonius himself may have. He was more interested in the solution and accepted any method that promised to lead him to his goal. This is the main difference between Getaldić and Anderson. Getaldić tried to solve this problem by the algebraic method within Viète's method and included it in *De resolutione et compositione mathematica*. However, Getaldić chose this method for different reasons. He solved this problem and others in the restored works of Apollonius by means of an algebraic method and included them in his *De resolutione et compositione mathematica*, because this method corresponded to the methodological concept of his work. Anderson, by contrast, did not try to solve the fifth problem by the algebraic method because the methodological concept of his work required it. He did it to arrive at a solution more easily and clearly.

Getaldić did not mention Anderson or Anderson's algebraic solution of the problem, though he adopted the algebraic method in his *De resolutione et compositione mathematica*. This indicates that Getaldić did not know of Anderson's treatise, because it is unlikely that he would have neglected to mention it.

At the end of the eighteenth century two restorations of Apollonius's work appeared: one was *Apollonii Pergaei inclinationum libri duo*, which Samuel Horsley published in Oxford in 1770, and the other was *A restitution of the geometrical treatise of Apollonius Pergaeus on inclinations*, which Reuben Burrow published in London in 1779.[14]

Horsley's approach to the solution of Apollonius's problems completely differs from that of Getaldić. While Getaldić favoured the synthetic solution, Horsley tried to solve the problems within the framework of Viète's analysis and synthesis. In his work he called these two components of the solution *Analysis* and *Compositio*. Horsley largely followed Getaldić's formulations as well as the numeration of the problems. The difference was in Horsley's division of the first problem from Getaldić's *Apollonius redivivus* into two problems, when the point is inside the given circle and when the point is outside the given circle. In the second book, apart from the general fifth problem in Getaldić's formulation, he solved a particular case of concentric circles.

In the second book Horsley mentioned Getaldić twice. He first mentioned him in connection with the algebraic method and the construction of a problem.[15] Horsley maintained that Getaldić had reached solutions by means of the algebraic method, and only afterward had he applied the geometric construction. Horsley's conclusion was correct and can be proved by several facts we have already pointed out. Horsley also mentioned Getaldić in connection with the difficulties he had in solving Apollonius's fifth problem on inclinations.[16] He said that Getaldić had not had time to solve it because he was too busy with State affairs, and that this gap was filled by Anderson in his *Supplementum Apollonii redivivi*. He outlined Anderson's method but was not acquainted with Getaldić's restitution of Apollonius's fifth problem.

Reuben Burrow mentioned Getaldić in the preface to his *Restitution* (1779) and stressed that only Getaldić and Horsley had studied these problems of Apollonius.[17] He had decided to restore Apollonius's work on inclinations because he was well acquainted with Pappus's mathematical collection and had formulated the problems himself. Burrow understood these problems in a way that differed from Getaldić, and consequently his formulations were unlike those of Getaldić. Only one of Burrow's problems corresponded to one of Getaldić's: Burrow's first problem and Getaldić's second. At the end of his solution to this problem Burrow said that it did not essentially differ from that of Getaldić. For the first time in the history of restoration of these problems, Getaldić's formulation had not been used as a basis. Burrow did not accept Getaldić's interpretation of Pappus's text translated by Comandino, and he formulated the problems as he understood them. Burrow's independent formulations show how intricate and even unintelligible Pappus's text was, which was also Getaldić's opinion. Getaldić said that it was more difficult for him to understand these formulations than to solve them. They also show that many different interpretations of Pappus's text are possible.

Burrow criticized Horsley for having used the algebraic method which hardly suited the original work of Apollonius. Burrow consistently used the ancient Greek geometrical analytical-synthetical method with the proof and the determination of a problem.

J. W. Camerer wrote about Apollonius's treatise on tangencies, Viète's restoration, and some lemmas of Pappus, in *Apollonii de tactionibus quae supersunt ac maxime lemmata Pappi in hoc libros Greace nunc primum edita*, published in Amsterdam in 1795. But, Camerer did not include those problems restored by Getaldić. Camerer's historical overview of the solutions to these problems only mentioned that Getaldić had supplemented Viète's restoration with his *Supplementum Apollonii Galli*. He was, therefore, acquainted with Getaldić's supplement and yet he did not include it in his work. He knew more than that, namely that Daniel Schwenter (1585–1636) had tried to solve Getaldić's problems on tangencies, more precisely the problems of three circles, in his *Geometriae practicae novae libri IV* (Book II, problem 21, 22, 23).

Getaldić's restoration of Apollonius's work on tangencies was included by John Lawson in his work *The two books of Apollonius Pergaeus concerning tangencies as have been restored by Franciscus Vieta and Marinus Ghetaldus* of which the first edition was published in Cambridge in 1764 and the second in London in 1771. On the basis of Viète's and Getaldić's works, Lawson completely gave Apollonius's treatise on tangencies and translated Viète's and Getaldić's solutions and formulations of the problems into English. Thus Lawson's book included almost the whole text of Getaldić in English translation, slightly adapted and polished since Lawson left out some less important sentences. Lawson listed Getaldić's solutions and problems in the same order as presented by Getaldić and put them forward at the beginning of his book as problems 1–6. They were followed by Viète's problems and solutions. Viète's eighth problem, which Getaldić solved independently, can be found also in Lawson's book, and here Lawson strictly followed Getaldić's solution, that is, he literally translated Getaldić's text into English. He added in a note that this construction was simpler than that of Viète. Unfortunately, nowhere in the book did he state clearly which were Getaldić's and which Viète's solutions so that one distinguishes them only if familiar with Getaldić's and Viète's work, or the group of problems studied by each of them. To these problems, which were considered the subject of Apollonius's treatise on tangencies, Lawson added six more problems as his addition. Lawson's problems were prompted by Getaldić's problems and were modified in the following way: to find a geometrical locus of the centres of circles each of which passes through m points, touches n straight lines, and p circles in such a way that $m+n+p = 3$. These problems were, therefore, a kind of enlargement upon Getaldić's problems.

In *Cursus mathematicus*[18] whose first volume was published in Paris in 1644, Pierre Herigone reviewed a number of older and more recent works on mathematics including Getaldić's *Apollonius redivivus*. He utilized some of them to explain his method of solving geometrical problems, which in his opinion was new, short and simple. The main characteristic of the method was the presentation of the solution by symbols instead of customary textual explanation. Symbols were listed and explained at the beginning of the work, they represented geometrical notions and operations such as intersecting, touching, etcetera. We can describe this method as a new kind of geometric expression. Herigone's idea is interesting especially because it appeared in the seventeenth century. Herigone applied his method to Euclid's *Elements* and *Data* and to the restored works of Apollonius. He used three restorations by Willebrord Snell, one by Viète dealing with tangencies and finally Getaldić's restoration of *Apollonius redivivus*. *Cursus mathematicus* was in Latin with a parallel French translation including the preface, introductory remarks, and formulations of the problems, which were given in the original Latin version and in the French translation. The rest of the book, namely the solutions, constructions and proofs of the problems were presented by means of Herigone's symbols.

What applied to the whole work, also applied to Getaldić's *Apollonius redivivus*. It was neither reprinted nor translated, but rather had a new form of expression. Through Herigone's publication, Getaldić's work, expressed in symbols, may have become more widely known than before. Judging by titles of Apollonius's restorations, one might conclude that Herigone included just one restoration by Getaldić, dealing with inclinations. This is not true. In the first volume of his *Cursus*, Herigone included the restoration of Apollonius's treatise on tangencies, but ascribed it completely to Viète. He listed sixteen problems of which the first six solutions were Getaldić's from his *Supplementum Apollonii Galli*, and he included Getaldić's drawings, though this was never mentioned in the text. Getaldić's six solutions are followed by Viète's restorations of Apollonius's ten problems numbered 7 to 16.

Getaldić's *Variorum problematum collectio*[19] was prepared for publishing after his return to Dubrovnik in the period from 1603 to 1606. The preface bears the date 1606 and the work was published in Venice the next year. Here Getaldić solved 42 geometrical problems, of which five were taken from Regiomontanus's *De triangulis planis et sphaericis* published in Nuernberg in 1533. Some problems were suggested to Getaldić by mathematicians Christoph Clavius and Christoph Grienberger, with whom he exchanged views all his life. Some were suggested by Jakov Restić, his maternal relation, and some were produced by Getaldić himself. We know which problems were taken from Regiomontanus's work, because Getaldić indicated their origin in the margins, but we cannot say who of the above mentioned persons contributed the other 37 problems.

Getaldić's work includes three groups of problems and each of them was solved in a different way. Common to all is that they were not solved by Viète's algebraic method, but by the ancient Greek geometrical methods or by a method directly based on them. The first and the largest group of problems was solved by the geometric synthetic procedure. The second group, problems 12–17, were solved by the analytic-synthetic geometric procedure. The third group, problems 1–2, 5–11, and 18–23, were solved using a procedure which was basically the same synthetic procedure used in the first group though with the addition of the so-called *consectarium*.

Contrary to expectations, these three groups were not presented one after the other, but only some problems solved in the same way followed each other. In other words, the problems were arranged in such a way that in certain points of construction they built upon solutions of the previous problems, or were considered variations of the previous problems. Getaldić grouped together the problems related to triangles, or to lenghts: the first 22 problems deal with triangles, some of them only with right-angled triangles; the next two problems, 23 and 24, deal with quadrangles;

while problems 25–42, with the exception of 26, require that a given length be intersected in a given ratio.

As I have pointed out, the synthetic procedure's aim is to find the desired magnitudes from the given magnitudes by means of a construction. If a geometrical analysis is indicated, it shows the way to the constructive solution of a problem. On the basis of symbolic algebra, Viète transforms geometric analysis and conducts it in the algebraic way so that it becomes algebraic analysis. From the algebraic analysis ensues a clear algebraic relation between the given and desired magnitudes, which later is called porism and is used in the construction of a problem. This procedure was emphasized in Getaldić's *De resolutione et compositione mathematica.*

The consectarium, which Getaldić used in *Variorum problematum collectio,* had its origin in Viète's works, among others, in his treatise *Effectionum geometricarum canonica recensio.* Here Viète used a synthetic procedure to solve a particular geometrical problem, and from the geometrical relations in this construction he deduced a relation between the given and desired magnitudes. To present this relation, Viète formulated a statement called a consectarium. This formulation can be expressed by means of an equation, though Viète did not use it.

In his *Variorum problematum collectio,* while solving the problems with the addition of a consectarium, Getaldić proceeded in the same manner as Viète. That is from a construction he deduced a relation between the given and the desired magnitudes. He called such a formulation a consectarium as had Viète. In contrast to Viète, Getaldić introduced a numerical example following the consectarium, which gave his method a particular importance. A comparison with the solutions of the same problems set forth in *De resolutione et compositione mathematica* leads to some important conclusions.

Getaldić's formulation of the consectarium is formally the same as porism in the algebraic analysis used in his *De resolutione et compositione mathematica.* It is a truth which contains the given and the desired magnitudes, but differs from porism in that is follows from synthesis and not the algebraic analysis. Consectarium has a general character and can be applied to any geometrical magnitudes given in a problem, which is the same as porism used in Getaldić's *De resolutione et compositione mathematica.* Since every consectarium was followed by a numerical example, and the same can be said of porism in his last work, the desired magnitudes were numerically derived from the given magnitudes by means of a consectarium.

In Getaldić's *Variorum problematum collectio* consectarium can be followed by numerical or geometrical solu-

tion, just as porism. Since porism in algebraic method is expressed in words as well as by an equation, the rise to Viète's species is more evident, and, being in general magnitudes, they can be numerical or geometrical. In consectarium Getaldić does not produce any equation that relates the given to the desired magnitudes, but the general formulation of the proposition following from the construction allows geometrical and numerical realizations, just as the porism. Getaldić places numerical realizations after the consectarium; a geometrical construction is also possible, but we need not consider it because it was put before the consectarium.[20]

Getaldić must have discussed an analysis of the problems to which a consectarium was added, just as he did for every synthetic solution. He did not mention it, because he reduced the solution of a problem to the formal structure as Viète did. Viète produced such a structure in his *Effectionum geometricarum canonica recensio.* Later he developed his algebraic analysis and symbolic algebra, particularly in *In artem analyticen isagoge* published in 1591. Getaldić did a similar thing and included the procedure with an indication of a consectarium in his earlier *Variorum problematum collectio,* which otherwise contained only geometrical solutions.

There is reason to believe that Getaldić did this on purpose. He conceived *Variorum problematum collectio* and *De resolutione et compositione mathematica* at the same time. The problems he solved here were solved again in *De resolutione et compositione mathematica.* That he occupied himself with both works simultaneously is shown in the preface to his collection where he said that he would solve some problems algebraically in *De resolutione et compositione mathematica.* In *Variorum problematum collectio,* Getaldić set forth the solutions he found by procedures that were not part of Viète's algebraic method, and in his *De resolutione et compositione mathematica* he wanted to show the possibilities of the analytic-synthetic method in general lines. He wrote *Variorum...* much earlier than the latter work. That is why Getaldić's *Variorum...* holds a central place in the earlier stage of his creative work. At the time of its publication, Getaldić considered it one of his most mature works, as he put it in the dedication, where he explained the reasons he dedicated this work to his benefactor Marin Gucetić, with whom he travelled around Europe. So it happened that problems 12 through 17 solved by the geometrical analytic-synthetic method were suitable for both works because they fitted in with the methodological conceptions. They were simply transferred into the work on mathematical analysis and synthesis, with a group of problems which did not belong to algebra and therefore could not be solved by Viète's algebraic method. In *Variorum problematum collectio*, they form a group of problems on acute-angled triangles.

While solving a problem, it is important to determine its conditions or limitations so that it can be solved. A problem can be solved only if conditions are set for the given magnitudes or their relations. In the analytic-synthetic method Getaldić used in *De resolutione et compositione mathematica*, these conditions were often obvious or could be deduced from porism. Possibly, in his *Variorum problematum collectio* Getaldić determined the conditions independently of porism, and in some problems they were not determined at all. Cyriaque de Mangin and Jacob Christmann reproached him for that and objected to his solution of the first problem in *Variorum problematum collectio*.

This problem was taken over from Regiomontanus's *De triangulis planis et sphaericis* (1533). Getaldić's solution was criticized by Cyriaque de Mangin in *Problemata duo nobilissima* (Paris, 1616)[21] and by Jacob Christmann in *Nodus gordius* (Heidelberg, 1612).[22] Getaldić solved this problem again in *De resolutione et compositione mathematica*, adding two conditions for the given magnitudes so that the problem could be solved. Though Getaldić did not include these conditions in *Variorum problematum collectio*, he rejected their criticism, saying that the two authors did not determine them either. According to Cyriaque de Mangin, the problem should have been determined, which is easily done by the analytical method, and it is also possible to obtain the analytically expressed condition as was required by Cyriaque de Mangin. Later Getaldić's friend Alexander Anderson discussed the solution of this problem with reference to Regiomontanus's and Getaldić's solution, in *Exercitationum mathematicarum decas prima* published in Paris in 1619.[23]

In *De resolutione et compositione mathematica* Getaldić tried to defend himself from Cyriaque de Mangin's and Christmann's attacks, but he did not mention this work of Anderson, nor his analysis. This is all the more interesting, because further on in his work Getaldić defended Viète, who was attacked by Cyriaque de Mangin, even though, as Getaldić himself put it, Anderson had already stood up in Viète's defence. Here Getaldić most probably referred to Anderson's *Animadversionis in Franciscum Vietam a Clemento Cyriaco nuper editae brevis Διακριδις*, published in Paris in 1617. We can thus assume that Getaldić did not know about Anderson's work from 1619 in which his solution was discussed. This is not an unreasonable assumption since we know that at the time Getaldić completely broke off his relations with natural philosophers, mathematicians, and friends abroad.

Michelangelo Ricci criticized Getaldić's solution to the sixteenth problem of the same collection in his *Algebra*, which remained in manuscript form in the library of the Mathematical Institute in Genoa.[24] Michelangelo Ricci was a student of Galileo Galilei and a friend and colleague of Stjepan Gradić of Dubrovnik. Ricci's work was written in the spirit of Viète and was probably a kind of textbook on the new algebra. Ricci mentioned only a few other authors and did not quote their works, so that the citing of Getaldić's solution was an exception. In the part of his manuscript, which describes the advantages of algebra in solving geometrical problems and the determination of the problem by means of an algebraic analysis, he directly drew upon Getaldić's *De resolutione et compositione mathematica*. At the beginning of this passage, Ricci discussed the porism and the determination of conditions for a problem. If the algebraic method is used to analyse the conditions set for a problem, it is possible to distinguish which cases can be solved and under which conditions. To support his argument, Ricci quoted the second book of Getaldić's *De resolutione et compositione mathematica* in which numerous problems were solved by the algebraic method. Ricci also stressed a thorough examination of all the conditions, even if the problem was solved geometrically. To explain his views better, he again used Getaldić's *De resolutione et compositione mathematica* and quoted the seventh problem of the fourth chapter of the fifth book, which Getaldić solved by the geometric method exclusively. Ricci also added that it was sixteenth problem in Getaldić's *Variorum problematum collectio*. Ricci said that he had chosen this problem because Getaldić had not studied the conditions under which the intersection of two lengths was possible. An analysis of Getaldić's solution has proved Ricci right.[25]

Not only in this problem were the conditions not determined. This deficiency was evident in other problems solved in both works. As an example I quote problem 9 in *Variorum problematum collectio*, where the conditions were not determined though the solution was followed by a consectarium identical to the porism in the solution of the same problem in *De resolutione et compositione mathematica* (problem 2, second book). However, in the latter work Getaldić correctly deduced the conditions from the analytical procedure or porism. In *De resolutione et compositione mathematica*, the conditions were determined later because Getaldić supplemented and improved his work for many years after the publication of his *Variorum problematum collectio* and because he conducted the analytical procedures in the framework of Viète's algebraic method, as he had not always done while writing *Variorum problematum collectio*.

Endnotes

1. M. Ghetaldus, *De resolutione et compositione mathematica*. Romae: Ex Typographia Reverendae Camerae Apostolicae, 1630.

2. M. Ghetaldus, *Nonnulae propositiones de parabola*. Romae: Apud Aloysium Zannetum, 1603.

3. J. Majcen, *Spis Marina Getaldića Dubrovcanina o paraboli i paraboličnim zrcalima* (Treatise of Marin Getaldić from Dubrovnik on parabola and parabolic mirrors). Rad Jugoslavenske akademije znanosti i umjetnosti, 1920, Vol. 223, Zagreb, pp. 1–43.

4. Among mathematicians who restored Greek mathematical works in the seventeenth century, I can mention Willebrord Snell, Vincenzo Viviani, and Stjepan Gradić.

5. F. Viète, Apollonius Gallus seu exsuscitata *Apollonii Pergaei* Περ`επαρων *geometria*. Paris 1600.

6. M. Ghetaldus, *Supplementum Apollonii Galli seu exsuscitata Apollonii Pergaei Tactionum Geometriae pars reliqua*. Venetiis: Apud Vincentium Fiorinam, 1607.

7. M. Ghetaldus, *Apollonius Redivivus seu Restituta Apollonii Pergaei Inclinationum Geometria*. Venetiis: Apud Bernardum Iuntam, 1607.

8. M. Ghetaldus, *Apollonius Redivivus seu Restitutae Apollonii Pergaei De Inclinationibus Geometriae, Liber secundus*. Venetiis: Apud Baretium Baretium, 1613.

9. Ž. Dadić, J. Balabanić, *Prva verzija Getaldićevih restauracija dvaju Apolonijevih problema dostavljena Galileju* (The first version of Getaldić's restorations of two of Apollonius's problems were sent to Galileo). Anali Zavoda za povijesne znanosti JAZU u Dubrovniku, 1982, Vol. 19–20, Zagreb, pp. 7–18.

10. S. Gradius, *Quaedam meditationes geometriae diversis temporibus a me Stephano Gradio factae*, manuscript, Vatican Library, Vat. 6921, pp. 103r–104r.

11. J. Balabanić, Ž. Dadić, *Pismo svicarskog matematicara Paula Guldina Marinu Getaldiću* (The letter of Paul Guldin, the Swiss mathematician, to Marin Getaldić), Anali Historijskog odjela Centra za znanstveni rad JAZU u Dubrovniku, 1978, Vol. 15–16, Dubrovnik, pp. 87–96.

12. A. Anderson, *Supplementum Apollonii redivivi*. Parisiis: Apud Hadrianum Beys, 1612.

13. A. Anderson, Αίτιολογία Pro zetetico Apolloniani problematis a se iam pridem edito in supplemento Apollonii Redivivi. Parisiis: Apud Oliverium de Varennes, 1615.

14. S. Horsley, *Apollonii Pergaei Inclinationum libri duo*. Oxford 1770, R. Burrow, *A Restitution of the Geometrical Treatise of Apollonius Pergaeus on Inclinations*. London: C. Etherington, 1779.

15. S. Horsley, Ibid, p. 103.

16. S. Horsley, Ibid, p. 113.

17. R. Burrow, Ibid, p. 5.

18. P. Herigone, *Cursus mathematicus, nova, brevi, et clara methodo demonstratus, tomus primus*, Paris 1644.

19. M. Ghetaldus, *Variorum problematum collectio*. Venetiis: apud Vincentium Fiorinam, 1607.

20. Ž. Dadić, Some methodological aspects of Getaldić's mathematical works. *Historia mathematica*, 1984, Vol. 11, No 2, New York, pp. 207–214.

21. C. Cyriaque de Mangin, *Problemata duo nobillissima*. Parisiis: Apud Viduam Da Vidis le Clerc, 1616.

22. J. Christmann, *Nodus gordius ex doctrina sinuum explicatus*. Heidelberg 1612.

23. A. Anderson, *Exercitationum mathematicarum decas prima*. Parisiis: Apud Oliverium de Varennes, 1619.

24. M.A. Ricci, *Algebra del Sig. Michel'Angelo Ricci che f| poi Cardinale d. S.R.C.*, manuscript, Biblioteca matematica della Universit... di Genova, Biblioteca Gino Loria A-XXIX-29.

25. Ž. Dadić, *Utjecaj Marina Getaldića na Michelangela Riccija*. (Marin Getaldić's influence on Michelangelo Ricci), Dijalektika, 1968, Vol. 3, No. 4, Belgrade, pp. 105–114.

Empowerment through Modelling: the Abolition of the Slave Trade

John Fauvel
Open University

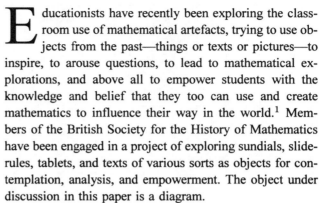

Educationists have recently been exploring the classroom use of mathematical artefacts, trying to use objects from the past—things or texts or pictures—to inspire, to arouse questions, to lead to mathematical explorations, and above all to empower students with the knowledge and belief that they too can use and create mathematics to influence their way in the world.[1] Members of the British Society for the History of Mathematics have been engaged in a project of exploring sundials, slide-rules, tablets, and texts of various sorts as objects for contemplation, analysis, and empowerment. The object under discussion in this paper is a diagram.

Only in the past few years have historians begun to take seriously the development of the important role of visual modelling of statistical and other data. Graphs, charts, and diagrams can be very powerful summaries of data and instruments of analysis, communication, and action and yet they are also poor relations, often regarded as inferior to linear reasoning. A well-known example is the pride with which Lagrange declared that there were *no figures* in his great *Analytical Mechanics* of 1788:

> There are no figures in this book. The methods that I demonstrate here require neither constructions, nor geometrical or mechanical reasoning, but only algebraic operations, subject to a regular and uniform development.

To some extent the devaluing of diagrams, as opposed to prose text or algebraic symbolism, is still the case today.[2]

We should be aware of this problem and ensure that this attitude is not inadvertently passed on to the next generation. Instead, we must enable today's pupils and students to realize the value and potential of graphical, non-linear display and argument as a part of their empowering mathematical repertory. What can we draw out from a fairly simple and accessible historical artefact in a classroom situation?

The artefact pictured comes from volume 1 of Thomas Clarkson's *The history of the rise, progress, and accomplishment of the abolition of the African slave-trade by the British parliament,* published in 1808. From a distance it looks like a blood system of arteries and capillaries, or perhaps some knarled deciduous trees in winter. But up close it turns out to be a set of rivers or streams, tributaries of estuaries labelled X and Y. To learn what this is about, we need a brief explanation of the book's context.

The abolition of the slave trade was one of the great radical causes of the late eighteenth century in Britain—in fact, *the* great moral crusade of the period—the 1790s version of Amnesty International or Friends of the Earth. After a prolonged campaign the House of Commons in

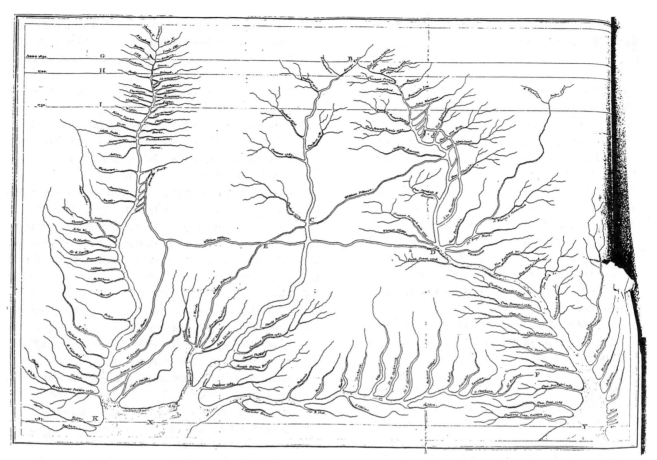

1807 passed the laws which made the slave trade illegal in British colonies, and the following year Thomas Clarkson (1760–1846), a prominent worker in the cause, published this history of the campaign and what led up to it. (Of course, making the slave trade illegal was not the same as preventing it, nor was it the same as abolishing slavery itself those things happened gradually over the next few decades; but it was a crucial and epoch-making achievement all the same.)

Clarkson's historical account is based on an explicit model or analogy:

> It would be considered by many, who stood at the mouth of a river, and witnessed its torrent there, to be both an interesting and a pleasing journey to go to the fountain-head, and then to travel on its banks downwards, and to mark the different streams in each side, which should run into it and feed it. So I presume the reader will not be a little interested and entertained in viewing with me the course of the abolition of the Slave-trade, in first finding its source, and then in tracing the different springs which have contributed to its increase. [...] In tracing the different streams

from whence the torrent arose, which has now happily swept away the Slave-trade,[3]

More than the historical *description*, the whole *analysis* is formulated in terms which can be represented in terms of the model. Clarkson's historical analysis is that the anti-slavery movement was brought about by the coming together, in 1787, of representatives of four *classes* of what he calls *forerunners and coadjutors,* separately concerned about the issue: namely, various English people; some English Quakers; American Quakers; and a fourth group which included Clarkson himself and William Wilberforce.

Clarkson gives a clear explanation of the diagram:

> The figure beginning at A and reaching down to X represents the first class of forerunners and coadjutors up to the year 1787, as consisting of so many springs or rivulets, which assisted in making and swelling the torrent which swept away the Slave-trade.
>
> The figure from B to C and from C to X represents the second class, or that of the Quakers in England, up to the same time. The stream on the right-hand represents them as a body, and that

on the left, the six individuals belonging to them, who formed the committee in 1783.

The figure from B to D represents the third class, or that of the Quakers in America when joined with others in 1774. The stream passing from D through E to X shows how this class was conveyed down, as it were, so as to unite with the second. That passing from D to Y shows its course in its own country, to its enlargement in 1787. And here I may observe, that as the different streams which formed a junction at X, were instrumental in producing the abolition of the Slave-trade in England, in the month of March 1807, so those, whose effects are found united at Y, contributed to produce the same event in America, in the same month of the same year.

The figure from F to X represents the fourth class up to 1787.

X represents the junction of all the four classes in the committee instituted in London on the twenty-second day of May, 1787.

The parallel lines G, H, I, K, represent different periods of time, showing when the forerunners and coadjutors lived. The space between G and H includes the space of fifty years, in which we find but few labourers in this cause. That between H and I includes the same portion of time, in which we find them considerably increased, or nearly doubled. That between I and K represents the next thirty-seven years. But here we find their increase beyond all expectation, for we find four times more labourers in this short term, than in the whole of the preceding century.[4]

This is, then, quite a sophisticated diagram, more subtle than it looks; there is a lot of information summarized in an accessible form (Clarkson was senior wrangler in 1783, that is, he placed first in the mathematical tripos at Cambridge, so he had received as good a mathematical education as any British person of his time). The reader can gain a global perspective from its overall characteristics, and can also focus in and find detailed information. This information is both about the particular writers and campaigners for the cause and also about the development of activity through time.

Clarkson went on to ponder lessons which the diagram afforded, about the Christian background and motivation of campaigners and the virtues of religious tolerance, and had various further reflections of an encouraging kind:

In looking between the first two parallels, where we see so few labourers, and in contemplating the great increase of these between the others, we are taught the consoling lesson, that however small the beginning and slow the progress may appear in any good work which we may undertake, we need not be discouraged as to the ultimate result of our labours; for though our cause may appear stationary, it may only become so, in order that it may take a deeper root, and thus be enabled to stand better against the storms which may afterwards beat about it[5]

Clarkson's grand model has become somewhat confused by this stage, as it turns into a tree in the heat of the moment! Nevertheless, what he goes on to say shows how deeply the model and his historical analysis are bound up with his political activity and the need to sustain the emotional strength of campaigners through reflecting upon the past:

... we discover the manner in which light and information proceed under a free government in a good cause. An individual, for example, begins; he communicates his sentiments to others. Thus, while alive, he enlightens; when dead, he leaves his works behind him. Thus, though departed, he yet speaks, and his influence is not lost. Of those enlightened by him, some become authors, and others actors in their turn. While living, they instruct, like their predecessors; when dead, they speak also. Thus a number of dead persons are encouraging us in libraries, and a number of living are conversing and diffusing zeal among us at the same time.[6]

Thus we begin to see what this has to do with empowerment. As a seasoned political campaigner, Clarkson clearly viewed his map as a means of rallying the troops and consolidating the achievements so far, by helping people to see how and where they were located in a tradition of radical protest.

There is an interesting footnote to our diagrammatic tale, an incident in its reception which highlights the clarity of judgement which the diagrammatic mode affords and forces on the analysis. The model turned out to be fiercely controversial in at least one quarter, namely the family of the other great anti-slavery campaigner William Wilberforce. Wilberforce's sons were greatly offended by Clarkson's chart, which they felt gave far too little prominence to their father's activities and influence. Thus the act of constructing such a diagram impels comparative judgements; as the Wilberforce family noticed, it cannot be a disinterested summary of facts but must carry critical evaluations of those facts—just as in history.

Let us move on now from the diagram, and put it in a wider historical framework. What is significant about what it represents? Why have I examined this diagram in such detail in the context of the history and pedagogy of mathematics? What, if any, is the benefit for today's schoolchildren or their teachers? Answers to these questions emerge in at least three contexts:

- the kinds of arguments used in the antislavery movement, and different ways in which these involve social mathematics
- the tradition of diagrammatic representations of concepts, and the use of these for further judgement or action
- the notion of "empowerment."

Arguments and Mathematics in the Antislavery Movement

Mathematics education can still learn much from the way mathematical ideas were used in past political and social arguments. Teachers need to be aware of what has or has not worked in the past, in order to see how successful equivalent ideas or practices might be in the present.[7] Several arguments were used in the anti-slavery campaigns which involve mathematics or mathematics education in one way or another. Three brief examples of these follow.

Clarkson on African calculators. One of the main arguments which anti-slavery campaigners had to contend with was that blacks were naturally inferior to whites. One way to defeat this prejudice was to produce examples of exceptional black achievements as counterexamples. Mathematical achievements were good for this purpose, as they were agreed to be objective and undeniable an example was given by Clarkson himself in an earlier essay. Here he discusses slave-trading practices in West Africa.

It is astonishing with what facility the African brokers reckon up the exchange of European goods for slaves. One of these brokers has perhaps ten slaves to sell, and for each of these he demands ten different articles. He reduces these immediately by the head into bars, coppers, ounces, according to the medium of exchange that prevails in the part of the country in which he resides, and immediately strikes the balance.

The European, on the other hand, takes his pen, and with great deliberation, and with all the advantages of arithmetick and letters, begins to estimate also. He is so unfortunate often, as to make a mistake; but he no sooner errs, than he is

detected by this man of inferiour capacity, whom he can neither deceive in the name or quality of his goods, nor in the balance of his account. Instances of this kind are very frequent: and it is now the general complaint of the captains sent upon the coast, that the African brokers are so nice in their calculations, that they can scarcely come off with a decent bargain.

I presume that instances of this kind will be received as proofs of the existence of their understandings, all arithmetical calculations being operations of the mind.[8]

Thomas Fuller. Another example used by anti-slavery campaigners to make the same point was a Virginia slave named Thomas Fuller (1710–1790), who was shipped from West Africa in 1724. Late in life he was discovered to have remarkable calculating abilities, and he was used throughout the nineteenth century as a counterexample to the charge that blacks were innately mentally inferior to whites.[9]

William Butler on sugar. My third example shows how anti-slavery concerns arose in William Butler's *Arithmetical questions* of 1795, a textbook devised to provide arithmetic exercises in a right-thinking context to influence the child's moral development. The book was designed for young ladies—Butler explicitly addressed the importance of female education. There were more than 300 questions, each focussing on some topic and making liberal reference to contemporary writers. One is about sugar:

No. 274. Sugar. Sugar is a very sweet agreeable saline juice, expressed from a kind of canes, or reeds, growing in great plenty in the East and West-Indies.[10]

(After a paragraph on the history of sugar, and how it reached the West Indies, he discusses its cultivation, in these terms.)

To cultivate the sugar-cane in the West-Indies, the wretched Africans are torn from their native land, in defiance of every principle of religion, humanity, and justice. When the European kidnapping manufacturers of human woe land on their coast; not only parents are dragged from their families, and children from their parents; and the most endearing ties of friendship, attachment, and relationship burst asunder; [...] but whole villages and towns are dispopulated or burnt, and every species of terror, dismay, and brutality, instantly pervades their unhappy country. [...]

As soon as the traffickers in human gore have gotten their complement on board the ship, it immediately makes off; when many of the miserable victims fall into such deep grief and despair, that they languish, sicken, and die in the passage: [...] Arrived at the destined plantations, the day hardly dawns when the whip resounds through those regions of horror; nor ceases, till darkness closes the scene, which day after day is renewed, till at length the miserable creatures expire beneath the lash, which in vain endeavours to rouse them to a renewal of their labour.

A celebrated French moralist well observes, that he cannot look on a piece of sugar without conceiving it stained with spots of human blood: [...]

The Abbé Raynal computes that, at the time of his writing, 9,000,000 of slaves had been consumed by the Europeans. Add 1,000,000 at least since, for it is about ten years, says Mr. Cooper, who published letters on this subject in 1787. Recollecting then that for one slave procured, ten at least are slaughtered, that a fifth die in the passage, and a third in the seasoning, and the unexaggerated computation will turn out, that the internal voracity of European avarice has been glutted with the murder of 180,000,000 of our fellow-creatures.

The English, it appears, not only outrage humanity on their own account, but engage in this disgraceful commerce to supply other nations. At Liverpool, the grand mart of this execrable trade, Dawson and Baker recently agreed to furnish 3,000 slaves annually to the Spanish islands, and upon this contract have received œ300,000 for those they have supplied.

Noble and generous, though hitherto unsuccessful efforts have, however, recently been made in the British parliament, to effect an abolition of this detestable and inhuman commerce.

After all this, the question itself is reached:

The island of Jamaica is said to produce annually 70,000 tons of sugar. What is their value at 19s. $4\frac{1}{2}d.$ per hundred weight? *Ans.* £1,356,250.[11]

This incongruous question is characteristic of Butler's style: powerful rhetorical material on injustices, the decay of liberties, and the like, followed by arithmetical questions which do not meet the point at issue. The problem

in this instance was taken to be a technical exercise in conversion between different monetary units and units of measure, as well as basic multiplication, when the discussion could have prompted a calculation on the wastage of slaves at points of the passage—the loss of a fifth, then a third, and so on. In a sense Butler does not go as far as he might in providing an empowering mathematics education, that is, mathematics enabling students to participate critically and with understanding in the issues of the day. But nonetheless his book is a remarkable attempt to teach arithmetic in a socially aware context.

Traditions of visual modelling

There is an important pedagogical point to make about the activity of introducing students and pupils to graphical objects from the past. People do not have an innate historical sense; it has to be learned. Children and some adults naturally assume that things are as they have always been—this is notably true of people's assumptions about mathematics. For example, my childhood mathematics lessons, perhaps like those of some readers, were done in exercise books consisting of squared paper, just the right size for multiplication and long division at one digit per square. I took this to be a timeless feature of the universe—whereas it was only some 60 to 70 years old by that time, having been introduced into British education in the late nineteenth century.[12] Similarly, graphic features which are familiar to everyone today through our print design culture have all been introduced and developed at some past stage, for some purpose. Exploring historical examples of these developments helps alert and educate children and others to the language, rhetoric, and analysis of graphic design, in mathematics and elsewhere.

It may be no coincidence that Clarkson's 1808 chart came at a time when there was increasing interest in exploring graphic modelling techniques. A good example is William Playfair's *Commercial and Political Atlas* of 1786. Playfair (1759–1823) was a brother of the Scottish mathematician John Playfair, and his book introduced various charts for presenting statistical data in a novel way, containing the first known use of time-series using economic data.

The diagram chosen for discussion in this paper was not of numerical or statistical data but what one might call conceptual data. There is growing support for an educational stance where diagrams of all sorts are seen as legitimate objects within the mathematics classroom—not only ones portraying numerical data. In the British National Curriculum for Mathematics, for example, level 1 speaks of "*creating simple mapping diagrams showing relation-ships and interpreting them*"; and that strand carries on up to level 10's "*interpreting various types of diagrams such as those used in analysis of critical path and linear programming*" (these are, one might notice, from Attainment Target 5, *handling data.*) So it is important to look at diagrams as objects in themselves, and to ponder educational issues about creating and interpreting them.[13]

A historical survey or taxonomy of different kinds of graphical representation and modelling would go back at least to Chinese and Babylonian and Egyptian maps and diagrams 4,000 or 5,000 years ago. Such representation is richly represented in the manuscript tradition of the high middle ages and in mystical traditions such as alchemy or kabbala, besides playing an illustrative role in many mathematical contexts. There is a fertile field here for exploration and promotion in a research or a teaching context.[14]

There has been an ideological devaluing of graphical modelling, for several easily perceived reasons. The development of print and typesetting produced a culture in which diagrams became separated from the natural flow of text rather than integrated with it as was possible in the manuscript tradition. Another reason lies in the development of mathematics where diagrams became seen as logically unrigorous, indeed deceptive and superfluous, epiphenomena of the underlying mathematical argument—that is the tradition in which Lagrange's pride at doing algebra, not geometry or constructions, can be seen to flourish. A third cultural analysis which is thought-provoking is Albert Biderman's perception[15] that the discrimination and segregation of graphics is analogous to that of women in a sexist culture.

Empowerment

The title of this paper refers to "empowerment." How does that arise from the present discussion? The aim here is to provide material and modes of consideration for helping today's schoolchildren to become confident users and moulders of the world around them, empowered to analyse, use, create, and communicate with graphical data as readily as in other ways; empowered also to become confident, critical, aware adults. This paper has given examples of how this happened in the past. To recapitulate, graphs and other diagrams were developed as a powerful way of revealing patterns to the inquirer, displaying information, patterns, relationships, and differences to others, and exploring further relationships, latent or not yet apparent in the raw data or previous information. Understanding the historical context in which graphs began to be used reveals how they em-

powered earlier generations. This knowledge in turn helps today's students to engage with the essence of the technique: the nature of the problem which it resolves, and the relation between that and the conceptual representation or transformation which the solution involved. Perhaps even more importantly, it also helps today's students to become similarly empowered, by learning to use skillfully the same or similar techniques, as well as learning the possibility of devising fresh exploratory data techniques purposefully for themselves.

References

1. A case in point is Alan Graham and John Fauvel (eds.), *Empowering students through the history of statistics,* Centre of Maths Education, Open University, 1995.
2. Albert D. Biderman, "The graph as a victim of adverse discrimination and segregation," *Information Design Journal,* 1980, 1: 232–241; Martin Gardner, *Logic Machines and Diagrams,* 2nd edition, (Brighton: Harvester Press, 1983).
3. Thomas Clarkson, *The history of the rise, progress, and accomplishment of the abolition of the African slave-trade by the British Parliament* (London: Longman, 1808), pp. 30, 32.
4. Ibid., pp. 259–261.
5. Ibid., p. 263.
6. Ibid., p. 264.
7. See also Marilyn Frankenstein, *Relearning Mathematics* (London: Free Association Books, 1989).
8. Thomas Clarkson, *An essay on the slavery and commerce of the human species, particularly the African* (London: Phillips, 1788), p. 125.
9. For further details, see John Fauvel and Paulus Gerdes, "African slave and calculating prodigy: bicentenary of the death of Thomas Fuller," *Historia mathematica,* 1990, 17: 141–151.
10. William Butler, *Arithmetical questions, on a new plan . . . for the use of young ladies,* (London, S. Couchman, 1795), p. 175.
11. Ibid., p. 177.
12. William H. Brock and Michael H. Price, "Squared paper in the nineteenth century: instrument of science and engineering, and symbol of reform in mathematical education," *Educational Studies in Mathematics,* 1980, 11: 365–381.
13. Further valuable discussion along these lines is contained in Edward R. Tufte, *The visual display of quantitative information* (Graphics Press, 1983), and *Envisioning information* (Graphics Press, 1990).
14. See also H. G. Funkhauser, "Historical development of the graphical representation of statistical data," *Osiris,* 1937, 3: 269–404; James W. Reidhaar, "An overview of nontabular methods for statistical presentation of data before this century," *Information Design Journal,* 1984, 4: 25–35.
15. Albert D. Biderman, "The graph as a victim of adverse discrimination and segregation," *Information Design Journal,* 1980, 1: 232–241.

The Calculus as Algebra, the Calculus as Geometry: Lagrange, Maclaurin, and their Legacy

Judith V. Grabiner
Pitzer College

Prelude: The Ways Mathematicians Think

Given a regular hexagon and a point in its plane: draw a straight line through the given point that divides the given hexagon into two parts of equal area.[1] Please stop for a few moments, solve the problem, and think about the way you solved it.

Did you draw an actual diagram? Did you draw a mental diagram? Were you able to solve the problem without drawing one at all? If you had a diagram, do you find it hard to understand how others could proceed without one? Were you motivated at all by analytic considerations? If you were not, do you understand how others were? Did you use kinematic ideas, such as imagining the line to be moving or rotating about a point? If not, can you understand how others might have? In solving the problem for the hexagon, were you consciously motivated by the analogy to the circle? Now that the analogy has been mentioned, do you think you were unconsciously so motivated? If not, do you understand how others were? Did you have any other method not already mentioned? (For instance, some people, often chemists, say they thought of folding the plane.) Finally, did you try to prove your solution? Did you nevertheless "know" that you were right?

The diversity of problem-solving approaches and of levels of conviction, even for a simple problem like this one, makes clear that there are many kinds of successful mathematical thinking. Distinctions like that between the diagram-drawers and the others, between those who use a geometric analogy and those who do not, between those who use kinematic ideas and those who do not, and between those who cannot imagine doing the problem analytically and those who naturally proceed that way, have been analyzed by Jacques Hadamard in his *Psychology of Invention in the Mathematical Field*.[2] One might, he said, distinguish between "geometric" and "algebraic" approaches to mathematics, for instance. Henri Poincaré did not find this distinction sufficient, so he divided mathematicians between what he called "intuitionalists" and "logicians."[3] For instance, Bernhard Riemann was intuitive, Karl Weierstrass logical. Hadamard explained Poincaré's distinction by citing the cases of Hermite (whom Hadamard saw as intuitive) and Poincaré himself (whom he saw as logical), saying, "Reading one of [Poincaré's] great discoveries, I should fancy ... that, however magnificent, one ought to have found it long before, while ... memoirs of Hermite ... arouse in me the idea, 'What magnificent results! How could he dream of such a thing?' "[4]

Hadamard explained the different types of mathematical thinking according to how deeply the unconscious is tapped in the process, by whether the thought is broadly directed in contrast to narrowly directed, and according to what kinds of mental pictures or other concrete representations are used.[5] Hadamard's general discussion can be nicely illustrated on a small scale by observing the diversity of responses one obtained when the Pólya hexagon problem is given to a group of mathematicians. Reflecting on Hadamard's discussion on the grander scale serves to set the scene for the contrast I want to draw between the work of two very different eighteenth-century mathematicians, Colin Maclaurin and Joseph-Louis Lagrange. Each contributed to the development of the calculus in the eighteenth century. But Maclaurin viewed the calculus as geometry, while Lagrange saw it as algebra. It will not be surprising if different readers of Maclaurin and Lagrange respond "of course!" or "how could he dream of such a thing?" to different arguments. Such differing responses illustrate our main point: there are many modes of creative mathematical thought.

Introduction to the Geometric and Algebraic Approaches

Figure 1 is a diagram from Maclaurin's *Treatise of Fluxions* of 1742.[6] Figure 2 is a diagram from Joseph-

FIGURE 2
This is not a misprint.

Louis Lagrange's *Théorie des fonctions analytiques* of 1797. This contrast underlines the distinction between the way these two men—Joseph-Louis Lagrange (1736–1813) and Colin Maclaurin (1698–1746)—thought about the calculus and its applications. For Maclaurin, the calculus was at heart geometric; for Lagrange, the calculus was algebraic. Maclaurin's great *Treatise of Fluxions* (1742) has over 350 diagrams; Lagrange's masterwork on the calculus, the *Théorie des fonctions analytiques,* search as one will, contains none—just pages of text and formulas.[7]

To see how contemporaries thought about these approaches, let us look at two illustrations from the early

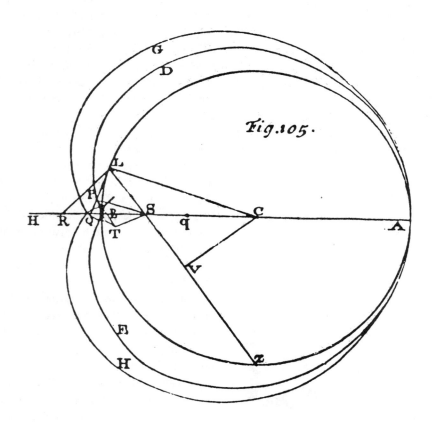

FIGURE 1

nineteenth century. First, here is an example of the Edinburgh tradition to which Maclaurin belonged, from the Scottish philosopher William Hamilton:

> The process in the algebraic method is like running a railroad through a tunneled mountain; in the geometrical like crossing the mountain on foot. The former carries us by a short and easy transit to our destined point, but in miasma, darkness and torpidity, whereas the latter allows us to reach it only after time and trouble, but feasting us at each turn with glances of the earth and of the heavens, while we . . . gather new strength at every effort we put forth."[8]

In contrast, followers of Lagrange saw mathematical truth arising from algebra, because of what Charles Babbage called "the accurate simplicity of its language."[9] Lagrange preferred algebra, not only for calculus, but even for mechanics. He wrote in 1797 that mechanics is just an analytic geometry with four coordinates, x, y, z, and t.[10] As is well known, Lagrange boasted in the preface to his *Analytical Mechanics* that "no diagrams are to be found in this work—only algebraic operations."[11] Thus, we avoid being misled by geometric intuition or by ideas about speeds and motion, and follow only the logic of pure, abstract systems of operations. The development of the calculus in the century after Isaac Newton and Gottfried Wilhelm Leibniz is the history of these two approaches: the calculus as geometry and the calculus as algebra, as exemplified in the present paper by Maclaurin and Lagrange.[12]

Let us now look at examples of the actual mathematical work of Maclaurin and Lagrange. The calculus is, of course, one subject, and Maclaurin and Lagrange had much in common. Nonetheless, as we shall see, their different approaches conditioned the problems they solved and how they solved them. The fact that successful mathematicians—even at the same time and in the same field—think in such different ways has implications for our work with students and for how we think about mathematical research.

Maclaurin

First, we will consider Maclaurin, whose calculus is Newtonian. He viewed the derivative—which he called the fluxion—as a mathematically-idealized velocity. Maclaurin wrote,

> The velocity by which a quantity flows. . .is called its *Fluxion* which is therefore always measured by the increment or decrement that would be generated in a given time by this motion, if it

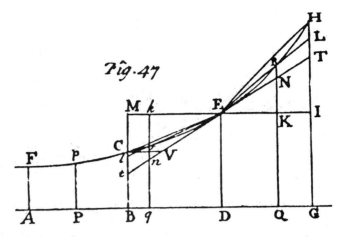

FIGURE 3

was continued uniformly from that term without acceleration or retardation.[13]

In Maclaurin's diagram reproduced above, for instance, the fluxion of the curve at E is measured by the line TI. Given this background, we will review a specific example of Maclaurin's approach to calculus: his theory of maxima and minima.

Long before the invention of calculus, it was well known that at the highest point of a smooth curve, the tangent is horizontal. The Greeks had treated not only tangents, but also the concavity and convexity of curves, and seventeenth-century mathematicians had gone beyond the Greeks to study many more curves, including those with points of inflexion and many branches. Maclaurin was interested in geometry; he knew about many kinds of curves; because of this, he wanted to give a systematic and complete theory of maxima, minima, concavity, and convexity for a wide variety of curves. He saw calculus as a means of doing so. Here are some of the curves that motivate and illustrate his theory (Figure 4).

Maclaurin defined the maximum not by inequalities, but by words. The words defined it in terms of velocities, but he immediately related his definition to graphs. Graphs are almost always intended to go with his verbal statements. For instance, he said,

> When for some time the variable quantity first increases till a certain assignable term, and then decreases . . . its magnitude at that term is considered as a maximum . . . . In problems of this kind. . .the variable quantity is represented by an ordinate of a curve . . . . The ordinate from a point of the curve is a maximum. . .when it is greater . . . than the ordinates which may be drawn from

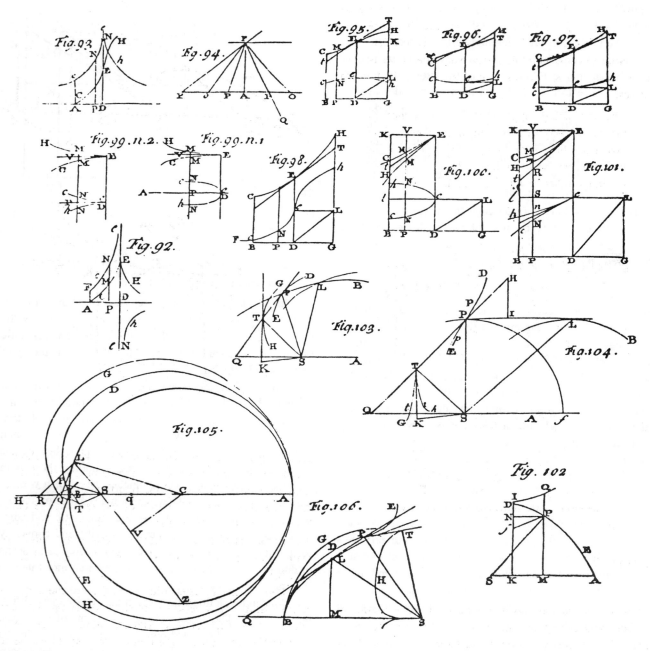

FIGURE 4

the parts of either branch of the curve adjoining to that point ... .[14]

Minimum is analogously defined. And what of concavity? Here Maclaurin began by following the ancient Greeks, saying that when a curve "has its concavity turned one way, the tangent at any point of it is on the convex side."[15]

But he continued by using the language of velocities. He gave a lemma which said that, if a curve is convex toward the base [what we call concave up],

the ordinate increases with a motion that is continually accelerated, and decreases with a motion that is continually retarded.[16]

This Lemma about acceleration and retardation is made plausible immediately by a diagram, where we look at successive values of the ordinate and see that the tangent is always under the curve.[17] (Look again at Figure 3.)

Similar considerations apply when the curve is concave down. Thus, for instance, for extrema of smooth curves:

> The arch being supposed to have its concavity turned one way, and the tangent at E being supposed parallel to the base, if the arch meet the base we may conclude that DE is a maximum" (or in the other case a minimum).[18]

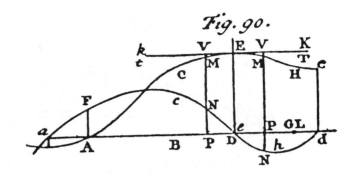

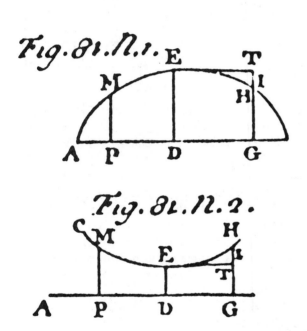

FIGURE 5

Thus he *defined* the maximum and minimum in terms of velocities, speaking of "increasing" or "decreasing" variables, but he *characterized* the difference in terms of *geometric* notions of concavity and convexity, illustrated by diagrams. The properties seem to flow from the underlying geometric intuition.

Using these properties, as one would expect, Maclaurin characterized the maximum, minimum, and point of inflexion of a smooth curve in terms of the values of the first and second fluxions. But he extended this method further to discover a new and general result. By a complicated geometric argument, drawing a succession of curves which we would call the graphs of $f(x)$, $f'(x)$, $f''(x)$, etc., and re-

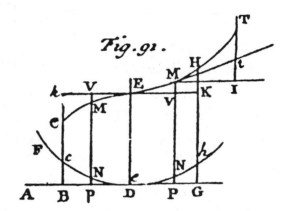

FIGURE 6

peatedly using convexity and concavity, Maclaurin showed that there will be a maximum or minimum when $0, 2, 4, \ldots$ fluxions vanish. The sign of the first fluxion different from zero determines on which side of the tangent the curve lies, because the sign of this fluxion determines whether the ordinate is being accelerated or decelerated. Similarly, there will be a point of inflexion when $1, 3, 5, \ldots$ fluxions vanish.[19]

Later in his *Treatise,* Maclaurin derived the same result by using Taylor series—a method which may be more familiar to us. Yet even this series derivation was geometrically motivated. In my exposition of that derivation, I shall use Lagrange's notation for derivatives to make the mathematics easier to follow.[20] Suppose we have a maximum; now let the first derivative $f'(x) = 0$. If h is small enough, then the diagrams we have looked at, and the lemma about acceleration and retardation quoted above, make clear that $f(x + h)$ and $f(x - h)$ will exceed $f(x)$ when $f''(x)$ is positive, and will be smaller than $f(x)$ when $f''(x)$ is negative. We see this from the geometry of the situation.

Maclaurin then used Taylor's series to express the same result. Maclaurin introduced Taylor's series in this

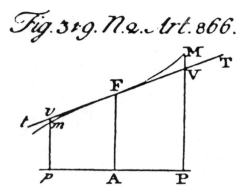

Fig. 319. N.q. art. 866.

FIGURE 7

context by giving the diagram in Figure 7 as well as the series. We give the series here in Lagrange's notation: Suppose

$$f(x+h) = f(x)+hf'(x)+h^2/2!f''(x)+h^3/3!f'''(x)+\cdots$$

where, in his diagram, AF is $f(x)$, PM is $f(x+h)$, and $AP = h$. Similarly, taking Ap in the opposite direction:

$$f(x-h) = f(x)-hf'(x)+h^2/2!f''(x)-h^3/3!f'''(x)+\cdots$$

Now, the geometric result that $f''(x) < 0$ at the maximum is confirmed by looking at the Taylor series; we see that, when $f'(x) = 0$ and when $AP = Ap = h$ is sufficiently small, the h^2 term (that is, the nonzero term with the smallest power of h), which here contains $f''(x)$, dominates the rest and thus f''(x) must be positive for a minimum, negative for a maximum. But thus translating the geometric inequalities into series now took Maclaurin beyond what is visible from the geometry, and again gave him the general result that there is a maximum or minimum when the first n derivatives for any odd n vanish; and then, if the next derivative is positive we get a minimum, otherwise a maximum. But, if n is even and the first n derivatives vanish, we get neither a minimum nor a maximum. Thus Maclaurin was led by geometric considerations of convexity and concavity and inequalities to formulate a series result which leads once again to the complete solution of the general problem of identifying maxima, minima, and points of inflection by computing higher-order derivatives.[21] The motivation for this set of results was far from being the calculus of polynomials, or even of functions defined by infinite power series. Maclaurin developed his theory of maxima, minima, points of inflexion, convexity and concavity, and orders of contact between curves because he wanted to study curves of all types, including those which cross over themselves, have many branches,

or have cusps.[22] Thus this analytic theory had a geometric motivation as well as a geometric mode of discovery. But the analytic result—the Taylor-series version—was highly influential, being adopted to treat the same problems, without any diagrams at all, by Euler and then by Lagrange.[23]

Finding relationships between functions represented by curves and their series expansions, motivated by inequalities derived from geometric diagrams, is characteristic of Maclaurin's *Treatise of Fluxions*. For instance, consider briefly what is now called the Euler-Maclaurin series, discovered independently and almost simultaneously by Euler and by Maclaurin (who published it first).[24] Suppose we have the curve $F(x)$, and a point a. Following the modern notation used by Herman Goldstine, we have:

$$\sum_{h=0}^{\infty} F(a+h) = \int_a^{\infty} F(x)dx + 1/2F(a)$$
$$+ 1/12F'(a) - 1/720F'''(a)$$
$$+ 1/30240F^{(v)}(a) + \cdots$$

How did Maclaurin get it? Characteristically, he started from a geometric diagram (Figure 8).[25] The left-hand side of Goldstine's version of the Euler-Maclaurin series is the sum of the successive ordinates AF, BE, CK, HT, etc. Maclaurin saw his result as showing that this sum is equal to the limit approached by the area $APNF$, plus 1/2 of AF, plus 1/12 of the fluxion of AF, etc. Maclaurin obtained the coefficients, which are now known as Bernoulli numbers, from representing the various quantities involved by Taylor series, integrating them term by term, equating like powers, and performing various substitutions. Still, the motivation was geometric.

Maclaurin applied his result in two ways: using various series to approximate the value of the area (what we call approximating the integral); and using the area's value to find the sums of various series. For instance, he used it to sum the powers of arithmetic progressions, to derive Stirling's formula for factorials, to obtain the Newton-Cotes numerical integration formula, and to get Simpson's rule as a special case.[26] In these applications we notice again the link between analytic results and their geometric motivation.

In fact Maclaurin based his understanding of the limit of the sum of a series on a geometric analogy that goes back to the ancient Greeks: the idea that the area under a curve is a limit of rectilinear figures. On the basis of this geometric model he gave the first clear definition of the sum of infinite series:

> There are progressions ... which may be continued at pleasure, and yet the sum of the terms be

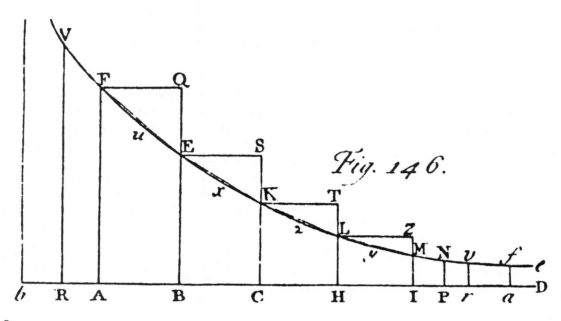

FIGURE 8

always less than a certain finite number. If the difference betwixt their sum and this number decrease in such a manner, that by continuing the progression it may become less than any fraction how small soever it be assigned, this number is the *limit of the sum of the progression,* and is what is understood by the value of the progression when it is supposed to be continued infinitely.[27]

The geometric motivation for this notion of the limit of the sum of a series is made clear in what immediately follows: "These limits are analogous to the limits of [geometric] figures which we have been considering, and they serve to illustrate each other mutually."[28]

Another geometrically-motivated topic for Maclaurin was elliptic integrals. Some integrals (Maclaurin used the Newtonian term "fluent" where we use "integral"), he observed, can be evaluated by being reduced to algebraic functions. Others cannot, but some of them can be reduced to finding circular arcs or logarithms. By analogy Maclaurin suggested that perhaps a class of integrals can be reduced to finding the length of an elliptic or hyperbolic arc.[29] His general approach was to use clever geometric transformations to reduce the integral that represented the length of a hyperbolic or elliptic arc to a "nice" form, and then reduce some previously intractable integral to the same form. For instance, he showed that the integral

$\int dx/\sqrt{x\sqrt{(1 + x^2)}}$ can be reduced to finding the length of an elliptic arc. In 1746 Jean d'Alembert used Maclaurin's work as a stepping-stone to his own analytic treatment, which in turn helped produce the influential work of Leonhard Euler and Adrien-Marie Legendre.[30] Thus Maclaurin initiated a very important type of investigation, again motivated by geometry, and foresaw its generality.

In physics Maclaurin created the subject of the gravitational attraction of ellipsoids and applied it to the problem of the shape of the earth. For instance, in 1740 he showed that an oblate spheroid is a possible figure of equilibrium under Newtonian mutual gravitation.[31] He devised some of the key methods used in studying the equilibrium of fluids,[32] including the methods of balancing columns and level surfaces.[33] His approach to the subject was geometric,[34] though his successors often began by putting his results in analytic form. Alexis-Claude Clairaut's classic *La Figure de la Terre* (1743) explicitly and repeatedly credited Maclaurin's work.[35] Lagrange began his own memoir on the attraction of ellipsoids by praising Maclaurin's prize paper on the subject as a masterwork of geometry, comparing the beauty and ingenuity of Maclaurin's work to that of Archimedes—before translating Maclaurin's geometric work into the language of the calculus.[36]

In every case that we have considered, Maclaurin contributed not to antiquarian geometry but to mainstream mathematics, and he did it in the same way. He started by defining a geometric problem, and used geometric insight to advance it. He also used the kinematic notion of fluxion as an idealized velocity, algorithmic tools of calculus, and nontrivial results from elsewhere in the calculus. In some cases—as when he used the Euler-Maclaurin formula to sum series—the final result is not geometric; in others (as for points of inflexion) it is. But always, he applied and advanced the calculus via his kinematic and geometric understanding of it.

Lagrange

Now, let us turn to calculus as practiced by Lagrange, beginning with the way he conceptualized the derivative. The contrast with Maclaurin is striking.

Lagrange wanted calculus to be "pure analysis," without appeals to intuitions about motion or geometry. Since he thought that algebra alone was rigorous, he believed that the way to make calculus rigorous was to banish the ideas of infinitesimals, geometry, and velocity, substituting instead "the algebraic analysis of finite quantities."[37] Lagrange also believed that expanding functions into infinite power series was an entirely algebraic process—as it appears to be, say, when one divides $1/(1 - x)$ to obtain $1 + x + x^2 + \cdots$. He seems to have based this opinion on Leonhard Euler's *Introductio in analysin infinitorum* of 1748, in which Euler successfully derived a wealth of power-series expansions by what appeared to Lagrange to be algebraic means.[38]

Because Lagrange thought that generating power-series expansions was purely algebraic, he thought it legitimate to define the derivative of a function as the coefficient of the linear term in the power-series expansion of the function, as follows. Note first that Lagrange's definition of function is algebraic: a function is an "expression de calcul" into which the variable enters in any way.[39] Suppose that

$$f(x + h) = f(x) + ph + qh^2 + rh^3 + \cdots,$$

where p, q, r, are all functions of x.[40] Then he defined $f'(x)$ as the coefficient of h, the function $p(x)$.[41] Analogously, he defined the second derivative as the coefficient of the linear term in the expansion of $f'(x + h)$, and so on recursively. The algebraic manipulation of power series then gave Lagrange the Taylor series

$$f(x + h) =$$
$$f(x) + hf'(x) + h^2/2!f''(x) + h^3/3!f'''(x) + \cdots.$$

Lagrange invented this notation for derivatives to emphasize that the derivative $f'(x)$, like $f(x)$, is a function of x. The notation also reminds us that the function $f'(x)$ is *derived* from $f(x)$.[42] (Lagrange's term "fonction derivée" is the origin of the term derivative.)

From the Taylor series, assuming implicitly that all the derivatives are bounded, Lagrange said that one can always find h small enough so that the h^n term (for any n) dominates the rest of the series. Of this result, which he called a "fundamental principle," he declared, "it is because of this . . . that the calculus is the most fruitful, especially in its application to the problems of geometry and mechanics."[43] (The reader will recall that Maclaurin used this result to derive the theory of extrema from the Taylor series.) As an example of Lagrange's own application of it, let us turn to his *Leçons sur le calcul des fonctions,* where he considered the result for the case $n = 1$. In this case, letting hH representing all the terms with derivatives of higher order than 1, he obtains:

$$f(x + h) = f(x) + hf'(x) + hH,$$

where H goes to zero with h.[44] And Lagrange meant something very precise when he said that H went to zero with h:

> Given any quantity D, h can be chosen sufficiently small so that $f(x + h) - f(x)$ is included between $h(f'(x) \pm D)$.[45]

Because of its importance, I have called this characterization of $f'(x)$ the *Lagrange property of the derivative.*[46] If we rewrite the Lagrange property with absolute value signs and modern notation, we get, for h sufficiently small,

$$\left| [f(x + h) - f(x)]/h - f'(x) \right| < D.$$

This is exactly Cauchy's delta-epsilon characterization of the derivative, to within an alphabetical isomorphism. And Lagrange was Cauchy's source.[47]

What did Lagrange do with this property of the derivative? First, he used it to prove that a function with a positive derivative on an interval is increasing on that interval. The proof is too complicated to give here.[48] For our present purposes it will suffice to say that it is entirely algebraic and proceeds by manipulating inequalities in a delta-epsilon way (though Lagrange did not distinguish between convergence and uniform convergence and assumed that, given D, his choice of h will work for any value of x). Still, it was an influential, pioneering effort.

The key point for us about Lagrange's lemma is this: the result is obvious to anybody who visualizes curves. Only someone who wanted to make calculus purely analytic and eliminate all geometric intuition would imagine that one needs to prove that a function with a positive derivative on an interval is increasing on that interval. Thus an algebraic view of the calculus was necessary for Lagrange even to think of this as a theorem, let alone prove it.

Once Lagrange had that result, he used it together with the intermediate-value property for continuous functions (which he had also tried to prove),[49] to give an algebraic proof for the mean-value theorem for derivatives, which he gave in this form:

$$f(x) + hf'(p) < f(x + h) < f(x) + hf'(q).$$

where $f'(p)$ is the minimum, $f'(q)$ the maximum, value of the derivative $f'(x)$ on the interval he was considering. The idea of his proof is this: If the minimum of $f'(x)$ is $f'(p)$ and its maximum is $f'(q)$, then the functions

$$g'(h) = f'(x + h) - f'(p) \, cr \text{and}$$

$$k'(h) = f'(q) - f'(x + h)$$

are both positive. Integrating with respect to the initial condition $g(0) = k(0) = 0$, he obtained the new functions

$$g(h) = f(x + h) - f(x) - hf'(p)$$

and

$$k(h) = hf'(q) - f(x + h) + f(x).$$

Since the lemma just proved shows that the functions g and k are increasing, $g(h)$ and $k(h)$ must be greater than 0, and thus

$$f(x) + hf'(p) < f(x + h) < f(x) + hf'(q).$$

The intermediate-value property for continuous functions then ensures that there is an X between p and q such that $f(x + h) = f(x) + hf'(X)$.[50] Reasoning in an analogous fashion but using higher-order derivatives, Lagrange found that

$$f(x + h) =$$
$$f(x) + hf'(x) + h^2/2f''(x) + \cdots + h^n/n!f^{(n)}(x),$$

the Taylor series with Lagrange remainder.[51]

Lagrange applied his remainders to solve a variety of problems. In particular he used the Taylor series with Lagrange remainder to prove his version of the Fundamental Theorem of Calculus, that the derivative of the area under the curve is the function itself. That is, he considered the function $y = f(x)$, and the function $F(x)$ that defines the area under $y = f(x)$ up to x. Then he showed that $F'(x) = f(x)$.

Lagrange's proof begins by observing (without a diagram!) that, for a monotonic function $f(x)$ where the area under $f(x)$ up to x is called $F(x)$,

$$hf(x) < F(x + h) - F(x) < hf(x + h). \quad (*)$$

Using his mean-value theorem and the Taylor series with second-order remainder, Lagrange obtained:

$$f(x + h) = f(x) + hf'(x + j)$$

for some j between 0 and h

$$F(x + h) = F(x) + hF'(x) + h^2/2F''(x + J)$$

for some J between 0 and h.

The inequality $(*)$ then yields

$$h[F'(x) - f(x)] + h^2/2F''(x + J) < h^2(f'(x + j)). \quad (**)$$

If one chooses h sufficiently small (and he explicitly calculated how small), the inequality $(**)$ cannot be true unless the $h[F'(x) - f(x)]$ term vanishes, so one must conclude that $F'(x) = f(x)$.

Lagrange also applied the Taylor series with Lagrange remainder to obtain the results Maclaurin got about extrema, and also to treat orders of contact between curves—again, without diagrams.[53] In addition, Lagrange applied his remainders to mechanics. For instance, he considered motion along a line such that $x = f(t)$. Between the time t and $t + \phi$, the distance traversed is

$$f(t + \phi) - f(t) = \phi f'(t) + \phi^2/2f''(t) + \phi^3/3!f'''(t + L\phi)$$

where L is between 0 and 1.[54] Lagrange then pointed out that the distance produced by the motion can be decomposed, via the right-hand side of the equation, into terms that represent the results of different partial motions, where the first term arises from a uniform motion, the second from a uniformly accelerated motion, and the third term represents all the other motions. For ϕ sufficiently small, he said that the motion composed of the first two terms gets very close to the actual motion.[55]

One more application of series to mechanics is Lagrange's explanation of an error that Nikolaus Bernoulli claimed to have found in Newton's *Principia* (Book II, Prop. X) on motion in a resisting medium; Lagrange said in effect that Newton had not considered terms of sufficiently high order.[56] Characteristically, Lagrange began his treatment of this subject by saying, "To discover the source of the error, we are going to reduce Newton's solution to analysis."[57] Thus, even in doing something which we think of as geometry or physics, Lagrange's method is self-consciously analytic. That Lagrange saw calculus as algebra—sometimes the algebra of inequalities, sometimes formal manipulations with power series—was the key to his success.

Why the Difference?

Contrasting the approaches to the calculus of Maclaurin and Lagrange raises a general historical question. Why were Maclaurin and Lagrange so different? The source of this difference is not just one of temperament, but arose from differences in education and cultural traditions. Maclaurin's mathematical philosophy, as Erik Sageng has shown, was shaped in part by the views of Sir Francis Bacon.[58] Maclaurin was a an initiator of the Scottish Enlightenment, whose ideas are linked with those of British philosophers of the empirical—and Newtonian—traditions. For instance, among Maclaurin's contemporaries in Edinburgh was David Hume, and among his successors was Adam Smith. In mathematics itself, Maclaurin's mentor was the classical geometer Robert Simson. Maclaurin's dissertation at Glasgow was a defense of Newton's ideas on physics, and Maclaurin's later mathematics was in the Newtonian tradition. Finally, Maclaurin taught at the University of Edinburgh, which, especially in the eighteenth and early nineteenth centuries, valued what we call general education, where students pursued classics and philosophy together with the exact sciences. Mathematics there, said Professor John Leslie, was "a branch of liberal education, [not] a mechanical knack."[59]

It was different for Lagrange, who was schooled in the major works of Continental mathematics and philosophy, notably those of René Descartes and Gottfried Wilhelm Leibniz. The prevailing Continental view of the philosophy of mathematics, arising from the influence of Descartes and Leibniz, was clearly expressed by the Enlightenment *philosophe,* the Marquis de Condorcet, when he said that algebra is "the *only* really exact and analytical language in existence."[60] Lagrange taught calculus, not as part of a liberal education in an institution like the University of Edinburgh, but first at the military school in Turin, and later at the *École polytechnique* in Revolutionary Paris, where most of his students were being educated to be engineers. Lagrange found his major mathematical influence in the works of Euler and was inspired by Euler's unparalleled algorithmic power. Calculus, of course, was still one subject. But Maclaurin and Lagrange learned it from different traditions, thought about it and taught it in different settings, and came to see its basic problems through different eyes.

Influence

Histories of the calculus sometimes deplore the influence of the approaches of both Maclaurin and Lagrange. Maclaurin, they say, looked backward toward ancient Greek geometry and away from modern analysis.[61] Lagrange did harm too, by overformalizing the subject.[62] But their immediate successors thought otherwise. Eighteenth- and early nineteenth-century mathematicians found much value in Mclaurin's work, even when they did not share his geometrical approach. As we have seen, Lagrange himself praised Maclaurin's treatment of extrema, and compared Maclaurin's work on the attraction of ellipsoids to that of Archimedes (before translating it into analysis).[63] Lagrange's contemporary Silvestre-François Lacroix, in his classic textbook on the calculus, praised and used Maclaurin's work on series.[64] The German mathematician Carl G. J. Jacobi worked out the remainder for the Euler-Maclaurin formula; Jacobi called the result simply the Maclaurin summation formula and cited it directly from the *Treatise of Fluxions.*[65]

As for Lagrange, his primary influence was on Augustin-Louis Cauchy, Bernhard Bolzano, and Karl Weierstrass. Cauchy used the Lagrange property of the derivative, which for Lagrange was a byproduct of his power-series definition of derivative, as the defining property. Once he had done so, all the proofs Lagrange had based on that property became legitimate.[66] Bolzano enthusiastically adopted Lagrange's philosophy of pure analysis; in fact, that is a major point of Bolzano's famous 1817 paper which gave a "Rein analytischer Beweis" (purely analytic proof) of the intermediate-value theorem for continuous functions.[67] Also, Bolzano's *Functionenlehre* (whose title is just a German rendering of Lagrange's *Théorie des fonctions*) bases many results on Lagrangian Taylor-series expansions.[68] Weierstrass adopted Lagrange's term "analytic functions" from the real-variable case to the complex, defined analytic functions as those with convergent power-series expansions, and then made use of some of Lagrange's techniques.

For the Newtonians, including Maclaurin, geometry played a key role in conceptualizing the calculus; Lagrangians later emphasized and exploited the algorithmic and abstracting power of algebra. But neither Maclaurin's nor Lagrange's understanding of calculus met the nineteenth-century standards of someone like Cauchy. Cauchy agreed with Lagrange that geometric intuition could not serve as a basis for proofs in analysis, but criticized Lagrange's overconfidence in what Cauchy called the "generality of algebra." This generality was used to permit jumps from results about the finite to the infinite, or from the real to the complex. Cauchy said instead that algebraic results hold only for particular values of the variables they contain and that infinite series can be used only when their convergence has been established.[69] Cauchy's was a more cautiously employed algebra, an algebra of inequalities rather than an algebra of power series, and it was constrained by the tradition of proof inherited from geometry. Thus Cauchy turned the calculus—the extensive body of results and techniques he inherited from his predecessors—from an overly algebraic subject toward the Euclidean model, where proofs rest on definitions and on explicit and intuitively reasonable assumptions. Cauchy thus synthesized the two ways of thought traced here.

Conclusion

Today we say that neither Maclaurin nor Lagrange was totally correct in his understanding of calculus. Nevertheless, each emphasized one way of thinking about it, and as a result each made distinct—and essential—contributions to the subject. The story I have told, like the Pólya experiment with which we began, makes clear that the question, "How do mathematicians think?" has many answers. As we teach, we need to be aware of the different ways our students think; we need to provide a range of examples and explanations to suit their different cognitive styles.

But the way mathematicians think is not just a matter of individual inclination. Rather, it is a product of education and culture. Further, the very multiplicity of ways that mathematicians think has great value. Partisans of individual points of view make discoveries that others could not see. It is thus worth recognizing and encouraging a wealth of approaches. We may wish to keep this in mind as we work to recruit non-traditional students into mathematical research, so that both we and they will value what their backgrounds have taught them. The story of calculus as geometry and calculus as algebra suggests that progress in mathematics is made by those who sharpen their thinking by exercising the courage of their sometimes idiosyncratic convictions.

References

1. G. Pólya, *How to Solve It: A New Aspect of Mathematical Method,* 2d edition (Princeton: Princeton University Press, 1971) p. 234.
2. Jacques Hadamard, *The Psychology of Invention in the Mathematical Field* (Princeton: Princeton University Press, 1945; Dover reprint, n.d.), p. 108n.
3. Hadamard, op. cit., pp. 108–109.
4. Hadamard, op. cit., p. 108n.
5. Hadamard, op. cit., pp. 112–115.
6. All diagrams in this paper are taken from Colin Maclaurin, *A Treatise of Fluxions in Two Books* (Edinburgh: Ruddimans, 1742).
7. Joseph-Louis Lagrange, *Théorie des fonctions analytiques* (Paris: Imprimérie de la République, An V [1797]; cp. the second edition (Paris: Courcier, 1813), reprinted in *Oeuvres de Lagrange,* pub. M. J.-A Serret, 14 volumes (Paris: Gauthier-Villars, 1867–1892), reprinted (Hildesheim and New York: Georg Oms Verlag, 1973), volume 9.
8. George Elder Davie, *The Democratic Intellect: Scotland and her Universities in the Nineteenth Century* (Edinburgh: The University Press, 1966), p. 127.
9. Preface to *Memoirs of the Analytical Society* (Cambridge: J. Smith, 1813). Attributed to Babbage by Anthony Hyman, *Charles Babbage: Pioneer of the Computer* (Princeton, Princeton University Press, 1982), p. 26.
10. Lagrange, *Fonctions analytiques,* in *Oeuvres,* vol. 9, p. 337.
11. Lagrange, *Mécanique analytique,* 2d edition, in Oeuvres, vols. 11–12.
12. Of course others shared these approaches; in particular, Lagrange's idea of the calculus as algebra owes much to Euler, especially his *Introductio in analysin infinitorum* of 1748. See Judith V. Grabiner, *The Origins of Cauchy's Rigorous Calculus,* (Cambridge, Mass.: The MIT Press, 1981), pp. 51–52.
13. Maclaurin, *Fluxions,* p. 57.
14. Maclaurin, op. cit., pp. 214–215. Maclaurin's text here refers to three diagrams, Figure 81, 82, and 83. Figure 81 shows two smooth curves and is our Figure 5; 82 and 83 show curves with various types of cusps.
15. Maclaurin, op. cit., p. 179; he refers immediately to his Figure 47—our Figure 3.
16. Maclaurin, op. cit., p. 179.
17. Maclaurin, op. cit., Figure 47, following p. 190, our Figure 3.
18. Maclaurin, op. cit., Figure 81, our Figure 5; see p. 217.
19. Maclaurin, op. cit., pp. 226–227.
20. Maclaurin used E for the ordinate of the curve (that is, the value of the function) at the given point. He used x for the increment we have called h, and he designated the fluxions of E by the appropriate number of dots over the E, divided by the fluxion of x to the appropriate power. The coefficient of the x^2 term, for instance, is
$$\ddot{E}/(\dot{x})^2.$$
21. Maclaurin, op. cit., pp. 694–696; compare section 261.
22. Maclaurin, op. cit., pp. 217–218. Look again at our Figure 4, above.
23. Leonhard Euler, *Institutiones calculi differentialis,* 1755, sections 253–255. In L. Euler, *Opera Omnia,* (Leipzig, Berlin, Zurich: Teubner, 1911–), Series 1, Vol XI. For Lagrange, see below.
24. James Stirling, congratulating Euler in a letter of 16 April, 1738, on Euler's publication (in *Comm Petrop* (22), 1738) of that formula, told him that Maclaurin had already made it public in the first part of the *Treatise of Fluxions,* printed and circulating in Britain in 1737. See Charles Tweedie, *James Stirling* (1922), p. 178, and A. P. Juškevič and

R. Taton, eds., *Leonhard Euleri Commercium Epistolicum* (Basel, Birkhäuser, 1980). On this early publication, see also Maclaurin, *Fluxions,* p. iii and p. 691n. Compare Herman H. Goldstine, *A History of Numerical Analysis from the 16th through the 19th Century* (New York, Heidelberg, Berlin: Springer, 1977), pp. 84–86.

25. For the derivation, see Maclaurin, *Fluxions,* p. 292ff; for the diagram, see Figure 146, following p. 310.

26. Maclaurin, op. ct., pp. 676–693.

27. Maclaurin, op. cit., p. 289.

28. Ibid.

29. Maclaurin, op. cit., p. 652ff.

30. See Alfred Enneper, *Elliptische Functionen: Theorie und Geschichte,* 2d. ed. (Halle: Nebert, 1896), pp. 526ff.

31. See Isaac Todhunter, *A History of the Mathematical Theories of Attraction and the Figure of the Earth, from the Time of Newton to That of Laplace* (London: Macmillan, 1873), p. 374; C. Truesdell, "Rational Fluid Mechanics, 1687–1765," introduction to Euler *Opera,* Ser. 2, vol. 12, p. xix.

32. For "Maclaurin ellipsoids" in twentieth-century classical dynamics, see S. Chandrasekhar, *Ellipsoidal Figures of Equilibrium* (New Haven: Yale, 1969), pp. 77–100.

33. Todhunter, *Attraction,* pp. 136–137.

34. See Todhunter, *Attraction,* pp. 145, 175, 409, 474. Compare Truesdell, "Rational Fluid Mechanics," pp. ix–cxxv, especially xix.

35. A.-C. Clairaut, *Théorie de la figure de la terre* (Paris: Duraud, 1743).

36. J.-L. Lagrange, "Sur l'attraction des sphéro‹des elliptiques," *Mémoires de l'académie de Berlin,* 1773, 121–148. Reprinted in *Oeuvres de Lagrange,* vol. III, p 619ff.

37. The full title of Lagrange's *Théorie des fonctions analytiques* is translated as "Theory of analytic functions, detached from any consideration of infinitely small or evanescent quantities, of limits or of fluxions, and reduced to the algebraic analysis of finite quantities." Amy Dahan Dalmédico gives a clear statement of Lagrange's program of using algebra "to obtain a priori the most general and uniform procedures and demonstrations independent of any specific geometric representation" in "L'ideal analytique de Lagrange," pp. 185–187 (esp. p. 185) in her "La méthode critique du ‹‹mathématicien- philosophe›› ," in Jean Dhombres, *L'école normale de l'an III: Leçons de mathématiques,* Paris, 1992, pp. 171–192.

38. On this point, see Grabiner, *Origins of Cauchy's Rigorous Calculus,* p. 51.

39. Lagrange, *Fonctions analytiques,* in *Oeuvres* 9, p. 15.

40. The notation is his except that I use h for the increment instead of his i.

41. Lagrange, *Fonctions analytiques,* in Oeuvres 9, p. 32.

42. For a full account of Lagrange's philosophy of the calculus, see Judith V. Grabiner, *The Calculus as Algebra: J.-L. Lagrange, 1736–1813* (Boston: Garland, 1990).

43. Lagrange, *Fonctions analytiques,* Oeuvres 9, vol. 9, p.29. Lagrange probably knew the result from Euler as well as from Maclaurin. For Euler, see *Institutiones calculi differentialis,* sections 253–254. Lagrange's knowledge of Euler's work is well known; see Grabiner, *Origins of Cauchy's Rigorous Calculus,* for example, pp. 118–120; for Lagrange's

admiration for Maclaurin's work on maxima and minima, see Maria Teresa Borgato and Luigi Pepe, "Lagrange a Torino (1750–1759) e le sue lezioni inedite nelle R. Scuole di Artiglieria," *Bollettino di Storia delle Scienze Matematiche* 1987, 7: 3–180, p. 154.

44. J.-L. Lagrange, *Leçons sur le calcul des fonctions,* new ed. (Paris: Courcier, 1806). In *Oeuvres,* vol. 10, pp. 86–87. (Compare *Fonctions analytiques,* in *Oeuvres* 9, p 77). The h and H notations are mine, in place of his i and I, but otherwise the presentation is Lagrange's.

45. Lagrange, *Calcul des fonctions, Oeuvres* 10, p. 87.

46. Grabiner, *Origins of Cauchy's Rigorous Calculus,* pp. 118–121.

47. A.-L. Cauchy, *Résumé des leçons données . . . l'école royale polytechnique sue le calcul infinitésimal* (Paris: Imprimérie royale, 1823); in *Oeuvres complètes d'Augustin Cauchy* (Paris: Gauthier-Villars, 1882–), Series 2, vol. 4, pp. 44–45. For an English translation, see Grabiner, *Origins of Cauchy's Rigorous Calculus,* pp. 168–170, and for Cauchy's debt to Lagrange's theory of the derivative, see *Origins,* chapter 5.

48. Lagrange, *Calcul des fonctions,* in *Oeuvres* 10, p. 86ff; compare *Fonctions analytiques, Oeuvres* 9, pp. 78–80. The *Calcul des fonctions* version is described in detail in Grabiner, *Origins of Cauchy's Rigorous Calculus,* pp. 123–126.

49. J.-L. Lagrange, *Traité de la résolution des équations numérique de tous les degrés,* 2d ed. (Paris: Courcier, 1808); in *Oeuvres,* vol. 8, pp. 19–20 and 134. This proof is not very successful logically; it uncharacteristically used the idea of motion, and was criticized on that account by Bolzano. See Grabiner, *Origins of Cauchy's Rigorous Calculus,* p. 73.

50. Lagrange, *Calcul des fonctions, Oeuvres* 10, p. 91ff; compare *Fonctions analytiques, Oeuvres* 9, pp. 80–81. A modern analyst would note that Lagrange, in these pioneering arguments, did not consistently distinguish between "less than," "less than or equal to," and "bounded away from."

51. Lagrange, *Calcul des fonctions, Oeuvres* 10, pp. 91–95; *Fonctions analytiques, Oeuvres,* vol. 9, pp. 80–85.

52. Lagrange, *Fonctions analytiques, Oeuvres* 9, pp. 238–239. In what follows, I have for clarity used j and J where Lagrange used the same letter for both, though he made clear in the text that the two quantities are not the same.

53. On extrema, see Lagrange, *Fonctions analytiques, Oeuvres* 9, pp. 233–237. On orders of contact, see *Fonctions analytiques, Oeuvres* 9, e.g. pp. 189, 198.

54. I use ϕ and L for Lagrange's theta and lambda. Lagrange, *Fonctions analytiques,* p. 341.

55. Lagrange, op. cit., pp. 341–342.

56. Lagrange, op. cit., pp. 365–376.

57. Lagrange, op. cit., p. 368.

58. Erik Lars Sageng, *Colin Maclaurin and the Foundations of the Method of Fluxions,* unpublished Ph. D. Dissertation, Princeton University, 1989, chapter II.

59. Davie, *Democratic Intellect,* p. 108.

60. Condorcet, *Sketch for a Historical Picture of the Progress of the Human Mind* (1793), tr. June Barraclough, in Keith Michael Baker, ed., *Condorcet: Selected Writings* (Indianapolis: Bobbs-Merrill, 1976), p. 238.

61. See, e.g., Morris Kline, *Mathematical Thought from Ancient to Modern Times* (New York: Oxford, 1972), p. 429; F. Cajori, *A History of the Conceptions of Limits and Fluxions in Great Britain from Newton to Woodhouse* (Chicago and London: Open Court, 1919), p. 187.

62. See, e.g., E. T. Bell, *The Development of Mathematics* (New York: McGraw-Hill, 1945), pp. 289–290; N. Bourbaki, *Elements d'histoire des mathématiques* (Paris: Hermann, 1960), pp. 217–218.

63. J.-L. Lagrange, "Sur l'attraction des sphéroïdes elliptiques," *Mémoires de l'académie de Berlin,* 1773, pp. 121–148; in *Oeuvres,* vol. 3, p. 619.

64. S.-F. Lacroix, *Traité du calcul différentiel et du calcul intégral,* 3 vols., 2d. ed. (Paris: Courcier, 1810–1819), vol. I, p. xxvii.

65. C. G. J. Jacobi, "De usu legitimo formulae summatoriae Maclaurinianae," *Crelles Journal* 18 (1834), 263–272.

66. This point is documented at length in Grabiner, *Origins of Cauchy's Rigorous Calculus,* esp. chapter 5.

67. B. Bolzano, *Rein analytischer Beweis der Lehrsatzes, dass zwischen zwey Werthen, die ein entgegengesetzes Resultat gewaehren, wenigstens eine relle Würzel liegt* (Prag, 1817).

68. For Lagrange's influence on Bolzano, see Grabiner, *Origins of Cauchy's Rigorous Calculus,* for example, p. 45, 52–53, 74, 95, 192n; compare Judith V. Grabiner, "Cauchy and Bolzano: Tradition and Transformation in the History of Mathematics," in E. Mendelsohn, ed., *Transformation and Tradition in the Sciences* (Cambridge: Cambridge University Press, 1984), pp. 105–124.

69. A.-L. Cauchy, *Cours d'analyse de l'école royale polytechnique,* 1821. In *Oeuvres,* Ser. 2, vol. 3, pp. ii–iii.

The Development of Algebraic Analysis from Euler to Klein and its Impact on School Mathematics in the 19th Century

Hans Niels Jahnke
University of Bielefeld

This paper will seek to describe the inner logic of a certain tradition in the history of calculus, that of algebraic analysis, and to investigate the concept of nineteenth-century arithmetic algebraic school teaching based on this tradition. To the frequent questions of why differential and integral calculus were not taught in German Gymnasia in the nineteenth century it gives a new answer. As will be shown, school teaching followed a mathematical paradigm in which there was neither room nor need for infinitesimal calculus. The paper closes with some conclusions about the teaching of calculus in the upper grades of gymnasia.

Genesis, meaning and conceptual structure of "Algebraic Analysis"

Since the term "algebraic analysis" is not generally known, it is necessary to start with its somewhat complicated history and to give some linguistic equivalents of it. The term "algebraic analysis" suggests how calculus was thought of and practiced in the eighteenth century, especially in the work of Leonhard Euler and Joseph-Louis Lagrange,[1] and the continuation of this tradition during the nineteenth century. Only in the nineteenth did the term receive a technical meaning and designate a certain class of textbooks treating the elementary and preparatory parts of infinitesimal calculus, especially the theory of infinite series. Ironically, this use of the term was presumably established by Cauchy's "Analyse algébrique"[2] which contributed more than any other to the final destruction of this tradition, which ended with the article "Algebraische Analysis" by Alfred Pringsheim and Georg Faber in the *Enzyklopädie der Mathematischen Wissenschaften*[3]. Nevertheless, themes and problems characteristic of this field have continued to today's mathematics. Some of the current research fields which originated from algebraic analysis are formal series, divergent series and generalized limits, reversion of series, recurrent series, infinite products, continued fractions, finite differences and difference equations, iteration, interpolation, generating functions, special functions, faculties, hypergeometric functions, gamma functions, operational calculus (differential and integral operators), fractional calculus, and the algebraic theory of real numbers.

The different eighteenth-century linguistic equivalents for what we call algebraic analysis are:
- analysis of the finite (and/or infinite)
- general or universal arithmetics
- combinatorial analysis (in Germany).

Thus, sometimes algebraic analysis included infinitesimal calculus, and sometimes not. It was both a subject and the algebraic view and treatment of the function concept.

During the eighteenth century the meaning and the scope of the theory underwent several changes. Leonhard Euler's famous *Introductio in analysin infinitorum,* vol. I, of 1748[4] provided its prototype. In this book functions were thought of mainly as finite or infinite symbolical expressions, i.e. infinite products, continued fractions, and power series.

At the end of the century different mathematicians elaborated algebraic analysis in different directions. The following two approaches are especially relevant. A) In his *Théorie des fonctions analytiques* (1797)[5] Joseph Louis Lagrange consistently treated infinitesimal calculus as a calculus of power series. The derivative of a function f is defined as the coefficient p in the power series expansion

$$f(x + i) = f(x) + pi + qi^2 + ri^3 + \cdots$$

Thus, setting $p = \frac{df(x)}{dx} = f'(x)$ and supposing that $f''(x)$ is derived from $f'(x)$ in the same way as $f'(x)$ is derived from $f(x)$, the above series will become the Taylor expansion

$$f(x + i) =$$
$$f(x) + if'(x) + \frac{i^2}{1 \cdot 2}f''(x) + \frac{i^3}{1 \cdot 2 \cdot 3}f'''(x) + \cdots.$$

B) At the end of the eighteenth century the term *combinatorial analysis,* which today is synonymous with combinatorics, designated an approach to calculus developed by a group of German mathematicians, who comprised the "Combinatorial School." The "spiritus rector" of this group was the Leipzig mathematician Carl Friedrich Hindenburg (1741–1808). The group sought to develop a system of "combinatorial operations" to determine and facilitate calculation of transformations of power series, infinite products, and continued fractions.[6]

To understand how the formal combinatorial approach including the use of divergent series, worked and was legitimized it is useful to look at some basic formulae and their interpretation. A crucial question is how equations including formal series can be interpreted. Consider first the geometric series

$$\frac{1}{1 + x} = 1 + x + x^2 + x^3 + \cdots.$$

Usually the $=$ sign means that both sides of the equality have the same numerical value for numbers x in the interval of convergence of the infinite series on the right-hand side. We call this the numerical interpretation of equality.

If infinite expressions are involved, this interpretation presupposes convergence. However, there is an aspect of this formula not covered by the numerical interpretation. The formula can be derived by a purely symbolical algorithm, that is polynomial division, comparison of coefficients, or by some other method. In any case, such a derivation operates on the symbols and does not regard numerical relations. Therefore, it can be asked whether a notion of symbolic or formal equality can be defined.

Second, how can the binomial formula

$$(1 + x)^m = \sum_{v=1}^{\infty} \binom{m}{v} x^v$$

be interpreted?[7] Again there is a numerical interpretation of this equality for "arbitrary" exponents m (including negative, rational, real and complex numbers) in the domain of those x for which the series on the right-hand side converges. If we confine ourselves to natural m, then the equality can also be considered as a purely algebraic or combinatorial relation. In this case we could speak of symbolic or formal equality.

A generalization of the binomial formula, the polynomial formula, provides a third example

$$(a_0 + a_1 x + a_2 x^2 + a_3 x^3 + \cdots)^m =$$
$$A_0 + A_1 x + A_2 x^2 + A_3 x^3 + \cdots.$$

In case of convergence it too has a numerical interpretation. For natural m there is a purely combinatorial proof and, therefore, a purely formal interpretation of the equality. This is true even if $a_0 + a_1 x + a_2 x^2 + a_3 x^3 + \cdots$ is an infinite power series, a so-called "infinitinom." This follows because the calculation of the A_i involves only finite segments of the series, and therefore $A_i = f(a_0, a_1, a_2, \ldots, a_i)$. The question arises whether this interpretation can extend to cases where m is a rational or negative number.

Before answering this question it is appropriate to say a few words about the relevance of the polynomial formula for algebraic analysis. Obviously, the formula determines the basic algebraic operations with power series: raising to a power and extracting a root in case of fractional exponents $m = p/q$, and division in case of negative exponents $m = -n$. In 1793 Heinrich August Rothe, a member of the Combinatorial School, proved a then famous theorem on the reversion of series. Given a power series

$$y = a_1 x + a_2 x^2 + a_3 x^3 + \cdots = P(x)$$

it is required to determine a series

$$x = b_1 y + b_2 y^2 + b_3 y^3 + \cdots = q(y)$$

such that $P(Q(y)) = y$. In the formative days of calculus, Isaac Newton had developed an algorithm for the recursive determination of the b_i.[8] This algorithm was basic for Newton's fluxional calculus. Rothe proved that the coefficients b_i could be independently calculated by means of the polynomial formula.[9] Thus, for the members of the combinatorial school, algebraic or combinatorial analysis was a closed, harmonic system of symbolic forms. In this system all elementary algebraic operations, namely addition, multiplication, division, exponentiation, extraction of roots of power series, and even the solution of arbitrary equations, are universally performable provided that infinite symbolic expressions are accepted as legitimate mathematical entities. The polynomial theorem was for Carl Friedrich Hindenburg and his adherents the most important theorem of analysis.[10]

The importance of the question of how formal, symbolic, and numerical equality can be related is obvious. At the beginning of the nineteenth century several more or less elaborate approaches to this problem had been developed.[11] It is well known that none of them was really successful. The influence of Cauchy's "Analyse algébrique" of 1821, a work that in spite of its title militated against the tradition of algebraic analysis, led to the exclusion of divergent series from mathematics for a long time and thus "destroyed" the concept of formal equality.

While there was no fully elaborated concept of formal equality, there was a sort of working definition of exponentiation of power series for fractional and negative exponents allowing an algebraic interpretation of equality even for these cases. If $m = p/q$, then $Q^{p/q} = R$ was to mean $Q^p = R^q$. This definition made it possible to interpret the polynomial formula for all types of exponents as a purely algebraic-combinatorial relation and to prove it without using infinitesimal methods.[12]

Formulae constituted the essence of algebraic analysis. This led to a global structure of the field which was very different from today's views of analysis. It is interesting for various reasons. This structure is listed in a well-known contemporary mathematical encyclopedia under the entry "Analysis."[13]

ANALYSIS OF THE FINITE according to G.S. Klügel's MATHEMATISCHES WÖRTERBUCH

I. The theory of functions or of the forms of quantities.
II. Introduction to the theory of series.
III. Combinatorics.
IV. Combinatorial Analysis in general.
V. Binomial and Polynomial Theorem.
VI. Products of unequal binomial factors.
VII. Faculties.
VIII. Logarithmic functions.
IX. Trigonometric functions.
X. Application of trigonometric functions to the decomposition of a function into trinomial real factors.
XI. Series, as continuation of the II.section.
XII. Equations in two or more variables.
XIII. Analysis of curved lines.
XIV. Calculus of finite differences.
XV. Connection between the Analysis of the Finite and the Differential Calculus:
  1. Through Taylor's theorem and some of its applications to the theory of series
  2. Through the general theorem of Lagrange for the reversion of series
  3. Through the determination of maxima and minima of a function
  4. In the geometry of curved lines, through the determination of tangents, normals and special points; through the formulae for different methods of generating lines by evolution, rotation etc.
XVI. Indeterminate or diophantine analysis, which may be viewed as the second main part of algebra.

From Klügel's survey we can conclude that *Taylor's theorem* was thought to belong to algebraic analysis, not to differential calculus. Shortly after 1900, Felix Klein argued that the 1812 syllabus for Prussian gymnasia, the so-called syllabus of Süvern, included differential calculus because Taylor's theorem was mentioned in it.[14] From the above survey it can be concluded that this assertion was historically not justified. It is also clear that the problems which are the main applications of calculus as taught in today's schools (extrema, tangents) were, in Klügel's classification, part of algebraic analysis.

The didactical conception of algebraic analysis

Throughout the nineteenth century the arithmetic-algebraic part of the school syllabus in the German gymnasia was a true, though reduced, image of algebraic analysis as exhibited in Klügel's survey.[15] In particular, calculations with series were treated as an application of combinatorics and the binomial theorem was the high point of the course.

The intrinsic meaning of the didactical version of algebraic analysis may be described as a universe ("organism") of symbolic expressions which can be treated by a uniform method. The extrinsic meaning consisted in elementary applications to geometry and physics. In 1834 the government decided to eliminate analytic geometry from the syllabus. Therefore the applications were reduced to calculations of simple geometrical quantities, such as areas and volumes, and to plane and spherical trigonometry. The movement to reform mathematics teaching did not start until 1872 with the demand for the reintroduction of analytic geometry. The demand was made in a speech by Émil Du Bois

Reymond.[16] Only around 1900 did a request for the introduction of the differential calculus into school teaching emerge.

To understand the nature of this school mathematics we must see it as a structure composed of theory and applications.

theory	applications
algebraic analysis	caculations of elementary geometrical and physical quantities, plane and spherical trigonometry

At first sight, the teaching of these subjects could seem the teaching of boring formal calculations. However, pedagogical and mathematical ideas were inherent in the combinatorial approach to algebra which are worth considering. On a general level, these ideas can be described by the catchword "insight" into structural properties of formulae. In contrast to the approach frequently followed today, algebra was not treated as a technique of applying rules to formulae. Rather, the guiding viewpoint was to build up a "network" of formulae connected by structural similarities and differences. The binomial formula is a good example with which to illustrate this idea. In the standard treatment

$$(a + b)^2 = a^2 + 2ab + b^2$$

is an isolated entity derived by applying rules such as the distributive law, the commutative laws for addition and multiplication, and so on.

The alternative view aims at building up sequences of connected formulae:

$$(a + b)^2 = a^2 + b^2 + 2ab$$

$$(a + b + c)^2 = a^2 + b^2 + c^2 + 2(ab + ac + bc)$$

$$(a + b + c + d)^2 = a^2 + b^2 + c^2 + d^2 +$$

$$2(ab + ac + ad + bc + bd + cd)$$

One can ask a pupil what will happen if a fifth letter occurs, or if one letter is omitted. Similarly, does the pupil know how the equality

$$(a - b)(b - c)(c - a) = -a^2b - b^2c - c^2a + ab^2 + bc^2 + ca^2$$

will change if another letter is introduced, or what is the law governing the distribution of positive and negative signs. In other words, the main aim was not to apply a rule, but to see the symmetries and regularities of a formula; a formula can be better understood if seen as a special case in a group of similar formulae. Thus the pedagogical and mathematical essence of the combinatorial approach to algebra consisted in what one can call a "functional principle for symbolic expressions," which is realized by conscious variations of form.

From these insights into the structure and pedagogical ideas of nineteenth-century school mathematics, it is possible to come to a new evaluation of the question of why differential and integral calculus were not taught in nineteenth-century gymnasia. This question was first posed at the end of the nineteenth century when the reform movement, under the influence of Felix Klein, demanded the introduction of this subject matter into the school curriculum. Impressed by the heated debates, historiographers of the time answered this question by saying that the absence of infinitesimal calculus was a consequence of the dominance of the classical languages and the suppression of mathematics in neohumanist gymnasia. However, the above analysis leads to a different conclusion. If it is accepted that school mathematics as described above was a complete and closed whole, then the question of why differential and integral calculus were not taught proves largely ahistorical. Infinitesimal calculus was by nature not part of the curriculum, and mathematics teachers themselves defended this position.

To elaborate my argument, I shall show that algebraic analysis was a complete paradigm for school mathematics. "Complete" means that it was possible to solve those problems, usually associated with infinitesimal calculus, for example the determination of extrema, within algebraic analysis. As noted above, at the beginning of the nineteenth century these problems were counted among the topics of algebraic analysis. Methods adapted to the needs of school teaching were developed during that century. One of them was the so-called "method of Schellbach."[17]

Karl Heinrich Schellbach (1804–1892) was a famous mathematics teacher and teacher-trainer in Prussia who exerted a great influence on mathematics teaching. Felix Klein claimed that "Schellbach's method" was "disguised infinitesimal calculus" and that Schellbach did not speak openly of infinitesimal methods because he feared the classical philologists.[18] For the latter assertion Klein gave no proof, and in regard to the former it is possible to show that Schellbach's method can be interpreted as an integral part of algebraic analysis and that it was didactically well founded.[19]

Let f be a given function with an extremal point at x_E (see figure). An arbitrary x is chosen sufficiently near to the extremal point x_E and the parallel to the x-axis is drawn through the point $(x, f(x))$. It intersects the graph of f at $(x_1, f(x_1))$, where x_1 is on the other side of x_E. Thus to every x there corresponds an x_1. The extremal point is characterized by the condition that x corresponds

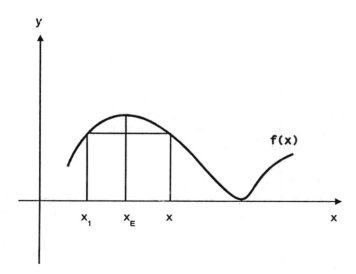

to itself when the parallel is tangent to the graph of f. By this condition we can calculate x_E in the following way.

The conditions on x and x_1 imply

$$f(x) - f(x_1) = 0$$

Schellbach claimed that it is "always possible" to factor $f(x) - f(x_1)$ and to write

$$(x - x_1) \cdot g(x, x_1) = 0.$$

For all x with $x \neq x_1$ it follows that:

$$g(x, x_1) = 0$$

For reasons of continuity this must also be the case for $x = x_1 = x_E$. Therefore x_E can be calculated from the equation

$$g(x, x) = 0.$$

This was Schellbach's simple procedure. Felix Klein claimed that this was "diguised infinitesimal calculus," because factoring $f(x) - f(x_1)$ and setting $x = x_1$ may be seen as equivalent to evaluating the limit

$$\lim_{x \to x_1} \frac{f(x_1) - f(x)}{x_1 - x}$$

without mentioning the conceptual difficulties involved.

The whole procedure, however, can be interpreted equally well as a purely algebraic calculation within the framework of analytic geometry. The extremal point is algebraically characterized by the condition that the intersection of the parallel with the graph of the function f becomes a single point which must be counted twice. "Schellbach's method" simply elementarized Descartes' procedure for calculating tangents and normals in his "La Géométrie",

published in 1636 as an appendix to the *Discourse de la Méthode*. Moreover, the method that a tangent to a curve results from the algebraic condition that the cut of a secant with a curve reduces for a tangent to a single point which has to be counted twice was also a common topic in contemporary university textbooks of analytic geometry. The schools also had a textbook on algebraic analysis which included a large part on analytic geometry and which gave Descartes's method in full generality and not only the special case considered by Schellbach.[20]

The application of this procedure requires deciding whether an extremum exists and where it lies. This investigation uses the concrete conditions of the problem at hand. For instance, to calculate the height h of a cone of maximal volume with a given lateral area k^2 one notes that for $h = 0$ and $h = \infty$ the volume is 0. For values of h between 0 and ∞ the volume is positive, and, therefore, it must take on a maximum.

Then Schellbach performed the following calculation.[21] Let the cone have generator s, height h and circular base with radius x. Then

$$\pi x s = k^2, \qquad s = \sqrt{x^2 + h^2},$$

and the sought volume

$$v(x) = \frac{1}{3}\pi x^2 h.$$

It follows that

$$h^2 = \frac{k^4 - \pi^2 x^4}{\pi^2 x^2}$$

and hence

$$V(x) = \frac{1}{3}\sqrt{k^4 x^2}.$$

This function becomes a maximum or minimum when the expression under the root is maximal or minimal. Thus, according to the method described above:

$$k^4 x^2 - \pi^2 x^6 = k^4 x_1^2 - \pi^2 x_1^6 \qquad \text{or}$$

$$k^4(x^2 - x_1^1) - \pi^2(x^6 - x_1^6) = 0$$

$$(x^2 - x_1^2)\left[k^4 - \pi^2(x^4 + x^2 x_1^2 + x_1^4)\right] = 0$$

In this case, it is possible to cancel the factor $x^2 - x_1^2$, because it corresponds to the solution $x = 0$ and $h = \infty$ with $V(x) = 0$. Therefore;

$$k^4 - \pi^2(x^4 + x^2 x_1^2 + x_1^4) = 0$$

For $x = x_1$ this leads to the solution

$$k^4 = 3\pi x^4, \qquad \text{or}$$

$$x = \frac{k}{\sqrt[4]{3\pi^2}}, \quad h = \frac{k\sqrt{2}}{\sqrt[4]{3\pi^2}}, \quad s = \frac{k\sqrt{3}}{\sqrt[4]{3\pi^2}}.$$

Clearly, this concrete investigation of a problem is often pedagogically preferable to a blind application of an algorithm requiring mere mechanical calculations of derivatives. Schellbach's method can be used to solve a concrete extremum problem involving a concretely given (algebraic or analytic) function. Questions of existence and uniqueness are treated by recourse to the concrete problem at hand.

Thus algebraic analysis provided an effective method for calculating extrema of concretely given algebraic and analytic functions. For these functions the factoring works, and Schellbach's method can be applied. These are the only functions that occur in school mathematics. Infinitesimal calculus becomes necessary only when more general functions have to be treated and when one wishes to give general conditions for the existence and uniqueness of extrema. In addition, the pedagogical superiority of a method which forces the pupils to study a problem concretely and does not use an unnecessary conceptual apparatus is obvious. Schellbach had good reasons for his method.

Conclusions

The foregoing analysis leads to three conclusions. First, there are obvious consequences for a concrete investigation of the Klein reform movement which is still missing. Such an analysis has to take into account that Klein's proposals met an existing tradition of teaching which was coherent and plausible for many mathematics teachers. Thus the whole reform can be understood as an interaction of two different mathematical paradigms: the elder tradition of algebraic analysis versus the reorganization of teaching due to the needs of the newly introduced infinitesimal calculus. Such a study would be important and instructive.

Second, two aspects of nineteenth-century school mathematics are of value for today's teaching . One is Euler's heuristic and experimental treatment of formulae. In today's schools students often learn to apply algebraic rules instead of learning to consider the shape of formulae and to recognize *symmetries, regularities,* and *structures.* The introduction of some elements of the Euler style of treating formulae into school teaching would benfit the students.

Third and final, algebraic analysis gives occasion to rethink the syllabus for the upper grades of school teaching. Instead of teaching a systematic differential calculus one could introduce a more general treatment of functions which develops and uses various methods for the investigation of individual functions. Nineteenth-century textbooks teach us that no systematic calculus is needed for the calculation of specific limits and tangents. Of course, to understand the algebraic procedure for calculating tangents might be as difficult for a pupil to grasp as the beginnings of differential calculus, and, obviously, we must not return to as extensive use of formulae as in the nineteenth-century. My point is of a more philosophical nature and aims at our attitude toward doing mathematics. In this regard, it would be healthy, I believe, to emphasize more the value of considering individual problems, instead of exclusively teaching powerful theories.

Acknowledgement. I would like to thank Abe Shenitzer, Toronto, and Ron Calinger, Washington, for valuable criticisms and for polishing the English of the present paper.

References

1. See Craig C. Fraser, "The Calculus as Algebraic Analysis: Some Observations on Mathematical Analysis in the 18th Century," *Archive for History of Exact Sciences,* 1989, 39: 317–335.
2. Augustin-Louis Cauchy, *Cour d'analyse de l'école royale polytechnique. Ire Partie. Analyse algébrique.* (Paris: Debure frères, 1821). References are to: Oeuvres complètes d'Augustin Cauchy, IIe série, tome III, Paris 1897.
3. Alfred Pringsheim and Georg Faber, "Algebraische Analysis". In *Enzyklopädie der Mathematischen Wissenschaften mit Einschlu ihrer Anwendungen,* II C 1 (Leipzig: Teubner, 1909–21), pp. 1–46.
4. Leonhard Euler, *Introductio in analysin infinitorum.* Tomus primus. (Lausanne: Bousquet, 1748). References are to: Leonhardi Euleri *Opera Omnia* I, 16.
5. Joseph Louis Lagrange, *Théorie des fonctions analytiques.* (Paris 1797). References are to: Oeuvres de Lagrange IX.
6. See Hans Niels Jahnke, *Mathematik und Bildung in der Humboldtschen Reform.* (Göttingen: Vandenhoeck & Ruprecht, 1990), chapter 4.
7. For the history of the binomial formula see Michel Pensivy, "Jalons historique pour une épistémolgie de la série infinie du binôme," *Sciences et techniques en perspective,* 1987/88, vol 14, and Jean Dhombres, "Quelque aspects de l'histoire des équations fonctionelles liés ... l'évolution du concept de fonction", *Archive for History of Exact Sciences,* 1986, 36(2): 91–181, on p. 150–163.
8. As a method for the solution of "affected literal equations" the algorithm appeared first in Isaac Newton, *De analysi per aequationes numero terminorum infinitas* (mscrpt. of 1669), published in: Derek T. Whiteside (ed.), *The Mathematical Papers of Isaac Newton,* vol. II (Cambridge: Cambridge University Press, 1968), 206–273, especially on p. 223–243. The method was made known to Leibniz by Newton's letter to Oldenbourg from June, 13, 1676, published in: H. W. Turnbull (ed.), *The correspondence of Isaac Newton, vol. II: 1676–1687* (Cambridge: Cambridge University Press, 1960), p. 20–47. For its relevance to Newton's calculus see Charles Henry Edwards, *The Historical Development of the Calculus* (New York: Springer-Verlag), p. 204–209.

9. Heinrich August Rothe, *Formulae de serierum reversione demonstratio universalis, signis localibus, combinatorio—analyticorum vicariis exhibita* (Leipzig: Sommer, 1793).

10. See Carl Friedrich Hindenburg, *Der polynomische Lehrsatz, das wichtigste Theorem der ganzen Anaysis: nebst einigen verwandten und anderen Sätzen, neu bearbeitet und dargestellt v. Tetens,... Zum Druck befoerdert und mit Anmerkungen, auch einem kurzen Abrisse d. combinatorischen Methode und ihrer Anwendung auf die Analysis versehen* (Leipzig: Fleischer, 1796).

11. See Hans Niels Jahnke, "Algebraische Analysis in Deutschland, 1780–1860," In Detlef Spalt (ed.), *Rechnen mit dem Unendlichen—Beiträge zu einem kontroversen Gegenstand* (Basel: Birkhäuser, 1990). pp. 103–121.

12. Consider for example the paper of Weierstrass's teacher Christoph Gudermann: *Allgemeiner Beweis des polynomischen Lehrsatzes ohne die Voraussetzung des binomischen und ohne die Hülfe der höheren Rechnung.* (Cleve: Schulprogramm, 1825). With some modifications this paper was published as part of Gudermann's larger paper "Theorie der Potenzial-oder cyklisch-hyperbolischen Functionen", *Journal für die reine und angewandte Mathematik*, 1830, 6: 1–39; 162–194; 311–356.

13. Georg Simon Klügel, *Mathematisches Wörterbuch oder Erklärung der Begriffe, Lehrsätze, Aufgaben und Methoden der Mathematik mit den nöthigen Beweisen und litterarischen Nachrichten begleitet in alphabetischer Ordnung. Erste Abtheilung: Die reine Mathematik, Bd. I bis III.* (Leipzig: Schwickert, 1803–8). Klügel (1739–1812) had become famous by his dissertation "Conatuum praecipuorum theoriam parallelarum demonstrandi recensio, quam publico examini submittent A. G. Kaestner et auctor respondens G. S. Klügel (Göttingen 1763) where he showed that all known attempts to prove Euclid's eleventh axiom were mistaken. Carl B. Boyer comments on Klügel's "Mathematisches Wörterbuch": "Portrays the status of the subject a century and a half ago." (*A History of Mathematics*, (New York: John Wiley & Sons, 1968), p. 680).

14. Felix Klein, *F. Vorträge über den mathematischen Unterricht an den höheren Schulen. Bearbeitet von R. Schimmack. Theil 1: Von der Organisation des mathematischen Unterrichts.* (Leipzig: Teubner, 1907), p. 109.

15. See Hans Niels Jahnke, Die algebraische Analysis im Mathematikunterricht des 19. Jahrhunderts. *Der Mathematikunterricht*, 1990, 36(3): 61–74.

16. Émil Du Bois-Reymond, *Kulturgeschichte und Naturwissenschaft.* (1877). Reference is to: S. Wollgast (ed.), E. Du Bois-Reymond, *Vorträge über Philosophie und Gesellschaft*, (Hamburg: Meiner, 1974), p. 105–158.

17. See Karl Heinrich Schellbach, *Mathematische Lehrstunden. Aufgaben aus der Lehre vom Gröten und Kleinsten.* Bearb. u. hrsg. v. A. Bode u. E. Fischer. (Berlin: Reimer, 1860).

18. l.c., 108.

19. Jahnke, l.c., *Die algebraische Analysis* ....

20. Wilhelm Gallenkamp, *Die Elemente der Mathematik. III. Theil. Die algebraische Analysis und die analytische Geometrie, insbesondere die Kegelschnitte enthaltend.* (Iserlohn: J. Bädeker, 1860).

21. l.c., 21/22.

The Mathematics Seminar at the University of Berlin: Origins, Founding, and the Kummer-Weierstrass Years

Ronald Calinger
Catholic University of America

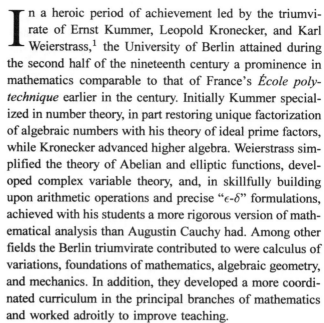

In a heroic period of achievement led by the triumvirate of Ernst Kummer, Leopold Kronecker, and Karl Weierstrass,[1] the University of Berlin attained during the second half of the nineteenth century a prominence in mathematics comparable to that of France's *École polytechnique* earlier in the century. Initially Kummer specialized in number theory, in part restoring unique factorization of algebraic numbers with his theory of ideal prime factors, while Kronecker advanced higher algebra. Weierstrass simplified the theory of Abelian and elliptic functions, developed complex variable theory, and, in skillfully building upon arithmetic operations and precise "ϵ-δ" formulations, achieved with his students a more rigorous version of mathematical analysis than Augustin Cauchy had. Among other fields the Berlin triumvirate contributed to were calculus of variations, foundations of mathematics, algebraic geometry, and mechanics. In addition, they developed a more coordinated curriculum in the principal branches of mathematics and worked adroitly to improve teaching.

The Mathematics Seminar at the University of Berlin was central to this creative endeavor. Founded in 1861, it was the first seminar on pure mathematics in Germany. Yet the English-speaking world knows little of its origins and work, and scholars have disagreed over who was the principal founder—Kummer or Weierstrass. This paper will examine the origins and founding of the Mathematics Seminar and its guiding research currents to 1883.

Long-Term Origins: The Research Model and Autonomous Discipline at the University of Berlin and Initial German Science Seminars

Fundamental to establishing a major mathematics seminar in Berlin was the development there of a new type of university—the modern research university—whose first stage had emerged in the German states chiefly at Göttingen University during the late eighteenth century. Founded in 1737, Göttingen in the kingdom of Hannover had a strong library, medical facilities, and museum, along with the *Societät der Wissenschaften,* which was added in 1751. In the 1790s leading German universities expected professors to conduct independent research that preserved or increased knowledge as well as to teach and guide students.[2] Soon after 1800 Göttingen University made explicit in its university regulations this double obligation of the professoriat.

After the Prussian defeat in 1806 at Jena-Auerstadt by Napoleon and the temporary loss of substantial amounts of territory, including Halle, by the Treaty of Tilsit (1807), King Friedrich Wilhelm III urged his state to "replace intellectually what it has lost physically."[3] During the intense

reform movement that followed, the University of Berlin opened as Prussia's "central university" in 1810. The main founder, Wilhelm von Humboldt (1767–1835), chief of the Section for Education and Religion in the Prussian Interior Ministry, designed Berlin to be the model of the modern research university. Sober, many-sided research in pursuit of a new ideal of *Wissenschaft* took place within disciplines increasingly differentiated. The potential for growth in knowledge was vast. The result has been called a *Geistesrevolution,* an intellectual revolution.

The foremost mission of Humboldt's university was to awaken a critical inquiry that would examine existing knowledge and discover new knowledge. In writing the statute for this and composing other university regulations, Humboldt drew upon philosopher Johann Fichte's plans for a university and was assisted by liberal Protestant theologian Friedrich Schleiermacher. Immanuel Kant's *Strife among Faculties* (1798), Friedrich Schelling's *Lectures on the Method of Academic Study* (1803), and speculative German *Naturphilosophie* argued for making the philosophical faculty—the arts and sciences faculty—central to research.[4] Schelling wanted the sciences considered separately from their practical applications. The requirement that all ordinary professors at Berlin possess a doctorate strengthened original research, for it meant that these faculty would have at least some experience in it. As a scholar and educator, Humboldt exemplified the pursuit of creative research and devotion to higher culture. Inspired by cultural nationalism and the concept of world history as an unfolding of consciousness, many literate Germans believed that rigorous studies of higher culture would release the spiritual and moral resources of their country along with the material.

The other mission of the University of Berlin, also conducive to research, was preparation for the professions and for state service (*Staatswissenschaften*). But these tasks, which had dominated eighteenth-century German universities, Humboldt made secondary to intellectual formation. Humboldt's reform relieved the philosophical faculty of its function of preparing students to enter the three higher faculties of law, theology, and medicine—a function soon assumed for all four faculties by reformed gymnasia. Humboldt's stress on original scholarship permanently broadened the mission of Prussian universities.

The university sustaining this twofold mission was to enjoy a degree of autonomy in governance, essentially by ordinary professors. The effort to rejuvenate learning in response to military defeat had fostered the ideal of academic freedom, and Humboldt wanted freedom as developed at Jena and Göttingen,[5] where faculty were not under supervision by theologians and censorship of lectures, research, and writings. But since the government funded the new university, government reviews and regulations existed, and the Prussian state reserved the right to appoint faculty. The infamous Carlsbad Decrees of 1819 constricted academic freedom in Berlin. They had German states appoint plenipotentiaries (university inspectors) and press censors.[6] Through the 1860s tensions continued with Prussian officials, particularly over the issues of free speech and constitutional government.

Its future strengthened by the university's commitment to research, mathematics profited also from the Prussian government's transference of teacher education in 1810 from the theological to the philosophical faculty. Previously professors in the philosophical faculty had taught mathematics to nonspecialists to complete a liberal education, and thus most of the mathematics faculty had been generalists. Now that education students had to take specialized courses in the subject over four years, mathematics came to be viewed as a major rather than an auxiliary discipline. Assigning professional preparation to the philosophical faculty seems integral to Humboldt's reforms. It moved the philosophical faculty toward an independent status equal to that of the formerly higher faculties of theology, law, and medicine.[7] In addition, the new disciplinary and professional nexus between the sciences and education extended among science and mathematics faculty—their authority now based on research—the high social ranking of scholars or *Gelehrte* in Prussian society. Secondary school teachers, who fostered the ideals of such studies, also attained that ranking.

The *Lehrplan* (1815/16) of the liberal *Staatsrat* Johann Wilhelm Süvern increased the need for mathematics school teachers in Prussia. Süvern, like Humboldt, was influenced by neohumanism. Reform in Prussia looked to autonomy or *Selbsttätigkeit,* embracing economic independence and cultural sophistication on the part of the individual, and neohumanism could address that last objective. Neohumanism would base general education mainly on ancient Greek and Roman literature and languages, history combined with geography, and the sciences.[8] Modern languages, seen as a practical skill, had a lesser place. Neohumanism included a Platonic concept of mathematics as elucidating hidden structures of things and providing methods that were exemplars for a higher order of rational contemplation.[9] For the new secondary schools or gymnasia, a term evoking the athletics and basic learning of classical Greece preparatory to the higher learning of Plato's Academy and Aristotle's Lyceum, Süvern's *Lehrplan* or curriculum stressed Greek and Latin, German, and mathematics.[10] The increased demand for secondary-school faculty prepared to teach mathematics supported the autonomy of mathematics at the University of Berlin.

Süvern's *Lehrplan* required mathematics to be taught six out of the thirty-two hour weekly schedule. This broke with the monopoly classical studies had enjoyed in the older, various *Gelehrtenschulen,* such as *Lycéen* and *Collegien,* renamed gymnasia in 1812. Opponents of Süvern's changes successfully limited the time for the natural sciences to a modest two hours a week, and it was difficult to implement the neohumanist *Lehrplan. Realschulen,* which gave more attention to science, competed with them. But gymnasia continued to administer the *Arbitur,* the graduation examination from secondary school.[11]

Within nineteenth-century German universities research methods and skills were chiefly developed in seminars, institutes, museums, and, after 1830, laboratories.[12] Most seminars were aligned with the humanities, and institutes with the sciences. These two were intended not to prepare teachers for schools in subject content but to build independent scholars. Propagation of research methods was their chief goal.[13] In seminars, lectures were complemented once a week with intellectual exercises to which students had to respond. This offered graduate students their only direct contact with a senior professor, who might systematically impart research methods and advise them.[14] Seminar students had to present a lecture on an assigned topic. Typically the faculty of an entire department and sometimes cognate areas as well would comment on the presentation. Seminars and institutes had separate funding lines for scholarships, for staff, and sometimes for library materials that included recent research publications.

Troubled by a lack of teachers qualified to teach the natural sciences, a subject important for a modernizing Prussia, the Prussian Cultural Minister Karl Altenstein founded in 1825 the first German science seminar.[15] Bonn was selected as the site, because it possessed necessary equipment and its science faculty had proposed a similar union three years earlier. Altenstein himself had proposed in 1824 that something like Berlin's Philology Seminar be created. Bonn's science faculty developed a three-year course of study in the natural sciences for seminar students planning to become teachers in gymnasia or municipal trade schools, but not other secondary schools or universities. The seminar curricular plan recommended the organization of student studies, time for each scientific discipline, and courses. After two semesters, students entered the seminar's upper level, in which essays employing sound research methodology were pivotal. Essays had to be in one of three forms: a critical review of a methodology, a de-scription of personal experiments, or a proposal of new research propositions. Perhaps because neohumanism postulated a unity between research and teaching, essays had to be presented orally and defended in class. The Bonn Seminar became highly successful. One of its *dozents* was Julius Plücker.[16]

Altenstein encouraged ventures at Halle and Königsberg similar to that at Bonn. By 1828 Heinrich Scherk had founded Halle's short-lived Mathematical Association, which had the character of a seminar in the natural sciences and mathematics. Ernst Kummer (1810–1893), who studied at Halle from 1828 to 1831, participated in the association. Scherk turned him from theology to mathematics, which Kummer considered a "preparatory science" for philosophy. For Scherk Kummer wrote a prize essay in 1831 on sines and cosines. When Scherk left Halle in 1833, the association ended.[17] Still, by the late 1820s the institutional structure for science seminars was developing in Prussia. By building an outstanding mathematics faculty, the University of Berlin provided the next and core element for the mathematics seminar.

The Mathematics Faculty at the University of Berlin to 1856

From its beginning in 1810, the University of Berlin recognized the importance of mathematics. A rich tradition in mathematics at the Berlin Academy as well as elements within romanticism, neohumanism, and French mathematics of the time influenced university founders and Prussian officials.

The learned mathematical tradition at the Berlin Academy went from Gottfried Leibniz to Leonhard Euler and Luigi Lagrange, and from them to Karl Gauss.[18] Gauss was elected a member in 1810 and Wilhelm von Humboldt attempted to draw him to Berlin. The Prussian Cultural Ministry subsequently attempted several times to hire him: in 1823/24 and again in 1828 the Ministry wanted him for a proposed polytechnic modeled after Paris's *École polytechnique*.[19] After Göttingen increased his salary, Gauss remained at his *alma mater.*

Although romantics were hostile to misapplying mathematics, that is, extending it to cover new fields without proper preparations or recognizing possible limits, which they saw as a fault of the late Enlightenment, a few Kantians developed critical mathematics to understand better a harmonious universe. In *Mathematische Naturphilosophie* (1822) Jakob Fries of Heidelberg, for example, reexamined the logical consistency of Euclid's postulates, especially parallelism. Gauss praised this work of Fries. At roughly the same time Gauss, Nikolai Lobachevsky, and Janos Bolyai were independently inventing non-Euclidean geometries.[20] In 1816 Fries had unsuccessfully sought a position in Berlin.

French achievements in mathematics sharpened the zeal for the discipline at the University of Berlin. Prussian officers attributed Napoleon's military victories largely to the training French officers had received at France's *École polytechnique,* whose faculty included Luigi Lagrange, Pierre-Simon Laplace, Gaspard Monge, and Joseph Fourier.[21] The Prussian military attempted to draw distinguished mathematicians to its military academy, founded in 1816, and succeeded in recruiting Martin Ohm and Gustav Peter Lejeune Dirichlet (1805–1859) in the 1820s.

After Gauss refused to come to Berlin, two men outside the university—geophysicist Alexander von Humboldt (1769–1859), brother of Wilhelm, and August Leopold Crelle (1780–1855)— undertook crucial measures to make it a European center for mathematical research. Both were imbued with neohumanistic ideals, had a thorough knowledge of French teaching methods, and were respected at the Prussian court. After two decades of service in Paris, where he had belonged to the circle of intellectuals who gathered around Joseph Fourier, Alexander von Humboldt returned to Berlin in 1827, determined to make the level of mathematics education in Berlin comparable to that in Paris. In Berlin he collaborated with Crelle, an engineer with a strong interest in mathematics. In 1826 Crelle had founded his *Journal für die reine und angewandte Mathematik,*[22] in which talented young authors published fresh and trend-setting papers in mathematics. In the years to come, new journals were to be crucial to the growth of mathematical research and the emerging professional community of mathematicians across national and multinational boundaries. While serving in the Education Section of the Prussian Ministry of Culture from 1828, Crelle studied extensively the French teaching of mathematics. The "true goal" of mathematics as he defined it went beyond applied mathematics, which the *École polytechnique* emphasized. Pure mathematics provided fundamental methodologies for coherent applications. Pure mathematics, wrote Crelle, must "develop . . . from within itself" and "be free to move and evolve in all directions"[23] and thereby enlighten human understanding.[24] For a better preparation of mathematics teachers, Crelle proposed in 1828 a mathematics seminar at the university with a library, reading room, and eighteen hours of lectures.[25]

In 1828 Berlin acquired Dirichlet from Breslau with the assistance of Alexander von Humboldt. Previously Dirichlet had studied in Paris under Fourier as well

as Sylvestre François Lacroix and Simeon-Denis Poisson. Having begun teaching at the Military Academy (*Kriegschule*) in Berlin, Dirichlet quickly moved to the university as well, remaining there for 27 years.

Led by Dirichlet, Berlin's mathematics faculty from 1828 to 1856 worked to improve examinations, introduce innovations in pedagogy, develop the curriculum beyond elementary university mathematics, and move into the forefront of research. Historian Gert Schubring has described this complex process as the modernizing of Berlin mathematical studies.[26] In this effort, Dirichlet was to be joined chiefly by Jakob Steiner (1796–1863) and in 1844 Carl Jacobi (1804–1851).[27] The faculty enjoined active learning through exercises and problem solving. In 1845 Ferdinand Joachimsthal (1818–1861), who had participated in Jacobi's Königsberg Seminar, became a *Privatdozent* in Berlin, where he developed impressive drill sessions. These Berlin faculty presented their research results in advanced classes, which were colloquia, and raised standards of rigor acceptable for proofs—an issue evident in Jacobi's remark:

> Dirichlet alone, not I, nor Cauchy, nor Gauss knows what a completely rigorous mathematical proof is. Rather, we learn it first from him. When Gauss says that he has proven something, it is very clear; when Cauchy says it, one can wager as much pro as con; when Dirichlet says it, it is certain . . . .[28]

A scholarly lifetime that included a cycle of 88 lectures and exemplary papers on number theory, foundations of mathematical analysis, and mathematical physics took early form in a careful study of Gauss's *Diquisitiones Arithmeticae* (1801). Dirichlet improved upon some proofs of Gauss and moved deeply into new areas of number theory, which remained his primary interest. He became a major interpreter of Gauss, and his lectures on number theory were the chief reason the *Disquisitiones Arithmeticae* was known beyond a small circle.

Dirichlet rose rapidly to prominence among European mathematicians. In an 1825 paper he provided an incomplete proof for the case $n = 5$ of the exasperating Diophantine equation known as Fermat's Last Theorem: there are no nontrivial integer solutions to the equation $x^n + y^n = z^n$, where n is any integer greater than 2. This proof, completed by Adrien-Marie Legendre, began a period of a broader, though frustrating, search by French and German mathematicians for a general solution. The case for $n = 4$ is simple, and Fermat had posed $n = 3$ as a challenge problem, which Euler and Gauss had subsequently solved. Thus, mathematicians now turned to $n = 7$ and

August Leopold Crelle (1780-1855)

Gustav Peter Lejeune Dirichlet (1805-1859)

Carl Jacobi (1804-1851)

greater integers.[29] Necessary techniques and conditions required for general solvability, however, were still missing.

Extending the work of Gauss and Jacobi, Dirichlet studied higher reciprocity laws on divisibility during the 1830s. This led him to search for a general theory of algebraic numbers having as its main theorem essentially unique factorization of complex numbers into prime numbers.[30] The genesis of the idea of applying unique factorization in proving theorems in number theory seems to be chapter 2 of Euler's *Vollständige Anleitung zur Algebra* (1770). Dirichlet did not achieve this prime factorization, but his correspondent Kummer did.[31]

A story has it that the acutely critical Dirichlet objected in 1843 to Kummer's preparatory theory. Reportedly Kummer's initial argument contained a gap, tacitly assuming that unique decomposition into primes applies to Dirichlet's class of algebraic numbers extending Gauss's complex integers. But that assumption is valid only for certain complex numbers and it lacked a proof. This was basically the same flaw as in Euler's 1753 proof for $n = 3$. This story, which mathematician Kurt Hensel first advanced in 1910, is questionable. No record of correspondence relating to it has survived, and, as historian Harold Edwards observes, it seems unlikely that Kummer would have made this error in a paper he was preparing for publication.[32]

In a series of papers from 1844 Kummer developed his bold new theory of ideal factors of complex integers.[33] Complex integers were also called cyclotomic integers. The name derives from Gauss's geometric interpretation of them based on his theory of the division of the circle (cycl- = circle and tom- = divide). In 1844 Kummer discovered that unique prime factorization does not hold for them. To reestablish unique factorization, Kummer had to develop a method within a set larger than the concrete integers. His larger set included adjoining "ideal numbers," which fill in the gaps that had prevented unique prime factorization. Kummer recognized that a proof of Fermat's Last Theorem for a class of primes requires that another condition be met. In order for the logic of his argument to hold, primes must be regular. Primes n are regular if and only if n does not divide numerators of Bernoulli numbers. These numbers comprise coefficients of the expansion $x(1 - e^x)^{-1} = \sum_{n=0}^{\infty} B_n x^n / n!$. The primes to 37 are regular, but many larger primes are irregular. In an 1847 article in Crelle's *Journal,* Kummer proceeded to prove for the most part Fermat's Last Theorem for regular primes.[34] He worked with regular primes $n < 61$,[35] knowing that 37 is irregular, but not yet recognizing that so is 59.[36] Kummer asserted that there are infinitely many regular primes but could not prove it. He later retracted this statement.

Dirichlet also improved the definition of functions, which was still in a formative stage, and showed more exactly how continuity and discontinuity relate to integrability of functions.[37] Fourier had cut a Gordian knot for mathematicians when he revealed that the requirement that any function be representable analytically by a simple formula is not significant. In articles in Crelle's *Journal* for 1826 and 1829, Dirichlet asserted that all continuous functions can be represented by convergent Fourier series. He examined discontinuities and offered his celebrated function that is discontinuous at each point in an interval. Dirichlet's function is:

$$f(x) = 1, \quad \text{when } x \text{ is rational, and}$$
$$= 0, \quad \text{when } x \text{ is irrational.}$$

His 1829 paper proved that the class of functions with a finite number of maxima and minima has convergent Fourier series. Integrability depends upon the functions' being piecewise continuous. After restating his views in the initial issue of *Repertorium der Physik* (1837), Dirichlet conceived of functions as correspondences,[38] as had Euler. Göttingen student Bernhard Riemann, in his *Habilitationschrift* (1854), carried forward Dirichlet's work on necessary and sufficient conditions to represent functions by Fourier series and on when they are integrable.

In mechanics Dirichlet invented the analytic tool that Riemann called "Dirichlet's principle."[39] This principle is a method to solve Dirichlet's problem in potential theory. For a continuous and uniform function in a region R, it assigns boundary values using differential equations. It seeks to minimize an integral, for example in electrostatics, and is thus part of the calculus of variations.

In 1828, Crelle had recommended that two other promising mathematicians—Carl Jacobi and Niels Abel—be hired to join Jakob Steiner at the new Berlin *Gewerbeinstitut,* the Vocational Training or Industrial Institute. In 1879 the Institute and the Royal Building Academy were to combine to form the *Technische Hochschule,* which today is the Technical University. Jacobi and Steiner had studied mathematics at the University of Berlin, and Abel had spent the winter of 1825 in Berlin on a Norwegian travel grant. All three became intimate friends and helped found Crelle's *Journal.* Their articles contributed to making it famous. Abel wrote seven papers for its first volume, Steiner five, and Jacobi one. Crelle had offered Abel the editorship, but a sufficient salary was not available. It took time for Crelle's recommendations to gain approval. Seeking a regular salaried position, Jacobi, who had been a Berlin *Privatdozent,* had transferred to Königsberg in 1826, later establishing there the research seminar in mathematics and physics, and Abel had returned home to Norway, where he died before an offer arrived in 1829. That year Steiner became senior master at the Industrial Institute with the strong backing of Wilhelm von Humboldt, whose son he had taught. In 1834, through the good offices of the Humboldt brothers and Jacobi, an extraordinary professorship of geometry was created for him at the University of Berlin, a post he held until his death in 1863.[40]

Concentrating on discovering relationships to establish its organic unity and coherence, Steiner was a principal founder of modern synthetic geometry. He first recognized that any point that can be constructed by a ruler and compass can be constructed by ruler alone if an auxiliary circle and its center are given in the plane of construction.[41] Criticized by Berlin colleagues who considered his approach elementary, Steiner in his teaching followed the Swiss educational reformer Johann Heinrich Pestalozzi (1746–1827), who sought to cultivate innate capacities of "head, heart, and hand." The teacher was to assist the student in learning, partly by posing questions, but not to follow an authoritarian rote imposition of information.[42] Students had to reflect independently upon materials, ask penetrating questions, and discover things for themselves with limited guidance from the teacher.[43]

Jacobi had a distinguished reputation based largely upon his having founded the theory of elliptic functions independently of Abel. Both Jacobi and Abel skillfully employed inverse functions to transform the theory of elliptic integrals of Adrien-Marie Legendre into the theory of elliptic functions. Both quickly discovered that these functions have a double periodicity and generalize trigonometric functions. Theirs was considered the most important mathematical discovery of its time.

Jacobi did not return to Berlin until June of 1844. In early 1843 poor health from diabetes exacerbated by the harsh climate of the region had sent him from the University of Königsberg to Rome, where he hoped to recuperate. He was joined there by his close friend Dirichlet. Jacobi lived his final seven years in Berlin. The retiring Dirichlet and the pugnacious Jacobi spent hours sitting together in silence in what Dirichlet's wife Rebecca called "Mathematik schweigen"—quiet mathematical inactivity.[44] In Berlin Jacobi was a member of the Academy of Sciences but not the university. The Cultural Ministry rejected his petition to become a faculty member. There were already three ordinary professors of mathematics, and the ministry refused to add a fourth. Academicians, however, had a statutory right to teach a course at the university.[45] Health problems, largely from his diabetes, permitted Jacobi to teach only sparingly.[46] In Berlin he lectured on function and number theory.[47] His students included Gotthold Eisenstein, Leopold Kronecker, and Bernhard Riemann. Jacobi died in 1851 of smallpox.

The end of the 1840s was an invigorating time. Senior mathematicians Jacobi, Dirichlet, and Steiner together with the well-known *dozents* Carl Wilhelm Borchardt, Ferdinand Gotthold Eisenstein, and Ferdinand Joachimsthal brought the university to an early peak in its mathematical reputation. Weierstrass was later to assert that no other university of the time could boast a mathematics faculty of their significance.[48]

In 1855 and 1856 the mathematics faculty underwent a critical transition. In 1855 Dirichlet left to succeed Gauss at Göttingen, and Crelle died. Dirichlet had long found it irritating that the mathematics and science faculty did not enjoy the status and income of the faculties of theology, law, and medicine. This recalled the practice at German universities before 1810 of treating as a lower faculty that of philosophy, of which mathematics and the natural sciences were a lesser part.

After his promotion to ordinary professor in 1839, Dirichlet had still lacked the full status of *Ordinarius,* since he had not written a dissertation in Latin, and his annual salary was modest. His salary in 1839 was 800 thaler against the average for an ordinary professor at Berlin of more than 1200 thaler.[49] He remained dissatisfied even after Humboldt and Jacobi obtained a near doubling of his salary to 1500 thaler in 1847 to keep him from moving to Heidelberg. Dirichlet, moreover, still had to teach an average of seven hours per week at the Military Academy. The lessons there were drudgery and interfered with his research. In 1855, Dirichlet announced that he would remain in Berlin only if his salary and teaching schedule were improved. Despite appeals from Humboldt and others, the Prussian Cultural Ministry did not respond until it was too late.

After Crelle died in 1855, Carl Wilhelm Borchardt, a former student of Jacobi and Berlin *Privatdozent* since 1848, became editor of Crelle's *Journal*. Borchardt continued to maintain its high standards for volumes 57 to 90.[50]

Steiner who might have succeeded Dirichlet as ordinary professor of mathematics was in ill health and was not promoted. Among the mathematical faculty Dirichlet recommended before leaving Berlin was Ernst Kummer of Breslau, whom he wanted to fill his post. Kummer, as Dirichlet noted, had begun his studies in Protestant theology, which helped his cause in Berlin. The energetic Kummer was admired for his research on complex numbers and his proofs of the laws of higher reciprocity, which he called his "archenemy." Although Kummer had never studied under Dirichlet, he was that mathematician's true disciple.

Ernst Kummer
(1810-1893)

Karl Weierstrass
(1815-1897)

He continued Dirichlet's program in number theory and the search for more rigorous proofs and improved pedagogy.

The University of Berlin confirmed Dirichlet's first choice. In the fall of 1855, Kummer arrived in Berlin followed by his favorite student from the Liegnitz Gymnasium, Leopold Kronecker (1823–1891). Kronecker, having received his doctorate from Berlin in 1845 while studying under Dirichlet, had been a mathematical amateur in the interim. Managing a family estate and selling his uncle's banking business in Liegnitz brought him a comfortable income. Kronecker was thus able to come to Berlin without a position and, spurred on by Kummer, to embark upon a period of creative research in number theory, algebra, and elliptic functions.

Dirichlet had also recommended Friedrich Richelot, Jacobi's successor at Königsberg, for a second position and praised the work of Otto Hesse of Königsberg, Georg Rosenhain of Vienna, and Karl Weierstrass (1815–1897) of Braunsberg Gymnasium on the Baltic in East Prussia. Richelot, Hesse, and Rosenhain were graduates of Jacobi's school. Weierstrass had just gained recognition as a result of his first paper, "On the Theory of Abelian Functions" in Crelle's *Journal* (1854). This paper solved the problem of inversion of hyperelliptic integrals by representing Abelian functions as quotients of power series that constantly converge. Crelle, who immediately recognized in the paper the mathematical genius of Weierstrass, considered him a continuer of the tradition of Abel, Jacobi, and Eisenstein. Borchardt traveled from Berlin to meet Weierstrass, and the two men became best friends. The nearby University of Königsberg quickly granted Weierstrass an honorary degree.[51] That was his status when Dirichlet spoke favorably of him to the authorities.

Dirichlet's response to Weierstrass had been guarded. When Crelle in January 1855 first pointed out to the Prussian Cultural Minister the merit of Weierstrass's research and requested a position devoted full time to research, the Minister asked Dirichlet to evaluate Weierstrass, but received no response to his initial letter. Responding to the second, Dirichlet recommended that Weierstrass have a suitable position devoted entirely to research.[52] But in March 1856 Dirichlet wrote to Kummer praising the work of Riemann in the theory of functions over that of Weierstrass.[53]

Kummer was determined to hire Weierstrass in Berlin. To block him from obtaining the vacant Breslau position, Kummer recommended geometer Ferdinand Joachimsthal, who had been a popular teacher while a *Privatdozent* at the University of Berlin. Joachimsthal was selected for the Breslau post, and Weierstrass, who had applied, was ranked

third. In 1856 Kummer acted to draw Weierstrass to the University of Berlin. In June he asked the Philosophical Faculty to recommend him and Borchardt to be extraordinary (*ausserordentlicher*) professors, but the vote was put off.

With help from Richelot, Humboldt obtained for Weierstrass the title of professor at Berlin's Industrial Institute with an annual salary of 1500 thaler.[54] Until September of 1856, when he traveled with Kummer to a science conference in Vienna, Weierstrass remained undecided about coming to Berlin. He strongly impressed the Austrians, who offered him a special professorship at any university in the Austrian portion of their Monarchy. To preclude that from happening, Kummer on 28 September wrote directly to the Cultural Minister.[55] Three days later the university offered Weierstrass a half-time *ausserordentlicher* professorship in addition to the post at the Industrial Institute, and he accepted.

In the fall of 1856, the forty-one-year-old Weierstrass came to Berlin from the Braunsberg Gymnasium. Crelle's *Journal* had just published the second of his famous papers on the theory of Abelian functions that first solved the Jacobian problem for hyperelliptic integrals. Kummer and Kronecker quickly became close friends with Weierstrass and Borchardt. With the arrival and collaboration of Kummer, Kronecker, and Weierstrass, the heroic period and the Mathematics Seminar were about to begin.

Funding Climate and Three Precursors

In the early 1860s funding for seminars and research institutes was sharply increasing at the University of Berlin. Budget comparisons in marks for the years 1811/12, 1834, and 1870 indicate that they enjoyed the greatest rate of increase of any major segment of the university budget.

Budgets	Seminars and Institutes	Faculty	University
1811/12	39,293	116,550	162,626
1834	78,444	193,650	299,538
1870	266,723	322,200	665,049 [56]

Inflation was not a significant factor. By 1870, the seminar and institute budget, only one-third of the amount for faculty salaries in 1811/12, was near to matching the entire budget for regular faculty salaries. From a base figure of 39,293 marks in 1811, state support of seminars and institutes roughly doubled every two decades in reaching 1870. Between 1870 and 1896 the increase was five-fold.

During the 1860s fields of emphasis for new seminars were shifting. Previously, faculty in the humanities and professions had established the important seminars, especially in philology, jurisprudence, and theology. The outstanding model was Friedrich August Wolf's and later August Böckh's philology seminar that had explicitly set the highest standards for its research. The Mathematics Seminar precedes a sharp upsurge in quasi-independent seminars and research institutes in medicine and the natural sciences that occurred after German unification in 1871, most of them in new or rapidly developing fields.[57]

If before 1860 few seminars had existed in the sciences or applications of mathematics at any German university, where did the Berlin founders look for guidance? They could consider the earlier Halle Association, in which Kummer had participated, or the Bonn Seminar. A careful reading of the purposes and terms of the initiative petition of the Berlin Mathematics Seminar and faculty contacts suggests three other seminars as direct precursors, they and Berlin's Philology Seminar having the most influence in planning for it.[58] They are the Königsberg Mathematical-Physical Seminar, begun provisionally by Jacobi in 1833,[59] the combined private seminar of Dirichlet and annual mathematics prizes in Berlin, and the mathematical pedagogy seminar established by Karl Schellbach at Friedrich-Wilhelm's Gymnasium in Berlin in 1856.[60]

The Berlin initiative petition praises the Königsberg Mathematical-Physical Seminar,[61] in which Jacobi, astronomer Friedrich Bessel, and electrodynamicist Franz Neumann had instilled a thorough research orientation and elucidated in detail the steps of proofs. The Königsberg Seminar had a broader base than Berlin's was to possess. It had two divisions: pure and applied mathematics, including mechanics and astronomy; and mathematical physics. Seminar students were selected from mathematics and physics students by examination or a record of academic achievement in science. Auditors could also attend. Seminarists investigated foundations or selected themes taken from one or several books, but not necessarily classics or journal searches as later in Berlin. Even papers in mathematical physics had to be purely theoretical or treat observations or measurements grounded in a mathematical theory. Königsberg seminarists circulated their papers before discussing them in class, and they kept journals and wrote annual reports.

Jacobi, who dominated the Königsberg Seminar, lived, according to Dirichlet, entirely in the world of thought.[62] In part he trained his students to achieve clarity in teaching. His enthusiasm for teaching and his own clarity at it, combined with his habitual finding of fertile connections in the discipline—his practice was to interweave textual materials with his original research—fired the imagination of his students. During Jacobi's tenure to 1844, the Königsberg Seminar was the most significant of the German seminars in the sciences. After Jacobi came to Berlin in 1844, he discussed the Königsberg enterprise and he championed Kummer, who was then at Breslau. Perhaps more important than common purposes or rules was the presence in the Königsberg and Berlin Seminars of faculty who were original research scholars and developed into talented teachers.

The import of Dirichlet's private seminar and mathematics prizes at the University of Berlin has been little recognized. Historian Wilhelm Lorey's *Das Studium der Mathematik an den Deutschen Universitäten* (1916) suggests that Kummer and Weierstrass thought about these when founding the Mathematics Seminar.[63] No less a figure than Richard Dedekind characterized Dirichlet's later seminars at Göttingen as "deep and penetrating,"[64] and at the Göttingen centenary honoring Dirichlet, Hermann Minkowski described him as an "inspired teacher."

The mathematical pedagogy seminar of Karl Schellbach (1804–1892) was a further precipitant to Berlin's Mathematics Seminar. As early as 1844 Schellbach had urged the Prussian Cultural Minister Friedrich Eichhorn to establish a mathematical institute separate from the university. The institute was to train students in the craft of instruction, which included sharpening their elocution, and to deepen their mathematical knowledge. Even with Jacobi's arrival in 1844 and staunch support, the proposal was not approved. Schellbach probably did not know that in 1838 the Cultural Ministry had rejected plans for a polytechnic institute that the influential Crelle supported. Schellbach continued his campaign to improve mathematical instruction. Among his allies was a past student, the future Kaiser Friedrich III.[65] In 1856 the mathematical pedagogy seminar was created for him, but at Friedrich-Wilhelm's Gymnasium rather than the university.[66]

Schellbach remained critical of mathematics teaching at the University of Berlin. His 1860 report to the Education Section of the Cultural Ministry complained that undergraduate students remained at a secondary-school level. Lectures were too difficult and proofs often too complex to be grasped well by the learner without further assistance. Schellbach believed that taking a seminar like Königsberg's could better prepare faculty to teach. His criticisms rankled some of the Berlin mathematics faculty. But Weierstrass, who was quite sensitive to the difficulties of learning, was impressed with Schellbach and even considered him a possible future editor of Crelle's *Journal*.

The triumvirate of Kummer, Weierstrass, and Kronecker pursued the same goals; and the Berlin Seminar was fundamental among their initiatives to improve mathematics education. In the winter semester of 1858, for example, Kummer had moved to modify the mathematics faculty's *Lehrfreiheit,* their freedom to choose the subjects they taught. He began a systematic coordination of courses offered by the regular mathematics faculty: himself, Ohm, and Weierstrass, together with the *dozents* Borchardt and Arndt. Steiner was ill and unable to teach. Kronecker did not become a member of the Berlin Academy until 1861, after which he taught. Kummer's action was a beginning stage in his efforts to develop a carefully coordinated, comprehensive curriculum and to improve lectures. As Schellbach's report showed, however, in 1860 the teaching of undergraduate mathematics at Berlin still had serious problems.

The Founding of the Mathematics Seminar

Scholars have disagreed over whether it was Kummer or Weierstrass who principally formulated the idea of the Berlin Mathematics Seminar and took the lead in founding it. Its formulation does not seem to reside with one person. Working by correspondence and negotiations with the Prussian Cultural Ministry, and making use of his faculty status and reputation for distinguished scholarship, Kummer took the public lead in establishing the seminar and had the able assistance of Weierstrass.

An initiative petition dated 6 April 1860 to the Prussian Cultural Minister Moritz von Bethmann-Hollweg proposing a mathematics seminar was the first formal measure leading to its founding.[67] Kummer and Weierstrass wrote and signed this petition. In the mathematical sciences more than in many other fields, teachers must do more than barely comprehend their subject. They proposed a seminar with a twofold purpose: to train its students better to be teachers, in part through developing a deeper understanding of materials and more effective elocution in lectures; and, more importantly, to give them direct, independent experience in pursuing new mathematical discoveries. Only research experience in generating new knowledge could illuminate "the foundations and clarity" of mathematics, which was vital to excellent teaching and learning of the subject.

The organizational principles enunciated in the initiative petition for the Berlin Seminar, along with its stress upon research as opposed to the art of pedagogy, derived in good part from the nature and statutes of Berlin's famous Philology Seminar and refinements of the Königsberg and Dirichlet Seminars. It was to address only pure mathematics, which goes to the heart of Jacobi's Seminar and reflects the growing autonomy of mathematics in Prussia.

Like the Philology Seminar, it was to be highly selective, annually accepting only a dozen or fewer students as regular members. Seminar faculty would choose them on the basis of records of high aptitude and diligence in mathematics.[68] The seminar was to consist of tutorials on selected themes, and was to proceed by a deftly crafted series of problems or exercises. Seminarists had to examine these themes through critical examinations of primary sources ranging from classical works to recent books and to articles in leading journals in German, French, and English. Journals were thought to offer the freshest thought. Appeal to the Philology Seminar and close collaboration with Crelle's *Journal* contributed to make this study of sources more sharply honed than had been that of its precursors. The seminar was to offer an annual prize. The Mathematics Seminar was also like the Philology Seminar in requiring a specialized library. Neither the royal nor the university library could address adequately its rapidly changing themes and questions. The specialized library was to begin as a small, select collection of classical works and leading journals and was to be added to yearly.

Behind the general university and governmental support of the proposal in the yearlong progress towards its final ministerial approval, the committed presence of Kummer is detectable. Only the skillful advocacy of a person of his stature at a time of favorable possibilities for funding could have gotten results. Kummer, an ordinary professor, had served as dean of the faculty in 1857/58. His reputation was international: he had received the Grand Prize of the Paris Academy of Sciences in 1857 for his theory of ideal prime factors. In 1860 Weierstrass was still a half-time associate professor with modest faculty support, as his case for promotion in 1864 was to demonstrate. Effective political leadership in founding the seminar, therefore, could come only from Kummer.

In his supportive response, dated 23 April 1861, to the initiative petition, Cultural Minister Bethmann-Hollweg wrote that he expected approval of the seminar.[69] But the guiding principles for the seminar still had to be jointly agreed upon, and he wished an outline of governing regulations. Academic freedom was still precarious in Wilhelmian Prussian universities. Tensions continued between state and various university interests. Ministries could interfere with proposals. This possibility troubled Kummer.

Modifications and changes of proposed regulations turned out to be minor. Except for the Prussian Finance Minister, who at first, Kurt -R. Biermann writes, did not think the seminar was needed,[70] no one raised substantial opposition to the petition and regulations. After a year, the Finance Minister relented. On 4 December 1862, he agreed to support the seminar, and Bethmann-Hollweg wrote in the margin of the petition "finaily!"[71] This action secured the financial base of the seminar. In the initiative petition, Kummer and Weierstrass had requested about 250 thaler yearly for the specialized library and a beginning annual budget of 500 thaler for the seminar.[72]

On 29 April 1861, less than a week after receiving the favorable first notice and long before they received final financial approval, Kummer and Weierstrass declared in a general letter:

> We here make known to students of the mathematical sciences that his Excellency Minister von Bethmann-Hollweg has decided upon the founding of a Mathematics Seminar at the Royal Friedrich Wilhelm's University. Its purpose is to improve mathematical studies through tutorials and introductions to selected works. We are empowered to establish and direct it.
>
> Consequently, those students who wish to participate as active members in the tutorials of the Mathematics Seminar, and have already progressed far in their studies so that they can successfully take part, are urged to communicate to one of the two signers of this letter their names, residences, and addresses.[73]

Seventeen students applied, and Kummer and Weierstrass admitted twelve.[74] On 8 May, just two weeks after they received Bethmann-Hollweg's notification, the seminar began.

Weierstrass quickly began work on the specialized library, which had been principally his idea. Throughout his career, Weierstrass believed that talented students learned best from original sources, reflecting upon the work of masters. He was not new to library projects. His home was next to the Berlin Library, and he assisted it. Later he was also to help obtain finances for specialized libraries for other Prussian seminars. Weierstrass now had bookcases set up in Auditorium 17 at the university for the Mathematics Seminar. A picture of this lecture room, where they gathered each Wednesday evening, was etched in the memory of the seminarists.

The first catalogue of the specialized collection, dated July 1861, includes these nine works:

1. Abel, Niels Henrik, *Oeuvres complètes,* 2 volumes, Christiania, 1839.
2. Cauchy, Augustin-Louis, *Exercises de mathématique,* 5 volumes, Paris, 1826–1830.
3. ——, *Leçons sur les applications du calcul infinitésimal à la géométrie,* 2 volumes, Paris, 1826–1828.
4. ——, *Résumés analytiques,* Turin, 1833.
5. ——, *Nouveaux exercices de mathématiques,* Prague, 1835.
6. ——, *Exercices d'analyse et de physique mathématique,* 4 volumes, Paris, 1840–1842.
7. Euler, Leonhard, *Institutiones calculi integralis,* 4 volumes, St. Petersburg, 1792–1845.
8. Monge, Gaspard, *Application d'analyse à la géométrie,* 5th edition, edited by J. Liouville, Paris, 1850.
9. Poisson, Simeon Denis, *Théorie mathématique de la chaleur,* with a supplement, Paris, 1835.

These books primarily covered developments from the integral calculus of Euler to the pursuit of rigor in Abel and Cauchy's version of mathematical analysis,[75] which Weierstrass soon supplanted with his own.

These nine works were a gift from the Berlin Academy. They had belonged to its series of academic publications by J. (III) Bernoulli, Bessel, Crelle, Dirichlet, Dirksen, Encke, Fischer, Hagen, Hansen, l'Huillier, Kummer, Lagrange, Lambert, Neumann, Pfaff, Poselger, Steiner, and Woepcke. This series was thus located within university confines. The Academy sent publications by three other notable authors: the *Canon arithmeticus* (1839) by Jacobi; *Resultate aus den Beobachtungen des magnetischen Verein in den Jahren 1838–1841* with separate papers by Gauss and Weber; and *Atlas des Erdmagnetism* (1840), published jointly by the same two men. The Academy continued thereafter to contribute a large proportion of the publications for the seminar library.

The seminar quickly purchased the *Journals* of Crelle and Liouville, the two chief mathematical journals of the nineteenth century. The seminar soon added the *Archiv der Mathematik und Physik* that had been founded in 1841 to improve rigor and clarify fundamentals, the *Jahrbuch über die Fortschritte der Mathematik,* the *Mathematischen Annalen,* and the *Zeitschrift für Mathematik und Physik* that dated from 1856. Major foreign journals were also purchased in time.

It was not until 7 October 1864 that the governing regulations for the Berlin Seminar gained official approval. The ten regulations cover the authority of the director, his responsibilities to the Prussian Ministry of Culture and the Philosophical Faculty, the selection of seminarists, the thematic approach of the seminar, and the specialized library.[76] The most important regulation was the eighth, which asserts that seminar tutorials will consist of an oral and a written part. The oral part includes a free and open discussion of selected mathematical problems and questions, which the director will set forth or perhaps the seminarists themselves will have raised, and a free discussion among seminarists of any prior investigations they have conducted on these questions and related publications. For the written component, the director is to see that exercises and assignments address a succession or series of problems from a selected area of mathematics, in such a way that study of them conveys a precise knowledge of the larger field. Themes pursued from the larger field will be determined by the director or possibly by the seminarists. Seminarists will submit written reports to the director, who will grade them.

Since only a dozen or fewer students were admitted to the seminar, other students, including Emil Lampe and Hermann Schwarz, founded the Mathematical Union in November 1861. The Mathematical Union was a student corporation. It sponsored lectures, discussions, and problem-solving contests to improve mathematical knowledge. It acquired its own library. The Union, which grew from twelve members in 1861 to eighty in the 1880s, became a model for student mathematical unions at other German universities. Georg Cantor, who served as its president in 1864, was later prominent in founding the German Mathematiker-Vereinigung.[77]

The labors of adjustment from Gymnasium teaching to research and teaching at a major university had already taxed Weierstrass's health. Previously in Braunsberg, he had worked to the point of exhaustion and occasionally suffered violent spells of vomiting following attacks of vertigo. Contemporaries considered this condition the result of brain spasms. In December 1861, the heavy demands put upon him by his research, especially on Abelian functions, and teaching responsibilities at the university and Industrial Institute led to a complete nervous collapse.[78] Weierstrass had to refrain from all academic work for a year. When he returned to the university, this formerly dexterous fencing master could lecture only while seated. An advanced student had to put materials on the chalkboard. This makes all the more impressive the exceptional reputation that Weierstrass earned as a teacher over the next decade.

In 1864 Weierstrass came up for advancement to ordinary professor. He, like Kummer, had chosen not to write books but to concentrate on articles. Though Crelle had recognized his genius in 1854, as did Kummer in hiring him two years later, Weierstrass's reputation was only beginning to emerge beyond select mathematical circles. The process of his promotion reveals that this extraordinary mathematician was not yet perceived as a leading member of the Philosophical Faculty.

In March 1864 Cultural Minister Heinrich von Muhler asked the senior Philosophical Faculty in March to report to

him as soon as possible on whether Weierstrass should be promoted to ordinary professor. The response, sent in May with the signatures of Dean Gustav Magnus, a physicist, and twenty other professors including Kummer, philologist August Böckh, and historian Leopold Ranke, indicates ambivalence toward Weierstrass on the part of senior members of the Philosophical Faculty.[79] The senior faculty wanted Weierstrass to surrender his position at the Industrial Institute and teach only at the university. They stressed university responsibilities and observed that his health had been adversely affected in the past. They raised as a possible obstacle to the promotion their opposition to increasing the number of ordinary professors. Between 1820 and 1870, the number of ordinary professors at the university increased from thirty to fifty.[80] The number in mathematics, however, fell from three in the 1840s to two in 1864—Kummer and Martin Ohm. The higher number was not to be restored for two more decades. But since Martin Ohm was almost blind and could not read or direct dissertations, the dean and senior faculty believed that the promotion of Weierstrass did not represent an addition. A second director of doctoral dissertations was very much needed.[81]

Kummer vigorously supported the promotion of Weierstrass on the basis of the quality of his research and teaching. He informed the faculty that no other mathematician would earn a more distinguished reputation. He also pointed to Weierstrass's aid in developing independent scholars in the Mathematics Seminar. Weierstrass won promotion, and over the next nine years the reservations among the senior faculty vanished. Kummer's projections were fulfilled. In 1870 the Philosophical Faculty elected Weierstrass to an annual term of dean. When Göttingen University attempted to hire him two years later, that faculty asked the Cultural Minister to end this "possibility [of his leaving Berlin] once and for all" and praised Weierstrass for possessing "a scientific capacity of the first rank."[82] In 1873 he was elected rector of the University of Berlin for a year in the face of such strong rivals as classicist Theodor Mommsen and physicist Hermann Helmholtz.

The Kummer-Weierstrass Years to 1883

The Mathematical Seminar, powerfully compounding departmental teaching, development of curricula, and research programs established during the years of Kummer and Weierstrass, brought students to the frontiers of mathematical knowledge. Mathematics faculty conducted research that was often significant and sometimes transforming in the theory of functions, analysis, algebra, algebraic geometry, and number theory. Teaching reflected these novel materials and posed fundamental questions. The University of Berlin became known for its new mathematics. Instrumental in bringing about this achievement were the format, teaching styles, and research programs of the seminar's director Kummer and even more its co-director Weierstrass, its prize competitions, and the role of seminarists in the Berlin school of mathematics.[83]

Under Kummer and Weierstrass the general enrollment in mathematics grew rapidly. During the 1860s undergraduate majors doubled. By 1872, after a brief decline during the Franco-Prussian War of 1870 and 1871, there were 125 majors. During the next decade that number more than tripled: in 1882, 450 students, or nearly twenty-five percent of all undergraduates matriculating into the Philosophical Faculty, concentrated in mathematics.[84] Initially these increases largely offset losses in the professional schools, and in the decade after 1875 they were part of an abrupt growth in university enrollment from 1824 to 5192.

To aspiring graduate students and postdoctoral scholars in mathematics, Berlin became the primary center for higher studies. Students came not only from all regions of Germany and the Dual Monarchy, Switzerland, Sweden, Denmark, Italy, and France but from the more distant lands of Russia and the United States.[85] Among them were Georg Cantor, Georg Frobenius, Georg Hettner, Wilhelm Killing, Felix Klein, Sophus Lie, Gøsta Mittag-Leffler and Hermann Amandus Schwarz.[86]

Speculating about what drew these students, Weierstrass pointed to the new mathematics embodied in the discipline's distinctive curriculum. During the early years, the triumvirate acted together to teach the integrated curriculum developed under Kummer's guidance. Weierstrass wrote in 1882 that students had

> the opportunity to [begin] with a two-year series of lectures on the most important mathematical disciplines in measured succession. Among topics covered, not a few will be read either not at all at other universities or not regularly. This has had the effect ... especially since lectures in mathematical physics received more complete treatment here than elsewhere ... [of drawing] young men ... not in trifling numbers.[97]

Quick to recognize the importance of mathematical physics, Weierstrass in his first Berlin course during the winter semester of 1856/57 had applied Fourier series and integrals to selected problems in physics. Graduate students gave two other important reasons for enrolling at Berlin—the seminar and Weierstrass.[88]

When the seminar began in 1861, its Wednesday evening meetings in Auditorium 17 dealt strictly with prob-

lems and exercises. This approach met requirement eight of the seminar's regulations. Drawing upon his research and publications, for example on the theory of surfaces, Kummer posed problems. In time, however, it also became the practice for each seminarist to pursue independent research on a designated topic and report findings to the seminar in a lecture. Soon Weierstrass was proposing most of the topics for the seminarists' research within the context of given themes. Topics were offsprings of his own fertile research, especially in the theory of functions and complex analysis. He also generously dispensed dissertation topics. The recitation format had the advantage of also facilitating improved elocution among the seminarists.

Kummer as a seminar teacher could draw on a fund of experience. When he arrived at Berlin, his teaching had already matured at the university level. His Berlin lectures were finely honed, clear, and vivid. His wry sense of humor and willingness to assist students in financial straits were also appreciated. As many as 250 auditors attended his lectures on algebraic geometry, analytic geometry, number theory, and mechanics. Only Joachimsthal had been as popular a mathematics teacher at the University of Berlin.

Kummer's mathematical research, which nurtured his teaching, may be divided into three periods. He had fixed on the theory of functions up to 1842, according to Emil Lampe, and number theory to 1859. From 1859 he concentrated on algebraic geometry, discovering Kummer's surfaces. Before 1861 Kummer's classroom lectures examined sound foundations and his own research. Students were increasingly required to base their study of the foundations of mathematics on "empirically given or presupposed" assumptions.[89] After the seminar formally began, Kummer confined his class lectures to foundations, while centering his Seminar offerings on results of his research. This allowed him to pose challenging new problems and to give seminarists the guidance of his own work.

Weierstrass, on the other hand, had come to Berlin without prior university faculty experience. His teaching style and lecture cycle developed and reached maturity here. His Westphalian accent seems to have had little effect upon the response to his teaching. Students appreciated his openness, his interest in their welfare, and his willingness to continue personal and mathematical discussions outside the classroom, often in pubs. While Kummer was the father of nine children, Weierstrass, a bachelor cared for by his two sisters, could devote his life to teaching and research.

Even so, some of Weierstrass's early Berlin lectures lacked order and seemed prepared the previous night or improvised. They showed students his brilliant mind at work, but were difficult to understand. This situation was quickly rectified, and, as his great lecture cycle encouraged by Kummer evolved during the 1860s, his reputation grew for presenting challenging, even inspirational, lectures aimed at serious students. Between 1857 and 1887, Weierstrass offered sixteen lecture cycles, generally of four semesters each. His teaching method and course content were steadily refined. In the "Academic Instruction" section of his Inaugural Lecture for the rectorship in 1873, Weierstrass revealed something of his method:

> Success [in teaching] depends ... to a great extent upon the teacher's leading the learner continually to some research. This, however, does not occur by chance through pedagogical direction, but chiefly as follows—through their arrangement of materials and emphasis, the teacher's presentation of lectures on a discipline lets the learner discern leading ideas appropriately. In these ways, the fully conversant thinker logically advances from mature and previous research and attains new results or better foundations than exist. Next, the teacher should not fail to designate boundaries not yet crossed by science and to point out some positions from which further advances would then be possible. A university teacher should also not deny the student a deeper insight into the progress of his own investigations, nor should he remain silent about his own past errors and disappointments.[90]

In his sixteen lecture cycles and research, Weierstrass emphasized four subjects. Elliptic functions had been important to him before his arrival in Berlin.[91] In a series of technical papers, he now sought to provide correct procedures to tackle them formally.[92] In summer semester 1857, Weierstrass also began to lecture on integral calculus, or general principles concerning the representation of analytic functions through infinite series, especially convergent Taylor series. The fourth subject that absorbed his attention was the application of variational calculus to selected topics in mathematical physics. Weierstrass's inaugural address (1857) to the Berlin Academy argued for a deeper understanding of relations linking mathematics and the natural sciences, especially physics—a subject that "mir allerdings sehr an Herzen liegt [to me certainly lies at the very heart of matters.]"[93] None of these subjects was developed to his satisfaction, and he regarded his great lecture cycle as inchoate rather than finished.[94]

By mid-1859 Weierstrass had shifted part of his research time from special functions to methods of the underlying theory of analytic functions.[95] Wishing to ensure that the techniques he employed with special functions were reliable, and responding to problems arising in his lectures on the theory of analytic functions, Weierstrass worked to introduce into the theory of analytic functions a higher standard of rigor and a systematic treatment offering greater generality and clarity. Kronecker also urged him to move in this direction.[96] Studies of the foundations of mathematical analysis by Cauchy, Dirichlet, and Riemann together with new results had increasingly revealed the lack of satisfactory foundations. A more critical attitude demanding stringent rigor was emerging and now came to the fore in Berlin. Proof of Cauchy's criterion, the definition of continuity, the question of what constitutes uniform convergence, the definite integral, and trigonometric series, all required a more thorough treatment. Weierstrass, for example, found Riemann's integral not sufficiently general.

In a flurry of activities in 1859 and 1860, the Berlin triumvirate concentrated on the foundations of mathematical analysis. Their internal discussions, which were not yet generally public, centered on Riemann's new theory of functions, based largely on his 1857 article on Abelian functions. Weierstrass nominated him for the Berlin Academy, praising the "originality and fecundity of his method" and the "essential and durable" influence of his thought.[97] Riemann came to Berlin in September 1859 with Dedekind and spoke with the triumvirate. In 1860 the Berliners were criticizing the level of rigor attained by the intuitive Riemann and began to develop provisional hypotheses on more satisfactory foundations.[98]

Kronecker wanted these built solely upon the arithmetic of natural numbers. While many mathematicians stubbornly opposed Kronecker's view, Weierstrass and Dedekind embraced it. They considered foundational arguments based on geometric continuity didactically useful, but not scientifically sound.[99] Weierstrass's studies of the theory of analytic functions of complex variables supported the arithmetic approach,[100] which Dedekind, influenced by Dirichlet, had pursued privately since 1858. Weierstrass, whose thinking about the theory of functions differed from Riemann's, later was to criticize "the disciples of Riemann ... [for] attributing everything to their master, while many [discoveries] had already been made and are due to Cauchy ..."[101]

A second research focus for Weierstrass on foundations was now unmistakable. His thought on the subject was crystallizing in his evolving great lecture cycle from his teaching of the private course "Introduction to Analysis" in 1859/60, which he repeated with modification as "Differential Calculus" in 1861. His 1861 catalogue of books for the Mathematics Seminar has 19 volumes by Euler, Abel, and Cauchy on calculus, although not those of Cauchy from the early 1820s on real analysis.

Hermann Schwarz, an Industrial Institute student, took and preserved accurate notes from his 1861 differential calculus class that reveal the novelty of Weierstrass's early Berlin ideas about the foundations of analysis as well as a practice among seminarists of publishing class notes, often with the instructor's assistance. Weierstrass precisely defines a function in the modern sense as a correspondence between two variable quantities—a definition he attributes to Fourier, Cauchy, and Dirichlet—and develops the epsilontic method. Using Cauchy's ϵ-δ formulation, he defines the notion of limit numerically without Cauchy's kinematics of "approaching indefinitely," which suggests continuous motion. The ϵ-δ technique, which he popularized, became essential to rigorous analysis.[102] Missing in Schwarz's notes are two other critical elements for the second phase of Weierstrass's repositioning of the foundations of analysis onto a more rigorous arithmetic base: a theory of irrational numbers and a continuous function that is nowhere differentiable.

Using epsilontics, Weierstrass and his disciples stringently demonstrated uniform convergence over the next two decades, supplied its classic definition, and established its fundamental importance in analysis. His definition is that

An infinite sequence of functions $\{u_n(x)\}$ is said to converge uniformly in an interval $[a, b]$ when, given any $\epsilon > 0$, there exists an N depending on ϵ, or N_ϵ, such that $|u_{n+k}(x) - u_n(x)| < \epsilon$ for $n > N_\epsilon$, $k > 0$, and for every x, where $a < x < b$.

Weierstrass and his students were then able to give conditions to integrate uniformly convergent series term-by-term.[103] Earlier, in the article "Note über eine Eigenschaft der Reihen ..." (1848), Dirichlet's student Philipp Seidel had examined convergent Fourier series representing a discontinuous function.[104] Near the point at which the function jumps, he had found values of x which converge arbitrarily slowly. Seidel's article introduced the uniform convergence concept,[105] while in the late 1840s the British mathematical physicist George Stokes had independently defined "quasi-uniform" convergence in a neighborhood. But they and Cauchy had failed to distinguish uniform convergence clearly from Cauchy's simple convergence. Weierstrass succeeded.

Weierstrass's "revolution in analysis," to borrow a phrase from Henri Poincaré, proceeded briskly during the 1860s. A serious illness prevented him from offering "Introduction to the Theory of Analytic Functions" in the winter semester of 1861/62, but he taught it in academic years 1863/64 and 1865/66, every other summer semester from 1868 to 1878, and every other winter semester from 1880/81 to 1884/85. French historian Pierre Dugac argues that Weierstrass developed his theory of irrational numbers from 1861 to 1863 and first presented it in these courses assisted by Georg Cantor, who arrived in Berlin in 1863.

The 1865/66 course was the source for part of Ernst Kossak's *Schulprogramm* "Elemente der Arithmetik," which appeared in print in 1872, the same year Cantor and Dedekind[106] published landmark works in developing the real numbers by using Cauchy sequences or Dedekind cuts. The *Schulprogramm* is a distinctive German publication. Each year one faculty member from each secondary school had to publish a dissertation. In Prussia, this practice had been introduced in 1824 to promote scientific research among secondary school teachers. Kossak's dissertation is confusing, setting notes from Weierstrass within a larger history of numbers without always differentiating its source. Weierstrass considered the book a mess and disowned it. A far better source for Weierstrass's theory of functions is Heinrich Heine's article "Die Elemente der Functionenlehre" in Crelle's *Journal* (1872).[107]

Collectively, these writings of Cantor, Dedekind, and Weierstrass arithmetically systematize continuity. Each author had his own reason for proposing his theory. Weierstrass drew upon arithmetic and the topology of the real line, especially with regard to the theory of irrational numbers, to construct his rigorous theory of analytic functions. Here are the roots of his famous quarrel with Kronecker, who attacked Cantor's concept of countability and saw errors "in the so-called present method of analysis." His attacks cooled relations with Weierstrass by 1875 and produced a complete break over the next decade.[108]

During the late 1860s and early 1870s, Weierstrass attacked intuitive geometric reasoning. Demonstrating its limitations in analysis was a critical phase to establishing the need for a shift from geometry to his strictly arithmetical reconstruction of the foundations of analysis. Seeking to free functions and their properties from constant reference to geometry,[109] he developed a striking counterexample to an erroneous, widely-held spatial intuition. Mathematicians then believed that continuous functions can be pictured as smooth curves that possess derivatives at most points, except possibly at isolated ones.[110] In a paper read to the Berlin Academy in July 1872, Weierstrass gave his example of a function continuous over an interval but that has no derivative at any point within the interval.[111] Here was a continuous curve with no tangent. Weierstrass's function with its sharp jumps is $f(x) = \sum_{n=1}^{\infty} b^n \cos(a^n \pi x)$, where

a is an odd integer, b a positive constant smaller than 1, and $ab > 1 + 3\pi/2$.

Weierstrass had likely presented this function earlier in the seminar. He must have been thinking on this problem for some time: he stated in a letter of 1873 that Riemann around 1861 indicated without proof that the function $f(x) = \sum_{n=1}^{\infty}(\sin n^2 x)/n$ "has no derivatives." This function had been studied in Berlin since at least 1864.[112] When Weierstrass's example of a continuous function, nowhere-differentiable, first appeared in print in 1875 in an article by his student Paul du Bois-Reymond, it shocked many mathematicians and shortly geometric preconceptions of functions became untenable. Moreover, it evoked a deeper investigation of the concept of continuity.

Also essential to Weierstrass's arithmetical reconstruction of the foundations of analysis was his criticism of Riemann for basing upon Dirichlet's principle, existence proofs for Abelian functions and the theory of functions. Again Weierstrass developed a counterexample. Dirichlet's principle assumes that within every bounded connected domain at least one function can be found to reduce a given integral to a minimum. In a paper read before the Berlin Academy in July 1870, Weierstrass found a case in which functions make an integral as small as we like while no continuous function exists to minimize it. Until David Hilbert in 1900 and 1901 effectively restored Dirichlet's principle, this finding constrained mathematicians not to employ it as a demonstrative argument.[114] Weierstrass based his theory of functions instead upon convergent power series and analytic continuation.[115] Contemporaries overwhelmingly adopted his simpler approach.[116]

These studies underpinned and were largely developed in Weierstrass's great lecture cycle. Firmly in place by the early 1870s, it consisted of these courses: "Introduction to the Theory of Analytic Functions," "Theory of Elliptic Functions," "Application of Elliptic Functions to Problems in Geometry and Mechanics," "Theory of Abelian Functions," "Application of Abelian Functions to the Solution of Selected Geometric Problems," and "Calculus of Variations."[117] During each four-semester cycle, Weierstrass clearly and systematically constructed a theory of elliptic and Abelian functions using only fundamental principles that he had proved.[118] Complex variable theory remained a major focus of his research: between 1857 and 1887 he gave 36 sets of lectures on elliptic functions, and most of the articles in the first six of his seven-volume *Mathematische Werke* deal with this theory.[119] When Borchardt died after editing Volume One of Jacobi's collected writings, published in 1881, Weierstrass also continued his work by editing the final six volumes.[120]

The work of a master logician, Weierstrass's complex variable theory was entirely analytic. He sought maximum generality that would allow him to deduce satisfactorily all consequences derivable from assumptions. At the same time, his treatment of elliptic functions revealed the problem-solving power of his method. Seminarists considered incontrovertible his logic in complex analysis. This was true even for those seminarists, notably Sophus Lie and Felix Klein, who later strenuously opposed his tendency to reduce all of mathematics to logical deduction.

The most distinctive features of Weierstrass's teaching in the seminar were his highly theoretical and systematic approach, especially to developing the theory of complex variables and arithmetical foundations of complex analysis, and his unmatched standard of rigor. His students beginning research into foundations did not limit themselves to "empirically given or presupposed assumptions," as Kummer specified. Weierstrass insisted on fresh strategies. He wanted to investigate thoroughly all analytical possibilities. In an October 1875 letter to Schwarz, he wrote: ". . . as long as he is working, the researcher must be allowed to follow every path he wishes; it is only a matter of systematic foundations."[121] This sounds like extending Jacobi's dictum that mathematics is "the science which must concern itself with everything."

Weierstrass uncompromisingly pursued a higher standard of completeness and clarity in proofs than previously achieved. His rigor soon became the model that seminarists and other European mathematicians concerned with method strove to achieve. This happened quickly in France, for example. In 1873 Camille Jordan of the *École polytechnique* introduced the French mathematical community to Weierstrass's exacting standards of rigor in analysis. The next year, Jean Gaston Darboux, the principal disseminator of Riemann's work in France, employed uniform convergence.

Because he constantly pursued firmer foundations and again and again found ways to make some point of an argument clearer, Weierstrass did not publish books detailing his conceptual discoveries. His reluctance to publish books did not bespeak an "aversion to printer's ink," as Felix Klein was to claim. Continuing suffering from bronchitis and phlebitis must also have impeded large publication projects. Weierstrass presented no systematic exposition of his analysis comparable to Cauchy's *Cours d'analyse* (1821) and *Résumé* (1823), which he may not have read.[122] His discoveries appeared instead in course lectures, seminar discussions, and articles. Accurate student notes and textbooks based upon his lectures were in demand in Europe and the United States. Much of this material remains scattered and unpublished to the present.[123] Schwarz and

Hettner, two of his most faithful seminar students, and Salvatore Pincherle prepared good, detailed sets of notes,[124] while Kossak, Adolf Hurwitz, and others incorporated lecture materials in their books.[125] Weierstrass unsuccessfully attempted to limit publication of unauthorized versions of his lecture notes, objecting chiefly to mistakes contained in some of them.

Besides including in the seminar topics from his great lecture cycle, Weierstrass turned to leading developments in modern geometry, an interest he had developed since coming to Berlin. He had worked with Steiner and come to admire him. He was one of the few people who could get along with the cranky Steiner, who hated analysis.[126] Weierstrass's special interest was in non-Euclidean geometries.

A cluster of developments had created the first favorable reception for non-Euclidean geometries. The posthumous publication from 1860 to 1865 of the correspondence between Gauss and Schumacher had first made public Gauss's high opinion of Lobachevsky's geometry. Guillaume-Jules Hoüel's French translation in 1866 of Lobachevsky's booklet *Geometrische Untersuchungen zur Theorie der Parallelinien* (1840) and a year later of Bolyai's twenty-four page "Appendix" to his father's *Tentamen* (1831) rescued the "Appendix" from near oblivion and improved the reception of both.[127] In 1868 Eugenio Beltrami demonstrated that Lobachevsky's geometry applies to geodesics on a pseudosphere (surfaces of constant negative curvature) and mapped the non-Euclidean plane on a disc. A thorough examination of each step in that construction produced the surprising conclusion that Euclid's and Lobachevsky's geometries are valid or invalid together, equally consistent in logic. Beltrami's disc with Lobachevsky's geometry, moreover, pictured physical reality as well as the Euclidean plane did. Essays on the foundations of geometry by Riemann and Helmholtz also appeared in 1868. Riemann's essay, which first presented his elliptic non-Euclidean geometry, caused a stir in Berlin. In 1870 Berlin seminarist Felix Klein extended Beltrami's findings by proving Euclid's geometry is a limiting case of Lobachevsky's with its parameter k or radius of curvature. When the curvature k becomes infinite, it gives the limit of straight lines and yields Euclidean geometry.

In 1864, a year after Steiner's death, Weierstrass had agreed to teach courses on modern geometry, giving a total of six until he completed that assignment in summer 1873. In 1870 Weierstrass offered the first university seminar on non-Euclidean geometries. Two years later, he gave another seminar on the foundations of geometry, which emphasized them. These seminars helped make non-Euclidean geometries respected in academic circles.

Faithful to the openness and discipline that characterized the seminar, Kummer and Weierstrass also improved the protocol for doctoral dissertations. As of 1870, examiners had to keep a written record of each oral for its strengths and weaknesses. Students who failed had to be informed in writing of areas needing improvement. In 1872, a year after German unification, Kummer and Weierstrass accepted the first mathematics dissertation written in German, settling a 350-year debate about using the vernacular in place of Latin in higher learning.[128] After 1879 no doctoral dissertation in mathematics was written in Latin.

By the early 1870s Weierstrass was considered the foremost theoretical mathematician in Europe and an exemplary teacher. His chief rival, Riemann, had died in 1866, but full recognition of the importance of Riemann awaited Hilbert. In 1872 Kummer largely turned over the Seminar to Weierstrass. The opinion of Charles Hermite was widely held. When Gøsta Mittag-Leffler arrived in Paris in 1873 for postdoctoral studies, Hermite told him: "You have come to the wrong place. You should study with Weierstrass in Berlin. *C'est notre maître tous.*"[129]

From its beginning the seminar held annual prize competitions that were separate from the older departmental prizes. Seminar competitions involved intense research on a selected topic and early identified for the director emerging talent among seminarists.[130] The prize was 40 thaler in 1862 and 50 thaler (150 marks) thereafter—a handsome sum for young men. Kummer was crucial to these competitions. He administered 24 of the 27 competitions held from 1862 through 1884, while Weierstrass administered two, in 1882 and 1883, and Kronecker one, in 1884.

In these competitions 49 seminarists received 56 prizes. Franz Josef Mertens, for example, won the 1863 prize for a paper that applied analysis to number theory using Dirichlet's methods.[131] Multiple prize winners included Hermann Schwarz, three times, and Eugen Netto and Ludwig Kiepert, twice. Kurt Biermann has traced the careers of 40 prize winners. Thirty-six became *dozents* in *Hochschulen,* university faculty, directors of schools or libraries, government officials, or industrial scientists.[132] The seminar thus quickly became a major source for mathematical faculty for German higher education.

Prize winners subsequently became the core of the Berlin school of mathematics, to which other seminarists and students from triumvirate classes also belonged. That school, with the triumvirate as its nucleus, promoted Berlin's new mathematics. Their work pushing back the existing boundaries of mathematical knowledge centered on Weierstrass's highly theoretical and systematic method, stringent rigor, and placement of complex analysis on arithmetical foundations. They and other Berlin school members joined Weierstrass to standardize and elaborate complex analysis. They established it so successfully throughout German higher education that by the 1880s the word *Funktionlehre* became a synonym for it. They also helped spread the new analysis abroad.[133]

The eighteen prize winners who had been undergraduates at the University of Berlin epitomized the Berlin school.[134] They were Heinrich Bruns, Georg Frobenius, Kurt Hensel, Georg Hettner, Ludwig Kiepert, Wilhelm Killing, Adolph Kneser, Johannes Knoblauch, Fritz Kotter, Emil Lampe, Franz Josef Mertens, Eugen Netto, Franz Rudio, Arthur Schönflies, Friedrich Schur, Hermann Schwarz, Maximilien Simon, and Paul Stäckel. All began as *dozents* in *technische Hochschulen* in Aachen, Berlin, Hannover, and Zurich. They later taught at such universities as Budapest, Cracow, Giessen, Göttingen, Halle, Heidelberg, Leipzig, Marburg, and Strassburg.[135]

Nine prize winners became extraordinary or ordinary professors at the University of Berlin itself. Among them, Eugen Netto helped Borchardt edit Volume One of Jacobi's *Collected Works* which appeared in print in 1881, while Hettner, Knoblauch, and Netto during term appointments in the 1880s taught differential calculus and introduction to analysis, application of elliptic functions to problems of geometry and mechanics, analytical geometry, mechanics, and potential theory. Georg Hettner, Fritz Kotter, and above all Johannes Knoblauch also helped edit the first two volumes of Weierstrass's *Mathematische Werke* (1894–95).[136]

Two prize winners were to lead the second generation of the Berlin school. After Kronecker died suddenly in 1891 and before Weierstrass retired in 1892, an elected, seven-member faculty commission chose their successors. Following the advice of Weierstrass, who was a member of the commission, they selected Frobenius, then teaching at Zurich Polytechnic, and Schwarz, who was at Göttingen. Schwarz had prepared a *Festschrift* for Weierstrass's seventieth birthday and "rescued" part of Riemann's work in analysis by providing rigorous demonstrations for boundary value problems instead of relying upon Dirichlet's principle. Also in 1892, Kurt Hensel, later the editor of the works of Kronecker, was named extraordinary professor. With this faculty transition in 1892, the heroic period ended.

While seminarists up to 1883 mainly advanced the Berlin school during their careers, Felix Klein and Sophus Lie among them became its chief critics. In their foundational studies, mathematicians at Bonn and Göttingen appealed to spatial intuition and physical experience. Klein,

who had attended both institutions, resisted the purist trend to treat arithmetic as the font of the whole of mathematical knowledge. He and David Hilbert rejuvenated the mathematical tradition at Georg-August Universität, that is, Göttingen University. From the late 1880s Klein diversified and reorganized its mathematics curriculum, succeeded in having Hilbert appointed editor of the influential *Mathematische Annalen,* and fostered applied research through a consortium formed with prominent German industrialists. After 1895 Göttingen displaced Berlin as the preeminent university in German mathematics.[137] Lie, Klein's ally at Leipzig, considered Weierstrass's arithmetical foundations of analysis to be an advance in logical acuity but not an exhaustive basis for mathematical foundations.[138] He blamed the "arrogance" and dominance of the Berlin school for the dearth of geometrical research in Germany. Klein's application to succeed Weierstrass in 1892 caused a mild sensation, when the faculty commission rejected him as a "charlatan" and "compiler."[139]

Still other names attest to the fecundity of the Berlin Seminar. Georg Cantor developed his new research field of set theory encompassing transfinite numbers, while Mittag-Leffler founded the international journal *Acta Mathematica,* which Weierstrass and Hermite vigorously supported. Products of the seminar who went outside the discipline of mathematics include physicist Max Plank and philosopher Edmund Husserl, who wrote his dissertation under Weierstrass on the calculus of variations.

Conclusion

The Berlin Mathematics Seminar had its origins and nurturing in the model research orientation of the University of Berlin, the growing autonomy of mathematics, and the building of a distinguished faculty there. These depended upon the confluence of often contingent mathematical, cultural, political, and economic factors. The Berlin Academy's mathematical tradition, neohumanist attempts to explore a purer reality through mathematics, and national rivalry with France's *École polytechnique* most encouraged the University of Berlin to have a strong mathematics program. Prussia's secularization of teacher education after 1810 and the Cultural Ministry's support of science seminars brought autonomy for mathematics, requiring the discipline to prepare teachers for new gymnasia. But not until external support after 1826 from Alexander von Humboldt and August Crelle with his *Journal* was a talented faculty recruited. By the late 1840s Dirichlet, Steiner, and Jacobi had developed a leading research program. A decade later the pathbreaking research and curriculum coordination of the triumvirate established the base for the seminar, while Schellbach's criticism in 1860 of undergraduate mathematical education at the University of Berlin further prompted it. Drawing upon three prior seminars—Berlin's in philology, Königsberg's in mathematics and physics, and Dirichlet's—and ably assisted by Weierstrass, Kummer guided the founding of the Mathematics Seminar through the university and the Prussian Cultural Ministry.

During the years of Kummer and Weierstrass, the seminar was the focal point of the mathematics faculty's efforts in research and the training of good teaching. Through thematic tutorials requiring independent research in primary sources and leading journals, seminarists gained clearer and deeper insights into the methods, foundations, and principal branches of mathematics. Seminarist lectures improved student elocution in preparation for teaching. These students had a specialized library at their disposal and participated in annual prize competitions. Seminar faculty brought them to the leading edge of knowledge in Berlin's new mathematics and such other fields as non-Euclidean geometries and prepared them to advance further. Weierstrass's complex analysis based upon an arithmetical foundation and his inculcation of stringent rigor were essential to the Berlin school of mathematics. Many of the seminarists spread Weierstrass's form of analysis, some through publishing his lecture notes, and they helped make it and the Berlin school dominant in mathematics across Germany during the late heroic period. But their rejection of geometric intuition for foundations and their purist stress on arithmetic and increasing specialization in research found strong critics in former seminarists Klein and Lie and, at the end of the century, the revitalized Göttingen mathematical tradition.

Acknowledgement. I am grateful to Joseph Brown (Rensselaer Polytechnic Institute), Gert Schubring (Bielefeld), and Thomas West (Catholic University) for reading drafts of this paper and making helpful comments to improve it.

REFERENCES

1. Adolf Kneser, a student of Kronecker and professor at Dorpat, characterized this time as the "heroic period." He limited it to the tenure of Weierstrass as ordinary professor from 1864 to 1892.
2. Frederick Gregory, "Kant, Schelling, and the Administration of Science in the Romantic Era," in *Science in Germany: Osiris,* ed. Kathryn Olesko, 2d ser., vol. 5(1989): 19 (hereafter *Science in Germany*).
3. As quoted in Notker Hammerstein, "Universitäten und gelehrte Institutionem...," in *Samuel Thomas Soemmerring und die Gelehrten der Goethezeit,* eds. Gunter Mann and Franz Dumont (New York: Fischer Verlag, 1985), p. 327.
4. Ibid., pp. 21–25, and see Hans Niels Jahnke, *Mathematik und Bildung in der Humboldtschen Reform* (Göttingen: Vandenhoek and Ruprecht, 1990).
5. Leonard Krieger, *The German Idea of Freedom* (Chicago: The University of Chicago Press, 1957), p. 157.
6. Charles E. McClelland, *State, Society, and University in Germany 1700–1914* (Cambridge: Cambridge University Press, 1980), pp. 218–219 (hereafter *State, Society, and University*).
7. For the professionalization of mathematics, see Gert Schubring, *Die Entstehung des Mathematiklehrerberufs im 19. Jahrhundert: Studien und Materialien zum Prozess der Professionalisierung in Preussen (1810–1870)* (Basel: Beltz, 1983, rev. 1991) and Gert Schubring, "Pure and Applied Mathematics in Divergent Institutional Settings in Germany: The Role and Impact of Felix Klein," in *The History of Modern Mathematics,* eds. David E. Rowe and John McCleary (New York: Academic Press, 1989), p. 176 (hereafter *Modern Mathematics*).
8. Neohumanism is a late-nineteenth century term, which gained acceptance through the writings of Friedrich Paulsen.
9. On the university level it rested on the dual pillars of *Lehrfreiheit,* the freedom of faculty to teach subject areas they chose rather than having them administratively set, and *Lernfreiheit,* the freedom of students not to be required to follow a strictly prescribed curriculum.
10. Gymnasia mathematics also included practical arithmetic. See Gert Schubring, "The Rise and Decline of the Bonn Natural Sciences Seminar," in *Science in Germany,* ed. K. Olesko, 5: 60 (hereafter "Bonn Natural Sciences Seminar").
11. Kathryn M. Olesko, "Physics Instruction in Prussian Secondary Schools before 1859," in *Science in Germany,* ed., K. Olesko, 5:95.
12. Joseph Ben-David, *The Scientist's Role in Society: A Comparative Study* (Englewood Cliffs, N. J.: Prentice-Hall, Inc., 1971), pp. 108–139.
13. See Walter Horace Bruford, *Deutsche Kultur der Goethezeit* (Konstanz: Akademische Verlagsgesellschaft Athenaion, 1965), p. 157.
14. Gert Schubring, "The German Mathematical Community," in John Fauvel, Raymond Flood, and Robin Wilson, *Möbius and His Band* (Oxford: Oxford University Press, 1993), p. 27.
15. Gert Schubring, "Bonn Natural Sciences Seminar," pp. 58 ff.
16. During his time at Bonn, Weierstrass does not appear to have studied under Plücker. See Gert Schubring, "Warum Karl Weierstrass beinähe in der Lehrerprufung gescheitert wäre," *Der Mathematikunterricht* 35 1(1989):22.
17. The official Halle seminar was not founded until 1839, five years after Greifswald established its Mathematical Society.
18. Its continuing influence may be seen in the "French colony" and the French gymnasium in Berlin. At that gymnasium and the Berlin Academy, the epigone Johann Philipp Gruson (1768–1857) now translated Euler and Lagrange into German and lectured on Euler.
19. Wilhelm Lorey, *Das Studium der Mathematik an den Deutschen Universitäten seit Anfang des 19. Jahrhunderts*

(Leipzig: B. G. Teubner, 1916), p. 41 (hereafter *Mathematik an den Deutschen Universitäten*).

20. Frederick Gregory, "Neo-Kantian Foundations of Geometry in the German Romantic Period," *Historia Mathematica* 10 (1983): 186.

21. See R. R. Palmer, *The Improvement of Humanity* (Princeton: Princeton University Press, 1985), pp. 325–326.

22. Crelle based it on the first purely mathematical journal, Joseph Diaz Gergonne's *Annales de mathématique pures et appliquées,* which began in 1810.

23. Gert Schubring, "The German Mathematical Community," p. 31.

24. Lorey, *Mathematik an den Deutschen Universitäten,* p. 45 and Lewis Pyenson, *Neohumanism and the Persistence of Pure Mathematics in Wilhelmian Germany* (Philadelphia: American Philosophical Society, 1983), p. 72.

25. Lorey, *Mathematik an den Deutschen Universitäten,* p. 43.

26. Gert Schubring, "Zur Modernisierung des Studiums der Mathematik in Berlin, 1820–1840," in Sergei S. Demidov *et al,* eds., *Amphora* (Basel: Birkhäuser Verlag, 1992), pp. 649–675.

27. On the mathematics faculty Enno Dirksen (1788–1850) and Martin Ohm (1792–1872) principally addressed elementary analysis, algebra, and mechanics, while Christian Ideler (1766–1846) lectured on mathematical geography, chronology, and foundations of French mathematical textbooks.

28. From a letter of 21 December 1846 to A. Humboldt as quoted in Kurt -R. Biermann, *Die Mathematik und Ihre Dozenten an der Berliner Universität 1810–1920* (Berlin: Akademie-Verlag, 1988), 2nd ed., p. 46 (hereafter *Die Mathematik und Ihre Dozenten*). Jacobi was critical of Cauchy for misplacing Abel's manuscript on elliptic integrals submitted for the Paris Academy prize competition in 1826. Cauchy, who chaired the referees for his manuscript, later claimed that illegibility hid the many new ideas in this long and difficult paper. Through pressure from Jacobi, this paper reappeared in 1841.

29. For histories of attempts to solve this problem, see Harold M. Edwards, *Fermat's Last Theorem: A Genetic Introduction to Algebraic Number Theory* 4th printing (Berlin: Springer-Verlag, 1993) and Paulo Ribenboim, *13 Lectures on Fermat's Last Theorem* (Berlin: Springer-Verlag, 1979). As noted in this paper, Euler's 1753 proof for the case $n = 3$ mainly by Fermat's method of infinite descent contained a gap. See Edwards, pp. 39 ff.

   Andrew Wiles of Princeton first provided a general proof for FLT in 1994. See David A. Cox, "Introduction to Fermat's Last Theorem," *The American Mathematical Monthly,* 100 (1994): 3– 14.

30. Harold M. Edwards, "The Background of Kummer's Proof of Fermat's Last Theorem for Regular Primes," *Archive for History of Exact Sciences* 14 (1974): 226 (hereafter "Kummer's Proof").

31. Ibid.

32. Ibid., pp. 219–236.

33. Kummer announced his breakthrough in a letter to Kronecker dated 18 October 1845.

34. Harold M. Edwards, "Kummer's Proof," pp. 226–228.

35. See Emil Grosswald, *Topics from the Theory of Numbers* (Boston: Birkhäuser, 1984), pp. 210–233.

36. André Weil, ed., *Ernst Eduard Kummer: Collected Papers* (Berlin: Springer-Verlag, 1975), p. 3.

37. See Thomas Hawkins, "The Origins of Modern Theories of Integration," in *From the Calculus to Set Theory 1630–1910: An Introductory History,* ed. I. Grattan-Guinness (London: Gerald Duckworth & Co. Ltd., 1980), p. 140 and Eberhard Knoblauch, "Von Riemann zu Lebesgue—Zur Entwicklung der Integrationstheorie" *Historia Mathematica,* 10, 3(1983): 318–343.

38. Harold M. Edwards, "Kummer's Proof," p. 226.

39. See Karl Weierstrass, "Über das Sogenannte Dirichlet'sche Princip" (1870) in idem, *Mathematische Werke* (New York: Johnson Reprint Corporation, 1967), II: 49–54. At the Dirichlet centenary at Göttingen University in 1905, Hermann Minkowski referred to the "other Dirichlet principle," which he described as the art of "overcoming problems with a minimum of blind calculation and a maximum of perceptive thought."

40. Lorey, *Mathematik an den Deutschen Universitäten,* pp. 47–48.

41. See Jacob Steiner, *Geometrical Construction with a Ruler* (1833), ed. Raymond Clare Archibald and trans. Marion Elizabeth Stark (New York: Scripta Mathematica, 1940) in *Scripta Mathematica Studies* No. 4: 10 and 13 ff.

42. See J. H. Pestalozzi, *How Gertrude Teaches Her Children,* ed. E. Cooke and trans. Lucy Holland and Francis Turner (London: G. Allen & Unwin Ltd., l915) 4th ed., p. 26.

43. In his later lectures at Berlin, Steiner often interrupted lectures to find whether students understood points and was sharp with those who did not. He thereby drove many from his classes.

44. Laurence Young, *Mathematicians and Their Times* (New York: North-Holland Publishing Company, 1981), pp. 184–185.

45. In his *Denkschrift* "Über die innere und aussere Organisation der hoheren wissenschaftlichen Anstalten in Berlin" (1810), Wilhelm von Humboldt asserted that every academician should have the right to lecture at the university. This was part of the Berlin Academy statutes up to 1945.

46. After the earlier loss of his family fortune in a bankruptcy, the Prussian government suspicion of him based on his constitutional discussions during the Revolution of 1848 and its temporary retraction of his salary bonus in 1849 made Jacobi's last years difficult.

47. Jacobi contributed to the history of mathematics by supplying notes on ancient Greek mathematics to Alexander von Humboldt for his book *Kosmos* (1845) and by examining the Diophantus manuscript in the Vatican Library. Partly through correspondence with Paul Heinrich Fuss (1798–1855), secretary of the St. Petersburg Academy and Euler's great-grandson, he also developed a detailed plan to publish the collected works of Leonhard Euler. Publication of Euler's prolific 873 writings did not begin until 1907 by the Swiss Society of Natural Sciences and continues to the present.

48. Biermann, *Die Mathematik und Ihre Dozenten,* p. 193.

49. McClelland, *State, Society, and University,* p. 208.

50. For these volumes it was widely called *Borchardt's Journal.*
51. Weierstrass was deeply moved when Richelot brought the honorary degree to Braunsberg and declared, "We have found in Mr. Weierstrass our teacher." See Pierre Dugac, *Elements d'Analyse de Karl Weierstrass* (Paris: Université de Paris VI, 1972), p. 9 (hereafter *Elements d'Analyse*).
52. Kurt -R. Biermann, "Die Berufung von Weierstrass nach Berlin," in *Festschrift zur Gedächtnisfeier für Karl Weierstrass 1815–1965,* eds. Heinrich Behnke and Klaus Kopfermann (Cologne: Westdeutscher Verlag, 1966), pp. 46–47 (hereafter *Festschrift . . . für Karl Weierstrass*).
53. Ibid., p. 9.
54. By 1856 the Industrial Institute had university characteristics. Its students were gymnasia graduates. The institution had academic freedoms, specialization, and faculty governance. See Eberhard Knobloch, "Mathematics at the Berlin Technische Hochschule/Technische Universität: Social, Institutional, and Scientific Aspects," in *Modern Mathematics,* eds., D. Rowe and J. McCleary, II: 251.
55. Dugac, *Elements d'Analyse,* p. 11.
56. Adolph Wagner, *Die Entwicklung der Universität Berlin 1810–1896* (Berlin: Druck von Julius Becker, 1896), pp. 57 ff.(hereafter *Universität Berlin*).
57. Prussia feverishly established these under Friedrich Althoff, Chief of the Higher Education Section of the Cultural Ministry from 1882 to 1907. Althoff encouraged specialization as opposed to the neohumanist preference for a general liberal education. He helped to establish: eighty-six medical clinics, institutes, and laboratories; seventy-seven institutes and seminars in philosophical faculties; nine law seminars; and four theology seminars. See McClelland, *State, Society, and University,* pp. 280 ff. Historian David Rowe has pointed out that Althoff supported Göttingen to lead in mathematics. He aided Berlin more for the arts, classics, and history.
58. Biermann, *Die Mathematik und Ihre Dozenten,* pp. 278–279.
59. Lorey, *Mathematik an der Deutschen Universitäten,* p. 114.
60. A fourth seminar, the Göttingen mathematical-physical seminar begun in 1850, appears not to have influenced Berlin planning.
61. Biermann, *Die Mathematik und Ihre Dozenten,* p. 279.
62. C. G. J. Jacobi's *Gesammelte Werke* ed. Carl Wilhelm Borchardt (Berlin: Verlag von Georg Reimer, 1881) I:1–28, esp. 22.
63. Lorey, *Mathematik an der deutschen Universitäten,* p. 120.
64. Umberto Bottazzini, *The Higher Calculus: A History of Real and Complex Analysis from Euler to Weierstrass,* trans. by Warren von Egmond (New York: Springer-Verlag, 1986) p. 265 (hereafter *The Higher Calculus*).
65. Friedrich III, who espoused constitutional liberalism, did not become kaiser until 1888. His reign lasted only 99 days. He died of throat cancer.
66. This seminar was important for younger mathematics lecturers. Rudolf Clebsch, Immanuel Fuchs, Leo Koenigsberger, and Franz Woepkke were among its participants.
67. The initiative petition is printed in full in Biermann, *Die Mathematik und Ihre Dozenten,* pp. 278–279.
68. An additional six associates might also be named.
69. Biermann, *Die Mathematik und Ihre Dozenten,* p. 99.
70. Ibid.
71. Ibid., pp. 99–100.
72. By 1870 its budget had grown to 1200 marks. See Wagner, *Universitat Berlin,* p. 61.
73. Biermann, *Die Mathematik und Ihre Dozenten,* p. 100.
74. According to regulation five, the number of seminarists could be no more than twelve, although as many as six associates could be named. This regulation was followed up to 1884. See Biermann, *Die Mathematik und Ihre Dozenten,* p. 280.
75. A solid study of Cauchy's work is Judith V. Grabiner's *The Origins of Cauchy's Rigorous Calculus* (Cambridge, Mass.: The MIT Press, 1981).
76. Biermann, *Die Mathematik und Ihre Dozenten,* pp. 279–281.
77. Joseph Warren Dauben, *Georg Cantor: His Mathematics and Philosophy of the Infinite* (Cambridge, Mass.: Harvard University Press, 1979), p. 160.
78. His teaching assignment alone was formidable. It included twelve hours of lectures every week at the Industrial Institute. See Heinrich Behnke, "Karl Weierstrass und Seine Schule," in *Festschrift . . . für Karl Weierstrass,* eds., Heinrich Behnke and Klaus Kopfermann, p. 25.
79. The 12 May 1864 letter is reproduced completely in Biermann, *Die Mathematik und Ihre Dozenten,* pp. 281–282.
80. McClelland, *State, Society, and University in Germany,* p. 204.
81. During the period from 1839 to 1850, the number of ordinary professors in mathematics had fallen from three—Enno Dirksen, Dirichlet, and Ohm—to two. When Dirksen died in 1850, his position was not filled. The number did not rise to three again until 1883, when an ordinary professorship was created for Kronecker.
82. Biermann, *Die Mathematik und Ihre Dozenten,* p. 102.
83. This account does not focus on Kronecker, who lectured at the university from 1862 as a member of the Berlin Academy but did not until 1883 become an ordinary professor with full faculty rights and Seminar codirector. The date of his appointment traditionally ends the Kummer-Weierstrass years. Kummer had resigned the previous year, and, with the appointment of his successor Lazarus Fuchs in 1884, Weierstrass, even though he remained director, had lost effective control of the seminar.
84. Biermann, *Die Mathematik und Ihre Dozenten,* pp. 103 and 291.
85. Weierstrass's letter dated 6 July 1882 is printed in Biermann, *Die Mathematik und Ihre Dozenten,* pp. 291–294. Graduate students from the United States still went mainly to Göttingen.
86. Weierstrass also privately tutored Sofya Kovalevskaya in 1870, but could not gain her admission to the doctoral program. After she wrote three strong papers on partial differential equations, Saturn's rings, and the reduction of certain Abelian integrals to simpler forms, he enthusiastically recommended her for a doctorate without examination and *in absentia* from Göttingen, which alone in Germany had the right to so confer mathematics doctorates. Although she never attended Göttingen, it granted her the doctorate *summa cum laude* in 1874. See p. 183 below and Reinhard Bolling, ed., *Briefwechsel zwischen Karl*

Weierstrass und Sofja Kowalewskaja (Berlin: Akademie-Verlag, 1993) and Ann Hibner Koblitz, *A Convergence of Lives: Sofia Kovalevskaia: Scientist, Writer, Revolutionary* (Boston: Birkhäuser, 1983), pp. 99–103 and 122–123.

87. Biermann, *Die Mathematik und Ihre Dozenten,* p. 193.

88. See Peter Ullrich, "Weierstrass' Vorlesung zur 'Einleitung in die Theorie der analytischen Funktionen,'" in *Archive for History of Exact Sciences* 40, 2 (1989), p. 145 and Thomas Hawkins, "Non-Euclidean Geometry and Weierstrassian Mathematics: The Background to Killing's Work on Lie Algebras,: in *Studies in the History of Mathematics,* ed. Esther R. Phillips (Washington, D. C.: The Mathematical Association of America, 1987), 26: 23 (hereafter *History of Mathematics*).

89. Kurt -R. Biermann, "Kummer," *Dictionary of Scientific Biography,* 18 vols., VII (1973): 523.

90. As quoted in Biermann, *Die Mathematik und Ihre Dozenten,* p. 107.

91. Weierstrass, like Riemann at Göttingen, was extending the work of Abel and Jacobi. As he told the Berlin Academy in his Inaugural Address in July 1857: "The theory of elliptic functions, . . . which I first became acquainted with...under my...teacher [Christof] Gudermann [at Munster Academy in 1840] . . . exercised a strong attraction on me, and has retained a definite influence on the entire course of my mathematical development." See Karl Weierstrass, *Mathematische Werke* (New York: Johnson Reprint Corporation, 1967), I: 223–224.

92. Ibid., p. 224.

93. Ibid.

94. Although Weierstrass offered six courses on modern synthetic geometry (see p. 170), this subject was not at the center of his research.

95. See Klaus Kopfermann, "Weierstrass' Vorlesung zur Funktionentheorie," in *Festschrift . . . für Karl Weierstrass,* eds., Heinrich Behnke and Klaus Kopfermann, pp. 75–96.

96. Umberto Bottazzzini, "The Influence of Weierstrass's Analytical Methods in Italy," in Sergei S. Demidov *et al,* eds., *Amphora,* p. 72.

97. Dugac, *Elements d'Analyse,* p. 14.

98. Bottazzini, *The Higher Calculus,* pp. 259 ff.

99. See Richard Dedekind, *Stetigkeit und irrationale Zahlen* (1872) in *Classics of Mathematics,* ed. Ronald Calinger (Englewood Cliffs, NJ: Prentice Hall, 1995), pp. 627–628.

100. Weierstrass, *Mathematische Werke,* 2: 235.

101. Bottazzini, *The Higher Calculus,* p. 263.

102. A function $f(x)$ has a limit l at $x = x_0$, if for $\epsilon > 0$, there exists a δ such that $|f(x) - l| < \delta$ for all x in the deleted interval $0 < |x - x_0| < \delta$. According to Munich mathematician Alfred Pringsheim, Weierstrass during his career "first gave the notion of the limit of a function all of the precision of which it was susceptible." See Dugac, *Elements d'Analyse,* p. 22.

103. See Saunders Mac Lane, *Mathematics: Form and Function* (New York: Springer-Verlag, 1986), p. 335.

104. See Philip L. Seidel, "Note über eine Eigenschaft der Reihen, welche diskontinuierliche Functionen darstellen," in Heinrich Liebmann, ed., *Ostwald's Klassiker* (Leipzig: Engelmann, 1900), No. 116: 35–45.

105. Michael Segre, "Peano's Axioms in Their Historical Context," *Archive for History of Exact Sciences,* 48, 3/4 (1994): 233.

106. For a biography of Dedekind, see Pierre Dugac, *Richard Dedekind et les fondements des mathématiques* (Paris: Vrin, 1976).

107. See Vol. 74 (1872): 172–188.

108. Felix Klein later commented that Kronecker wanted "to ban irrational numbers and reduce mathematical knowledge to relations between whole numbers alone." See Bottazzini, *The Higher Calculus,* p. 290.

109. Weierstrass was so critical of geometric intuition that he considered geometric drawings for the purpose of demonstrations to be in poor taste. If he clarified a point by drawing a diagram, he was careful to erase it completely.

110. See K. Volkert, "Zur Differentzierbarkeit Stetiger Funktionen—Ampere's Beweis und Seine Folgen," *Archive for History of Exact Sciences* 40 (1989): 37–112.

111. "Über continuierliche Functionen eines reellen Argument die für keinen Werth des letzteren einen bestimmten Differential-quotienten besitzen," in Weierstrass, *Mathematische Werke,* 2: 71–74.

112. See Erwin Neuenschwander, "Riemann's Example of a Continuous 'Nondifferentiable' Function" in *The Mathematical Intelligencer* 1(1978): 40–42 and L. L. Segal, "Riemann's Example of a Continuous 'Nondifferentiable' Function Continued," in Ibid., p. 81.

113. See Karl Weierstrass, "Über das Sogenannte Dirichlet'sche Princip" in *Mathematische Werke,* II: 49–54. Dedekind had communicated this principle to Weierstrass from Dirichlet's summer 1856 lectures at Göttingen.

114. Hermann Helmholtz stated, however, that for physicists "Dirichlet's principle continue[d] to remain a demonstration." Felix Klein found physicists "less concerned with mathematical details; for them 'evidence' is sufficient." See Bottazzini, *The Higher Calculus,* p. 302.

115. Detlef Laugwitz argues that Weierstrass may not have known directly of the early 1820s classic writings of Cauchy, because Weierstrass's work on delta functions and convergence factors for his approximation theory of 1885 appear clumsier than Cauchy's. See Detlef Laugwitz, " 'Das letzte Ziel ist immer die Darstellung einer Funktion': Grundlagen der Analysis bei Weierstrass 1886, historische Wurzeln und Parallelen," *Historia Mathematica* 19 (1992): 341–355. See also S. Mac Lane, *Mathematics: Form and Function,* p. 336.

116. The method of Riemann and that of Weierstrass and Dedekind dominated mathematical analysis during the final third of the nineteenth century. Henri Poincaré characterized Riemann's method as "above all a method of discovery"; and that of Weierstrass and Dedekind as "above all a method of demonstration." See Dugac, *Elements d'Analyse,* p. 64.

117. According to Oskar Bolza, Weierstrass began lecturing on the calculus of variations in 1865. His three best known courses were given in the summer semesters of 1875, 1879, and 1882. For Weierstrass's contributions to this field, see Herman H. Goldstine's *A History of the Calculus of Variations from the 17th through the 19th Century* (New York: Springer-Verlag, 1980), pp. 190–250.

118. As Weierstrass continuously revised and improved his lecture courses, students enrolled in increasing numbers. His course on the theory of elliptic functions drew 50 students in the winter semester of 1868/69 and 100 in the spring semester of 1885. Higher enrollment at the university accounts only partially for this. In the winter semester of 1884/85, as many as 250 students attended Weierstrass's "Introduction to the Theory of Analytic Functions," spilling over from the Barracks Auditorium to the garden outside. Weierstrass had to move the class to the amphitheater of the Chemical Institute, but within a few sessions attendance fell to under 200 and the class was moved to smaller quarters.

119. See Weierstrass, *Mathematische Werke,* 7 vols.

120. C. G. J. Jacobi's *Gesammelte Werke,* vols. II–VII (1881–1891).

121. See Weierstrass, *Mathematische Werke,* 2: 235.

122. See reference note 98.

123. Roger Cooke, Review of Reinhard Bolling, ed., *Briefwechsel zwischen Karl Weierstrass and Sofja Kowalewskaja* (1993) in *Historia Mathematica* 22 (1995): 73–77.

124. Pincherle based his "Saggio di una introduzione alla teoria della funzioni analitiche secondo i principii del prof. C. Weierstrass" on his teacher's Abelian functions class in Berlin in academic year 1877/78.

125. Pincherle, who was a secondary school teacher in Pavia, wrote his "Saggio . . ." for the *Giornale di Matematiche* (1880). This article introduced into Italy the basic principles of Weierstrass's theory of analytic functions. See Umberto Bottazzini, "The Influence of Weierstrass's Analytical Methods in Italy," in Sergei S. Demidov *et al,* eds., *Amphora,* pp. 67–90.

126. Weierstrass collected and edited Steiner's writings. See Jacob Steiner, *Gesammelte Werke,* ed. Karl Weierstrass (Berlin: Reimer, 1881–1882) 2 vols.

127. See Jeremy Gray, "The Discovery of Non-Euclidean Geometry," in *History of Mathematics,* ed. E. R. Phillips, pp. 48–51.

128. Breaking with academic tradition, Paracelsus had lectured in German at Basel University in the late 1520s.

129. Dugac, *Elements d'Analyse,* p. 18.

130. Biermann, *Die Mathematik und Ihre Dozenten,* p. 50.

131. Hermann Schwarz won the prize in 1863 for a "clear, precise" paper on fifth-degree surfaces that contained "only a few minor errors" in its Latin style. Kummer announced the winner in a 12 June 1863 letter that appears in Biermann, *Die Mathematik und Ihre Dozenten,* p. 108.

132. Four died young, and nine others could not be traced. See Biermann, *Die Mathematik und Ihre Dozenten,* p. 108.

133. Weierstrass and Richard Dedekind primarily developed this new analysis.

134. For further information on the Berlin school, see Gottfried Richenhagen, *Carl Runge (1856–1927): Von der reinen Mathematik zur Numerik* (Göttingen: Vandenhoeck & Ruprecht, 1985).

135. American historian Thomas Hawkins has described Wilhelm Killing's exhaustive Weierstrassian analytical study of the foundations of geometry in the article "Non-Euclidean Geometry and Weierstrassian Mathematics: The Background to Killing's Work on Lie Algebras," in *History of Mathematics,* ed. E. R. Phillips, pp. 21–37.

136. See Weierstrass, *Mathematische Werke.* Knoblauch also edited volumes III (1903) and V (1915).

137. At the close of the nineteenth century, German mathematics was undergoing a major transformation. Research in pure mathematics shifted from the traditional fields of the theory of functions, invariant theory, and algebraic geometry to new fields like set theory and point-set topology. Abstract structures fascinated mathematicians. Much of the new mathematics, for example tensor analysis and matrix algebra, soon was crucial to handle problems in relativity theory and quantum mechanics. In addition, pressures from the second industrial revolution in Germany increased interest in potential applications in ship design, aerodynamics, and financial investments. In this environment of widening applications, the line between pure and applied mathematics was blurring, and the old neohumanist ideology no longer guided the mathematics profession.

From 1895, Klein together with David Hilbert (1862–1945) revitalized the Göttingen mathematical tradition at Georg-August University. These two mathematical universalists redirected the research lines of German mathematics away from narrow specialization and encouraged vibrant, integrating relations with other disciplines. Chiefly as a result of their intellectual and institutional work, Göttingen gained primacy in German mathematics. See David E. Rowe, "Klein, Hilbert, and the Göttingen Mathematical Tradition," in *Science in Germany,* ed. K. Olesko, 5: 186–213.

138. Bottazzini, *The Higher Calculus,* p. 290; David Rowe, "Klein, Hilbert, and the Göttingen Mathematical Tradition," *Science in Germany,* ed. K. Olesko, 5: 186–213; and I. M. Yaglom, *Felix Klein and Sophus Lie* (Basel: Birkhäuser, 1988).

139. David Rowe, "'Jewish Mathematics' at Göttingen in the Era of Felix Klein," *Isis,* 77 (1986): 433.

S. V. Kovalevskaya's Mathematical Legacy: The Rotation of a Rigid Body

Roger Cooke
University of Vermont

The lasting contributions of the Russian mathematician Sof'ya Vasil'evna Kovalevskaya to mathematics consist of two results. The first is her existence theorem for analytic solutions of certain partial differential equations (the Cauchy-Kovalevskaya theorem), which formed the major part of the doctoral dissertation that she wrote under the direction of Weierstrass in 1874; the second is her discovery of a physical configuration for which the equations of motion of a rigid body about a fixed point under the influence of gravity can be integrated in closed analytic form. The first of these is well-known, since the Cauchy-Kovalevskaya theorem continues to be an important tool in establishing the existence or nonexistence of analytic solutions of partial differential equations. In this paper I shall discuss the context and significance of her second achievement and consider its continuing influence in the late twentieth century.

To reconstruct precisely the present state of any nineteenth-century mathematical topic is in a sense impossible. No mathematical problem is understood exactly as it was understood at the time of Kovalevskaya's death a century ago. If a nineteenth-century mathematician could be resurrected and acquainted with contemporary work, that person would find the subject bewildering, not so much because many old problems have been solved as because the framework in which problems are posed has been generalized, and the new framework has generated a new set of problems. Where earlier mathematicians studied particular functions or geometric figures, twentieth-century mathematicians are inclined to think in terms of classes of things and deduce the properties of members from their membership in a class. For example, the classical theorem of Karl Weierstrass says that any continuous function on an interval $[a, b]$ can be approximated by a polynomial. The modern version, known as the Stone-Weierstrass theorem, says that, if an algebra $\mathfrak{A}$ of real-valued continuous functions on a compact Hausdorff space X separates points and for each point x of the space contains a function that does not vanish at x, then $\mathfrak{A}$ is dense in the space of continuous real-valued functions on X.

This change of language has affected all areas of mathematics, including those studied by Kovalevskaya. Thus where Kovalevskaya studied a particular system of differential equations intended to describe the rotation of a rigid body about a fixed point under the influence of gravity and used theta functions to express the solutions, a twentieth-century analog might be the study of flows on r-gonal Jacobians, to use the title of a paper by Emma Previato,[1] dedicated to the memory of Kovalevskaya. Theta functions, which were defined by very specific formulas in the nineteenth century, are today defined by structural properties and are consequently more abstract. In summary, the difference between the mathematics of the nineteenth century and that of the twentieth century is only partly a matter of old problems having been solved. It entails even more a difference in point of view, reflected in ever higher levels of abstraction.

This difference will appear in the example discussed below—Kovalevskaya's analysis of the problem of a rigid body rotating about a fixed point. Typically a problem in applied mathematics consists of many parts, which for purposes of discussion are considered in three groups: (1) analysis of the physical problem and abstraction of the appropriate mathematical problem corresponding to it; (2) solution of the mathematical problem; (3) application of the mathematical solution to the original problem, usually involving new mathematical complications because of the difficulty of numerical computation. We shall consider each of these problems in order, although the solution of the first problem was complete before Kovalevskaya began her work.

The Equations of Motion—Special Cases

The first work on the rotation problem was carried out by Leonhard Euler over many years. Euler derived various systems of equations to describe the motion of such a body, finally settling on a system of nine equations for the motion of a rigid body on which no force is acting (the "Euler case").[2] His three principal equations are:

$$dx + \frac{cc - bb}{aa} yz\, dt = 0,$$
$$dy + \frac{aa - cc}{bb} xz\, dt = 0,$$
$$dz + \frac{bb - aa}{cc} xy\, dt = 0.$$

In these equations a^2, b^2, and c^2 are the moments of inertia of the body about its three principal axes, t is time, and x, y, and z are the components of the angular velocity of the body measured in a coordinate system through the center of mass of the body.

In analyzing this physical problem Euler essentially worked alone. To solve a system of equations of this type requires ability in manipulating formulas so as to obtain a single equation containing only one dependent variable, which can be solved explicitly for the derivative of one variable and then integrated to give the solution.

Joseph-Louis Lagrange also considered this problem. He derived the following equations for the case in which the fixed point is taken at the origin and lies on one of the principal axes of the body passing through the center of gravity (if in addition the moments of inertia about the other two principal axes are equal, this case is known as the "Lagrange case"): [3]

$$\frac{d}{dt}\left(\frac{dT}{dp}\right) + q\frac{dT}{dr} - r\frac{dT}{dq} + km\zeta'' = 0,$$
$$\frac{d}{dt}\left(\frac{dT}{dq}\right) + r\frac{dT}{dp} - p\frac{dT}{dr} - km\zeta' = 0,$$
$$\frac{d}{dt}\left(\frac{dT}{dr}\right) + p\frac{dT}{dq} - q\frac{dT}{dp} = 0.$$

Here $T = \frac{1}{2}(Ap^2 + Bq^2 + Cr^2) - Fqr - Gpr - Hpq$, where p, q, and r are the components of the angular velocity of the body; A, B, and C are the principal moments of inertia of the body (what Euler denoted a^2, b^2 and c^2); F, G, and H are the three products of inertia; m is the mass of the body; k is the distance from the fixed point to the center of gravity of the body; and $(-km\zeta'', km\zeta', 0)$ is the torque on the body.

These studies by Euler and Lagrange illustrate the nature of the relationship between mathematics and physics. When one has constructed a mathematical model of a physical situation using differential equations, the solutions of the resulting mathematical problem are of great interest. If the mathematical problem is intractable, the physicist may well retrace the situation and model it differently (so that the simple 1-2-3 scheme we are following is not entirely realistic). In the cases of Euler and Lagrange the integrals that express the solution involve the square root of a cubic polynomial, and hence cannot be expressed in a finite number of algebraic and exponential functions. They lead to a new class of functions, known as elliptic functions. Three mathematicians, Adrien-Marie Legendre, Niels Henrik Abel, and Carl Gustav Jacobi, clarified the nature of elliptic functions. In his 1825 treatise on the subject Legendre used the Euler and Lagrange cases of these equa-

[1] Emma Previato, "Flows on r-gonal Jacobians," *Contemporary Mathematics*, **64** (1987): 153–180.

[2] L. Euler, "Du mouvement de rotation des corps solides autour d'un axe variable," *Mémoires de l'Académie des Sciences de Berlin*, **14** (1758): 154–193.

[3] J.-L. Lagrange, *Mécanique Analytique*, Paris: Ve Courcier, 1811, pp. 243–276.

tions as applications of elliptic functions.[4] After Abel died in 1829 and Legendre in 1833, Jacobi carried the work to completion by introducing theta functions, with which elliptic functions can be elegantly expressed. In 1849 Jacobi achieved a beautiful set of formulas for the change of variable from coordinates fixed in the rotating body to coordinates fixed in space in the case considered by Euler.

Jacobi's work prompted the Prussian Academy of Sciences in 1852 to propose a prize of 100 ducats for the solution of these equations:[5]

It is known that the number of cases in which the differential equations of analytical dynamics can be integrated in finite form, or even reduced to quadratures, is quite limited; and in view of the repeated efforts which the greatest mathematicians have applied to this subject, it is very probable that most of the mechanical problems whose solution has not yet been achieved in the form mentioned are by their nature not susceptible of integration by quadratures, and other analytic forms will be needed to handle them successfully. Since Jacobi has recently given a beautiful representation in series form of the rotation of a rigid body on which no accelerative force acts, it seems worthwhile to attempt to give a wider expression to the application of series, and with their aid to handle cases of rotational motion which have not yet been reduced to quadratures. One such case is offered by the problem of rotation of a heavy body, for which reduction to quadratures has been attained only in one special case due to Lagrange. The Academy therefore makes the complete solution of this problem the subject of a competition.

Despite this offer, which remained in effect from 1852 to 1858, the prize was not awarded. Weierstrass worked on the problem about this time, since Kovalevskaya alludes to such work in letters to Gösta Mittag-Leffler.

As its statement shows, the prize was motivated by the thought that the recent progress in understanding of elliptic functions would apply to the general equations of motion. However, these functions apply only *after* the problem has been reduced to quadratures. In a sense it was premature

to hope to apply the new knowledge of algebraic integrals to the physical problem: One needed the integrals first, and they were not available. Pure mathematics had already gone beyond mere elliptic integrals and was now capable of handling very general algebraic integrals. Therefore, seemingly, if one could only set up the proper integrals, the mathematical techniques developed by Jacobi and others would solve many important physical problems. To see how that process works is the subject of the next section.

The Euler equations were given a final form by the English mathematician R. B. Hayward:[6]

$$A\frac{dp}{dt} = (B-C)qr + Mg(z_0\gamma' - y_0\gamma''),$$
$$B\frac{dq}{dt} = (C-A)rp + Mg(x_0\gamma'' - z_0\gamma), \qquad (1)$$
$$C\frac{dr}{dt} = (A-B)pq + Mg(y_0\gamma - x_0\gamma').$$

Here γ, γ', and γ'' are the components of a unit vector pointing downward, expressed in terms of the principal axes of the body. (The primes simply distinguish components; they do *not* indicate differentiation.) As such they must be related to the angular velocity of the body by

$$\frac{d\gamma}{dt} = r\gamma' - q\gamma'',$$
$$\frac{d\gamma'}{dt} = p\gamma'' - r\gamma, \qquad (2)$$
$$\frac{d\gamma''}{dt} = q\gamma - p\gamma'.$$

The six equations (1) and (2) are the mathematical problem that must be solved in order to describe the motion of the body.

Solving the Differential Equations

Reduction to Quadratures. Again, the problem of solving a system of differential equations consists of two parts: manipulating the equations to reduce the problem to evaluating certain integrals; and performing the resulting integration. Consider the problems encountered in reducing the system of six equations (1) and (2) to quadratures. The nineteenth-century technique for this reduction can be understood well by looking at a simple and standard equation, that of simple

[4] A.-M. Legendre, *Traité des fonctions élliptiques et des intégrales eulériennes*, Paris: Husard-Courcier, 1825, Vol. 1: 366–410.

[5] The full statement of the problem was: "to integrate the differential equations for the motion of a body rotating about a fixed point, on which no accelerative force except gravity acts, by means of regularly progressing series which represent explicitly as functions of time all quantities required for the knowledge of the motion."

[6] R. B. Hayward, "On a direct method of estimating velocities, accelerations, and all similar quantities with respect to axes movable in any manner in space," *Transactions of the Cambridge Philosophical Society*, **10** (1858): 1–22.

harmonic motion:

$$x'' + x = 0,$$

where the primes denote differentiation with respect to a variable t, the only independent variable of this problem. (It usually represents time in the physical situation.) Making the substitution $y = x'$, we can write this equation as a system of two equations: $x' = y$, $y' = -x$, which can then be written as

$$dt = \frac{dx}{y} = \frac{dy}{-x}. \tag{3}$$

The second of these equations does not contain t, and can therefore be solved separately. Because the divergence of the vector $(y, -x)$ is zero, that is,

$$\frac{\partial y}{\partial x} + \frac{\partial(-x)}{\partial y} = 0,$$

this equation is exact, and its general solution can be found by elementary techniques. This solution is

$$x^2 + y^2 = C, \tag{4}$$

where C is a constant. The function $f(x, y) = x^2 + y^2$ is an invariant of the motion, called in the nineteenth century an *integral* of the system (3). I shall use only the term *invariant* to avoid confusion with the more familiar use of the term *integral*. In general an integral of such a system is any function of the variables x and y that remains constant as a function of the independent variable t when x and y are chosen as functions of t satisfying the system.

Equation (4) can be solved for $y = x'$, yielding the first-order equation

$$x' = \sqrt{C - x^2}, \quad \text{i.e.,} \quad dt = \frac{dx}{\sqrt{C - x^2}}.$$

Thus the problem is reduced to quadrature.

This technique, when suitably generalized, applies to the system of six equations (1) and (2), which can be written in the form

$$dt = \frac{dp}{P} = \frac{dq}{Q} = \frac{dr}{R} = \frac{d\gamma}{\Gamma} = \frac{d\gamma'}{\Gamma'} = \frac{d\gamma''}{\Gamma''}.$$

This system has two helpful elements of degeneracy: First, the independent variable t appears only as dt (the system is said to be *autonomous*). For that reason, t can be ignored, so that we really need solve only the other five equations. With a system of five equations, we need five independent invariants to get a complete solution.

In finding them we are helped by the other bit of degeneracy: The divergence of the coefficients is zero; in fact P is independent of p, Q is independent of q, etc., so that naturally

$$\frac{\partial P}{\partial p} + \frac{\partial Q}{\partial q} + \frac{\partial R}{\partial r} + \frac{\partial \Gamma}{\partial \gamma} + \frac{\partial \Gamma'}{\partial \gamma'} + \frac{\partial \Gamma''}{\partial \gamma''} = 0.$$

Because of this fact, a technique known as the Jacobi last-multiplier method makes it possible to produce a fifth independent invariant by integration once four independent invariants are known. We therefore need four independent invariants. Now three independent invariants are suggested immediately by the fact that $(\gamma, \gamma', \gamma'')$ is a unit vector, and by the laws of conservation of energy and conservation of angular momentum; because the torque is horizontal, the vertical component of the angular momentum is constant. Thus the general problem is solvable if one can find just one more independent invariant. This fact explains the success achieved in the Euler case, since the assumption that the torque is zero allows one to conclude that the two horizontal components of angular momentum are also constant, thus providing a full set of five independent invariants immediately. Similarly in the Lagrange case the derivative of the function T with respect to r is an invariant. From the point of view of physics, the Kovalevskaya case is the only other case in which a fourth algebraic invariant can be found. Once the problem has been reduced to quadratures, there remains the still-formidable task of performing the resulting integration. This task can be performed, provided the invariants are algebraic.

Algebraic Integrals.   Abelian integrals—integrals whose integrands are rational functions of two variables that are related by a polynomial equation—formed one of the most prominent areas of research in mathematics during the nineteenth century, and the clarification of the nature of algebraic functions and their integrals was one of the most significant achievements of the century. A broad outline of this large corpus of work will have to suffice to show its significance.

Many problems of mathematical physics, for instance the motion of a simple pendulum, lead to integrals that cannot be evaluated in the eighteenth-century sense of this term, that is, expressed in terms of a finite number of elementary algebraic and exponential functions. The pendulum problem leads to the square root of a cubic polynomial, and hence to elliptic integrals. The problem of rotation of a rigid body leads to integrals involving more complicated expressions. In the pendulum problem, for small oscillations one can replace the equation by an approximation that has elementary solutions, but there is a loss of realism in the solutions. As for the rigid body problem, as the Prussian Academy noted in announcing the 1852 prize competition, the only cases that could be handled in the eighteenth and early nineteenth centuries were degenerate cases in which either no force acts on the body (the Euler case was later known as the Euler-Poinsot case, because of a geometric discussion of it given by Louis Poinsot in 1834), or in which the body has a certain amount of symmetry (the Lagrange case). In both cases the integrals are elliptic. Although elliptic integrals are the simplest non-elementary integrals, they were poorly understood before the work of Legendre, Abel, and Jacobi between 1780 and 1840. Euler had strenuously attempted to avoid them on occasion, writing several papers on the special cases in which an integral that is apparently elliptic can be reduced to elementary functions. But the general case of the Euler equations is so complicated that the integrals are not even elliptic. In this problem, as in many others, such as the three-body problem, physics presented mathematicians with a real need for basic research.

From the 1780s to the 1820s Legendre advanced the study of elliptic integrals. Euler and Legendre, in turn, stimulated Abel and Jacobi to study still more complicated algebraic integrals; and their study, emphasizing hyper-elliptic integrals (those involving the square root of a polynomial of degree five or higher), attained complete generality, applicable to all algebraic integrals. In this work Jacobi discovered that elliptic functions, which are doubly periodic meromorphic functions having poles at a lattice of points in the complex plane, can be expressed as quotients of a remarkable class of entire functions called *theta functions*. Jacobi applied these functions to the physical problems that lead to elliptic functions, giving a thorough discussion of the Euler-Poinsot case of the motion of a rigid body in terms of them.

Jacobi's success in handling elliptic functions with theta functions led to a search for the appropriate generalization of them to handle more complicated algebraic integrals. The discovery of this generalization required the work of Ernst Minding, Gustav Göpel, Johann Rosenhain, Bernhard Riemann, and Karl Weierstrass, using a variety of approaches. The basic problem was to find an expression for the inverse of the transformation defined by a complete set of integrals of first kind. This problem was known as the *Jacobi inversion problem*, and its solution involved theta functions of several variables. The problem arose out of one of the greatest theorems of the early nineteenth century—Abel's Theorem.[7] This theorem shows that any sum of a finite number of such integrals can be reduced to a fixed number, depending only on the underlying algebraic relation between the variables, whose limits are algebraic functions of the limits of the original integrals. For elliptic integrals, this number is one, and so the problem of finding the limit of integration is determinate. For more complicated integrals there is more than one limit of integration, and so the problem becomes indeterminate. In order to make the problem a determinate one, Jacobi considered a system of such sums of integrals, with different integrands in different sums. The problem of determining the limits of integration then again becomes determinate.

To see the problem and its solution in an illustrative case, let us look at the simplest non-elliptic case, the inversion of a set of two hyper-elliptic integrals of "genus two". This situation poses a transformation $(x, y) \mapsto (u, v)$ expressed by two integrals of *first kind*, i.e., integrals having no singularities in the extended complex plane:

$$u = \int_a^x \frac{1}{\sqrt{P(s)}}\, ds + \int_b^y \frac{1}{\sqrt{P(s)}}\, ds,$$
$$v = \int_a^x \frac{s}{\sqrt{P(s)}}\, ds + \int_b^y \frac{s}{\sqrt{P(s)}}\, ds, \tag{5}$$

in which $P(s)$ is a polynomial of degree 5 or 6, and the problem is to express x and y in terms of u and v. The solution was achieved by Rosenhain in a letter to Jacobi published after Rosenhain's death.[8] The solution required the invention of theta functions of two variables. One way of expressing it follows. Let $Z(x)$ be any rational function, say, having poles at the points $\gamma_1, \ldots, \gamma_k$. Let $\xi_1, \ldots, \xi_k$ be such that $Z(\xi_j) = X$, where X is any fixed complex number. Finally, define the theta function of two variables $\theta(s, t)$ by the following relation:

$$\theta(s, t) = \sum_{n=-\infty}^{\infty} \sum_{m=-\infty}^{\infty} e^{\pi i \left(M^* T M + 2 M^* \binom{s}{t} \right)},$$

where $M = \binom{m}{n}$, $T = \begin{pmatrix} \tau_{11} & \tau_{12} \\ \tau_{21} & \tau_{22} \end{pmatrix}$, and $\mathrm{Im}\,(T)$ is a positive-definite matrix. Choose c so that

$$\theta \left(\int_c^u \frac{ds}{\sqrt{P(s)}} - \int_a^u \frac{ds}{\sqrt{P(s)}} - \int_b^v \frac{ds}{\sqrt{P(s)}}, \right.$$
$$\left. \int_c^v \frac{s\, ds}{\sqrt{P(s)}} - \int_a^u \frac{s}{\sqrt{P(s)}} - \int_b^v \frac{s\, ds}{\sqrt{P(s)}} \right) \equiv 0.$$

Then

$$\frac{[Z(x) - X][Z(y) - X]}{[Z(a) - X][Z(b) - X]} =$$

---

[7] See the article by R. Cooke, "Abel's Theorem," in *The History of Modern Mathematics*, edited by David Rowe and John McCleary, Boston: Academic Press, 1989, pp. 387–421.

[8] *Journal für die reine und angewandte Mathematik*, **40** (1850): 320–360.

$$\prod_{j=1}^{k} \frac{\theta\left(\int_c^{\xi_j} \frac{ds}{\sqrt{P(s)}} - u, \int_c^{\xi_j} \frac{s\,ds}{\sqrt{P(s)}} - v\right)\theta\left(\int_c^{\gamma_j} \frac{ds}{\sqrt{P(s)}}, \int_c^{\gamma_j} \frac{s\,ds}{\sqrt{P(s)}}\right)}{\theta\left(\int_c^{\gamma_j} \frac{ds}{\sqrt{P(s)}} - u, \int_c^{\gamma_j} \frac{s\,ds}{\sqrt{P(s)}} - v\right)\theta\left(\int_c^{\xi_j} \frac{ds}{\sqrt{P(s)}}, \int_c^{\xi_j} \frac{s\,ds}{\sqrt{P(s)}}\right)}.$$

Because of the possible choices of Z, a, and b, there is more than one possible solution to this problem. In solving the problem in a particular case one must therefore make choices so that variables significant to the problem at hand appear naturally and prominently. In other words, solving this problem is not merely a matter of following a recipe. This point should be kept in mind in the discussion of Kovalevskaya's application of the technique.

Even though it was premature to think of applying the theory of Abelian integrals to the equations of the rotation problem when the Prussian Academy proposed the contest in 1852, this beautiful technique was not unapplied for long. Physicists and mathematicians immediately began seeking applications for it, and their search was soon rewarded in the dissertation of Carl Neumann at the University of Königsberg in 1856, which was subsequently published (1859). The English translation of the Latin title of the work is "On a certain problem of mechanics that can be reduced to hyper-elliptic integrals of first kind." Neumann considered the following problem: *Describe the motion of a point constrained to remain on the sphere $x^2 + y^2 + z^2 = 1$ when acted on by a force whose potential is $ax^2 + by^2 + cz^2$.* Here a, b, and c are any real numbers, positive or negative, satisfying $a < b < c$. Neumann gave the attraction exerted by a homogeneous ellipsoid on an internal point as a specific example of such a problem. Using the Hamilton-Jacobi equations, he deduced the system

$$\frac{\partial \varphi}{\partial \lambda_1} = u_{11} \frac{d\lambda_1}{dt} + u_{12} \frac{d\lambda_2}{dt},$$

$$\frac{\partial \varphi}{\partial \lambda_2} = u_{21} \frac{d\lambda_1}{dt} + u_{22} \frac{d\lambda_2}{dt},$$

where

$$u_{11} = \left(\frac{\partial x}{\partial \lambda_1}\right)^2 + \left(\frac{\partial y}{\partial \lambda_1}\right)^2 + \left(\frac{\partial z}{\partial \lambda_1}\right)^2,$$

$$u_{22} = \left(\frac{\partial x}{\partial \lambda_2}\right)^2 + \left(\frac{\partial y}{\partial \lambda_2}\right)^2 + \left(\frac{\partial z}{\partial \lambda_2}\right)^2,$$

$$u_{12} = u_{21} = \frac{\partial x}{\partial \lambda_1} \cdot \frac{\partial x}{\partial \lambda_2} + \frac{\partial y}{\partial \lambda_1} \cdot \frac{\partial y}{\partial \lambda_2} + \frac{\partial z}{\partial \lambda_1} \cdot \frac{\partial z}{\partial \lambda_2}.$$

Neumann reduced these equations to a pair of integrals similar to (5). Using the same letters as above so as to emphasize the similarity:

$$0 = \varepsilon_1 \int_{l_1}^{\lambda_1} \frac{ds}{\sqrt{P(s)}} + \varepsilon_2 \int_{l_2}^{\lambda_2} \frac{ds}{\sqrt{P(s)}},$$

$$t = \varepsilon_1 \int_{l_1}^{\lambda_1} \frac{s\,ds}{\sqrt{P(s)}} + \varepsilon_2 \int_{l_2}^{\lambda_2} \frac{s\,ds}{\sqrt{P(s)}}.$$

Here ε_1 and ε_2 denote ± 1. By this reasoning Neumann produced a system remarkably similar to that of Kovalevskaya 30 years later. Being interested in physical applications, he proceeded to discuss various cases that arise from different data, giving a geometric description of the motion of the point in four different cases. Discussions of the ambiguous signs resulting from ε_1 and ε_2 and of the way in which the constants of integration l_1 and l_2 can be determined followed. Finally, he invoked Rosenhain's work on theta functions of two variables to give a detailed description of the motion. In view of Kovalevskaya's work, this part of his paper is the most interesting today. He used the theta functions

$$\varphi_{kh}(iv, iw, p, q, \alpha) =$$
$$\sum_{m=-\infty}^{\infty} \sum_{n=-\infty}^{\infty} (-1)^{ms_k + ns_h} p^{\frac{1}{4}m_k^2} q^{\frac{1}{4}n_h^2} e^{m_k iv + n_h iw + m_k n_h \alpha}$$

and eventually obtained explicit expressions for x, y, and z as functions of time.

Although Neumann's work involved a rather artificial problem of mechanics, it had interest for its own sake (and still does today). Principally, however, it stimulated interest in the ways in which Abelian integrals can be applied to the differential equations of physics. The major difficulty was the absence of ways of reducing equations to quadratures, so that some new approach was needed. Weierstrass looked at this problem from the point of view of analytic function theory, probably shortly after the 1852 prize competition. Ignoring the fact that time in physics is a real variable, he regarded all the functions as analytic functions of complex-variable time and asked what restrictions (special cases) lead to physical functions that are single-valued analytic functions of time—not necessarily entire functions, since a pole or essential singularity will not have any effect on the physical interpretation of the variable unless it occurs on the real axis. From this point of view he found that general solutions of the Euler equations could *not* be single-valued functions of time. None of this work was published. We know of it from a letter Kovalevskaya wrote to Mittag-Leffler from Berlin on 21 November 1881:[9]

> This work [on the refraction of light] was quite well along when I had the weakness to allow myself to be distracted by another question, which has troubled my head since almost the beginning of my mathematical studies, and in which, for a time, I feared others would surpass me. The problem involves solving the general case of rotation of a heavy body about a fixed point by means of Abelian functions. Weierstrass had once previously suggested that I work on this problem, but all my attempts at the time were fruitless; and Weierstrass' own investigations showed that the differential equations of this problem cannot by satisfied by single-valued (eindeutig) functions of time. This result compelled me to abandon this problem for a while, but since then the beautiful, still-unpublished research of our master on the stability of the solar system and the analogy with other problems of dynamics have renewed my zeal and given me the hope of satisfying the conditions of this problem by Abelian functions *whose arguments are nonlinear functions of time*... The route I followed consisted of expressing the variables of the problem by theta functions of two variables which for certain values of the constants reduce to the elliptic theta functions that arise in the particular case of Lagrange, then trying to choose them so as to be able to integrate the differential equations between the theta functions and time. The calculations this led me into were so difficult and complicated that I cannot yet say if I will reach the desired end by this route... Weierstrass is consoling me that even in the worst case I could always reverse the problem and try to find out which forces lead to a rotation whose variables can be expressed by Abelian functions—a poor problem, to be sure, and far from having the same interest as the one I have set myself; but I shall have to settle for it if I have bad luck, relying on the example of Neumann, who chose an analogous problem for his doctoral dissertation.

The new approach thus assumes the solutions can be expressed as theta functions of two variables (a reasonable assumption) and adjusts the parameters of the theta functions so that the equations are satisfied. This approach allows mathematicians to circumvent the natural barricade caused by the absence of quadratures. It does not make the problem easy, however. The number of parameters involved, and the number of possible adjustments are prodigious, and Kovalevskaya was certainly not exaggerating when she described the complexity of her computations. That she was able to deal with these complexities is ample testimony to her profound knowledge of this area of mathematics. Before discussing details of this work, we consider how she came to be an expert in Abelian functions.

Kovalevskaya's Education. Kovalevskaya began the study of higher mathematics with Leo Königsberger and Paul Du Bois-Reymond in Heidelberg in 1869 and continued as a private student of Weierstrass in Berlin starting in 1870, obtaining the doctoral degree *in absentia* from the University of Göttingen in 1874. What was the content of this education? The courses Weierstrass gave from 1870 to 1873 give some clue.

Summer 1870	Analytic Function Theory
Winter 1870–71	Theory of Elliptic Functions
Summer 1871	Modern Synthetic Geometry
	Selected Geometric and Mechanical Problems Solvable Using Elliptic Functions
Winter 1871–72	Theory of Abelian Functions
Summer 1872	Introduction to Analytic Function Theory
	Calculus of Variations
Winter 1872–73	Theory of Elliptic Functions
Summer 1873	Elements of Modern Synthetic Geometry
	Selected Geometric and Mechanical Problems Solvable Using Elliptic Functions
Winter 1873–74	Theory of Abelian Functions

Weierstrass undoubtedly discussed many topics with Kovalevskaya that were not in his regular lectures. But he would at least have given her copies of the lectures he was reading to his other students at the time. We know that he did so on some occasions, since in one of his early letters to her (14 January 1872) he discusses the topic for their next meeting, which is a problem in Abelian integrals.[10] This letter is especially noteworthy since he asks her to

---

[9] See R. Cooke, *The Mathematics of Sonya Kovalevskaya*, Berlin-Heidelberg-New York: Springer-Verlag, 1984, pp. 95–96.

[10] P. Ya. Kochina, *Briefe von Karl Weierstrass an Sofie Kowalewskaja*, Moscow: Nauka, 1973, pp. 11–12.

"The route I followed consisted of expressing the variables of the problem by theta functions of two variables... The calculations this led me into were so difficult and complicated that I cannot yet say if I will reach the desired end by this route."

A page of computations in Kovalevskaya's hand.
(Courtesy Institut Mittag-Leffler)

return pages on which his observations are written because he needs them for his next lecture.

By 1874 Kovalevskaya had become one of the world's foremost experts on algebraic integrals, and one can only regret that her talent and expertise lay unused for a decade because the social conditions of the time posed insuperable barricades to women's participation in the world of scholarship. The basic theorems about algebraic integrals were proved before Kovalevskaya, but in her dissertation she showed great facility in this area by reducing a class of hyper-elliptic integrals to elliptic integrals. This topic—degeneracy of algebraic integrals—was of considerable interest because one is always interested in simplifying formulas to the maximum extent possible. Poincaré later wrote a paper on the same subject and praised Kovalevskaya's contribution to it very highly in his analysis of his own

Berlin, 14 January 1872

Dear Madame,

To my regret I shall not be able to receive you tomorrow morning, since I have been ill with a cold for several days.

I am therefore sending you my notes on the topic that will be the subject of our next conversation. These notes are sufficiently complete that you will be able to assimilate them even without my being there. Please be so kind as to return *these* papers, which I need for my next lecture.

In order to rectify the general formulas in a simple case, please note that if

$$y = \sqrt{R(x)},$$

and $R(x)$ is an entire function of degree $(2\rho + 1)$—for ρ has exactly this meaning in the present case—the simplest function

$$H(xy, x'y')$$

is the following, in which $P(x)$ denotes a divisor of $R(x)$ of degree ρ:

$$H(xy, x'y') = \tfrac{1}{2}\left(1 + \tfrac{P(x)}{P(x')}\tfrac{y}{y'}\right)\tfrac{1}{x-x'}$$

The latter becomes infinite, if

$$P(x) = (x - a_1)(x - a_2)\cdots(x - a_\rho),$$

for the points

$$x = x', \quad a_1, \ldots, a_\rho$$
$$y = y', \quad 0, \ldots, 0$$

and moreover each pole is of first order. In addition its expansion in powers of $x - x'$ for x in a neighborhood of x' and y in a neighborhood of y' begins with

$$-\tfrac{1}{x-x'}.$$

Finally it becomes 0 for $x = \infty$, $y = \infty \ldots$

Letter from Weierstrass to Kovalevskaya, 14 January 1872

works.[11] The theory of Abelian functions was essentially complete by the time Kovalevskaya became a student of Weierstrass. She learned the subject in Weierstrass's presentation, which differed from that of Riemann, though Weierstrass certainly knew the relation between the two and made sure that Kovalevskaya did also.

The solution of the Jacobi inversion problem leads to the conclusion that algebraic integrals can be *evaluated*, provided the class of allowable functions is expanded to include theta functions of many variables. Jacobi's 1849 paper is a splendid example of the application of this principle to the case of elliptic functions, where only theta functions of one variable are required for the solution. One might then naturally expect to see theta functions of several variables applied to solve more complicated problems of significance to physics. This step, however, was long in coming, despite every encouragement from various Academies of Science, such as the 1852 contest of the Prussian Academy. The reason is not difficult to find: the equations must be "reduced to quadratures," that is, transformations must be performed so that the problem becomes purely a matter of evaluating integrals. That step was missing in the general case of the equations of motion of a rotating rigid body, and also in many other systems, such as the three-body problem. Kovalevskaya's contribution to this problem was to discover a new and very general case in which such a reduction was possible, then to carry out the highly non-trivial details of the transformations, so as to obtain a system of the form (∗), which can then be solved by known techniques.

[11] H. Poincaré, "Sur la réduction des intégrales abéliennes," *Comptes Rendus*, **99** (1884): 853–855.

Kovalevskaya's Discovery. From the letter written to Mittag-Leffler in 1881 we learn that Kovalevskaya had become interested in the rotation problem early in her career. Weierstrass must at least have mentioned it in his course on physical problems leading to elliptic integrals, given in the summer of 1873. In any case as a well-educated mathematician she would certainly have heard of it. Kovalevskaya hoped to solve the general case of this problem, that is, to find the general solution of the equations (1) and (2). The first step is to reduce the equations to quadratures when trying to solve the system in closed form. Realizing the difficulty of this step, Kovalevskaya did not really try to do that. She followed the path Weierstrass had opened, trying to express the functions p, q, r, γ, γ', and γ'' as meromorphic functions of time and to see what constraints such an assumption placed on the parameters A, B, C, x_0, y_0, z_0. As she stated in her memoir,[12]

> Up to now these equations have been integrated [completely] in only two particular cases:
>
> 1) the case of Poisson (or Euler), where
>
> $$x_0 = y_0 = z_0 = 0,$$
>
> [Kovalevskaya seems to be confusing Poinsot and Poisson here—RLC]
>
> 2) the case of Lagrange, where
>
> $$A = B, \quad x_0 = y_0 = 0.$$
>
> In these two cases the integration can be carried out using a function $\vartheta(u)$ whose argument is an entire linear function of time.
>
> The six quantities p, q, r, γ, γ', γ'' are single-valued functions of time in both these cases and have only poles as singularities for finite values of the independent variable.
>
> Do the integrals of these differential equations retain this property in the general case?
>
> If so, these differential equations could be integrated using series of the form
>
> $$p = t^{-n_1}(p_0 + p_1 t + p_2 t^2 + \cdots)$$
> $$q = t^{-n_2}(q_0 + q_1 t + q_2 t^2 + \cdots)$$
> $$r = t^{-n_3}(r_0 + r_1 t + r_2 t^2 + \cdots)$$
> $$\gamma = t^{-m_1}(f_0 + f_1 t + f_2 t^2 + \cdots)$$

[12] S. Kovalevskie, "Mémoire sur un cas particulier du problème de la rotation d'un corps solide autour d'un point fixe, où l'intégration s'éffectue à l'aide des fonctions ultra-élliptiques du temps," *Mémoires Présentés par Divers Savants*, **31** (1890): pp. 1–2.

$$\gamma' = t^{-m_2}(g_0 + g_1 t + g_2 t^2 + \cdots)$$
$$\gamma'' = t^{-m_3}(h_0 + h_1 t + h_2 t^2 + \cdots),$$

where n_1, n_2, n_3, m_1, m_2, m_3 are positive integers; and in order for these series to represent the general system of integrals of the differential equations under consideration, they must contain *five* arbitrary constants.

The need for only five arbitrary constants is explained by the fact that the system is autonomous, that is, does not involve time explicitly, and so time $t = 0$ can be moved to any desired starting value. In these last equations Kovalevskaya was assuming it was moved to a pole of the solutions. By substituting these expressions into the equations, anyone can find, as Kovalevskaya did, that $n_1 = n_2 = n_3 = 1$ and $m_1 = m_2 = m_3 = 2$. Kovalevskaya's argument follows. In general these conditions cannot be met, but that they can be met when $A = B = 2C$ and $z_0 = 0$. This is the Kovalevskaya case. Charles Hermite accurately described it in a letter to Kovalevskaya in 1886 as a beautiful discovery. By studying the poles of the solution (the asymptotic behavior of the solutions at a pole), Kovalevskaya found the only possible new case in which the solutions can be single-valued analytic functions of time.

By a suitable choice of axes and units it is possible to eliminate most of the parameters and assume $A = 2 = B$, $C = 1$, $y_0 = 0 = z_0$. One is then left with only one parameter, which Kovalevskaya took to be $c_0 = MGx_0$. Equations (2) are unaffected by all these normalizations, but equations (1) become

$$2\frac{dp}{dt} = qr,$$
$$2\frac{dq}{dt} = -pr - c_0\gamma'', \qquad (1')$$
$$\frac{dr}{dt} = c_0\gamma'.$$

Once she discovered this case, which nature had hidden from all eyes before hers, Kovalevskaya had little difficulty finding a fourth invariant of the system, namely

$$\{(p + qi)^2 + c_0(\gamma + i\gamma')\}\{(p - qi)^2 + c_0(\gamma - i\gamma')\}.$$

She had therefore made possible the reduction to quadratures by use of the Jacobi last-multiplier method. The first phase of her task was therefore complete. She had found a case in which the reduction to quadratures was possible. It remained to carry out the actual reduction, then to solve the equations for the resulting invariants.

Thus the second phase of her project began, that of transforming the equations $(1')$ into a form suitable for quadratures. Again the transformations involved are extremely complicated. Kovalevskaya found a long sequence of changes of variable, leading finally to variables that she denoted s_1 and s_2 but that we shall call ξ and η here, in terms of which the equations can be written as

$$0 = \frac{d\xi}{\sqrt{\mathcal{P}(\xi)}} + \frac{d\eta}{\sqrt{\mathcal{P}(\eta)}},$$
$$dt = \frac{\xi\, d\xi}{\sqrt{\mathcal{P}(\xi)}} + \frac{\eta\, d\eta}{\sqrt{\mathcal{P}(\eta)}}, \tag{6}$$

where $\mathcal{P}$ is a polynomial of degree 5. A comparison of equations (6) with equations (5), shows that ξ and η can be expressed in terms of two variables u and v of which u is constant and v is time. Thus ξ and η are linear functions of time. This completes the second phase of the process and initiates the third phase—carrying out the inversion of equations (6).

The fine detail of this part of the work occupies nearly three-fourths of Kovalevskaya's 62-page paper. She expressed p, q, and r as quotients of linear combinations of six algebraic functions whose worst irrationality is the square root of a polynomial of degree 2 or 5 in the variables we have called ξ and η. She then managed to express these functions as quotients of theta functions of two variables, each argument of which is a linear function of time. Thus the third and final part of the mathematical analysis was completed.

Let us now respond to two questions about this work:
1) Why did it deserve the Bordin Prize?
2) What lasting impact has it had?

Significance of Kovalevskaya's solution

Kovalevskaya's work on this problem contains several achievements, any one of which is evidence of a very high mathematical ability and creativity:

1) By asking new questions about the system of Euler equations, she discovered a special case that was capable of closed-form solution;

2) She created the transformations needed to reduce the system to quadratures;

3) She applied the techniques for handling systems of algebraic integrals in a special case in such a way as to exhibit significant variables of the problem explicitly in terms of algebraic functions, each of which was a quotient of theta functions of linear functions of time.

These achievements gained the admiration of jury of the Paris Academy, which awarded the Bordin Prize to Kovalevskaya in 1888 in the following words.[13]

> The author has not merely added a result of very high interest to those that were bequeathed to us by Euler and Lagrange; he [the identity of "l'auteur" was officially unknown to the jury— RLC] has made a profound study of the result due to him, in which all the resources of the modern theory of theta functions of two independent variables allow the complete solution to be given in the most precise and elegant form. One has thereby a new and memorable example of a problem of mechanics in which these transcendental functions figure, whose applications had previously been limited to pure analysis and geometry.

The continuing influence of Kovalevskaya's techniques can be discovered in the following techniques: 1) analysis of the poles or asymptotic behavior of solutions of a differential equation to determine whether turth invariant that she discovered has a geometric interpretation that provided the basis for a paper by Nikolaĭ Egorovich Zhukovskiĭ, in which several important characteristics of the motion in the Kovalevskaya case were pointed out.[14] The complexity of Kovalevskaya's mathematics turns out to be only a reflection of the complexity of nature in this case: the Kovalevskaya case remains to this day a mysterious motion, not noticeably simpler than the rotation of a completely arbitrary rigid body.

The Kovalevskaya Legacy

We now know what to look for in seeking examples of the influence of Kovalevskaya's ideas in twentieth-century mathematics. This influence, in the case of the rotation problem, will be expressed in 1) studying the singularities of solutions to determine whether an equation can be integrated in closed form, and 2) the use of theta functions to solve differential equations. As mentioned above, the language of mathematics has changed greatly since Kovalevskaya's day. Nowadays, for example, a mathematician is unlikely to ask whether a system of differential equations

[13] *Comptes Rendus*, **CV** (1888): p. 1042.

[14] N. E. Joukowsky, "Geometrische Interpretation des von Sophie Kowalevski behandelten Falles der Bewegung eines schweren starren Körpers um einen festen Punkt," *Jahresbericht der deutschen Mathematiker-Vereinigung*, **4** (1897): 144–150.

can be integrated by quadratures. The question will more likely be framed in an abstract form, such as: Can a flow be linearized on an Abelian variety?

Still, even when the change in language is taken into account, there are certain constant criteria of mathematical profundity. For example, in judging the lasting effect of any new mathematical technique, it is useful to ask: *Was the technique, after being published, used to solve any other outstanding problems?* In the case of Kovalevskaya's theta-function technique, the answer is yes, but with considerable delay. Nearly ninety years passed before a firm positive answer could be given to this question. The Kovalevskaya case of the Euler equations was mentioned earlier in standard works on mechanics, and this is another indication of its value. We may perhaps understand both Kovalevskaya's legacy and the power of twentieth-century mathematics, which lies in generality and synthesis, by looking at another paper, very similar to hers, which also waited a long time before being put into the proper perspective in the twentieth century.

The paper is Carl Neumann's doctoral thesis, which Kovalevskaya herself mentioned as an example of the approach to the problem suggested by Weierstrass.[15] In view of this work of Neumann it was not accurate to say, as the Bordin jury did, that the applications of theta functions of two variables had previously been limited to analysis and geometry. Now the problem considered by Neumann seemed rather artificial and limited in scope, as Kovalevskaya's characterization of it shows. In that respect it resembles the Kovalevskaya case of the rotation problem. Recently, however, it has been discovered that Neumann's problem can be interpreted as the problem of several coupled harmonic oscillators, and is more fundamental than at first appears.

The question—can a given flow be linearized on an Abelian variety—was posed and solved by M. Adler and P. van Moerbeke for a variety of flows in a recent paper,[16] and the solution was achieved by analyzing the asymptotic behavior of the solutions at their poles, just as Kovalevskaya had done, as the authors explicitly observed. Thus it seems that the technique that led Kovalevskaya to discover her case of integrability continues to be used, though in a new language that she would not immediately recognize.

In general, applications of Kovalevskaya's theta function techniques were rather slow in coming because of the complexity of the systems in which they can be applied. Nevertheless, by the late 1960s they began to appear. Emma Previato gives a good example in the Korteweg-de Vries equation for wave motion

$$u_t - \frac{1}{4}u_{xxx} - \frac{3}{2}uu_x = 0.$$

This equation was proposed by D. J. Korteweg and G. de Vries in an 1895 paper written a few years after Kovalevskaya's work, and Previato gives an expression for a solution of it in terms of a theta function (of one variable).

After Kovalevskaya's work[17] some 85 years passed before anyone made similar use of theta functions of two variables, in problems of dynamics, but then a whole series of such papers appeared. By the 1950s theta functions were again a topic of interest, their basic theory having been given a new cast in the context of Abelian varieties by S. Conforto, Karl Ludwig Siegel, and André Weil in the 1940s.[18] In the meantime another fundamental equation of mathematical physics had appeared, the Korteweg-de Vries equation written above.

According to Previato, until the late 1960s this nonlinear equation was thought to have no solutions that could be expressed exactly in terms of known functions. Then in the 1970s B. A. Dubrovin, V. B. Matveev, and S. P. Novikov in a series of individually and jointly written papers, established connections between the Korteweg-de Vries equation and Abelian varieties. At the same time, Henry McKean and P. Moerbeke were proving some very similar results on Hill's equation, which is important in the theory of lunar motion. The work of Dubrovin, Matveev, and Novikov was summarized in a joint paper,[19] in which they showed that the set of complex solutions of a stationary higher-order Korteweg-de Vries equation is a certain Abelian variety. Moreover the potential function for such a problem can be expressed in algebraic form using theta functions of many variables. The work of McKean, Moerbeke, and others is summarized in the proceedings of the Global Analysis Sem-

[15] C. Neumann, "De problemate quodam mechanico, quod ad primam integralium ultraellipicorum classem revocatur," *Journal für die reine und angewandte Mathematik*, **56** (1859): 54–66.

[16] M. Adler and P. Van Moerbeke, "Kowalewski's asymptotic method, Kac-Moody Lie algebras, and regularization," *Communications in Mathematical Physics*, **83** (1982), No. 1: 83–106. This paper and the paper of Dubrovin, Matveev, and Novikov mentioned below were pointed out by Ann Hibner Koblitz in *A Convergence of Lives*, Boston: Birkhäuser, 1983.

[17] See note 11 above.

[18] See Jun-Ichi Igusa, *Theta Functions*, Berlin: Springer-Verlag, 1972.

[19] B. A. Dubrovin, V. B. Matveev, and S. P. Novikov, "Nonlinear equations of Korteweg-de Vries type," *Uspehi Matematicheskih Nauk*, **31** (1976): 55–136 [Russian].

inar in Calgary in 1978,[20] in which it is shown that Hill's equation is actually contained in the Neumann equations discussed above. Thus twentieth-century algebraic geometry has unified the work of Kovalevskaya, Neumann, and Korteweg and de Vries.

Of course, it would be difficult to recognize that these theorems involve techniques Kovalevskaya had pioneered. The linguistic disguise, in terms of flows on a Jacobian variety, is perfect. I think that, if Kovalevskaya could see one of the papers just referred to, she would have to read as far as the appearance of theta functions before realizing that this was a topic familiar to her. Dubrovin, Matveev, and Novikov, however, did not forget the person who first used these techniques. At the end of the introduction to their paper they wrote,

> Incidentally, as a concluding remark we point out that in the classical nontrivial cases of Jacobi (geodesics on a 3-dimensional ellipsoid) and of Sonya Kovalevski (the case of a heavy gyroscope) Abelian varieties also occur as level surfaces of commuting integrals: the direct product of one-dimensional ones for Jacobi's case, and non-trivial two-dimensional Abelian varieties for the Kovalevski case... Here we quote from a letter written by Sonya Kovalevski in December 1886 [The letter is actually undated—RLC]:
>
>> He (Picard) reacted with great disbelief when I told him that functions of the form
>>
>> $$y = \frac{\theta(cx + a, c_1 x + a_1)}{\theta_1(cx + a, x_1 x + a_1)}$$
>>
>> can be useful in the integration of certain differential equations.

The analysis of the present authors shows that for 90 years after the work of Sonya Kovalevski, until the 1974 papers on the K-dV equations, Picard's skepticism was justified.

It is fascinating to speculate on what Kovalevskaya would make of all the papers that carry on the work she began a century ago. Like any nineteenth-century mathematician, she would have to learn an entirely new mathematical notation in which the concepts familiar to her are special cases of more general entities that are now the main focus of attention. Yet she would have one enormous advantage over the current generation of mathematicians who focus on generality and are sometimes at a loss when faced with concrete particular cases. If she could be with us today,

she most likely would repeat her experiences in Heidelberg and Berlin, sequester herself with a few papers and monographs in order to become familiar with the new language, and emerge a few weeks later, ready to show the mathematical world new things that can be done with the legacy it has inherited from her and her contemporaries.

References

1. M. Adler and P. Van Moerbeke, "Kowalewski's asymptotic method, Kac-Moody Lie algebras, and regularization," *Communications in Mathematical Physics*, **83** (1982): 83–106.
2. R. L. Cooke, *The Mathematics of Sonya Kovalevskaya*, Springer-Verlag, Berlin, 1984.
3. ——. "Sonya Kovalevskaya's place in nineteenth-century mathematics," *Contemporary Mathematics*, **64** (1987): 17–52.
4. ——. "Abel's Theorem," In: *The History of Modern Mathematics*, edited by David Rowe and John McCleary, Academic Press, Boston, 1989.
5. B. A. Dubrovin, V. B. Matveev, and S. P. Novikov, "Nonlinear equations of Korteweg-de Vries type," *Uspehi Matematicheskih Nauk*, **31** (1976): 55–136 [Russian].
6. L. Euler, "Du mouvement de rotation des corps solides autour d'un axe variable," *Mémoires de l'Académie des Sciences de Berlin*, **14** (1758): 154–193.
7. R. B. Hayward, "On a direct method of estimating velocities, accelerations, and all similar quantities with respect to axes movable in any manner in space," *Transactions of the Cambridge Philosophical Society*, **10** (1858): 1–22.
8. J. Igusa, *Theta Functions*, Springer-Verlag, Berlin, 1972.
9. C. G. J. Jacobi, "Dilucidationes de æquationum differentialium vulgarium systemate earumque connexione cum æquationibus differentialibus partialibus linearibus primi ordinis," *Journal für die reine und angewandte Mathematik*, **23** (1842): 1–104.
10. ——. "Sur la rotation d'un corps," *Comptes Rendus*, **29** (1949): 97–106 = *Journal für die reine und angewandte Mathematik*, **39**: 293–350 = *Journal des mathématiques pures et appliquées*, **14**: 337–344.
11. A. H. Koblitz, *A Convergence of Lives. Sofia Kovalevskaia: Scientist, Writer, Revolutionary*. Birkhäuser, Boston, 1983.
12. ——. "Changing views of Sofia Kovalevskaia," *Contemporary Mathematics*, **64** (1987): 53–78.
13. D. J. Korteweg and G. De Vries, "On the change of form of long waves advancing in a rectangular canal and on a new type of long solitary waves," *Philosophical Magazine*, **39** (1895): 422–443.
14. S. V. Kovalevskaya, "Mémoire sur un cas particulier du problème de la rotation d'un corps solide autour d'un point fixe, où l'intégration s'effectue à l'aide des fonctions ultra-élliptiques du temps," *Mémoires présentés par divers savants*, **31** (1890): 1–62.
15. A.-M. Legendre, *Traité des fonctions élliptiques et des intégrales eulériennes*, Huzard-Courcier, Paris, 1825.
16. C. Neumann, "De problemate quodam mechanico, quod ad primam integralium ultraellipticorum classem revocatur,"

---

[20] H. P. McKean, "Integrable systems and algebraic curves," in: *GLobal Analysis* Springer Lecture Notes in Mathematics, No. 755 (1979), pp. 83–200.

Journal für die reine und angewandte Mathematik, **56** (1859): 54–66.

17. H. Poincaré, "Sur la réduction des intégrales abéliennes," *Comptes Rendus*, **99** (1884): 853–855.

18. ——. *Œuvres*, 11 Vols. Paris: Gauthier-Villars, 1951–1956.

19. K. Weierstrass, *Werke*, 7 Vols., Georg Olms Buchhandlung, Hildesheim, 1894–1927.

20. N. E. Zhukovskĭ. "Geometrische Interpretation des von Sophie Kowalevski behandelten Falles der Bewegung eines schweren starren Körpers um einen festen Punkt," *Jahresbericht der deutschen Mathematiker-Vereinigung*, **4** (1897): 144–150.

Aspects and Problems of the Development of Mathematical Education at Technical Colleges in Germany during the Nineteenth Century

Susann Hensel
University of Jena

This paper addresses a rather significant debate surrounding the mathematical education of engineers in German technical colleges during the 1890s. The history of technology includes several debates about the type of mathematical teaching most appropriate for engineers. Among these, the occasionally heated discussions around the turn of the twentieth century are particularly striking and attracted public attention. Moreover, the issues they raised anticipated many recent problems in this field, and thus merit special interest.

First let me sketch the most important factors that shaped the scientific and social context for these debates. The *École Polytechnique* in Paris, which influenced, to a degree, the establishment of mathematical teaching at the German technical colleges, is a place to start. The *École Polytechnique* was founded in 1794 in order to provide the young French Republic with needed military and civil servants. It was to supplement the training by older technical schools, such as the *École des Ponts et Chaussées,* the *École du Génie,* or the *École des Mines.* An extensive education in mathematics and science was the basic founding principle. Throughout the first three decades of the *École Polytechnique,* for instance, mathematics covered on the average roughly 50%, mechanics and physics about 23% of the curriculum during the first year. Because of the destruction of *Ancien Régime* learned institutions during the revolution, many leading scientists such as Gaspard Monge (1746–1818), Pierre-Simon Laplace (1749–1827), and Joseph-Louis Lagrange (1736–1813) taught at the *École Polytechnique.*

Attempting to manage the increasing demands posed by the advancing industrial revolution, the German states founded polytechnical schools during the first quarter of the nineteenth century to train the technical staff. In this process the *École Polytechnique* in Paris, to a certain extent, served as a model; it also influenced the establishment of American technical colleges in the nineteenth century, especially West Point.

In the German states, the polytechnical schools had primarily taken over the basic intention to make mathematics, science, and mechanics the fundamental subjects of engineering education. However, they were no mere images of the *École Polytechnique* but followed rather a particular polytechnical concept, emphasizing more the practical side of technical education. Also there was no education for military staff, and both the scientific instruction and the professional subjects were housed under a single roof. Until the middle of the nineteenth century, the level of science and mathematics taught in German polytechnical schools was quite low compared with the Paris school. Around

1850 there were hardly any known mathematicians working at the polytechnics, and even these mathematicians were mainly teachers, not researchers.

Consequently, Germans valued their engineers and their technical colleges far less than the French valued their *polytechniciens* and the *École Polytechnique*. Indeed, the fight for a social acknowledgement, equal to that of other academic groups, concerned Germany's engineering community until the turn of the century. Obtaining the status and rights of a university for the technical colleges, an essential part of this process, shaped the development of the technical colleges and their relationship to the universities.

At German universities, the centers for scientific research, mathematics had developed increasingly along "pure" lines since the first quarter of the nineteenth century and even more strongly since the mid-nineteenth century. Mathematicians also applied mathematics to astronomy, physics, and mechanics, but they did not consider technical applications as belonging to their fields. Most university professors were concerned with "pure" science. This was legitimated by the universities' principle of *Lehr- und Lernfreiheit*, the freedom of teaching and learning. This principle strongly influenced the training of future mathematics teachers for secondary schools and among them, candidates for professorships at technical colleges too. Owing to this kind of training, which stressed the unity of research and teaching, many teachers later became active researchers. To be sure, not all mathematicians should be treated alike. In particular, Felix Klein forged a faction, beginning in the 1890s, which resisted specialization and sought to intensify the relationship between mathematics and other scientific and technological disciplines.

Clearly, some aspects of the technological sciences also had a crucial impact on the development of the mathematical education of engineers. By about 1860 industrial expansion had rendered obsolete the trial-and-error approach to solving technological problems. This was especially true in machine building, which was the core of the German industry, and is thus focused on here. In response to this growing pressure, a new direction in mechanical engineering arose that aimed to ground it in theory. Ferdinand Redtenbacher (1809–1863), Franz Reuleaux (1829–1905), Franz Grashof (1826–1893), and Gustav Zeuner (1828–1907) were the foremost representatives of this new direction, which sought to develop a technological science to overcome Germany's lack of capital and its relative inexperience in machine building. They wanted to turn inventing and constructing into a deductive science in order to make engineering highly calculable and more manageable. The theoretical conceptions were based in particular

on the rigorous deductive and axiomatic method of mathematics. Thus during the 1860s and 1870s, mathematics, no longer just an auxiliary tool, played an outstanding role in establishing the scientific foundations of engineering.

In order to understand these new engineering conceptions, it was necessary to think mathematically and to know something about mathematical methods. Yet the typical mathematical education at polytechnical schools could not meet the new demands. Consequently, leading engineers wanted to establish higher standards in mathematics for engineers. To do so they pursued a policy of appointing mathematicians from universities at polytechnic schools. In 1865 the "German Association of Engineers" (VDI), the largest engineering organization, adopted two related objectives: to elevate the education of engineers and to promote the status of engineers in German society. The programmatic paper by VDI's president Franz Grashof, entitled "Principles of the Organization of Polytechnics," spelled out these goals clearly. The principal aim was to develop the polytechnics into colleges with university status, including the right to award academic degrees "Dr. Ing." and "Diplomingenieur" comparable to those conferred by the universities.

Following this intervention of the VDI, the above-mentioned initiatives of leading engineers became generally applied policy at technical colleges. As a consequence many prominent mathematicians and scientists were appointed to technical colleges from universities in the 1870s and 1880s: at Berlin, Heinrich Weber (1842–1913) and Paul du Bois-Reymond (1831–1889); at Darmstadt, Aurel Voss (1845–1931) and Ludwig Kiepert (1846–1910); at Dresden, Axel Harnack (1851–1888); at Karlsruhe, Ernst Schroeder (1841–1902); at Munich, Felix Klein (1849–1925), Alexander von Brill (1842–1935), Jacob Lüroth (1844–1910), and Walther von Dyck (1856–1934).

Although this produced a higher scientific level in mathematics instruction at technical colleges, there was the other side of the coin, in particular for the mathematical education of engineers. Most mathematicians had a strong orientation towards "pure" mathematics—an orientation they brought with them to the technical colleges. There, the recruitment of the mathematics faculty was placed in charge of mathematicians and scientists, who wanted mathematics taught according to the new "methods of modern analysis." They were thus expected to offer calculus courses based on a rigorous approach to foundations of analysis. In spite of some formal statements, mathematicians were not obligated to consider the special needs of a technical college, either in their research or teaching. From the last third of the nineteenth century this followed mainly because technical colleges had adopted the universities' principle of *Lehrfreiheit,* the freedom of teaching. Thus the mathematical education for engineering students became similar in scope, content, and methods to that for prospective mathematics teachers at universities. Before the turn of the century, mathematicians who took into account the special needs of technical colleges were rather exceptional.

Now, what comprised the mathematical education of engineers in "higher mathematics" (according to the contemporary usage, this meant differential and integral calculus and analytical geometry), about which the debates of the 1890s centered? During the first two years students had to take courses in "higher mathematics", descriptive geometry, science, mechanics, and drawing. The proportion of mathematical subjects, including mechanics, was about 55% in the first year, and around 40% in the second. During the third and the fourth years, students could take elective mathematical courses. First-year calculus covered fundamental notions, theorems, and rules up to the introduction of the definite integral. The second year included the further treatment of integrals as well as introduction into the theory of differential equations and applications of calculus to geometry. Beginning in the 1880s, catalogues of courses showed a variety of special lectures on research specialties of mathematicians. These lay outside the standard curriculum for engineers and reflected the contemporary spectrum of mathematical research. Among others they covered elliptic functions, complex variable theory, and number theory. Some of these special courses on Fourier series, determinants, or vector analysis were recommended to students, in their third or fourth year, depending on their major subject. This suggests an increasing realization at technical colleges of the unity of research and teaching.

A crucial feature of many courses in calculus was the inclusion of elements of the Weierstrassian "rigorous" treatment or "arithmetization of analysis." According to statements made by contemporary mathematicians and remarks in several textbooks written by mathematicians of technical colleges, Weierstrassian rigor played an essential role. Depending on the point of view of the individual professor, courses typically covered such topics as the construction of number systems, the introduction of real numbers by means of Cantor's fundamental sequences, and analytical proofs for nearly all theorems or results from the theory of functions. Students were taught to use uniform convergence, uniform continuity, and theorems about continuous functions. Applications of mathematical methods, however, were confined to geometry and some physical fields, but almost no problems resulting from the technological sciences. Numerical and graphical methods of solution also played a minor role up to the 1880s.

As research and teaching in mathematics had been turned to the more abstract side, engineering sciences developed in a theoretical direction too, minimally related to real problems rooted in technical practice. In research the theoretical foundation of the technological sciences was not accompanied by a sufficient development of the empirical component. As a result, in the last decade of the nineteenth century, the rapidly increasing demands of industry clashed more and more with the mainly theoretical education at the technical colleges. This led to intense discussions among engineers focusing on engineering education.

Industrial demands and these discussions brought a shift in the further development of the technological sciences. No longer did it follow the theoretical path based on mathematics that Redtenbacher, Reuleaux, Grashof and others had foreseen. It now favored the experimental side more. Engineering laboratories for research and teaching were founded or extended at all technical colleges. A crucial stimulus for this change came from the 1893 Chicago World's Fair, where German engineers, industrialists, and government officials became acquainted with the American system of technological education.

In searching for ways to integrate the new experimental subjects into the curriculum, the so-called auxiliary sciences came under discussion in 1894 and had consequences for mathematics. Whereas mathematics had been considered a fundamental subject for the technological sciences and the study of engineering up to the second third of the nineteenth century, now its importance was increasingly questioned. Because mathematics had failed to provide a reliable basis for solving engineering problems and an education that equipped engineers to solve evolving problems posed by industry, stronger conflicts arose between mathematicians and engineers during the last decade of the century.

The engineering community began to adopt a pragmatic point of view on the value of different sciences for training engineers. In particular, the most outspoken group of engineers in the fight for a practical education was led by Alois Riedler (1850–1936), a professor of machine building at the technical college in Berlin-Charlottenburg. He and his followers introduced a new "chapter" within the methodological debate in the technological sciences which contemporaries labeled the "antimathematical movement" of engineers. This group sharply criticized the content, scope, and methods of engineers' mathematical education. They felt that it was more appropriate for engineering students to skip the intricacies of rigorous mathematical reasoning corresponding with the standards of contemporary mathematics and pay instead particular attention to exercises and applications as well as to graphical and numerical methods of solution. At the most extreme, this group called for large reductions in hours for mathematics instruction, restriction to elementary mathematics, and the exclusion of calculus from the regular curriculum at technical colleges.

These engineers further criticized the university monopoly on educating teachers of mathematics and science and demanded the right to educate all future professors for technical colleges. They argued that theorists should be divested of their crucial influence on appointments in mathematics. Some thought mathematics professorships should be occupied only by those who had studied an engineering subject in graduate school. A more radical group even wanted only engineers as instructors for the mathematical courses. Headed by Riedler, in 1897 the engineering departments succeeded in attaining the right to control the scope and content of the mathematical training. Consequently, hours for mathematics and other theoretical subjects were reduced at some technical colleges, the reductions in Berlin being most significant.

A faction of more progressive engineers strove to develop new mathematical methods for the technological sciences. They pointed, for example, to the use of projective geometry in Karl Culmann's graphical statics. F. Grashof and August Foeppl (1854–1924) pioneered using partial differential equations and vector analysis in engineering. They solved eigenvalue problems to prevent resonance catastrophes, and used linear algebra to calculate electrical networks. Of course, these engineers did not support the extreme demands of the strictly praxis-oriented engineers.

Mainly through the efforts of the VDI did the latter faction win out. Recognizing the dangers implicit in the extreme demands of Riedler's group, the VDI chose to advocate a moderate position regarding mathematics education. In its Aachen resolution of 1895, the VDI called for changes in the quality of mathematics instruction rather than cut backs in the quantity. Mathematics teaching was to be improved by diminishing its abstract character and utilizing more illustrations, graphics, exercises, and applications. Mathematicians were expected to dedicate more attention to mathematical needs of the technological sciences. Therefore, the resolution called for future technical-college teachers to spend three or four semesters studying there.

Questions about engineering education dominated the 1895 VDI meeting in Aachen—which was also a key event in the entire methodological discussion within the technological sciences. First, because of these concentrated efforts for an improvement of engineering education, funds for engineering labs were made available to the technical colleges by all German states, beginning in 1896. Second, engineers with their strenghtened status-consciousness stressed social demands for the development of the technical colleges. Third, aims of different factions among engineers became more pointed. In particular, discussions, once individual statements, took on the character of confrontations of group interests and ideals. These were largely coloured by tensions between universities and technical colleges. Fourth, the debate then went in several different directions. Machine-building engineers had begun the debate, but the debate spread to almost all branches of engineering. Only electrical engineering took no part in these discussions, which penetrated into the public sector through numerous articles in scientific journals, newspapers, and books.

In 1897 all 33 teachers of mathematics, mechanics, and descriptive geometry at the nine technical colleges in Germany issued a joint declaration reacting to the demands made by the engineers. They agreed to abolish failures of previous educational practices in mathematics. In particular, they supported the idea of adopting teaching methods better suited to needs of engineers. The declaration specified the latter aim in remarking that the mathematics professors were to work towards a firm and skilled use of mathematical methods by the students. They were further expected to find a sound combination of analytical and geometrical methods and to offer elective courses in mathematics and mathematical physics which were of special importance to the various engineering subjects. But mathematicians also maintained the need for a two-year basic course in mathematics and stressed that mathematics had to be taught, without exception, by professional mathematicians.

The antimathematical movement approached its climax in 1897 when about 50% of all professors of engineering at technical colleges made a public declaration emphasizing their adherence to the views of the strict "practitioners" among engineers. At this stage the professional and social interests of engineers frequently dominated the debate.

One aspect of the debate was whether mathematics should be a fundamental or auxiliary subject. A popular slogan of these engineers was "technical colleges into the hands of technicians!" As a consequence, the important problems of how to improve mathematics education and increase its practical efficiency through a closer cooperation between mathematicians and engineers were sometimes forgotten.

Nevertheless, by the end of the first decade of the twentieth century, the debate with its frequently antagonistic points of view had largely been settled. More tolerant engineers then pointed out the importance of mathematics for the development of the technological sciences and consciously tried to intermediate and find a compromise. They included distinguished engineers such as August Foeppl, Aurel Stodola (1859–1942), Hans Lorenz (1865–1940), and Carl J. Bach (1847–1931), and important industrialists such as the general director of the Maschinenfabrik Augsburg-Nürnberg AG, Anton von Rieppel, and the director of the Farbenfabriken Friedrich Bayer, Leverkusen, Henry Th. von Böttinger. Furthermore, a growing number of engineers realized the need for stronger education in mathematics, new mathematical methods, and better cooperation with mathematicians. The VDI and the German Ministry for Culture (Kultusministerium) played a key role in encouraging necessary changes in mathematical education while preventing possible longterm damages to mathematics at the technical colleges. Thus, mathematicians succeeded in maintaining the mathematical chairs at the technical colleges.

At the same time, beginning around the turn of the century, more mathematicians at technical colleges were adjusting their courses to the particular needs of engineering students. Mathematicians increased the use of numerical and graphical methods for solving mathematical problems and stressed the theory and use of calculating devices as slide rules, integraphs, planimeters, or harmonic analyzers. Also mathematicians made efforts to enrich their courses and textbooks with application problems from physics, electrophysics, technical mechanics, and other fields closely related to professional subjects. "Applied mathematics" began to play an increasing role in the curricula of engineers in the first decade of the century. By the 1920s the once-heated debate had largely abated.

References

1. For details see Susann Hensel, Karl-Norbert Ihmig, Michael Otte, *Mathematik und Technik im 19. Jahrhundert in Deutschland. Soziale Auseinandersetzung und philosophische Problematik* (Göttingen: Vandenhoeck & Ruprecht, 1989), pp. 1–111.

2. Ambroise Fourcy, *Histoire De L'Ecole Polytechnique* (Paris: Belin, 1987) p. 376. For the history of the early *Ecole Polytechnique,* see also Felix Klein, *Development of Mathematics in the 19th Century,* Translation of the 1926/27 original by M. Ackerman (Brookline, MA: Math Sci Press, 1979), pp. 59–84; Peter M. Molloy, *Technical Education and the Young Republic: West Point as America's Ecole Polytechnique, 1802–1833,*(Ph.D. dissertation: Brown University,

1975), pp. 1–150; Friedrich Klemm, "Naturwissenschaften und Technik in der französischen Revolution," *Deutsches Museum, Abhandlungen und Berichte, München, Oldenburg, Düsseldorf,* 1977,45, no. 1; M. Paul, *Gaspard Monges 'Geometrie descriptive' und die Ecole Polytechnique* (Bielefeld: Institut für Didaktik der Mathematik der Universität Bielefeld, 1980); Terry Shinn, *Savoir scientifique et pouvoir social. L'École polytechnique, 1794–1914* (Paris: Presses de la foundation nationale des sciences politiques, 1980).

3. See Karl-Heinz Manegold, Wilhelm Treue (eds.), *Documenta Technica. Darstellungen und Quellen zur Technikgeschichte,* Reihe 1: Darstellungen zur Technikgeschichte: Moritz Rühlmann, *Vorträge über die Geschichte der Maschinenlehre und der damit in Zusammenhang stehenden mathematischen Wissenschaften,* Nachdruck der Ausgabe Leipzig und Braunschweig 1881–1885 (Hildesheim, New York: G. Olms, 1979), p. 404); also Walter Purkert, Susann Hensel, "Zur Rolle der Mathematik bei der Entwicklung der Technikwissenschaften," *Dresdener Beiträge zur Geschichte der Technikwissenschaften, Dresden,* 1986, 11:3–53, on pp. 25–27.

4. On the French influence on American technical schools, see, e.g., James G. McGivern, *First Hundred Years of Engineering Education in the United States (1807–1907)* (Spokane: Gonzaga University Press, 1960), pp. 9–61; George S. Emmerson, *Engineering Education: A Social History* (New York: Crane, Russak; England: Newton Abbott, 1973), pp. 134–156; Molloy, *Technical Education*; Nathan Reingold, Marc Rothenberg (eds.), *Scientific Colonialism* (Washington, D.C.: Smithsonian Institution, 1987), pp. 157–158; Peter Lundgreen, "Engineering Education in Europe and the U.S.A., 1750–1930: The Rise to Dominance of School Culture and the Engineering Professions," *Annals of Science,* 1990, 47:33–75, on pp. 46–56.

5. For the characteristics of this "German way" see, e.g., Lundgreen, "Engineering Education", pp.41–46; Gisela Buchheim, Rolf Sonnemann (eds.), *Lebensbilder von Ingenieurwissenschaftlern. Eine Sammlung von Biographien aus zwei Jahrhunderten* (Basel, Boston, Berlin: Birkhäuser-Verlag, 1989), pp. 47–58; Gisela Buchheim, Rolf Sonnemann (eds.), *Geschichte der Technikwissenschaften* (Basel, Boston, Berlin: Birkhäuser-Verlag, 1990), pp. 195–203.

6. See D.A. Binder, "Europäische Technische Studienanstalten im Jahre 1853. Karl Koristkas Reisebericht," *Mitteilungen der Österreichischen Gesellschaft für Geschichte der Naturwissenschaften,* 1984, 4, no. 2/3: 49–100; Purkert, "Zur Rolle der Mathematik," pp. 30–32.

7. See Karl-Heinz Manegold, *Universität, Technische Hochschule und Industrie. Ein Beitrag zur Emanzipation der Technik im 19.Jahrhundert mit besonderer Berücksichtigung der Bestrebungen Felix Kleins* (Berlin: Duncker & Humblot, 1970).

8. As for technical applications, we refer here to problems posed by or related to engineering practice or research. But even the connections to astronomy, geodesy, or physics were comparatively rare and became loc꞉ꞏ toward the turn of the century, giving way to a rather abstract direction of research. See also Lewis Pyenson, *Neohumanism and the Persistence*

of Pure Mathematics in Wilhelmian Germany (Philadelphia: American Philosophical Society, 1983), p. 22.

9. For an excellent early evaluation of this development in mathematics see Felix Klein, "Über die Beziehungen der neueren Mathematik zu den Anwendungen. Antrittsrede, gehalten am 25.10.1880 in Leipzig," *Zeitschrift für den mathematischen und naturwissenschaftlichen Unterricht,* 1895, 26:535–536.

10. See, e.g., David E. Rowe, "Klein, Hilbert, and the Göttingen Mathematical Tradition," *OSIRIS,* 2nd series, 1989, 5:186–213.

11. For example, F. Redtenbacher initiated the appointments of Alfred Clebsch (1833–1872) and Wilhelm Schell (1826–1904) to the Karlsruhe polytechnic in 1856 and in 1861, respectively. F. Reuleaux supported the appointments of Karl Weierstrass (1815–1897),in 1856, and Elwin Bruno Christoffel (1829–1900), in 1869, to the Berlin Gewerbeinstitut.

12. Franz Grashof, "Principien der Organisation polytechnischer Schulen," *Zeitschrift des Vereins Deutscher Ingenieure,* 1865, 9:721–723.

13. For details see Susann Hensel, "Zu einigen Aspekten der Berufung von Mathematikern an die Technischen Hochschulen Deutschlands im letzten Drittel des 19. Jahrhunderts," *Annals of Science,* 1989, 46:387–416.

14. For example, in 1882 Axel Harnack was considered, among others, a candidate for the vacant mathematical chair at the technical college at Aachen. His evaluation which the mathematics faculty reported to the Prussian Kultusministerium may serve to illustrate the criteria for the mathematics professors at technical colleges: "Professor Harnack is as distinguished a teacher as he is a scholar. As a mathematician, Harnack has distinguished himself with a number of works showing his acquaintance with the various areas of analysis. Professor Harnack's *Differential-und Integralrechnung,* published in 1881, is a splendid work which documents the professor's teaching abilities. In this work, Harnack makes the rigorous methods of modern analysis accessible to students by a clear and readable presentation." (quoted in Hensel, "Zu einigen Aspekten der Berufung", pp. 412–413, translation by the author) This book was written as a text accompanying Harnack's teaching at the technical college at Dresden. It was characterized as belonging to the "rigorously artihmetizing direction" (G. Bohlmann, "Uebersicht über die wichtigsten Lehrbücher der Infinitesimalrechnung von Euler bis auf die heutige Zeit," *Jahresberichte der Deutschen Mathematiker Vereinigung,* 1899, 6:93–110, on p. 105.).

    Paul Stäckel (1862–1919), who had been mathematics professor at technical colleges for several years, wrote in his evaluation of the mathematical education on behalf of the International Mathematical Commission on the Teaching of Mathematics in 1915: "The mathematical education for engineers became more and more similar to the studies for majors in mathematics, and moreover, a strong arithmetizing direction gained predominance in mathematics. The professors felt obligated to dedicate a considerable part of their lectures in higher mathematics to a foundation of notions and theorems, but the efficiency of the lectures suffered from that practice." Paul Stäckel, *Die mathematische Ausbildung der Architekten, Chemiker und Ingenieure an den deutschen Technischen Hochschulen* (Leipzig, Berlin: B.G. Teubner, 1915), [Abhandlungen über den mathematischen Unterricht in Deutschland, veranlasst durch die Internationale Mathematische Unterrichtskommission, 4, no. 9], on p. 27, translation by the author.

15. Z. d. V. D. I., 1895, 39: 1095–1096.

16. *Zeitschrift für Architektur und Ingenieurwesen,* 1897, 22: 242–244.

17. Ibid., pp. 609–611.

18. Paul von Lossow, "Zur Frage der Ingenieurerziehung," Z. d. V. D.I., 1899, 43:355–361, on p. 361.

19. For the especially important role of Carl J. Bach in this regard, see Susann Hensel, "Der Technikwissenschaftler Carl Julius Bach (1847–1931) und die Frage der mathematischen Ausbildung der Ingenieure," *NTM-Schriftenreihe für Geschichte der Naturwissenschaften, Technik und Medizin, Leipzig,* 1991/92, 28, no. 2:231–257.

20. To provide a substitute for the lack of textbooks with applications to technical problems, in 1902 Robert Fricke (1861–1931), aided by an electrical engineer, translated John Perry's *Calculus for engineers* into German. Beginning in the 1910s, numerous calculus textbooks were published containing practical examples for application supplemented by books of problem sets. Examples are: Otto Dziobek, *Vorlesungen über Differential-und Integralrechnung* (Leipzig, Berlin: B.G. Teubner, 1910); Georg Helm, *Die Grundlehren der höheren Mathematik. Zum Gebrauch bei Anwendungen und Wiederholungen* (Leipzig: Akademische Verlagsgesellschaft, first ed. 1910, 2nd ed. 1914); Friedrich Dingeldey, *Sammlung von Aufgaben zur Anwendung der Differential-und Integralrechnung* (Leipzig, Berlin: B.G. Teubner, first part 1910; second part 1913). In 1908 Eugen Jahnke (1861–1921) started editing *Teubners Leitfäden für den mathematischen und technischen Hochschulunterricht,* This series dealt, in an "easy-to-read" form, with various mathematical methods which were increasingly needed in particular fields of engineering research; the books were addressed mainly to engineers.

21. For the essential role of Carl Runge in strenghtening a more "applied" direction in mathematics, see Gottfried Richenhagen, *Carl Runge (1856–1927): Von der reinen Mathematik zur Numerik* (Göttingen: Vandenhoeck & Ruprecht, 1985).

American Mathematics Viewed Objectively: The Case of Geometric Models[1]

Peggy Aldrich Kidwell
Smithsonian Institution

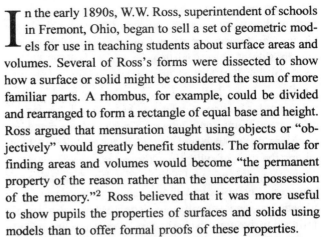

In the early 1890s, W.W. Ross, superintendent of schools in Fremont, Ohio, began to sell a set of geometric models for use in teaching students about surface areas and volumes. Several of Ross's forms were dissected to show how a surface or solid might be considered the sum of more familiar parts. A rhombus, for example, could be divided and rearranged to form a rectangle of equal base and height. Ross argued that mensuration taught using objects or "objectively" would greatly benefit students. The formulae for finding areas and volumes would become "the permanent property of the reason rather than the uncertain possession of the memory."[2] Ross believed that it was more useful to show pupils the properties of surfaces and solids using models than to offer formal proofs of these properties.

The belief that "objective" presentations can replace proofs in mathematical demonstrations is outside the tradition of rigorous mathematical thinking that dates from ancient Greece. Historians seeking to trace the origins and internal development of mathematics have properly focused their attention on written texts. From ancient times, however, people have used mathematical objects. The general increase in material abundance that has characterized the western world from the eighteenth century, accompanied by rapid changes in mechanical, electrical, and electronic technology, has brought with it a profusion of objects fulfilling mathematical functions. Examining these yields valuable insight into the place of mathematics in society at large; the history of mathematics teaching; the role of mathematics in business, engineering, and the sciences; the development of mathematical communities; and even the practice of mathematical research.

Most mathematical objects fit into two broad classes—those that assist in carrying out routine mathematical operations and those that illustrate mathematical structures and principles. The first category includes such aids to computation as the abacus, slide rule, calculating machine, cash register, and electronic calculator. Such instruments help people solve routine problems of arithmetic, geometry, and sometimes more advanced areas of mathematics, without necessarily encouraging any understanding of the principles involved. They are used by mathematicians but are more widely sold for engineers, business people, office workers, and the general public. Aids to calculation have been made from antiquity and, especially in the case of electronic calculators and computers, are manufactured in large numbers to this day.

The number and variety of objects made to illustrate mathematical structures and principles is far smaller than the plethora of computational devices. These objects include geometric models, mathematical paintings and sculp-

tures, and pedagogical devices such as logic machines, a form of abacus known as the numeral frame, and blocks for teaching arithmetic known as Cuisenaire rods. Of course, this division of objects is suggestive, not absolute. For example, in the first half of this century manufacturers like Keuffel & Esser built large slide rules for teaching purposes, not for routine calculations. Pedagogical devices like the numeral frame could be used for calculations, and logic machines do, to an extent, automate the process of determining the truth of syllogisms. Moreover, present-day electronic computers are used both for mundane calculation and to represent complex structures such as geometric surfaces. Nonetheless, one generally can distinguish between devices designed to avoid mathematical labor and those intended to point up mathematical truths. In this volume, it seems appropriate to concentrate on the second sort of object. This essay examines American use of geometric models in the nineteenth and early twentieth centuries as a result of a series of pedagogical reforms.

Geometric models are and have been made for classroom use, to satisfy the aesthetic sense of makers, and to illustrate research presentations. Many are found in museum collections and in cases that line the halls of mathematics buildings at schools and universities. Surviving models, along with manuals and catalogs issued by those who made and sold these models, offer tangible testimony to transformations in American mathematics. At the same time, they are part of more general stories of object-oriented pedagogy and the visual representation of scientific and technical ideas.

Geometric models entered the American classroom along four intertwining but distinct paths. In each instance, new models were associated with new educational ideals and institutions. From the 1820s American communities began to establish large numbers of common schools that offered general elementary instruction. Reformers wrote at length about the curriculum and teaching techniques appropriate for these classrooms. Many agreed that young children would benefit from object-oriented pedagogy, including simple geometric models. From the mid-nineteenth century, a second market for models developed, with the founding of professional schools for engineering education. Engineers had traditionally learned their trade through apprenticeships. However, the success of Napoleon's army inspired the establishment of West Point in 1801.[3] West Point could not meet the country's demand for engineers, and several civilian engineering schools were established.[4] Seeking to emulate French and German practice, these schools purchased models for teaching descriptive geometry and engineering drawing, as well as other apparatus.

Third, as drawing became a routine tool of engineers, Massachusetts legislators voted that it should be taught to future industrial workers as well. From the 1870s, courses in industrial drawing were offered in Massachusetts. The practice spread to other states. The art curriculum developed in connection with this reform emphasized use of models. Finally, in the 1880s and 1890s, American universities increasingly sought to mold themselves on the pattern of their German counterparts. American students took degrees in Germany, and schools purchased German laboratory apparatus to teach subjects ranging from the physical sciences to psychology.[5] They also bought German mathematical models. A few American makers copied or sought to extend these collections.

Sellers of models consistently affirmed their pedagogical value. It is difficult to know, however, the extent to which models were used in the classroom. One suspects that they often served as much to assert the educational ideals and enhance the prestige of those who bought and exhibited them as to fill a pedagogical function.

Simple Geometric Solids

In the first half of the nineteenth century, civic activists, particularly in the northern states, advocated common schools, that is to say schools providing elementary education for the general population. Following doctrines of Swiss reformer Johann Heinrich Pestalozzi (1746–1827) and German pedagogue Friedrich Froebel (1782–1852), many educators stressed the value of objects as aids to understanding.[6] During the 1830s, Henry Barnard (1811–1900), the first secretary to the State of Connecticut's Board of Commissioners of the Common Schools, prepared a list of questions to be asked by inspectors visiting a common school. Several queries were about instructional apparatus. Did the school have a numeral frame? a set of geometrical solids? a cube that could be divided to illustrate the process of taking a cube root?[7]

The devices Barnard recommended were among those described by Josiah Holbrook in his *Apparatus Designed for Families, Schools, Lyceums and Academies* (Boston, 1833). According to Holbrook, for any school or family investing in educational apparatus, geometric models were "probably the first articles to be procured, whatever may be the second."[8] Models helped a child learn the name of simple shapes. At the same time, blocks marked with a grid of lines on the surface made it easier to learn mensuration, that is, the calculation of surface areas and volumes. For those studying art and engineering-drawing, models of prisms, cylinders, cones and other shapes were useful.

In his 1833 book, Holbrook described other apparatus for teaching mathematics. His "arithmeticon," a device for teaching multiplication, used balls arranged on a square grid. He also noted that stiff paper, properly cut and folded, could be used to make models of the Platonic solids. Flat figures made from strips of tin could help teach relations between areas of polygons (as two pieces arranged to form either a square or a rhombus with sides of equal length). Such devices were not new and would be produced later. However, they never acquired the popularity of geometrical solids, the numeral frame, or the cube root block.[9]

As early as 1826, Holbrook established an exchange for the sale of school apparatus. During the 1840s, his sons, Alfred and Dwight, established the Lyceum Village in Berea, Ohio and sold apparatus made there. In about 1854, Dwight Holbrook moved to Hartford, Connecticut and founded the Holbrook School Apparatus Manufacturing Company.

In 1857, the solids and numeral frame mentioned by Henry Barnard sold as part of a standard collection called "Holbrook's School Apparatus," a set that also included an orrery, a tellurian, a terrestrial globe, a hemispheric globe, a magnet, and a textbook. In 1857, the set sold for twenty dollars; the geometric solids alone cost $1.25.[10] Barnard thought so highly of this apparatus that he served as an agent for Holbrook in the 1850s.[11] Similarly, John C. Spencer, Superintendent of Common Schools for the state of New York, specifically recommended Holbrook's apparatus in an 1841 set of instructions for those administering commons schools in his state.[12]

Several other dealers in the United States sold simple wooden models. An 1835 list of apparatus for schools sold by Claxton and Wightman of Boston includes a set of wooden models for use in teaching solid geometry.[13] Mid-century catalogs of Benjamin Pike, Jr., of New York and James W. Queen & Co. of Philadelphia included sets of geometric solids.[14] An 1869 catalog of W. Mitchell McAllister, another Philadelphia dealer, lists models much like those of Holbrook.[15] It is quite possible that Pike, Queen, and McAlister simply marketed apparatus made by the Holbrook family. Sets of shapes sold by Queen in 1859 and later by J.W. Schermerhorn and Company of New York (1871), by Baker, Pratt & Co. of New York (1879) and by A.H. Andrews of Chicago (1881) included models of both polygons and solids.[16] Such sets of solids and flat surfaces sold at least as late as 1965.[17]

Nineteenth-century catalogs of teaching apparatus include not only wooden polygons and polyhedra, but various other devices. In the 1850s Benjamin Pike, Jr., of New York

offered a cone, cut to show the various conic sections. Numerous other dealers did likewise. In 1859 James W. Queen & Company of Philadelphia sold flat sheets of pasteboard, marked so they could be cut out and folded into Platonic solids. Holbrook had recommended such sheets, but I have found no evidence that they became a common product. Queen also offered models of polyhedra inscribed in glass cubes.[18] In 1871 J. W. Schermerhorn of New York sold a set of rods which fit together to form the edges of polygons. This "goniograph," which resembles apparatus sold for kindergartens, came with rods of wood, metal or whalebone. Apparently it found few buyers as other catalogs do not list it.[19]

After the Civil War, some makers emphasized models for teaching mensuration. In 1873, for example, Isaac Harrington of Huntington, Connecticut, patented an "improvement in apparatus for teaching mensuration." Harrington's apparatus consisted of a set of wooden blocks or flat shapes, hinged or doweled so that they could be transposed into other shapes whose areas were known to students. He proposed "ocular demonstrations" for finding the areas of a rhombus and a regular polygon, and the volumes of a right triangular prism, any regular right prism, and so forth. Mathematicians Hubert Anson Newton of Yale University and Charles Davies of Columbia commended Harrington's invention, and his "geometrical blocks" were offered for sale in the National School Furniture catalog for 1872.[20]

Harrington's patent did not guarantee him a full share of the market. During the 1880s, a set of blocks designed by Albert H. Kennedy (1848–1940) of Rockport, Indiana, was used in some schoolrooms. In 1884 Kennedy also patented an "educational appliance" designed to demonstrate the relative volume of a cylinder, a sphere inscribed in the cylinder, and a cone of the same diameter and height. Versions of this apparatus, which may have been influenced by Kennedy's patent, were produced well into this century.[21] About 1891, shortly after Harrington's patent expired, W.W. Ross published the manual for the "dissected surface forms and geometrical solids" mentioned at the outset of this paper.[22] Ross collected recommendations for his blocks from educators throughout Ohio. The examples in the Smithsonian collections come from Wesleyan University in Connecticut. There are also a few of Ross's models at Hood College in Frederick, Maryland.[23] These instances indicate that Ross found buyers outside Ohio. Such extensive collections of models for teaching mensuration apparently did not become common, however.

Models for Engineering Drawing

A second tradition of geometric modelmaking emerged from the program of mathematics instruction Gaspard Monge (1746–1818) established at the École Polytechnique in Paris in the early nineteenth century. As a professor at a school for military engineering (École du Génie Militaire) at Mézières in the 1770s, Monge had systematically developed descriptive geometry, finding techniques for representing three-dimensional objects in two dimensions. These methods were classified as a military secret until after the French Revolution. During the 1790s, Monge acquired considerable influence over French technical education. He believed that instruction in descriptive geometry should occur at all levels of school, promoting habits of exact thinking needed in an industrial nation.[24] By 1801, students at the newly established École Polytechnique, a school for training military engineers that was a model for nineteenth century American engineering schools, spent 20% of their first two years of class time studying descriptive geometry. This fraction would decrease, with increasing time devoted to analysis. Nonetheless, descriptive geometry remained entrenched at the École Polytechnique long after Monge departed.[25]

Monge not only asserted the importance of descriptive geometry but argued that it was important to consider the generation of surfaces by the motion of lines. Indeed, he sought to classify surfaces by their mode of generation, not by the form and degree of the corresponding analytic expression. This approach in time gave rise to differential geometry. More immediately, it led Monge to pay special attention to ruled surfaces, that is, surfaces like the cylinder which may be considered as resulting from the motion of one or more straight lines. Monge made models of at least two surfaces which were preserved at the Conservatoire National des Arts et Métiers in Paris.[26]

Théodore Olivier (1793–1853), a graduate of the École Polytechnique who taught descriptive geometry at the École Centrale des Artes et Manufactures and then at the Conservatoire, went one step further. To suggest how surfaces were generated, Olivier designed a set of models of ruled surfaces which included several moveable models. For example, a ring of threads joining two parallel circles might form a cylinder. If one circle were rotated, the cylinder became a hyperboloid of one sheet. This was called a "warped" surface, as two successive threads were not in the same plane. Olivier models also were designed to draw attention to the curves of intersection of surfaces, with these curves marked by special grommets.

Olivier gave a nearly complete set of his models to the Conservatoire in 1849. He died four years later, leaving his personal collection of nearly forty models. Not long thereafter, William M. Gillispie, Professor of Civil Engineering at Union College in Schenectady, New York, and a former student of Olivier, visited Paris. He persuaded Mme. Olivier to sell him her husband's model collection. Gillispie put the models on exhibit at Union College. When he died in 1869, his wife sold the collection to the college for $800.[27]

According to William C. Stone, Professor of Mathematics at Union College, the models there were largely made by the firm of Pixii, Père et Fils, under Olivier's direction. Fabre de Lagrange, successor to Pixii, continued this modelmaking. In 1857 24 mathematically minded friends of Harvard University subscribed $1,175 for the purchase of copies of the Olivier models made by Fabre de Lagrange. Benjamin Peirce and Joseph Lovering were to purchase the models for the "Mathematical Cabinet of Harvard College." Once the models arrived, Peirce pledged to give a series of 10 to 12 lectures about them, which subscribers and their friends could attend without charge. It is not clear whether Peirce gave these lectures, but the models survive in the Harvard Collection of Scientific Instruments.[28] Olivier models also were purchased by West Point, the Stevens Institute of Technology, and the Columbia School of Mines.[29] Duplicates of the Olivier models were used to teach drawing at Princeton.[30]

Descriptive geometry also attracted the attention of German scholars and modelmakers. Ferdinand Engel (1805–1866), a Prussian carpenter, studied mechanical drawing on his own until the excellence of his drawings persuaded his father to allow him to attend the Royal School of Arts in his native Magdeburg. After completing his studies, Engel taught drawing privately, and prepared drawings for textbooks on descriptive geometry and optics. He also made a model of the wave surface in biaxial crystals that won him a prize at the famous Crystal Palace Exhibition, the World's Fair held in London in 1851.

When the Revolution of 1848 failed, Engel, who had liberal sentiments, fled Prussia for England and then the United States. In 1855 he published an English edition of *Axonometrical Projections of the Most Important Geometrical Surfaces: Drawings in Descriptive Geometry*. This volume includes a catalog of 38 models in wood and plaster which are shown in the drawings. Engel's notion of important geometrical surfaces was similar to that of Monge and Olivier, but he made solid models and not ruled surfaces. He included quadric surfaces such as the ellipsoid, hyperboloid, paraboloid, and cone, as well as helicoids, screws, and compounds of cubes. As one would expect, he also made models showing optical surfaces, like the one that had won him a prize in 1851. Prices ranged from 30 cents for a simple wooden cube to $12 for a plaster hyperboloid of one sheet, shown with its lines of curvature, to $100 for a wooden model of one of Fresnel's wave surfaces. In general, wooden models cost more than plaster ones of the same surface.31

Engel's book contained testimonials from Wolcott Gibbs of New York and the German mathematician G.P. Lejeune Dirichlet. Some of Engel's models survive in the Physics Collection of the University of Mississippi.[32] However, he apparently did not find commercial success. Engel's abilities as a draftsman did attract the attention of Alexander Dallas Bache, who hired him to make drawings of instruments and apparatus of the US Coast and Geodetic Survey. In this capacity, Engel also made plaster models of variations in the earth's magnetic field, based on data Bache had collected in Philadelphia between 1840 and 1845. One of these models survives at the American Philosophical Society and another at the Smithsonian.[33]

Olivier and Engel were not the only designers of models for descriptive geometry. Models constructed in Prussia at the Polytechnisches Arbeits-Institut in Darmstadt under the direction of J. Schroeder were of special importance. Schroeder and his students constructed a few moveable string models like those of Olivier, and some models of intersecting lines and solids. To help students master mechanical drawing, they also made a series of models showing the projection of various solids onto horizontal and vertical planes. To aid those drawing actual machines, they also made models of steam engines, turbines, water-wheels, and the like.[34] For a general market, they made wooden models of polyhedra.

In the United States, Schroeder's models of polyhedra were sold by James W. Queen and Company from at least 1881.[35] Small sets were available for elementary instruction, and larger ones for advanced work. Universities such as Johns Hopkins, Columbia, and Michigan purchased Schroeder models of polyhedra. Johns Hopkins, the Stevens Institute of Technology, and Princeton purchased his models for descriptive geometry. Columbia president Frederick A.P. Barnard also purchased some of his models of machines for the Columbia School of Mines.[36]

Models for Art Education

Instruction in drawing that relied heavily on models was not confined to engineering schools and universities. Although Josiah Holbrook had suggested in 1833 that his geometrical forms could be used to teach drawing, few early nineteenth-century American teachers followed this sugges-

tion. After the Civil War, however, the growing importance and prosperity of American industry, combined with exposure to British and European drawings at World's Fairs, led educators to take a new interest in teaching industrial drawing. In 1869 the legislature of the Commonwealth of Massachusetts passed a law establishing drawing as a subject to be taught in public schools. All towns of 10,000 or more inhabitants had to offer courses in industrial and mechanical drawing.

To carry out this program, the city of Boston hired the Englishman Walter Smith (1836–1886). Smith had trained in London at the South Kensington Museum and had directed several English art schools. He received a joint appointment as Director of Art Education for Massachusetts and Superintendent of Drawing in Boston.[37] By 1876 Smith had founded a school for training drawing masters, the Massachusetts Art Normal School (later the Massachusetts College of Art). He also had patented a drawing instrument and had begun to write a series of textbooks on drawing. Smith began his course with lessons on drawing plane and solid geometrical figures. Indeed, students were not only to draw these shapes, but to "refer all things drawn to regular geometrical forms." Drawings were not to be made by copying other drawings, but by examining objects themselves. Hence it was important for students to have at hand a set of geometrical models.[38]

From at least 1876, a set of 12 painted tin forms known as "Melville's Geometric Drawing Models" had been available from New York instrument dealers.[39] Smith designed four sets of what he called "American Drawing Models." The first set had 30 pieces and included not only common geometrical solids, but also a flight of four steps, a cross, and three vases. Several shapes were shown in skeleton form—a circular ring, a triangular frame, and a skeleton cube, for example. Smith's second set consisted of wooden vases. The third had four large models for use in front of a class. The fourth set, intended for primary schools, had 12 common geometric figures. The models were made in Worcester, Massachusetts, at the Washburn Machine Shop of the Free Institute of Industrial Science (later Worcester Polytechnic Institute).

Smith's first three sets of models, especially the first, proved popular and were offered by dealers such as Devoe, Andrews (1881), and the Boston firm of Frost & Adams (1891). They were manufactured at Worcester Polytechnic Institute as late as 1897.[40]

In 1886 Smith lost his positions in Massachusetts and returned to England. However, three Massachusetts educators, John B. Clark, Walter Perry, and Mary Dana Hicks continued to develop texts sold by his publisher, Prang, as *The Prang Course.* These textbooks, particularly *The Use of Models,* emphasized the importance of models in art education. Students were to handle models of geometric solids, shape them in clay, make paper models, and eventually draw each one in perspective.[41]

Art educators said little about descriptive geometry or the mathematical properties of solids. Their use of models deserves mention as another example of the relationship between modelmaking and eductional reform. Moreover, the Washburn Shops merit attention. Production at these shops parallels, in a limited way, modelmaking under Schroeder at Darmstadt. At the same time, Worcester was the training ground for A. Harold Wheeler, who would become one of the most prolific modelmakers in the twentieth century United States.

Models for the Professional Mathematician

By 1890 models had become a well-established part of geometry teaching in American high schools and colleges. Florian Cajori's *The Teaching and History of Mathematics in the United States,* published that year, includes a survey of teaching practices in 168 American colleges and universities as well as in 181 high schools and academies. In answer to the question "To what extent are models used in geometry?", most respondents indicated that models were used, particularly to introduce forms in solid geometry and, at the college level, to teach descriptive geometry. The replies Cajori printed suggest that practices were not uniform—some colleges could not afford models, and some college professors thought models were better suited to elementary instruction. Cajori himself believed that "to teach plane geometry to very young students, or solid and spherical geometry to students of any grade, without the aid of models, is a great mistake."[42]

Johns Hopkins University, the first American university to attempt to emulate mathematical practices of German research universities, is not listed in this survey. However, Cajori described the curriculum at Johns Hopkins in some detail elsewhere in his book. Here he mentioned "a magnificent set of geometrical models for the study of surfaces," acquired in 1884 and used particularly by William E. Story, a Harvard graduate who had obtained his doctorate from the University of Leipzig in 1875.[43] Story listed these models in full in the *Johns Hopkins University Circular* for January, 1885. They included a collection of polyhedra and quadric surfaces made from wood, as well as string models for descriptive geometry, made by Schroeder of Darmstadt and mentioned earlier. Johns Hopkins also had an extensive collection of paper and plaster models distributed by another Darmstadt firm, that of Ludwig Brill. Several Brill models were quadric surfaces, illustrating the same principles as models made by Engel and Schroeder. Others were more complex, including surfaces of third and fourth degree, as well as surfaces of rotation of constant negative or positive curvature.

The models made by Brill and his successor, Martin Schilling, are significant for several reasons. First, like Monge and Olivier's models for descriptive geometry, they were designed by mathematicians. Secondly, they were associated with the educational ideals of German universities, particularly those of mathematician Felix Klein. Schools that purchased these models asserted their admiration of German methods, just as those that had bought Olivier models proclaimed their admiration of French techniques. Thirdly, Brill models were purchased by many American mathematics departments for use in teaching relatively advanced topics. These were not models for art education, elementary geometry, or engineering drawing; they were models for mathematicians. This section traces the early history of the Brill models and discusses their reception in the United States. This large topic has already attracted some attention from mathematicians and historians of mathematics, as indicated in later notes.

The story of the Brill models begins with the mathematician Alexander Brill (1842–1935), a native of Darmstadt who studied architecture at Karlsruhe and Giessen before obtaining a PhD in mathematics under Alfred Clebsch at Giessen in 1864. In 1867 Brill was appointed as a *Privatdozent* at the University of Giessen, and in 1869 was named professor at the technical school in Darmstadt.[44] In 1874 he brought out a series of paper models of second order surfaces. These models were inspired by a model of an elliptic paraboloid made from half-circles which the German-educated mathematician Olaus Henrici of London had sent to a meeting of mathematicians in Goettingen.[45] Brill's models, which represented surfaces by delicately interlaced circles or quadrilaterals, were not as sturdy as wooden or even plaster models, but cost considerably less.

In 1875, Brill moved to the Technische Hochschule München.[46] That same year, geometer Felix Klein joined the faculty. Klein shared Brill's belief that mathematics students should have models at their disposal. In the inaugural address Klein had delivered at Erlangen in 1872, he had stressed the advantages students at technical schools received from building models and making drawings.[47] Brill and Klein soon directed their students in designing new models. Sales proved sufficient to enlist the assistance of Alexander Brill's brother Ludwig, who operated the family printing business in Darmstadt. Ludwig Brill distributed the models and issued catalogs describing them.

Ludwig Brill issued models in series, with a total of 21 series released between 1877 and 1892 Particularly in the early years, one series might contain models designed by several students and relating to such diverse topics as Kummer surfaces, surfaces of rotation of constant curvature, and the path of a heavy point on a sphere.[48] In general, the models in the first 10 series were designed by students at Munich. Several of these students, such as Walther Dyck, Karl Rohn, and Ludwig Schleirmacher, went on to careers in mathematics teaching. Another, Anton von Braunmuehl, became an historian of mathematics. Student Rudolph Diesel, who designed a series of plaster models of second-order surfaces that proved particularly popular, is now better remembered as the inventor of the pressure-ignited heat engine that bears his name.[49] Not all models were designed in Munich. The plaster models that made up Brill's Series 9 were copied from originals at the University of Berlin which had been described by Ernst Kummer in papers of 1862, 1866, and 1872.[50] Carl Rodenburg designed 21 models of surfaces of the third order, published as Series 7. Rodenberg, who had his doctorate from G}ttingen and was professor of mathematics at the technical school in Darmstadt, went on to teach in Hannover.[51] Series of models published in later years were more often associated with a single mathematician and related to only one topic. Klein left Munich in 1880, going first to the University of Leipzig and then, in 1886, to Goettingen. Alexander Brill left for Tübingen in 1884. Both remained actively interested in models.

Models distributed by Ludwig Brill and later Schilling diffused gradually in the United States. As noted earlier, Johns Hopkins acquired Brill models in 1884. In 1889, Story moved from Hopkins to Clark University in Massachusetts. As Roger Cooke and V. Frederick Rickey have shown, Clark University soon had its own collection of Brill models. Story and Henry S. White presumably used these in courses on the theory of surfaces and twisted curves, as Story had at Hopkins. The Clark mathematics department took pride in these tangible symbols of its affiliation with the German educational system. Clark prepared an extensive exhibit for the Columbian Exposition of 1893, including photographs of its educational apparatus. Several photographs showing the collection of models survive at Clark.[52] Not all early Brill models were at large universities. Judy Green and Jeanne Laduke have published an 1886 photograph of Helen A. Shafer's senior mathematics class at Wellesley College in Wellesley, Massachusetts. Professor Shafer and 10 students admire two Brill mod-

els of ruled surfaces that sit on the table in front of them. A cabinet in the back of the photograph holds additional models of quadric, cubic, Kummer, and optical surfaces.[53]

Clark University sought to impress visitors to the Columbian Exposition with photographs of its Brill models. Those who wanted to learn more could see the models themselves at the German Educational Exhibit. Moreover, as Karen Parshall and David Rowe have indicated, Felix Klein visited Chicago to attend the International Mathematical Congress held in conjunction with the fair.[54] Both at the fair and in a series of lectures at Northwestern University soon thereafter, Klein used models to illustrate his talks. Apparently this combination of exhibition and exposition led several American colleges to purchase Brill models. For example, in the summer of 1893, President Seth Low of Columbia College visited Chicago to see his school's exhibit at the fair. Low made several purchases from the German educational exhibit, including a collection of mathematical models which he presented to Columbia's Department of Mathematics.[55] Edwin M. Blake, who had recently received a doctorate from Columbia and taught mathematics there, attended the Evanston Colloquium in August. It is not clear whether he or other Columbia mathematicians influenced Low's purchase.

Wesleyan University also purchased numerous Brill models at the fair. Much of this collection is now at the Smithsonian. Evidence from college catalogues suggests further diffusion of models. The *Catalogue* of the University of Illinois for 1899–1900 boasted that the mathematics department was "supplied with eighty-five of Brill's models," including examples in paper, plaster, and string.[56] The 1895–1896 Yale University *Catalogue* mentions that the university's "laboratories, museums, and collections" include "various collections of models, casts and photographs used in the teaching of mathematics."[57] No modelmaker's name is given, but a fine collection of Brill models survives at Yale. The *Catalogue* also mentions that fortnightly meetings of the Mathematical Club occasionally feature "descriptions and models of new apparatus."[58] Neither the collection of models nor the Mathematical Club was mentioned in the 1890-1891 catalogue. Similarly, the Cornell University *Register* for 1895–1896 notes that "A collection of models has been begun, which will be very useful in the study of surfaces of functions, of lines and systems in space, and of hyper-geometry."[59] These models were not mentioned in the catalogue for 1894–1895.

Mathematicians outside Ivy League schools and state universities also purchased models. Not only Wellesley, but Amherst College had several models by Brill.[60] Nearby Mt. Holyoke College has a collection of models by Schilling,[61] and Bowling Green State University in Ohio acquired a few models by Schilling.[62]

Martin Schilling of Halle and then Leipzig took over distribution of models from Ludwig Brill in about 1900.[63] By 1903 Schilling offered 30 series with a total of 321 models.[64] His 1911 catalog has 41 series of 377 models.[65]

Schools that did not want to order models from Germany sometimes had copies made locally. The Smithsonian collections contain several plaster models of optical surfaces and bar linkages by Brill and Schilling from the Department of Mathematics of the University of Michigan. Michigan also acquired moveable ruled surfaces, similar to those of Brill, that were made by Eberbach of Ann Arbor. Presumably this was the Ann Arbor firm of Christian Eberbach and his son Ottmar. The company was founded in 1843 as a pharmacy, but in 1874 became an importer and manufacturer of scientific instruments. The firm Eberbach established remained in business well into this century, although it always concentrated more on apparatus for scientific laboratories than on mathematical models.[66]

Alexander Brill and Felix Klein argued that students might benefit from making models. Few American teachers seem to have attempted this at any but the most elementary level, however. Three American modelmakers deserve mention. A. Harold Wheeler (1873–1950), a graduate of Worcester Polytechnic Institute mentioned earlier, built polyhedra. Wheeler taught mathematics in the high schools of Worcester. Between about 1910 and 1950, he designed and made models, and taught his students to do so. They used inexpensive materials such as plastic, string, balsa wood, and especially paper. Wheeler devised and indeed patented methods of making models that replaced glued parts by ingeniously cut paper folds or slits in paper or plastic. Wheeler exhibited models at the Massachusetts Institute of Technology, Wellesley College, and Brown University. A large number from Brown University and his personal collection are now at the Smithsonian. Wheeler also corresponded with mathematicians including Donald (H.S.M.) Coxeter and Raymond C. Archibald. Moreover, he described his work at mathematical meetings ranging from gatherings of local teachers to the 1924 International Mathematical Congress in Toronto.[67]

Wheeler attended the 1934 Century of Progress Exhibition, in Chicago, and saw models exhibited by Saul Pollock of Indiana State Teacher's College in Terre Haute, Indiana. Pollock (born 1904), who had received his doctorate from the University of California, made moveable models of ruled surfaces in aluminum frames as well as ruled surfaces showing the intersection of solids. In principle, these models are similar to those of Olivier.[68] Examples

of the style, perhaps copies made by Wheeler, survive in the Smithsonian collections. Pollock's models did not exert the influence on twentieth century Americans that Olivier's ruled surfaces did in nineteenth-century engineering education.

Another midwestern mathematician, Richard P. Baker (1866–1937), also made and sold models in the 1930s. A native of England with an undergraduate degree from Oxford and a doctorate from Chicago, Baker taught mathematics at Iowa State University. He systematically attempted to produce models to compete with those of Schilling. In a 1931 catalog, Baker listed over 500 items, including not only models for mathematics, but for physics, chemistry, and crystallography. Baker's models included familiar figures from solid geometry, ruled surfaces, and quadric and cubic surfaces. He also made Riemann surfaces, models relating to statistics, and various three-bar links. The models are made from wood, thread, painted sheet metal, and plaster.[69] The Smithsonian collections include a few models Baker sold to Brown University, as well as a collection donated by his daughters, mathematician Frances E. Baker and her sister Gladys E. Baker.

The depression years of the 1930s were a poor time for selling to universities, and the rhetoric of mathematical pedagogy no longer encouraged use of models. Baker's models apparently did not sell widely, and, in 1932, Schilling discontinued his distribution of models for lack of business.[70] Modelmaking continued, and indeed continues to this day. However, the pedagogical reforms of recent years have not featured this particular kind of mathematical object.

Conclusion

An "objective" examination of the history of American study of mathematical structures reveals an intimate connection between geometric models and pedagogical reform. The extension of American education during the nineteenth century, with new schools for elementary education and engineering, new curricula in art education, and the adoption of German methods of graduate instruction in mathematics all brought with them new geometric models. These reforms occasionally were associated with eminent European mathematicians such as Gaspard Monge and Felix Klein. In the twentieth century, American mathematicans such as Wheeler, Baker, and Pollock took up modelmaking, but none succeeded in making models a permanent part of American mathematics education. Wheeler's extensive study of polyhedra did influence Coxeter's research on icosahedra, but Coxeter ultimately turned to England for models.[71] Some recent mathematicians have made models to demonstrate new kinds of surfaces, such as Robert Connelly's flexible polyhedra. Increasingly, however, they rely on quite a different object, the electronic computer, to generate images of important surfaces. These objective presentations, however, are quite a different story.

References

1. This paper arose as part of a continuing effort to catalog the geometric models in the collections of the Section of Mathematics of the Smithsonian's National Museum of American History. I am grateful to the many donors who made this collection possible, as well as to Uta C. Merzbach, Curator Emeritus of Mathematics, who assembled much of it. Smithsonian colleagues who have assisted in this work include Deborah J. Warner, Bernard Gallagher, Jim Roan, and Margaret Sone. Interns Melissa Collier, Kristen Haring, and Rushel McCarney, as well as David L. Roberts of Johns Hopkins University, deserve special thanks. Finally, I thank various mathematicians who have generously answered questions about models known to them. Further information of this sort is welcome.

2. W. W. Ross, *Mensuration Taught Objectively with Lessons on Form* (Fremont, Ohio: no publisher, about 1891), p. 3.

3. S.E. Ambrose, *Duty, Honor, Country: A History of West Point* (Baltimore: Johns Hopkins University Press, 1966), p. 19.

4. M.A. Calvert, *The Mechanical Engineer in America, 1830–1911, Professional Cultures in Conflict* (Baltimore: Johns Hopkins University Press, 1967). There were at least two civilian schools in the United States that offered training for engineers well before mid-century. The Literary, Scientific and Military Academy of Norwich, Vermont, founded in 1820, was, as its name suggests, patterned after West Point. The Rensselaer Institute (later the Rensselaer Polytechnic Institute), founded in 1824, introduced a civil engineering curriculum in 1835. However, neither of these schools initially offered serious competition to West Point. See S. Rezneck, *Education for a Technological Society: A Sesquicentennial History of Rensselaer Polytechnic Institute,* Troy, New York: Rensselaer Polytechnic Institute, 1968.

5. The enthusiasm of American educators for German practice is suggested by the activities of President Seth Low of Columbia University. At the Columbian Exposition of 1893 in Chicago, Low purchased books, mathematical models, a calculating machine, and scientific apparatus from the German educational exhibit. He donated these to various departments of the university. See Seth Low to the Trustees of Columbia College, January 8, 1894, Columbia College Collection, Rare Book and Manuscript Library, Columbia University, New York. On psychological apparatus, see A.B. Davis and U.C. Merzbach, *Early Auditory Studies: Activities in the Psychological Laboratories of American Universities,* Washington: Smithsonian Institution Press, 1975.

6. D. J. Warner, "Commodities for the Classroom: Apparatus for Science and Education in Antebellum America", *Annals of Science,* 1988, 45:387–397. S. G. Kohlstedt, "Parlors, Primers, and Public Schooling: Education for Science in Nineteenth-Century America," *Isis,* 1990, 81:425–445. E.W. Stevens, Jr. "Technology, Literacy and Early Industrial Expansion in the United States," *History of Education Quarterly,* 1990, 30:523–544.

7. Barnard's inquiries, originally published in the first volume of the *Connecticut Common School Journal* (1838), were republished as "Inquiries Respecting a School," *American Journal of Education,* 1856, 1:687.

8. J. Holbrook, *Apparatus Designed for Families, Schools, Lyceums and Academies* (Boston: Allen and Ticknor, 1833), p. 6.

9. On the cube root block, see D.L. Roberts, "Romancing the Root: An Episode in the Promotion of Geometric Aids for Arithmetic Instruction," *Rittenhouse,* 1993, 7:33–36.

10. F.C. Brownell, *The Teacher's Guide to Illustration: A Manual to Accompany Holbrook's School Apparatus* (Hartford: Holbrook School Apparatus Co., 1857), p. 19.

11. R.E. Thursfield, *Henry Barnard's "American Journal of Education"* (Baltimore: Johns Hopkins University Press, 1945), p. 36.

12. New York (State of), *Laws, Statutes of the State of New York Relating to Common Schools . . .,* (Albany: Thurlow Weed, 1841), p. 167.

13. Claxton and Wightman, "A Card," Boston, October 1, 1835, Collection of Business Americana, Archives Center, National Museum of American History, Washington, D.C.

14. B. Pike, Jr., *Pike's Illustrated Descriptive Catalogue of Optical, Mathematical and Philosophical Instruments* (New York: B. Pike, Jr., 1848), 2:267–269. J.W. Queen and S.L. Fox, *Illustrated Catalogue of Mathematical, Optical and Philosophical Instruments and School Apparatus* (Philadelphia, J.W. Queen & Co., 1859), pp. 52–53.

15. W.M. McAllister, *A Priced and Illustrated Catalogue of Meteorological and Philosophical Instruments and Chemical and School Apparatus* (Philadelphia: McAllister, 1869), pp. 12–16.

16. J. Johonnot, *School-Houses* (New York: J.W. Schermerhorn & Co., 1871), pp. 84–85. Baker, Pratt & Co., *Illustrated Catalogue of School Apparatus* (New York: Baker, Pratt & Co., 1879), pp. 111-115. A.H. Andrews, *Catalogue of School Merchandise* (Chicago: A. H. Andrews), 1881, pp. 100–103.

17. Welch Scientific Company, *Welch Mathematics Catalog,* (Skokie, Ill.: Welch Scientific Company, 1965), p. 12.

18. Queen & Fox, *Illustrated Catalogue,* 1859, pp. 52–53.

19. Johnnot, *School-Houses,* 1871, pp. 84–85.

20. I. Harrington, "Improvement in Apparatus for Teaching Mensuration", U.S. Patent 137075, March 25, 1873. C. Anderson, *Technology in American Education 1650–1900* (Washington, D.C.: U.S. Department of Health, Education and Welfare, 1962), p. 45.

21. A.H. Kennedy, "Educational Appliance", U.S. Patent 296018, April 1, 1884.

22. W.W. Ross, *Mensuration Taught Objectively with Lessons on Form,* pp. 29–32.

23. Personal Communication, Rushel McCarney, March, 1993.

24. G. Monge, *Geometrie Descriptive: Leçons Donné aux Écoles Normales . . .* (Paris: Baudouin, an VII (1798 or 1799)), p. 1.

25. P.J. Booker, *A History of Engineering Drawing* (London: Chatto & Windus, 1963), pp. 86–113. L.J. Daston, "The Physicalist Tradition in Early Nineteenth Century French Geometry," *Studies in History and Philosophy of Science*, 1986, 17:269–295. E. Glas, "On the Dynamics of Mathematical Change in the Case of Monge and the French Revolution," *Studies in History and Philosophy of Science*, 1986, 17: 249–268.

26. Conservatoire National des Arts et Mètiers, *Catalogue des Collections* (Paris: Dunod Editeur, 1882), p. 31.

27. W.C. Stone, *The Olivier Models*, Schenectady, N.Y.: Friends of the Union College Library, 1969.

28. Harvard University Faculty of Arts and Sciences, "Subscription of Twelve Hundred Dollars For the purchase of a copy of the Olivier Models," UAIII 28.56 hd, Harvard University Archives; Harvard College Papers, Second Series, vol. 24, pp. 242–242, UAI 5.125, Harvard University Archives; and J.M. Shaw, "The Silk-Thread Models of Monge and Olivier: Teaching, the Study of Surfaces, and Generality in Synthetic Geometry," unpublished paper, 1990. I thank Will Andrewes for graciously providing information about these models and a copy of Shaw's paper.

29. D.C. Arney, "Resources in the History of Mathematics at the United States Military Academy," *Historia Mathematica*, 1988, 15:368–369. F. de R. Furman, ed. *A History of the Stevens Institute of Technology* (Hoboken, N.J.: Stevens Institute of Technology, 1905), p. 7. Columbia College, *Catalogue, 1870–1871*, p. 94.

30. College of New Jersey, *Catalogue*, 1881–1882, p. 78.

31. "Christian Gottlieb Ferdinand Engel," *American Journal of Science*, 1868, 45:282–284. C.G.F. Engel, *Axonometrical Projections of the Most Important Geometrical Surfaces...* (New York: H. Goebler, 1855).

32. "University of Mississippi," Files of the Division of Physical Sciences, National Museum of American History.

33. R.P. Multhauf with D. Davies, *A Catalogue of Instruments and Models in the Possession of the American Philosophical Society* (Philadelphia: American Philosophical Society, 1961), p. 13–14. R.P. Multhauf and G. Good, *A Brief History of Geomagnetism ...* (Washington, D.C.: Smithsonian Institution, 1987), p. 87.

34. E.A. Davidson, "Industrial and Scientific Education as exemplified in the Paris International Exhibition of 1866," *American Journal of Education*, 1872, 23:701–704.

35. J. W. Queen & Co., *Priced and Illustrated Catalogue and Descriptive Manual of Mathematical Instruments and Materials for Drawing, Surveying and Civil Engineering* (Philadelphia: J. W. Queen & Co., 1881), p. 59.

36. W.E. Story, "List of Models of Mathematical Surfaces Belonging to the Johns Hopkins University," *Johns Hopkins University Circular*, January, 1885, #36:36–38. There are a few examples of models of polyhedra by Schroeder from the University of Michigan at the Smithsonian. Furman, *A History of the Stevens Institute of Technology*, p. 7. College of New Jersey, *Catalogue*, 1881–1882, p. 78. F.A.P. Barnard to G.M. Ogden, October 9, 1868, Papers of Governeur Morris Ogden, Rare Book and Manuscript Library, Columbia University, New York, N.Y.

37. R.G. Boone, *Education in the United States* (New York: D. Appleton & Company, 1889), p. 229; and D. Korzniak, *Drawn to Art: A Nineteenth-Century American Dream* (Hanover and London: University Press of New England, 1985), pp. 153–166.

38. W.J. Smith, *Teacher's Manual for Freehand Drawing in Intermediate Schools* (Boston: L. Prang & Co., 1876), pp. 246–247, 262–267.

39. J. Johnnot, *School-Houses*, p. 84. F. W. Devoe & Co., *Priced Catalogue of Surveying and Mathematical Instruments ...* (New York: Devoe, 1876), p. 199.

40. F.W. Devoe & Co., *Priced Catalogue...*, 1876, p. 200–201. A.H. Andrews & Co., *Illustrated Catalogue of School Merchandise* (Chicago: A.H. Andrews & Co., 1881), p. 96–97. Frost & Adams, *Descriptive Catalogue of Artist's Materials...* (Boston: Frost & Adams, 1891), p. 107–108. Worcester Polytechnic Institute, *Annual Catalogue*, 1897, 96–97.

41. P.C. Marzio, *The Art Crusade: an Analysis of American Drawing Manuals, 1820–1860* (Washington, D.C.: Smithsonian Institution Press, 1976), pp.65–66.

42. F. Cajori, *The Teaching and History of Mathematics in the United States* (Washington: Government Printing Office, 1890), p. 359. Cajori surveyed normal schools as well as high schools and colleges, but apparently did not ask those teaching at these schools about geometric models.

43. F. Cajori, *Teaching and History of Mathematics*, p. 269.

44. On A. Brill, see: J.B. Pogrebyssky, "Alexander Wilhelm von Brill," *Dictionary of Scientific Biography* (New York: Scribners, 1970), 2:465. S. Finsterwalder, "Alexander von Brill Ein Lebensbild," *Mathematische Annalen*, 1936, 112:653–663.

45. W. Dyck, *Katalog mathematischer und mathematisch-physikalischer Modelle, Apparate und Instrumente* (Munich: K. Hof- u. Universitaetsbuchdruckerei, 1892), p. 258.

46. J.C. Poggendorff, *Biographisch-Literarisches Handworterbuch...* (Leipzig: J.A. Barth, 1898), III:193.

47. D.E. Rowe, "Felix Klein's 'Erlangen Antrittsrede : A Transcription with English Translation and Commentary," *Historia Mathematica*, 1985, 12:135, 139.

48. L. Brill, *Catalog mathematisher Modelle fuer den hoeheren mathemateschen Unterricht...* (Darmstadt: L. Brill, 1892), p. 5.

49. W. R. Nitske and C. M. Wilson, *Rudolph Diesel: Pioneer of the Age of Power* (Norman: University of Oklahoma, 1965), pp. 46–47.

50. L. Brill, *Catalog*, 1892, p. 19.

51. L. Brill, *Catalog*, p. 14. J.C. Poggendorff, *Biographisch-Literarisches Handwoerterbuch...* (Leipzig: J. A. Barth, 1898) III:193 & 1130; and Poggendorff, 1904, IV:1260.

52. R. Cooke and V.F. Rickey, "W.E. Story of Hopkins and Clark," *A Century of Mathematics in America*, edited by R. Duren with the assistance of R.A. Askey, H.M. Edwards, and U.C. Merzbach (Providence, R.I.: American Mathematical Society, 1988), 3:51–61.

53. J. Green and J. LaDuke, "Women in the American Mathematical Community: The Pre-1940 Ph.D.'s," *The Mathematical Intelligencer*, 1987, 9:11–23. The photograph discussed is reproduced on page 17 of this article by Green and

LaDuke. The original is in the Wellesley College Archives in Wellesley, Massachusetts.

54. K.V.H. Parshall and D.E. Rowe, "Embedded in the Culture: Mathematics at the World's Columbian Exposition of 1893," *Mathematical Intelligencer*, 1993, 15, #2:40–45.

55. Columbia College in the City of New York, *Fifth Annual Report of President Low to the Trustees* (New York: Columbia College, 1894), p. 129.

56. *Catalogue of the University of Illinois* (Urbana, Illinois: 1899–1900), p. 119.

57. Yale University, *Catalogue of Yale University* (New Haven: Yale University, 1895–1896), p. 181.

58. Yale University, *Catalogue of Yale University* (New Haven: Yale University, 1895–1896), p. 226.

59. Cornell University, *Register* (Ithaca, N.Y.: Cornell University, 1895–1896), p. 130.

60. Personal Communication, Duane Bailey, Department of Mathematics, Amherst College, April 12, 1993.

61. Personal Communication, Cecile Girard, Department of Mathematics, Mt. Holyoke College, April 12, 1993.

62. Personal Communication, V.F. Rickey, March 16, 1993.

63. W. Barth and H. Knoerrer, "Algebraic Surfaces," *Mathematical Models,* ed. Gerd Fischer (Braunschweig/Wiesbaden: Fried. Vieweg & Sohn, 1986), 2:8 gives the date 1900. Fis-

cher, in the same volumes, 1:IX, says that Schilling took over Brill in 1899.

64. *Catalog mathematischer Modelle fuer den höheren mathematischen Unterricht veroeffentlicht durch de Verlagshandlung von Martin Schilling in Halle a.S.* (Halle a. S.: M. Schilling, 1903), pp. VII–XIII.

65. *Catalog mathematischer Modelle für den höheren mathematischen Unterrict veroeffentlicht durch die Verlagshandlung von Martin Schilling in Leipzig* (Leipzig: Martin Schilling, 1911).

66. "Ralph Horace Miller," *National Cyclopaedia of American Biography,* (New York: James T. White & Co., 1950), 33:151.

67. David L. Roberts, "Albert Harry Wheeler (1873–1950): A Case Study in the Stratification of the American Mathematics Community, forthcoming.

68. *Official Handbook, of Exhibits in the Division of the Basic Sciences: Hall of Science* (Chicago: A Century of Progress, 1934), p.39–40, 43.

69. R.P. Baker, *Mathematical Models* (Iowa City, 1931).

70. G. Fisher, *Mathematical Models,* IX–X.

71. H.S.M. Coxeter, P. Du Val, H.T. Flather, and J.F. Petrie, *The Fifty-Nine Icosahedra,* New York: Springer-Verlag, 1982 (reprint of 1938 article), p.4–5.

The Social and Intellectual Shaping of a New Mathematical Discipline: The Role of the National Science Foundation in the Rise of Theoretical Computer Science and Engineering

**William Aspray,
Andrew Goldstein, and
Bernard Williams**
*Rutgers University, and
R&D Publications*

In the second half of the twentieth century, the domain of mathematics has widened, partly through the rise of new mathematical disciplines such as mathematical statistics, operations research, and computer science. The formation of these new disciplines was partly intellectual, but it was also shaped by social factors, such as professionalization, institutionalization, patronage, politics, and interactions with other parts of society. This paper explores the development of one of these new mathematical disciplines, theoretical computer science and engineering, and the role played in its development by its chief patron in the United States, the National Science Foundation (NSF).[1]

In the United States, much of the early research on computer science was conducted in a few research universities. For an area of computer science to advance, someone had to fund computer time as well as faculty research, graduate students, and support staff. The federal government, rather than industry or academe, provided most of the funding. Among the federal agencies involved, NSF was the second largest supporter of computer science and engineering but almost the only supporter of theoretical computer science, the most highly mathematical part of the discipline. NSF valued this theoretical research because it appeared more "scientific" than most parts of computing, whereas industry and other federal agencies disdained it because they could not see its practical application. (As is shown at the end of this paper, however, interesting practical applications did arise eventually.)

Computer Science and its Federal Patrons

Attempts to establish a science of computing were clearly an outgrowth of the development of the technology of the electronic, stored-program, digital computer, which was a by-product of the Second World War. Although there were results in numerical analysis and logic predating the computer that became part of computer science, most concepts and issues were created anew as researchers labored to organize and apply computing systems more effectively and to discover their limits.

By the late 1940s a few researchers, such as Alan Turing and John von Neumann, were investigating subjects later considered part of computer science, but the intellectual discipline did not take shape until the 1960s. The 1950s were devoted first to finding hardware to carry out the arithmetic, storage, and logical control functions in the computer, and afterwards to finding ways of automating the programming process through the development of high-level programming languages, operating systems, and utilities. In the 1950s a little theoretically oriented research

was carried out on numerical analysis, automata theory, self-learning systems, and a few other topics, when the main task was to build an effective working machine.

The 1960s was a decade of transformation with major technical advances, such as timesharing, graphical user interfaces, and networking. Most graduate computer science programs in existence today were formed, and enrollments skyrocketed. Professional organizations, such as the Association for Computing Machinery and the IEEE Computer Society developed model curricula, which were widely adopted in American colleges and universities. Through efforts to develop curricula and extensive discussions in the professional journals, such as the *Communications of the ACM,* computer researchers struggled to identify a set of theory, methods, and facts to define a science of computing. In this context theoretical studies began to expand and take on importance as the scientific core of this new discipline.

After World War II, first the military and then other parts of the federal government recognized the importance of computers in helping to meet their agency's mission and to become a vital tool in American society. By 1960 almost every branch of the government invested in computers, and some supported research on computing. Federal support of academic research was critical to the development of the discipline.

Many federal agencies sponsored extramural research programs in computing. In 1954, the year in which NSF awarded its first grant in computing, the Army, Navy, Air Force, Atomic Energy Commission, and National Security Agency were all funding computer developments and applications that supported their missions.[2] The National Bureau of Standards, which since the war had been the leading federal agency in computing, was diminished because of a political scandal over a battery additive. They were further diminished by staff resignations following attacks from Senator Joseph McCarthy, and by a significant loss of military funding for its programs.[3] The Atomic Energy Commission had just begun a campaign to establish itself as the national center for advanced computation. (Perhaps because of AEC Commissioner John von Neumann's early death in 1957, the AEC labs never ascended to the national leadership position in computing they sought although they became the most advanced users of supercomputers.) The Office of Naval Research and the Air Force Office of Scientific Research were providing modest research grants to academic researchers.

After NSF gave its first support, two other federal agencies began to fund computing at substantial levels. One was the National Aeronautics and Space Administration (NASA), which began funding of computing soon after its formation in 1958. It supported research on in-

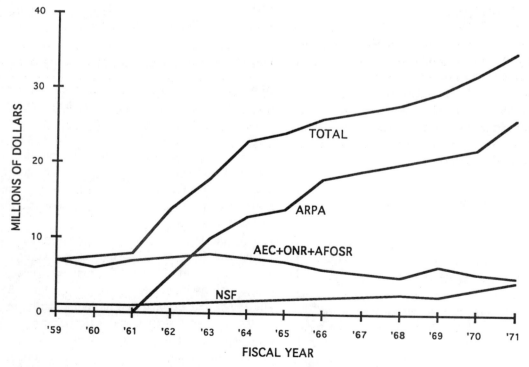

TABLE I.   Federal Support of Research in Computer Science and Engineering.
Source: NSF, Director's Program Review-Office of Computing Activities 12/15/1970, p. 14 (redrawn).

formation systems, artificial intelligence, computer-aided design, data-display systems, and programming languages, but most of its sponsored research focused narrowly on producing products needed by NASA.

The other new sponsor was the Advanced Research Projects Agency (ARPA, later known as DARPA when "Defense" was prefixed to the agency's name in the early 1970s). President Dwight Eisenhower formed the agency in 1958 after the Soviet launching of the Sputnik satellite and because he believed that interservice rivalry created inefficiency and waste in the development of high technology for the nation's defense. DARPA's first task was the military space program, but it soon became involved in advanced research on materials, reconnaissance satellites, nuclear test verification, and other projects.

DARPA began to support computing research in 1962, when it formed its Information Processing Techniques Office (IPTO) under Joseph C.R. Licklider, a psychologist from Massachusetts Institute of Technology. The agency massively funded computing because it believed this technology to be the solution to the military's desire for advanced command-and-control systems. DARPA projects fundamentally contributed to timesharing, networking, graphics, and artificial intelligence. The three most prominent academic computer research programs in the United States, MIT, Stanford, and Carnegie Mellon, were products of DARPA largess.

As Table I shows, DARPA's support of computer science research dwarfed that of all other federal agencies almost immediately. Within a year of its first interest in computing, its support exceeded the total support from all other federal agencies combined. NSF support grew throughout the 1960s and 1970s. (See the black bars in Table II. The other bars in the table represent NSF's other forms of support for computing.) By the 1970s NSF's support was about one-half that provided by DARPA.

National Science Foundation Support for Computing

Before turning to an examination of NSF support for theoretical computer science and engineering, this section reviews generally NSF programs in support of computing and computer science. The National Science Foundation was established in 1950 to provide federal support for scientific research and education. Many parties wanted the federal government to continue its support of science after the Second World War, given that science had been mobilized effectively for national interests during the war. However, political wrangling over the nature of the new

organization delayed enactment of the founding legislation for five years from the time Vannevar Bush set out his vision of a National Research Foundation at the request of President Franklin Delano Roosevelt.

Given the nascent state of computing, the Foundation's original organizational structure made no provision for computing, but rather was organized along lines of traditional scientific disciplines such as physics and chemistry. The first computer grants were made through the mathematics program, and this is how things stood for the next thirteen years. An Office of Computing Activities was first established within the Foundation in 1967 as a result of President Lyndon Johnson's Great Society program, which called for the use of computers to enhance education. Various programmatic reorganizations occurred within NSF in the 1970s and 1980s, and computer science and mathematics were again grouped together briefly—although as more equal partners the second time around. Only in 1986 was a separate directorate established for computer science and engineering.

At NSF the computing program had two missions— to develop a science of computing and to make computers available as a tool in the service of other sciences. These two missions, which were not always mutually supportive, were carried out through programs in computer facilities, computer science research, information science, and computer education.[4] The programmatic emphasis varied markedly over time, but in most periods all four activities were supported. This work was pursued almost exclusively through research grants to academic researchers and institutions. Rather than supporting only top researchers and institutions, NSF attempted to build up a wide national base of computing in research universities, four-year and junior colleges, and even high schools. The management style was participatory, using strong advisory panels and peer review of proposals.

Until 1967 NSF's chief computing activity was a program that provided computer facilities to more than 400 universities. During the next five years the emphasis was on "computer education," an ambiguous term that meant three things: instruction in computer science and computer engineering, the application of computers to science and engineering teaching, and the development and implementation of general-purpose instructional tools driven by computers. During most of the 1970s support emphasized computer science research. In the early 1980s there was a renewed interest in computer facilities programs. Over the entire postwar period, NSF provided the leading federal effort to develop large bibliographic and data management systems, which began with research on automatic language

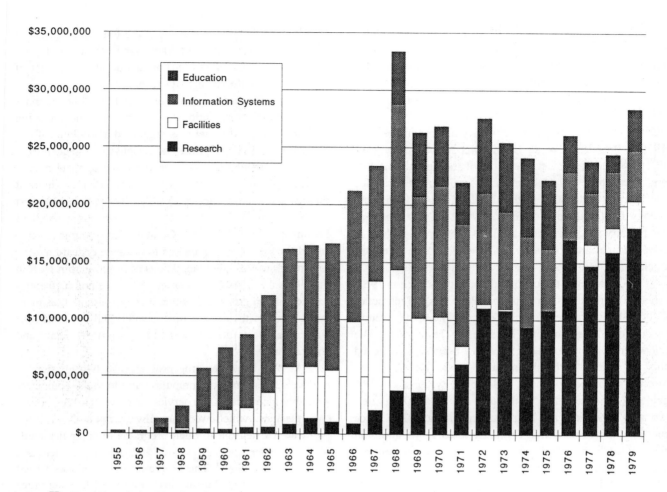

TABLE II Total Foundation Funding for Computing and Computer Science.

translation in the 1950s and evolved to support information science by the 1970s.

We shall examine here only one of these programs, NSF support for computer science research. In 1967 the Foundation responded to a presidential mandate for increased funding for computer-aided education by launching a tri-fold program of support for computers in colleges and universities. The effort included funding for computer-aided education, continued funding for computer facilities infrastructure, and added funds for research in computer science. The creation of the Office of Computer Activities (OCA) at this time formalized the Foundation's role as a patron for computer research, which the agency had assumed ten years earlier. In the late 1950s and early 1960s, mathematicians and electrical engineers had occasionally come to the Foundation with proposals to study aspects of computers. By 1966 the Foundation had awarded 155 grants through the mathematics and engineering directorates, for computer science research totaling over $6,000,000. OCA greatly increased the support of computer

science research. Over the next two years, 101 new grants (amounting to almost $6,000,000) were awarded.

As Table III shows, the Foundation's research program started in the late 1960s and grew in fits during the 1970s. In response to the Mansfield Amendment, the military agencies shed non-military research projects to other agencies.[5] This resulted in a doubling of NSF's annual budget for research in 1972. Annual funding passed the $10 million mark as the number of grants increased from 97 to 167. A similarly abrupt infusion of funds occurred in 1976, when there was a 60% increase in the number of grants and total annual funding for research.[6] After scaling back in 1977, growth in funding for the remainder of the decade barely kept pace with inflation. Over the 26 years between 1955 and 1980, the Foundation provided $150 million to hundreds of researchers at more than 230 institutions. These researchers worked on 2,700 different grants at an average of $57,000 per grant.[7]

The Foundation's pattern of support for computer research differed from those of other funding agencies. Al-

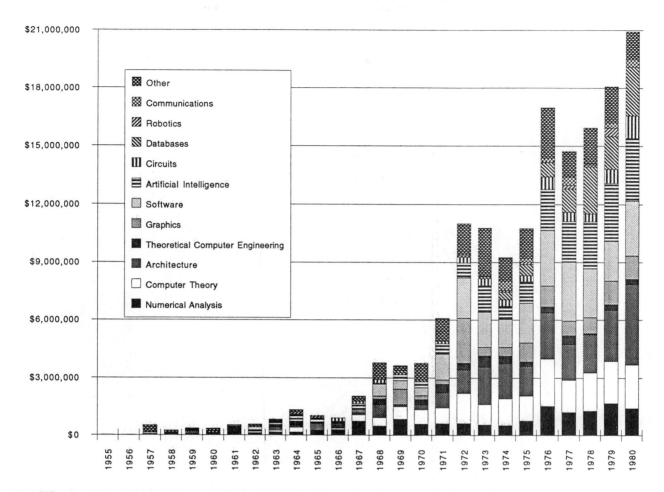

TABLE III   Foundation Funding for Computer Science Research.

though NSF supported technological development, including computer architecture, circuits and components, artificial intelligence, and robotics, its grants were intended primarily to promote theoretical research of an abstract and mathematical nature.

The Foundation supported scientists working in every area of computer science research. Table III identifies twelve distinct funding categories, which, except for robotics, were supported with more than $1 million in research grants.[8] Software and architecture research garnered the most funds. Theoretical computer science was the third most heavily supported area, even though the average grant was only 84% that of the overall average (perhaps reflecting a lack of need for computer hardware). Artificial intelligence and numerical analysis were next in total funding. Graphics, databases, circuits and components, theoretical computer engineering, communications, and robotics received less support. Table IV indicates how the distribution of support among the categories varied over time.

The Foundation staff did not usually set a research agenda for funding.[9] They relied on the scientific community to set the agenda, through the recommendations of advisory panels, the proposals scientists submitted, and the reviews the scientific community gave those proposals. Because the program's direction was driven by the scientific interests of applicants, in trying to understand NSF's significance in the rise of theoretical computer science and engineering it is more appropriate to consider examples of research sponsored by the Foundation than to analyze the formal programs established by Foundation staff. Through a carefully chosen set of examples, we simultaneously give an overview of the historical development of theoretical computer science and engineering and the influence of NSF support on that research. Many other examples could have been chosen, but these case studies indicate the breadth of topics the Foundation supported, the character and significance of the work within those topics, and the nature of the relationship between the Foundation and its grantees.

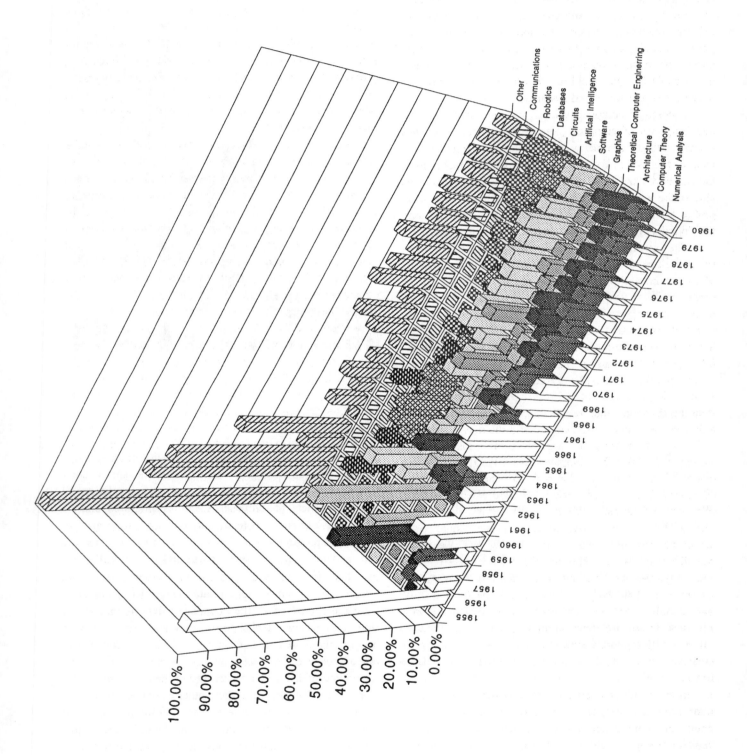

TABLE IV   Foundation Funding Percentages by Category of Research

NSF Support for Theoretical Computer Science

This and the next section examine NSF research support of two areas—theoretical computer science and theoretical computer engineering. These were the most mathematical of the computer science and engineering subdisciplines receiving Foundation support. Our approach is to give some general remarks followed by case studies. These cases were chosen to include some researchers who received the largest number of NSF grants and to show the evolving character over time of the sponsored research.

Theoretical computer science was one of the earliest subdisciplines of computer *science* to coalesce and was correspondingly central to the definition of computer science in the 1960s. Except for some modest support from the Air Force and Navy in the 1950s, the Foundation was almost the sole supporter of this field. Our investigation probes both the formation of a science of computing and the Foundation's critical role in that effort.

Theoretical computer science originated in an interest at the turn of the century in logic and the foundations of mathematics. Powerful tools of mathematical logic were constructed in a search for a secure foundation for mathematics. One important outgrowth of this movement was the new field of recursive function theory, which investigated the ability to construct sets of mathematical objects according to purely "mechanical" rules from some other given set of mathematical objects. In 1936 the British mathematician Alan Turing conceptualized his answer to one of the outstanding questions of recursive function theory in terms of a theoretical machine that would build the new mathematical objects from the given ones.[10] Turing's machine was subsequently taken by computer scientists as an abstract model of the computer. This was the start of the theory of computability, which became the computer scientist's equivalent of recursive function theory. Although the results of the two fields were coextensive and results could readily be translated from one to the other, the orientation was different: the recursive function theorists were interested in gaining results about the foundations of mathematics and the relationship between various kinds of mathematical objects, whereas the computability theorists were interested in investigating the computing power of various real and abstract computing machines. In the 1950s and 1960s, however, this distinction was not yet so clearly drawn.

In the 1950s mathematical logicians/theoretical computer scientists worked on recursive functions/computability, and the Foundation supported some research in this area. An example is the work of Martin Davis on Hilbert's Tenth Problem. In his famous address at the Second International Congress of Mathematicians held in Paris in 1900, David Hilbert posed a set of 23 unsolved problems of mathematics, which set a research agenda for many mathematicians in the twentieth century.[11] Hilbert's Tenth problem asked whether there is a general solution to Diophantine equations that can be determined by a finite number of operations. Davis, a mathematician at NYU's Courant Institute, used the methods of computability to attack this problem.[12] He did not achieve a solution, but his results laid the groundwork for the solution by Yuri Matiyasevic, who in 1971 showed that Hilbert's Tenth Problem is formally undecidable.[13]

The research of Arthur Burks and his colleagues at the University of Michigan illustrates the result of NSF support in the other major area of theoretical computer science active in the 1950s, known as cellular automata theory. In a series of path-breaking studies in the 1950s, Institute for Advanced Study mathematician John von Neumann developed automata theory based on Turing's work on theoretical machines and research on neural nets by Warren McCulloch and Walter Pitts.[14] Burks, who was von Neumann's colleague and collaborator for a short time in the 1940s, carried these ideas forward from the 1950s until the 1980s with frequent Foundation support.

Burks's research was originally supported by the Burroughs Corporation, which undertook several theoretical projects in its effort to enter the digital computer market. At Burroughs's request, Burks formed a Logic of Computers research group at the University of Michigan in 1949.[15] The group designed a computer with an architecture heavily influenced by mathematical logic.[16] Their machine was the predecessor of the B5000, one of the leading Burroughs computer products manufactured in the 1960s.[17] The company continued its support of the Logic of Computers Group until 1957, when it decided the theoretical orientation did not provide enough practical benefit to justify continued support.

A 1957 grant from the Foundation—$31,700 for a two-year study of switching systems—enabled Burks to keep the group intact. This began the group's more than three decades of research in theoretically oriented computer science. Some funding came from the Air Force, Army, and Navy in the first years and from NIH in the 1960s, but the Foundation was the main supporter.[18]

Some of Burks's work extended von Neumann's research on self-reproducing automata.[19] Von Neumann invented cellular automata, which are two-dimensional arrays of abstract "cells" (modeling nerve cells) governed by mathematical axioms and act together as a purposeful formal system (for example, like the nervous system), but he died before completing this work. Burks completed and amended von Neumann's work. He used these cellular automata to describe computing elements such as switches, delays, counters, and memory. Burks and other members of the Logic of Computers Group used these to build biological models, such as a model by Bernard Zeigler and Roger Weinberg of living cells, and models of heart fibrillation in dogs that Burks developed with his long-time collaborator John Holland and with Henry H. Swain of the University of Michigan medical school.[20]

Learning was one of the group's major research interests. Assuming that learning is best understood as a process of gradual adaptation, they developed cellular automata known as classifier learning systems.[21] These large networks of multiply interconnected cells could be simulated on conventional serial computers, such as the VAX PDP-11. Beginning with a random configuration of connections, but with carefully chosen classifiers (rules of operation), the system passed bits of data from cell to cell within the network. Each cell processed its input in a way determined by the particular classifier associated with the cell to produce an output that was passed along the network. Feeding the output of certain cells back as the input of others, the classifier processed its data over multiple cycles. By using various mechanisms (with names such as the genetic algorithm and the bucket-brigade algorithm) to strengthen the connection between certain cells and weaken others, the classifier system evolved dynamically. It "learned" to produce good outputs more often than bad ones.

Burks has noted how these classifier systems could control robots, with robot "brains" that use input from sensory equipment as the initial data and output from the classifier system as instructions for the robot's navigational system. Performance was measured by collision avoidance, and the classifier adjusted accordingly.

The Logic of Computers Group emphasized the similarities of its mathematical models to natural systems. Recent applications of classifier systems include models of the stock market and the control of oil flow through unreliable pipes.[22] Some models involved genetic mechanisms that randomly combined or "mutated" classifiers and allowed only the "fittest" systems to survive. Burks's final PhD student, Chris Langton, took this analogy the furthest, creating what he called "artificial life"—elaborate computer simulations of biological organisms without physical embodiment.[23] These systems greatly interested researchers in the 1980s investigating complex non-linear systems from the physical and biological realms. This research program begun by von Neumann and Burks is today carried forward today by the Santa Fe Institute, an interdisciplinary research

institute founded by Nobel Laureates Murray Gel-Mann and Kenneth Arrow and Princeton physicist Phil Anderson. Burks's former student and colleague, John Holland, guides much of the institute's research.

Burks's work represents a theoretical research discipline studied continuously from the advent of the computer to the present. During most of this time, cellular automata theory was pursued only as a theoretical discipline, with few practitioners and low levels of funding. In the 1980s, however, interest in cellular automata grew rapidly because the subject informed efforts to build massively parallel computers and to write programs for using these new kinds of machines effectively.

Although they seem promising, it is too soon to foretell the significance of cellular automata. Other areas of theoretical computer science, including many studies in the 1950s supported by the Air Force on learning systems, died out in the 1960s because the systems were too simplistic and did not model the physical world adequately. A new set of theoretical research problems emerged to take their place in the mid-1960s. Foremost among them was computational complexity theory.

The classical work in theoretical computer science, which evolved from Turing's original ideas, said little about practical issues of computing. Turing machines were highly abstract mathematical fictions. They were powerful in some respects (for example, unlimited memory) but restrictive in others (for example, narrow channel). By the 1960s when many computers were in operation and many practitioners were attempting to solve difficult problems from business and science, interest grew in a more practical computer theory that would "measure the amount of 'computing work' done and classify computations by their complexity."[24] This was the origin of computational complexity theory.

Juris Hartmanis was a leading pioneer in this discipline. After receiving a doctorate in mathematics from California Institute of Technology in 1955, he joined General Electric Research Laboratory in Schenectady, New York. While there, he studied the classical theory of automata. By 1962 he discarded these issues and began to develop a method to classify functions according to the time necessary to calculate them. Inspired by Martin Davis's work on computability, Hartmanis and Richard Stearns published a landmark paper in 1965 which introduced a classification system for complexity of computational problems.[25] It opened a new field of research that addressed questions of practical computing differently from Turing machines.

Hartmanis left General Electric in 1965 to become the head of a new computer science program at Cornell University, gaining more opportunity to pursue theoretical research.[26] At Cornell he quickly built both a strong faculty in theoretical computer science and one of the nation's leading computer science departments.[27] Hartmanis worked to consolidate the field that he had helped to found. In the early 1970s, for example, he published two papers with his colleague John Hopcroft spelling out premises of computational complexity.[28]

Even with departmental administrative responsibilities, Hartmanis remained in the forefront of computer theory research with steady Foundation support. Beginning with a 1967 NSF award to study "Automata-Based Computational Complexity," Hartmanis explored implications of the complexity measures that he and Stearns introduced. Their paper identified the time taken by a computer or the space the computer used as measures of a computation's complexity. He extended this concept by adding a new measure of complexity: the amount of tape a Turing machine requires to complete a calculation.[29] His work in the 1970s focused on recasting results from recursion theory as results that had significance for complexity theory. Of particular concern were so-called isomorphism problems, which examine programming systems that are computationally isomorphic under the operation of particularly important classes of functions, such as polynomial time computable functions. With two graduate students, Ted Baker and Leonard Berman, Hartmanis posed two conjectures that became centerpieces of complexity theory research.[30]

In 1963 while Hartmanis formulated his ideas on complexity theory, he learned about the work of MIT doctoral student Manuel Blum, whose dissertation took complexity theory in a different direction. Whereas Hartmanis defined complexity in terms of concrete models (automata and the resources they use), Blum's took an axiomatic approach that defined complexity as a machine-independent concept depending only on the problem and the algorithm used to solve it.

Blum's thesis on computational complexity of recursive functions was partially supported by NSF.[31] Its main result was the "speed-up theorem."[32] Blum showed the existence of a class of computational problems which have no optimal solution. Given any program to solve a problem in this class, another program existed that would do the computation faster. This surprising and highly unintuitive result stimulated Blum's fellow graduate student, Schmuel Winograd, to employ similar techniques to find a method (called the Straussman method) for multiplying matrices that improved on the traditional row-by-column approach.[33]

After graduation Blum joined the University of California-Berkeley, where Lofti Zadeh encouraged him to apply to the Foundation to pursue complexity research. The Foundation awarded Blum $94,700 in 1970 to study computability and computational complexity. This began a string of Foundation grants that supported Blum's research and his graduate students into the 1980s.

With this support Blum continued his research in computational complexity. Most importantly he examined the use of probability theory to improve the performance of algorithms.[34] The output of each step of a computation of a deterministic algorithm is completely predictable from its inputs. Probabilistic algorithms introduce random numbers as inputs in order to make the output unpredictable. By introducing random variables one can speed up a calculation or even make a calculation possible when it could not be carried out deterministically.

Blum applied his results from complexity theory to make computer simulations of brain processes. Endeavoring to build an inductive inference system, he worked in the late 1970s on a system capable of evaluating a sequence of symbols and inductively inferring a rule that generated them. In the 1980s his attention turned to more modest models of cerebral functions. He investigated how organisms with rudimentary brains, such as ants, can exhibit complex behavior. The finite state automata that he developed could successfully maneuver through any maze in a flat plane. But when the mazes were placed on planar graphs, on which there did not always exist a north-south-east-west orientation, the automata did not have complete success at maze negotiation. Discoveries by naturalists in the late 1970s and 1980s that pigeons can detect magnetic north and that protozoa can sense direction he took as confirming his modeling results.[35]

Blum's impact on the field was not limited to his research. He also achieved acclaim for being a fecund producer of doctoral computer scientists. His students, including Len Adelman, John Gill, Choffe Goldwasser, Russell Impagliazzo, Sylvio Micali, Gary Miller, and Steven Rudich, have gone to premiere computer science departments and have become leaders in developing the field. Gill, for example, conducted a theoretical study of randomness in computing that served as the foundational base for subsequent research in probabilistic algorithms.[36] Miller developed a probabilistic algorithm for determining whether a number is prime, which surpassed all deterministic algorithms and had important consequences for cryptanalysis research in the 1980s.[37]

Blum's nondeterministic methods reflect a major trend in computational complexity research since the 1970s. This work derives in part from research performed under NSF support by Blum's Berkeley colleague, Richard Karp. By applying complexity theory, Karp was able to find non-deterministic methods to improve the performance of algorithms for combinatorial problems—one of the most important and computationally demanding categories of problems faced by computer scientists.

After receiving his doctorate from the Harvard Computation Laboratory in 1959, Karp worked at the IBM Research Center in Yorktown Heights, New York. He studied combinatorial problems, including the traveling salesman problem and a combinatorial method to design logic circuits. The former, famous problem of computer science seeks to find the shortest path a salesman can travel to a given set of cities. With just a small number of cities in the route, the problem grows quickly unmanageable. Many hoped that high-speed computers would solve the problem, but even the fastest machines were overwhelmed by moderately complex trips, given that the computation required to determine an optimal path grows exponentially as the number of cities increases. Computer scientists sought an algorithm for calculating optimal solutions to this and a large collection of related problems which had the characteristic that the amount of calculation grew at less than an exponential rate as the problem became larger.

In the early 1970s, by then at Berkeley, Karp examined perhaps the most important unsolved problem of complexity theory: noted short-hand as P=NP? In P-type problems (polynomial-time), the amount of computation required to solve them increased only modestly (as a polynomial function) as the size of the problem increased. This was not necessarily true of NP-type (non-polynomial time) problems, such as the traveling salesman problem. It was known about NP-type problems only that a solution to them could be checked in a reasonable amount of time; the amount of computation required to find an optimal solution, however, might grow inordinately, for example, at an exponential rate, as the input variable, for example, number of cities the salesman traveled to, grew. In a 1971 paper, Karp's colleague Stephen Cook identified a particular NP-problem known as the satisfiability problem and showed that, if it could be proved to be a P-type problem, then every other NP-type problem would also be a P-type problem.[38] This result would mean that practical algorithms would exist for all these NP-type problems.

Cook defined a problem to be NP-complete if it had this same property as the satisfiability problem. Thus, if any NP-complete problem is a P-type problem, every NP-problem is then a P-type problem. With ingenuity, Karp demonstrated that a large set of important combinatorial problems were NP-complete. Following his lead, computer scientists proved the NP-completeness of hundreds of common combinatorial problems.

In 1973 when Karp became head of the Computer Science Division at Berkeley, his administrative duties precluded active research on NP-completeness research. He received his first Foundation grant the following year, namely $106,000 to search for problems that are NP-complete and to develop algorithms to handle difficult combinatorial problems. The NSF grant files show that the agency gave the grant in part because they thought that such a productive researcher should be freed from administrative duties to pursue research while his intellectual abilities were at their zenith.

Research on NP-problems intensified efforts to find efficient, if not optimal, algorithms for many problems. Karp noticed that these algorithms were efficient even though based on heuristic methods. He wondered why they are so effective. Most computer scientists, when using heuristics to develop an algorithm, did worst-case analysis; they contrived a problem intended to thwart the algorithm as a means to test it. Karp instead considered algorithm performance in realistic settings; he wanted to study the operation of algorithms with data that matched probability distributions found in actual applications. For this purpose he developed a probabilistic analysis of algorithms.[39] Applying his results he increased the speed of combinatorial optimization algorithms by increasing the confidence that the fastest of the known algorithms produces nearly optimal results. His influential results increased the efficiency of computational mathematics, particularly linear programming. Following his lead, many researchers contributed to these improvements.

The value of Karp's method can be seen in its application to the traveling salesman problem.[40] Strictly optimal solution methods not using excessive amounts of computer time existed for up to about 60 cities. Heuristic solution methods that gave near-optimal solutions existed for up to 300 cities. Karp's probabilistic approach enabled him to find near-optimal solutions for up to many thousands of cities.

Karp's work represents the last major trend in theoretical computer science supported by the Foundation prior to 1980. As Table V shows, by 1980 the Foundation had supported all major areas of theoretical computer science. Given above is a representative sample of the research supported by the 400 awards NSF made in this area. This research included recursive functions, cellular automata, and computational complexity. Its application included computer science (parallel computing), other sciences (biological modeling), and practical disciplines (cryptography).

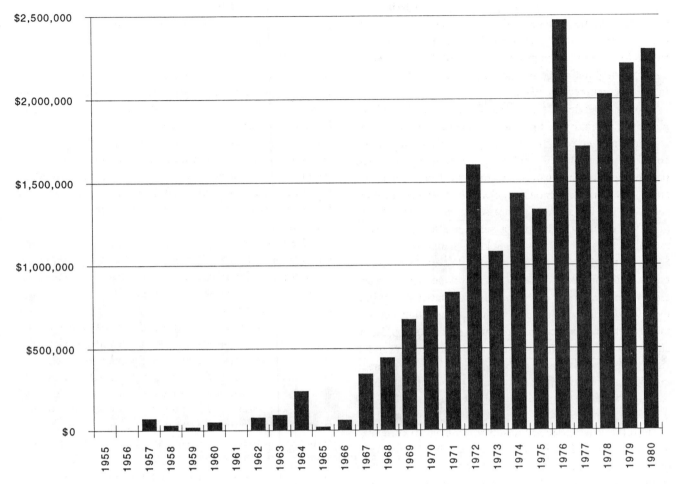

TABLE V Foundation Funding for Research in Theoretical Computer Science.

This research support helped cement the theoretical underpinning of computer science, develop new talent, and build strong computer science departments.

NSF Support for Theoretical Computer Engineering

The Foundation funded research that investigated topics at the boundary between engineering, mathematics, and computer science, which we call "theoretical computer engineering." It includes information and coding theory, systems theory, network theory, control theory, and related work.

Foundation support for theoretical computer engineering has been modest; the $5,000,000 distributed over 112 grants between 1954 and 1980 is only a little more than 3% of the total that the Foundation awarded for computer research through 1980. (See Table VI.) Only numerical analysis grants were smaller on average. In the early days of Foundation-sponsored research, however, theoretical com-

puter engineering was more significant. In 1961 the Foundation committed $300,000, nearly 60% of its computer research budget, to this field—mainly for systems theory research. Of all theoretical computer engineering categories, systems theory gained the most Foundation support during the 1960s. Most of these funds went to the University of California at Berkeley, where Lofti Zadeh and Charles Desoer worked.

In the 1970s, as the Foundation's budget for other areas of computer science ballooned, support for theoretical computer engineering still increased slightly. Emphasis shifted during this decade from systems theory to information and coding theory. As applications to data communications and cryptography grew during the 1970s, the Foundation increased its support. Between 1975 and 1980, 70% of funds for theoretical computer engineering went to information and coding theory.

Lofti Zadeh's career exemplifies the first phase of Foundation support for theoretical computer engineering. His research began before the Foundation was founded,

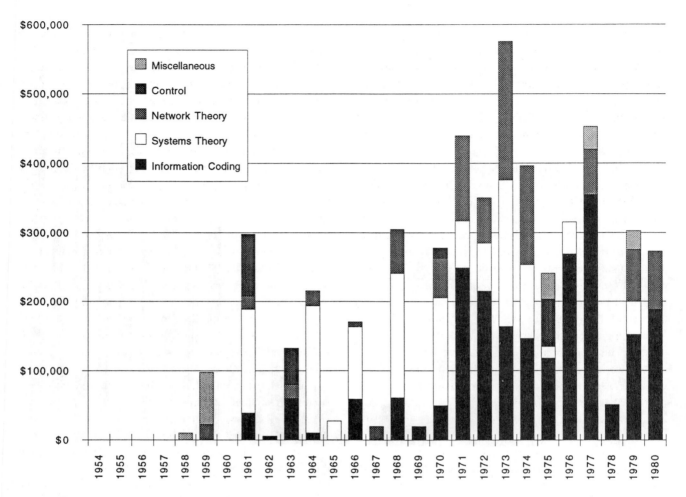

TABLE VI   Annual Funding for Theoretical Computer Engineering Research.

when he was an electrical engineering doctoral candidate at Columbia University in the late 1940s. There he studied systems analysis, creating mathematical theories to explain the behavior of systems of electrical devices. His 1949 dissertation introduced the concept of time-variant transfer functions (to replace time-invariant ones). His concept permits analysis of complicated systems, such as long-distance communications with multiple amplifiers and filters, by a few simple measurements.[41] In 1949 he accepted a faculty position at Columbia University, where he continued to work on systems analysis and related areas.

During the 1950s Zadeh substantially contributed to systems theory and became a leading practitioner in this engineering field. In 1950 he and John R. Ragazzini generalized a theory of time-series prediction proposed by Norbert Wiener two years earlier.[42] Two years later, in a work on sampled-data systems, they introduced a z-transformation that made analysis of such systems similar to the analysis of more familiar linear time-invariant systems.[43] In both cases, they explored uncertain events with probabilis-

tic methods—a tool that figured prominently in Zadeh's later research.

Zadeh's first Foundation support in 1957 resulted in the book *Linear Systems Theory* (1963), coauthored with Charles Desoer.[44] Its state-space approach became the standard technique for optimizing linear control systems. Zadeh also developed the theory of non-linear systems. By showing how they could be organized hierarchically, he facilitated the design of non-linear filters and predictors that were more effective than their linear counterparts. This work was applied to military equipment for signal processing, fire control, and trajectory calculations.

Although he early incrementally improved systems analysis, Zadeh was pessimistic about the applicability of existing systems analysis approaches to highly complex systems.[45] As early as 1961 he began to think that the richness of humanistic systems stemmed from their imprecise judgements and shades of interpretation—an imprecision unknown in his electromechanical systems. He became convinced that an adequate approach to the analysis of complex systems would involve the flexibility to handle intermediate, indefinite quantities. Instead of models of complex systems circumscribed by rigid quantifiability, he wanted to introduce variables that have imprecisely defined values. Zadeh's reputation as an innovative theorist enabled these ideas to get a hearing. With NSF support, he formalized his ideas on a new system of control, which he described as "fuzzy."

Zadeh first described fuzzy sets in 1965 and thereafter worked exclusively in that area.[46] In the early 1970s, he refined its concepts, seeking to construct a system that functioned analogously to human reasoning. The unsharp boundaries of fuzzy sets allowed for uncertainty, a quality Zadeh was unable to integrate satisfactorily into system models using conventional control theory. His 1968 paper in *Journal of Mathematical Analysis and Applications* integrated probabilistic measures of events into his fuzzy approach.[47] Working with Richard Bellman in 1970 he developed a theoretical underpinning for fuzzy sets.[48]

As his understanding of fuzzy sets deepened, Zadeh discovered unanticipated applications. Originally, he wanted to develop systems that mimicked human reasoning. In a 1972 paper in *IEEE Transactions on Systems, Man, and Cybernetics* he proposed using fuzziness to improve control of electromechanical systems.[49] Within a year he expanded the possibilities of fuzzy control through the introduction of linguistic variables, natural-language words such as "hot" or "tepid" or "chilly" that embraced an imprecisely defined range of quantitative values. Linguistic variables simplified control of electromechanical tasks because they made it easier to describe the way the task was to be performed. By 1976 Danish engineers were using fuzzy set theory to govern the operation of cement kilns, and today that method of control dominates this industry.[50]

Several other industries, including consumer electronics, automobiles, and home appliances such as washing machines and air conditioners, adopted fuzzy control in the following years. Almost all applications occurred outside of the United States, many in Japan.[51] Skepticism of US industry about applications of fuzzy control systems mirrored the skepticism of US academic researchers about the theory.

From its inception, fuzzy logic (as the field came to be called) was controversial,[52] and this controversy illustrates the weakness of the grant system employed by NSF. Many researchers in systems analysis denigrated the idea of fuzzy logic. With its reliance on the peer review system, NSF had a low chance of finding sympathetic reviewers. If an application was submitted by an established researcher, reputation alone sometimes carries a proposal. Young re-

searchers, no matter how talented, do not gain Foundation support for controversial research topics. For example, Enrique Ruspini, who is now at the Stanford Research Institute and contributed to fuzzy logic, failed to win support from the Foundation at the beginning of his career.[53]

Zadeh, however, continued to enjoy generous support. The Army Office of Research, NASA, and primarily NSF provided the funding into the 1980s. Zadeh continued to work prolifically on fuzzy logic. He adapted fuzzy logic principles to expert systems and natural language processing. In these efforts, he took advantage of the flexibility of fuzzy systems to bestow them with commonsense knowledge by creating techniques for expert systems to represent imprecise information and combine evidence.[54] In 1986, for example, he proposed a way to represent meaning in natural language systems. With these advances, he led a growing community of researchers investigating fuzzy logic.[55] The Japanese practice of incorporating fuzzy logical control into consumer products is now making academic and industrial computer engineers in the United States reexamine their attitudes towards fuzzy logic.

Critics often ask why the federal government should support theoretical investigations such as those in computer engineering theory. Implicit in this criticism is the view that these studies do not address practical national needs. Zadeh's work illustrates the potential practical value of this theoretical research. So does our next case case regarding Martin Hellman's Foundation-supported research on applications of information theory to cryptology.

In 1948 when Claude Shannon published his seminal paper on information theory, information theory and cryptography were closely associated because of the importance of transmitting secure messages during the Second World War.[56] Research on data compression, error correction, and error detection continued after the war, but public work on cryptography waned. Most cryptographic research was classified and pursued by government organizations, such as the National Security Agency (NSA). Only in the late 1960s was cryptography supported outside these federal organizations.

One pioneer, Martin Hellman, helped legitimize cryptography as an academic research subject. Upon obtaining his doctorate from Stanford in 1968, he worked for a year at International Business Machines's research center in Yorktown Heights, New York, just as encryption research began there. He did not actively participate in this research but was aware of the effort. The next year Hellman left for Massachusetts Institute of Technology, where he discussed information theory with Peter Elias, a former colleague of Shannon. Elias showed Hellman Shannon's 1949 paper, which made clear to Hellman several connections between cryptography and information theory.[57] Hellman suspected that Shannon's information theory was motivated by his work on cryptography. When he returned to Stanford in 1971, he began to consider cryptography.

Hellman explored several areas of information theory. With support from the Foundation, he examined error-detecting codes, data compression, and multi-access channels—so-called "aloha" channels developed in 1970 by Norman Abramson, which were predecessors to the now-popular Ethernet.[58] This work stimulated ideas about encryption, which he studied in his spare time. He soon published a small paper on the subject and, pleased with his results, thought about concentrating his research on cryptography.[59] Many colleagues advised him against this. The National Security Agency had a virtual monopoly on cryptography research, and its results were classified.[60] A researcher working outside NSA could expect to inadvertently duplicate its research or have new results appropriated and classified by the agency. In either case, prospects of advancing this field did not appear promising. Undaunted, Hellman approached NSF and described his projected work. With Foundation permission to change research topics in mid-grant, he turned to cryptography full time. In 1976 with Whitfield Diffie, he published a landmark paper entitled "New Directions in Cryptography."[61]

This paper suggested an alternative to the basic system by which all coded messages were encrypted. A message to be coded was traditionally operated on by a key cipher, which transformed the elements of the message. The recipient of the coded message then used the same key to unscramble the message into its original form. Only someone with the correct key could decipher the message, but anybody who knew the key could do so. The Digital Encryption Standard, created by IBM and accepted as the standard for electronic information encryption by the National Bureau of Standards in 1976, was based on this principle.

This system had two important shortcomings. It was necessary to distribute copies of the key to everyone who should have access to the coded messages, while at the same time denying access to people for whom the message should remain secret. The military had traditionally communicated keys by courier, but this was infeasible for large groups of people communicating on electronic networks. The other problem was authentication. The conventional systems offered no way to prevent the reader, who knows the key, from sending himself an encrypted message and falsely claiming it was sent by the transmitter.

Hellman and Diffie proposed an encryption scheme that resolved both of these issues. They used two different keys on the message: one to encode, the other to decode. The keys were inverses of one another; that is, applying one key would undo the coding of a message done with the other key. They were chosen so that it was very difficult to derive one key by knowing the other. One key could be made public, allowing anyone to encrypt a message decipherable only to the person who had the partner key, which was kept private. Because keys were not inverses of themselves, knowledge of the public key was of no use in decoding. Anyone wanting to send a message simply looked up the recipient's public key in a directory, encoded the message using the public key, and felt confident that only the recipient with the private key could decipher it. The plan eliminated the complications of distributing shared keys. It also guaranteed the authenticity of a message. If a person encrypted something with a private key, no one was able to add to or change the message in the code in which it had originally been ciphered.

The coding system was only as foolproof as the code was unbreakable. The question of the security of a public key system was of paramount importance, and the call to investigate it generated substantial research in related disciplines. A system for creating pairs of keys was invented that made it easy to generate random inverse pairs but very difficult to derive the private key from the public one. Computer scientists studied problems with this property when studying the NP class of problems. Prior to Hellman's paper, the practical value of computational complexity was unknown. The public key system posed a problem of practical significance, which invigorated the field and gave it focus. It was essential to prove that the calculation of a private key from a public one belonged to that class of intractable problems in order to have confidence in the coding system's security.

Hellman contributed to this work, but not entirely by choice. His research agenda was circumscribed by agencies funding him. Because NSA controlled military spending in cryptography, funding that Hellman received from the Air Force Office of Scientific Research could not be used for cryptographic research.[62] Only later was Hellman allowed to spend Air Force money on mathematics research related to cryptography. A believer in freedom of research, NSF program director Frederick Weingarten resisted NSA pressure to curtail NSF funding for Hellman's cryptography work.[63]

Some problems Hellman could not work on directly at all. He wanted to appraise the security of IBM's Data Encryption Standard, for example.[64] He believed that NSA had deliberately weakened it during the standards adoption process in order to ensure that it could break any code encrypted with this standard.[65] NSF could not fund work related to a standard, however, and Hellman's funding from the military could not be used for cryptographic research. Seeking to maintain his funding base, he could approach the question only obliquely, doing NSF-supported research on the security of codes. He sought short-cuts to exhaustive, trial-and-error methods to break codes and derived a set of cryptanalytic time-memory-processor tradeoffs that cut the resource costs of code-breaking.[66]

There was some delay between the proposal and implementation of public key cryptosystems. In 1978 MIT researchers Ronald Rivest, Adi Shamir, and Leonard Adleman proposed the first function that could implement these public key codes.[67] Their research, also supported in part by NSF, resulted in the creation of the RSA cipher, a public key cryptosystem based on exponentiation in modular arithmetic. That same year, Hellman and his student Ralph Merkle proposed a different public key cryptosystem based on the so-called "knapsack" problem, which was considered computationally too complex to solve.

Despite these successes, business clients were slow to employ the public key systems.[68]

This is now changing, however, as the explosive growth of electronic communications over networks or by radio transmission increases the desire and need for privacy and authenticity. Banks, in particular, have accepted public key cryptosystems to conduct electronic funds transfers securely. Widespread implementation of public key cryptosystems in the mid-1980s coincided with the cessation of Foundation support for Hellman's work. The Foundation had thus supported basic theoretical research and an important initial application of it. Once the research was complete, the Foundation left the development and implementation to others.

Facilities for Theoretical Computer Science

Although theoretical computer scientists did not make extensive computations in their research, they did frequently use computer communications facilities. The most practical design implemented for communication among a large number of computers separated over long distances is packet switching, which uses low-cost, high-speed computers and memory buffers to route messages through the network. Messages are broken into packets, each containing the address of the recipient. The packets are routed through the network independently of each other, then reassembled at the destination. With this method, transmission is hundreds of times more efficient than another allocating a communications link for the entire duration of

a single message transmission. Conceived in 1964, packet switching was proven in a wide-area network known as the ARPANET, which was implemented by DARPA in 1969 to connect researchers at government laboratories, commercial defense contractors, and selected academic computer science departments.[69]

Large companies such as IBM and General Electric established private networks, and publicly available commercial packet network services were introduced by TELENET in 1975, soon followed by TYMESHARE. The computer science community quickly saw potential benefits of computer communications and developed a view of building a networked nation, which NSF computer program staff shared. John Pasta, head of the NSF Division of Computer Research, provocatively asserted in 1975 that "as we see it, the university computer center as it is now constituted, is doomed," for he anticipated that large, stand-alone, centralized facilities relying on one or two mainframe computers would be replaced by local networks of interactive minicomputers, connected by packet switching to other clusters of computers dispersed across the nation and around the world.[70] NSF's Computer Science and Engineering Advisory Panel recommended in 1974 that the Foundation provide access for qualified computer scientists to an international computer network, which the Foundation staff accepted as their highest priority.[71]

The federal Office of Management and Budget stymied Foundation efforts to connect computer scientists through networks because it thought such a plan too expensive. The OMB believed that building computer networks should be left to the private sector.[72] The NSF staff was dismayed at this attitude. If it could not build a system, it was still going to study the question. It countered with a research program into practical economics and organizational strategies for sharing computer resources across networks connecting independent institutions. The problem was exacerbated by the power of ARPANET. It was so effective a research tool that it created tensions between computer scientists who had access to it and those who did not; "because communication over the network was convenient, researchers with ARPANET access tended to interact among themselves and ignore those without access."[73]

Many theoretical computer scientists did not have access to ARPANET, and they eventually provided NSF with an opportunity to sponsor computer networks. In 1977 Lawrence Landweber, head of the Computer Science Department at the University of Wisconsin, organized Theorynet, a low-budget electronic mail facility for theoreticians. With NSF funding, Theorynet provided members with a mailbox on a central computer at the University of

Wisconsin, accessed by terminals over dial-up phone lines or through the public packet-switched networks. Theorynet substantially increased collaboration among the more than 100 theoretical computers scientists around the world who used it.

In order to overcome a Theorynet limitation, NSF entered into the full-scale networking business. Theorynet was limited to message transfers and did not allow the more sophisticated on-line computing or use of remote computing facilities that was useful, especially in experimental computer science research (not so much in theory). In 1979 Landweber proposed that NSF support his request for hardware and pay the connection fees to link the University of Wisconsin computer to ARPANET. The ensuing series of proposals culminated in the CSNET, a network available to all researchers in computer science departments across the United States that did enable file transfers and remote log-ins. By piggybacking CSNET on ARPANET, the Foundation established it at a cost of less than three million dollars. NSF skirted OMB opposition by burying the project inside a larger program for funding equipment in computer science departments. By this time a national shortage of trained computer faculty was recognized, and NSF took this situation to justify undertaking a more aggressive networking initiative.

The success of CSNET paved the way for the more expensive NSFNET in mid-1980s. By the early 1980s computer research and development was regarded as essential to US national competitiveness. The Foundation was authorized to establish five national supercomputer centers and a communications network to connect academic, government, and industrial laboratories to these supercomputers. The NSFNET, originally established to provide access to the supercomputers, was subsequently expanded to interconnect most academic and industrial research centers engaged in government-funded research in all scientific areas, not just computer science. The NSFNET became the backbone of the international network of computer networks known as the Internet, whose technology was developed by DARPA. Thus, NSF's modest effort to make it possible for theorists to communicate by electronic mail led the Foundation into a full-scale national networking effort.

Industrial Reaction and NSF Response

Foundation staff were proud of the progress made in the 1960s and 1970s to develop a science of computing. A measure of how far computing had come as a science was its theoretical basis. As NSF staff member Thomas Keenan remarked on NSF's emphasis on theoretical investigations

rather than empirical research, systems development, or applications:

> Computer science had achieved the title 'computer science' without much science in it, early ... [my colleague Kent Curtis and I] decided that to be a science you had to have theory, and not just theory itself as a separate program, but everything had to have a theoretical basis. And so, whenever we had a proposal ... we encouraged, as much as we could, some kind of theoretical background for this proposal—not [for example] just software, and not just write a program, but there should be some basis for it.[74]

Of course, theoretical computer science and engineering were among the most theoretical subdisciplines of computer science, and NSF had been the leading sponsor of these subjects. This desire to define a science of computing and a theory-based discipline of computer engineering fit the Foundation's budgets, which could not afford capital-intensive systems and applied computer research—as supported by DARPA and the industrial sector.

Although the Foundation's programs made some headway in defining a science of computing, other serious problems loomed for computing that had to be addressed. By 1977 academic and industrial circles widely believed that a national crisis prevailed. The annual production of US computer science doctorates had begun to fall off.[75] Some government and industrial computer positions went unfilled, but this was minor compared to the acute shortage in academia. In previous decades, academic researchers had made fundamental contributions to systems design and other experimental research, such as timesharing and networking. These materially benefited industry, the military, and the nation as a whole. But that experimental line of research in the academic sector was perceived as having been replaced with theoretical studies having little practical benefit.

To address this issue, the Foundation convened a study group in 1978 under the direction of Jerome A. Feldman of the University of Rochester. The committee found that the prevailing cause was a movement of faculty and graduate students into industry. In effect the community was eating its "seed corn" for the next generation of computer scientists.[76] The attraction of industrial positions was partly the higher salaries, but mainly the availability of good laboratory equipment. The NSF computing staff had been forced by the Office of Management and Budget to terminate its facilities programs to colleges and universities in the early 1970s, and universities permitted their computing facilities to become obsolete.[77] The Feldman Report recommended that NSF make multimillion dollar awards to 25 schools over a five-year period, so that each of these schools could become a "center of excellence" with good facilities, a critical mass of researchers, and an active experimental research program.

The Association for Computing Machinery, one of the major professional organizations, dissented from these recommendations because they believed the remainder of the academic community would suffer if funding was concentrated in twenty-five institutions. In preparing the Foundation's response, NSF Director Richard Atkinson queried 25 leading industrial computer research directors on how to make the most effective use of Foundation support. The research directors favored sponsorship of experimental research facilities, large research projects, and opportunities to attain "critical mass." IBM's Chief Scientist, Lewis Branscomb, concurred by giving his lowest priority "to the small project grant which tends to proliferate the relatively abstract theoretical projects that have characterized much of academic computer science in the past."[78] Thus these theoretical research grants, esteemed by staff and some of the academic community, were disdained by some in the industrial community.

In 1980 the Foundation launched its Coordinated Experimental Research Program, a scaled-down version of the "Centers of Excellence" program recommended in the Feldman Report. It was consonant with the Foundation's budget reality. The original plan was for schools to submit proposals for large, single-focused, multi-investigator research projects in keeping with the objective of establishing a critical mass. The program quickly devolved, however, into a facilities program. It is too early to fully evaluate the program, but it seems to have partly met its objectives. Cornell and Wisconsin used CER grants to develop experimental research strength to complement their existing strength in theory. CER also helped transform the Universities of Washington and Massachusetts into prominent research institutions in computer science by providing equipment and funding that attracted good students and faculty. Clearly, CER lessened the premium NSF placed on theoretical studies in computer science.

Conclusions

The areas discussed in this paper such as information theory and computational complexity, are typical of the fruitful mathematical subjects that arose as a result of a new technology. At first the technology existed largely independently of theory, and there was no pressing practical reason

to develop a theory. Some academic researchers believed, however, there was value in constructing a science of computing. Although it is not fully explored in this paper, there seem to be strong social reasons behind this drive to mount a science. Apparent social reasons are the greater status society proffers those who study an applied science instead of merely building or operating machines, the greater government funding for applied scientific research rather than mere tinkering, and the justification of research efforts to colleagues and university administrators.

Most government agencies support academic researchers only to carry out assignments that advance agency objectives. Atypically, at NSF the academic research community was influential in determining the agency's objectives and making decisions about how they were carried out—through selected scientific advisory committees and peer review. Thus, NSF's goals largely followed the academic research community's goals. Not surprisingly, NSF supported the development of theoretical computer science and engineering.

Congress and the White House have only modestly supported the academically defined goals for the NSF program, as witnessed by the modest budgets allocated to programs such as university computer facilities or individual research grants in comparison to those given to defense agencies for their research. When NSF goals resonated with national interests held by Congress or the White House, NSF funding rapidly increased. This happened when President Johnson promulgated his computer education initiative in 1967. It happened again, to a certain extent, with CER in 1980, and again in the mid-1980s, when the NSF budget was vastly increased to support supercomputer centers that were viewed as improving national competitiveness.

Although NSF also sponsored computer programs in education, facilities, and information science, theoretical computer science and engineering gained most from the research program.

NSF's supporting role in the creation of model curricula in the 1960s may have been beneficial, and the Theorynet electronic-mail system did enhance communication among theory researchers. But it was the individual research grants that grew and sustained a research community and hence enabled a body of results to be produced. The value of this support is seen in the contributions made by such researchers as Martin Davis on computability theory, Arthur Burks on cellular automata, Juris Hartmanis on automata theory, Manuel Blum on computational complexity, Richard Karp on NP-completeness, Lofti Zadeh on fuzzy logic, and Martin Hellman on applications to cryptology.

References

1. Theoretical computer science includes such subjects as automata theory, decidability theory, computational complexity, programming semantics, and neural networks. Theoretical computer engineering includes information and coding theory, systems theory, network theory, control theory, and related subjects.

2. For background on this subject, see Kenneth Flamm, *Creating the Computer* (Washington, DC: Brookings Institution, 1988).

3. For background on this subject, see William Aspray and Michael Gunderloy, "Computing and Numerical Analysis at the National Bureau of Standards, *Annals of the History of Computing*" (January 1989): 3–12.

4. For more information about these program areas, see: W. Aspray and B.O. Williams, "Computing in Science and Engineering Education: The Programs of the National Science Foundation:, IEEE *Electro/93 Proceedings*, 1993; ———, "Arming American Scientists: The Role of the National Science Foundation in the Provision of Scientific Computing Facilities," *Annals of the History of Computing*, forthcoming; Bernard O. Williams, "From Machine Translation to Information Science: Programs of the National Science Foundation" (forthcoming).

5. The Mansfield Amendment was an amendment to the 1970 US military procurement law, named after Senator Mike Mansfield of Montana. The amendment made it unlawful for the defense department to sponsor basic scientific research unless it was clearly related to military objectives. As a result, military and even federal agencies not covered by the amendment stripped themselves of basic research projects of questionable relation to their missions, and the NSF became the recipient of many of these projects and the funds associated with them.

6. The reason for this increase is uncertain. It could be a response to the energy crisis or a willingness of the liberal Congress elected in 1974 to support the Foundation. Perhaps it was the first measure of John Pasta and Kent Curtis's success at increasing research support.

7. The average Foundation grant was remarkably consistent over time. In the early 1960s, investigators were given close to $40,000 for their work. This figure increased to $60,000 by 1980. This suggests, after adjusting for inflation, some decline in the real value of Foundation research grants over time. It is important to remember, however, that computing power, purchased either as hardware or as time leased on someone else's computer, grew cheaper year after year, in opposition to the trend of prices of items that constitute the Consumer Price Index. Because computing power represented a significant cost in research budgets, standard inflation measures are an unreliable adjustment for determining the purchasing power of the grants.

8. Historically, the organization of computer fields changed over time. For example, in the 1950s and early 1960s the Foundation staff regarded theoretical computer science as including numerical analysis, computational mathematics, formal languages, and intelligent systems, as well as areas we today regard as part of theoretical computer science, such as complexity and automata theory. In the coding of grants and the preparation of these charts, we have taken a presentist approach that carves up computer science into subdisciplines as it was in the late 1980s. Thus, our category of theoretical computer science does not include numerical analysis, computational mathematics, formal languages, or intelligent systems.

9. Foundation staff occasionally set its own research agenda, for example a program in networking in the late 1970s. The way in which NSF staff divided up its programs and allocated budgets also shaped the research agenda somewhat.

10. Turing, A.M. "On Computable Numbers, with an Application to the Entscheidungsproblem," *Proc. of the London Math. Soc.*, series 2, 42 (1936–37): 230–265.

11. Constance Reid, *Hilbert* (New York: Springer Verlag, 1970) and Ronald Calinger, ed., *Classics of Mathematics* (Oak Park, Ill.: Moore Publishing, 1982), pp. 655–678.

12. For related work, see Martin Davis, Hilary Putnam, and Julia Robinson, "The Decision Problem for Exponential Diophantine Equations," *Annals of Mathematics* series 2, 74 (1961): 425–436.

13. Y. Matiyasevic, "Diophantine Representation of Enumerable Predicates," *Izv. Akad. Nauk. SSSR*, ser. Matem. 35 (1971): 3–30.

14. W.S. McCulloch and W. Pitts, "A Logical Calculus of the Ideas Immanent in Nervous Activity," *Bulletin of Mathematical Biophysics* 5 (1943): 115–133. See William Aspray, *John von Neumann and the Origins of Modern Computing* (MIT Press, 1990), Chapter Eight, for a discussion of von Neumann's theory of automata and for citations to his various published and unpublished papers on the subject.

15. Original members of the Logic of Computers Group were Burks, Jessie Wright, and Don Warren. They were joined in 1955 by John Holland and Cal Elgat.

16. The computer was the Burroughs Truth Function Calculator. See A. Burks and Warren Wright, "An Analysis of a Logical Machine Using Parenthesis-Free Notation," *Mathematical Tables and Other Aids to Computation* 8 (1954): 52–57.

17. Private correspondence, Burks to Andrew Goldstein, 30 March 1992.

18. Oral history interview with Burks, conducted by Andrew Goldstein, July 29, 1991, Archives, IEEE Center for the History of Electrical Engineering.

19. See John von Neumann, *Theory of Self-Reproducing Automata* Edited and completed by Arthur W. Burks (Urbana: University of Illinois Press, 1966).

20. Burks, oral history interview. Also see Bernard P. Zeigler and Roger Weinberg, "System Theoretic Analysis of Models: Computer Simulation of a Living Cell," *Journal of Theoretical Biology* 29 (1970): 35–56.

21. See, for example, L.B. Brooker, E.E. Goldberg and J.H. Holland, "Classifier Systems and Genetic Algorithms," *Artificial Intelligence* 40 (1989): 235–282; also J.H. Holland, *Adaptation in Natural and Artificial Systems* (Ann Arbor: University of Michigan Press, 1975).

22. See, for example, J. Holland, Keith J. Holyook, Richard E. Nisbett, Paul R. Thagard, *Induction: Processes of Inference, Learning, and Discovery* (Cambridge, MA: MIT Press, 1986): Chapter 4.

23. See, for example, C. Langton, "Studying Artificial Life with Cellular Automata," *Physica D* 2D (October-November 1986): 120–149; C. Langton, ed., *Artificial Life* Santa Fe Institute Studies in the Sciences of Complexity (Reading, MA: Addison-Wesley, 1989).

24. Juris Hartmanis, "Observations About the Development of Theoretical Computer Science," *Annals of the History of Computing* 3 (January 1981): 46.

25. J. Hartmanis and R. E. Stearns, "On the Computational Complexity of Algorithms," *Trans. of the Amer. Math. Soc.* 117, 5 (May 1965): 285–306.

26. After computational complexity theory became well established, a number of companies took an interest. Most of the fundamental work was carried out in the academy, however.

27. Hartmanis brought to Cornell John Hopcroft, Bob Constable, David Gries, Ken Brown, John Dennis, Ellis Horowitz, George Mor, Alan Shaw, Bob Tarjan, Peter Wegner, and others. See Alan Borodin, "Juris Hartmanis: Building a Department—Building a Discipline," in Alan L. Selman, ed. *Complexity Theory Retrospective* (New York: Springer Verlag, 1990).

28. The papers are Hartmanis, J., and J. Hopcroft, "What Makes Some Language Theory Problems Undecidable," *Journal of Computer and System Sciences* 4, 4 (August 1970): 368–376, and Hartmanis, J., and J. Hopcroft, "An Overview of the Theory of Computational Complexity," *Journal of the ACM* 18 (1971): 444–475. An appraisal of their significance is found in Alan Borodin, "Juris Hartmanis: Building a Department—Building a Discipline," in Alan L. Selman, ed. *Complexity Theory Retrospective* (1990), 22.

29. P. Fischer, J. Hartmanis, and M. Blum, "Tape Reversal Complexity Hierarchies," *IEEE Conference Record of the 1968 Ninth Annual Symposium on Switching and Automata Theory,* (October 1968), 373–382.

30. The Hartmanis-Baker conjecture and the Berman-Hartmanis conjecture are described in Paul Young, "Juris Hartmanis: Fundamental Contributions to Isomorphism Problems," in Alan L. Selman, ed. *Complexity Theory Retrospective* (New York: Springer Verlag, 1990), 28–58.

31. NSF grant GP-2495. Other support came from the Joint Services Electronics Program, NIH, NASA, and the US Air Force. The thesis title was "A Machine-Independent Theory of the Complexity of Recursve Functions, MIT, 1964. See published versions: Blum, Manuel, "A Machine-Independent Theory of the Complexity of Recursive Functions," *Journal of the ACM* 14 (1967): 322–336.; "On Effective Procedures for Speeding Up Algorithms," *Journal of the ACM* 18 (1971): 290–305.

32. Manuel Blum, "A Machine-Independent Theory of the Complexity of Recursive Functions," *Journal of the ACM* 14 (1967): 322–336.

33. See, for example, S. Winograd, "On the Time Required to Perform Multiplication," IBM Research Report RC-1564, 1966.

34. See M. Blum, "Independent Unbiased Coin Flips From a Correlated Biased Source—A Finite State Markov Chain," *Combinatorica* 6 (1986): 97-108; M. Blum, Alfredo De Santis, Giuseppe Persiano, "Noninteractive Zero-Knowledge," *SIAM Journal of Comp.* 20 (December 1991): 1084–1118.

35. See, for example, K.-P. Ossenkopf and R. Barbeito, "Bird Orientation and the Geomagnetic Field: A Review," *Neurosci. Behavi. Rev.* 2 (1978): 255–270; P. Semm, D. Nohr, C. Demain, and W. Wiltschko, "Neural Basis of the Magnetic Compass: Interactions of Visual, Magnetic and Vestibular Inputs in the Pigeon Brain," *Journal of Comparative Anatomy A* 155 (1984): 283–288.

36. J.T. Gill, "Probabilistic Turing Machines and Complexity of Computation" Ph.D. dissertation, Department of Mathematics, University of California, Berkeley, 1972. Testimony to this dissertation's significance can be found in Karp, Richard M., "Combinatorics, Complexity and Randomness—Turing Award Lecture," *Communications of the ACM* 29, 2 (February 1986): 108.

37. Gary Miller, "Riemann's Hypothesis and Tests for Primality," *Journal of Computer and System Sciences* 13, (1976): 300–317.

38. See S. A. Cook, "The Complexity of Theorem-Proving Procedures," *Proceedings of the Third ACM Symposium on Theory of Computing* (1971): 151–158.

39. Karp investigated a number of related research questions that we will not describe: design of a probabilistic algorithm to test for pattern matching, randomized algorithms in space-bound computation, combinatorial enumeration, parallel computation of matchings and maximal independent sets, data structures, and load balancing and on-line computation.

40. Richard M. Karp, "Probabilistic Analysis of Partitioning Algorithms for the Traveling-Salesman Problem in the Plane," *Mathematics of Operations Research* 2 (August 1977): 209–224.

41. See, for example, a published version in L. Zadeh, "Frequency Analysis of Variable Networks," *Proceedings of the IRE* 38 (March 1950): 291–299.

42. Lofti Zadeh and John R. Ragazzini, "An Extension of Wiener's Theory of Prediction," *Journal of Applied Physics* 21 (July 1950): 645–655.

43. Zadeh, Lofti, and John R. Ragazzini "Optimum Filters for the Detection of Signals in Noise," *Proceedings of the IRE* 40, 10 (October 1952): 1223–1231.

44. Lofti Zadeh and Charles A. Desoer, *Linear system theory; the state space approach* (New York, McGraw-Hill, 1963).

45. Lofti Zadeh, telephone interview with Andrew Goldstein, July 24, 1991.

46. ——, "Fuzzy Sets," *Information and Control* 8, 3 (June 1965): 338–353.

47. ——, "Probability Measures of Fuzzy Events," *Journal of Mathematical Analysis and Applications* 23 (1968): 421-427.

48. R.E. Bellman and Lofti Zadeh, "Decision-Making in a Fuzzy Environment," *Management Science* 17 (1970): B-141-B-164.

49. Lofti Zadeh, "Outline of a new Approach to the Analysis of Complex Systems and Decision Processes," *IEEE Transactions on Systems, Man, and Cybernetics* SMC-3, 1 (January 1973): 28–44.

50. Zadeh interview, July 24, 1991. Fc. other actual and potential applications of fuzzy control, see B.R. Gaines and L.J. Kohout, "The Fuzzy Decade: A Bibliography of Fuzzy Sys-

tems and Closely Related Topics," *International Journal of Man-Machine Systems* 9 (1977): 1–68; also *Proceedings of the 1977 IEEE Conference on Decision Control, including the 16th Symposium on Adaptive Processes and a Special Symposium of Fuzzy Set Applications,* where authors detail applications to curve fitting, image analysis, inference, medical diagnosis, failure diagnosis, electromagnetic interference source identification, heat exchange, traffic control (on the Japanese railroad), and "phases of measurement, evolution and control of a communication-and-formation process of Morality concepts."

51. Kevin Self, "Designing with Fuzzy Logic," *IEEE Spectrum* 27, 11 (November 1990): 42–44, continued on 105.

52. See Jim Bezdek, "Editorial: Fuzzy Models—What are They and Why?" *IEEE Transactions on Fuzzy Systems* 1 (February 1993): 1–6.

53. Zadeh interview, July 24, 1991. For Ruspini's contributions, see for example, E. Ruspini, "A New Approach to Clustering," *Information and Control* 15 (1969): 22–32; and "New Experimental Results in Fuzzy Clustering," *Information Science* 6 (1973): 273–284.

54. Lofti Zadeh, "A Computational Approach to Fuzzy Quantifiers in Natural Languages," in Cercone, Nick, ed. *Computational Linguistics* (New York: Pergamon Press, 1983).

55. Lofti Zadeh, "Test-Score Semantics as a Basis for a Computational Approach to the Representation of Meaning," *Literary and Linguistic Computing* 1, 1 (January 1986): 24-N35.

56. C. Shannon, "A Mathematical Theory of Communication," *Bell System Technical Journal* 27 (1948): 379–423, 623–658.

57. C. Shannon, "Communication Theory of Secrecy Systems," *Bell System Technical Journal* 28 (October 1949): 656–715.

58. Norman Abramson, "Development of the ALOHANET," *IEEE Transactions on Information Theory* IT-31 (March 1975): 119–123.

59. See M.E. Hellman, "An Extension of the Shannon Theory Approach to Cryptography," *IEEE Transactions on Information Theory* IT-23 (May 1977): 289–294. Presented at the IEEE International Symposium on Information Theory, Notre Dame, Indiana, October 27–31, 1974.

60. John Cherniavsky, interviewed by William Aspray, Washington, D.C., September 28, 1991.

61. Whitfield Diffie and Martin Hellman, "New Directions in Cryptography," *IEEE Transactions on Information Theory* IT-22, 6 (November 1976): 644–654.

62. A discussion of NSA's policies on cryptography research can be found in James Bamford, *The Puzzle Palace : a Report on America's Most Secret Agency* (Boston: Houghton Mifflin, 1982).

63. Richards Adrion, interviewed by William Aspray, Amherst, MA., 29 October 1990.

64. See Hellman, Martin, R. Merkle, L. Schroeppel, W. Washington, W. Diffie, S. Pohlig, and P. Schweitzer, "Results of an Initial Attempt to Cryptanalyze the NBS Data Encryption Standard" Electrical Engineering Department, Stanford University, Stanford, CA, SEL 76-042, September 9, 1976; and Diffie, W., and Martin Hellman, "Exhaustive Cryptanalysis of the NBS Data Encryption Standard," *Computer* 10, 6 (June 1977): 74–84.

65. Martin Hellman, telephone interview with Andrew Goldstein, July 31, 1991. Also see Hellman, M., "DES will be Totally Insecure within 10 Years," *IEEE Spectrum* 16 (July 1979): 32–39.

66. Martin E. Hellman, "A Cryptanalytic Time-Memory Trade-Off," *IEEE Transactions on Information Theory* IT-26, 4 (July 1980): 401–406.

67. Ronald Rivest, Adi Shamir, and L. Adleman, "On Digital Signatures and Public Key Cryptosystems," *Communications of the ACM* 21 (February 1978): 120–126.

68. See Martin Hellman, "Commercial Encryption," *IEEE Network Magazine* 1, 2, (April 1987): 6–10 for a discussion of issues involved in commercial acceptance of encryption standards.

69. Howard L. Resnikoff, "Implications of Advances in Information Science and Technology for Institutions at Higher Learning," *EDUCOM Bulletin,* v.16 no.1, Spring, 1981 pp. 2–10 ; also Arthur L. Norberg and Judy E. O'Neill with contributions by Kerry J. Freedman, *A History of the Information Processing Techniques Office of the Defense Advanced Research Projects Agency* Charles Babbage Institute Report, University of Minnesota, Minneapolis, October 1992.

70. John Pasta, "Overview, Trends, Issues," Directors Program Review Computer Research, NSF, May 20, 1975, p. 13.

71. Walter Sedelow, "Networking Research: Some Issues," Directors Program Review, Computer Science, May 20, 1975, NSF.

72. A.M. Noll, "Program Review and Evaluation of NSF Office of Computing Activities," May 29, 1973, copy in the office files of Thomas Keenan, CISE, NSF.

73. Douglas Comer, "The Computer Science Research network CSNET: A History and Status Report, Communications of the ACM, vol. 26 no 10, October, 1983 p. 747–753.

74. Keenan oral history interviewed by William Aspray, Washington, D.C., 28 September 1990. Charles Babbage Institute archives, University of Minnesota.

75. US computer science doctorates awarded were 244 in 1976, but fell to 216 in 1977 and 196 in 1978. By 1976 this trend was already clear from the number of students entering doctoral programs and taking qualifying examinations.

76. Jerome A. Feldman and William R. Sutherland, eds., "Rejuvenating Experimental Computer Science: A Report to the National Science Foundation and Others," *Communications of the ACM* 22 (September 1979): 497–502.

77. Kent K. Curtis et al..,"John R. Pasta, 1918-1981: An Unusual Path Toward Computer Science," *Annals of the History of Computing* 5,3 (July 1983): 224–238; John Pasta, "Introduction and Background, Director's Program Review Computing Activities, December 15, 1970, pp. 5–6, NSF.

78. Branscomb's letter was reproduced in Kent Curtis, "Computer Science" *Program Reports* 4, 2 (April 1980): 21–25, National Science Foundation.

III

Integration of History
with Mathematics Teaching

Fundamentals and Selected Cases

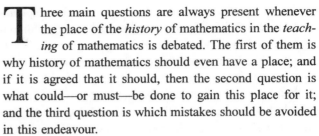

History of Mathematics and the Teacher

Torkil Heiede
*Royal Danish School
of Educational Studies*

hree main questions are always present whenever the place of the *history* of mathematics in the *teaching* of mathematics is debated. The first of them is why history of mathematics should even have a place; and if it is agreed that it should, then the second question is what could—or must—be done to gain this place for it; and the third question is which mistakes should be avoided in this endeavour.

To the first question there are many answers, all of them personal, see for example my 1992 paper in *The Mathematical Gazette* with its long list of references[1] or the even longer list in its Danish version in *Normat*.[2] Here the following general—and personal—remarks may suffice.

In many ways our time is ahistoric, that is: most people seem to live ahistoric lives. We focus so much upon living in the present moment, in the happiness or unhappiness of the present now, without thinking either on the past or on the future, that we deny their importance or even their existence and reality. Maybe since everything seems to change so rapidly—too fast for us to handle it—we grab at the moment and ignore change, so that each of us becomes like Zeno's arrow in mid-air, frozen, unmoving, and dead, because the past is over and done with and might never have been, and the future is unsecure and unsafe and may never come.

Many of us, especially many young people, have a deep contempt for both the past and the future, and both for everyone older and for everyone younger than ourselves. This might be a result not only of the rapid changes in almost everything but also simply of the way children and young people are educated in our time. They are sent to day nurseries, kindergartens, schools, colleges, and universities, and in all these institutions they are forced to be together mostly with others of precisely their own age for many hours every day, year after year. Such segregated schooling may make it all to easy for them to feel such an unreasoned and unreasonable contempt for all who are not of their own age—in both senses of the word "age."

But if one finds this wholly synchronic outlook not only narrow and constricting but even deadly, if one's outlook is more diachronic, then Zeno's arrow has not only a place—it also has a velocity, it moves, and somewhere in its past there may be a bow, and somewhere in its future there may be a target, that is: the arrow has a history. Then everything man touches has a history: the tree, the axe, the house, architecture, every other trade and art, every conviction, every subject— therefore also mathematics—and the history of a subject becomes an inseparable part of the subject itself. To say that if we teach mathematics properly we must somehow include its history is just another way of saying that mathematics is a living subject. One explanation of the fact that so many people— particularly children and young people in schools and colleges—find mathematics dull, boring, uninteresting, even hateful, could be that they were taught—or are now being taught—mathematics without its history, that is mathematics as if it were dead.

In his lecture at the History and Pedagogy of Mathematics session at the Sixth International Congress on Mathematical Education in Budapest in 1988, Robert L. Hayes stated:

> I believe that it is a grave mistake and error of strategy to attempt to teach mathematics without reference to its cultural, social, philosophical, and historical background.[3]

Yes—a grave mistake because mathematics is a living subject, and an error of strategy because when it is taught as if it were dead, it is killed for the students.

With this we can turn to the second of the three questions: what can be done to make the *history* of mathematics appear in the *teaching* of mathematics? Clearly the wish that the history of mathematics should in some way be present whenever mathematics is taught makes great demands on all mathematics teachers, and maybe quite new demands since math teachers are often as ahistoric in outlook as most everyone else. Of course one must try to ensure that the history of mathematics gets an important place in the education of future teachers, but this is not enough; something must also be done for the teachers who are already teaching. There are maybe 40 times as many of them as there are new teachers coming out every year, at least if you put the working life of a teacher at 40 years and trust in proportionality.

In the following I will survey the situation in Denmark, my own country. Not only because I know best our problems and the attempts to handle them, and not at all because I believe they are typical; they are surely not. But I hope that by learning of the concrete situation in my country, my readers will see a sidelight being cast on the problems in their own countries, which will throw them in relief and make them, and maybe thereby also the more general problems, easier to grasp. This is indeed what has often happened to me.

There is, however, a more timely reason for describing the situation in Denmark just now. The Government of Denmark has recently made obligatory that the history of mathematics appear in the teaching of mathematics in the upper grades of Danish schools, which has of course made the whole problem more pressing and has called for action here and now.

In 1988 the Danish Ministry of Education issued new regulations for the teaching of all subjects in the gymnasium, that is grades 10 to 12, which are now taken by 35% of the 17- to 19-year-olds, after nine or ten years in what is in Danish called the folkeskole, literally: the folk school. These regulations signal a wind of change through the gymnasium with respect to history both as a subject in itself and also as an intrinsic part of other subjects, among them mathematics which is taken as a major subject by two-thirds of all the students in the gymnasium and as a minor subject by the rest.

The new regulations have been published by the Ministry as a series of booklets, one for each subject in the gymnasium, the booklet which deals with mathematics having the title Matematik.[4] With respect to the content of mathematics as a major subject, the new regulations differ little from previous ones; this content is mainly the same as it has been for most of this century. In grades 10 and 11 it consists of the following five themes:

Numbers	Differential calculus
Geometry	Statistics and probability
Functions	

In grade 12 there are, for those who take mathematics in all three gymnasium years, three additional themes:

> Vector geometry in two and three dimensions;
> Integral calculus and differential equations;
> A theme in which mathematics and computers can cooperate.

What is really new in the regulations is what they say about the way in which these five plus three themes should be taught, namely with regard for what the Ministry calls the three aspects of mathematics. To take them in reverse order, the third aspect is called

The inner structure of mathematics;

one could also call it the deductive aspect, thereby meaning the whole edifice of definitions and proofs and theorems and examples and counterexamples, etc. The second of the three aspects is

The model aspect of mathematics;

one could also call it the aspect of mathematics that is applicable, in physics, chemistry, technology, geography, economics, the humanities; it is possible to interpret mathematical concepts, and to use mathematical structures, to describe or even to explain non-mathematical phenomena.

These two aspects are not very surprising, even if there is far more to them than can be conveyed if one just said pure and applied mathematics. But it is rather surprising that they are mentioned on a par with one more aspect of mathematics which is even mentioned before them, and that this first aspect is

The historical aspect of mathematics.

This is a new departure, and what it implies for the teaching of mathematics is specified by the Ministry as follows: (The original text is of course in Danish, and as far as I know there is no official English version. This is my own translation.)

Students must obtain knowledge of elements of the history of mathematics and of mathematics in its connection with culture and society.

In combination with the treatment of each of the (five plus three) main themes, it will be natural to put it into perspective by incorporating a summary of its history, also touching on the historical environment of each theme.

The historical aspect can also be covered in special teaching sequences, for example on the basis of historical source material. It may also be advantageous to treat—to a certain extent — features of the epoch, the culture, and the society, in which the mathematical theme in question evolved. Examples of such teaching sequences could be the evolution of the number system, solutions of equations in Babylonian or Greek mathematics, the evolution of the concepts of limit and derivative, elements of the history of probability and statistics.

The historical aspect can also be treated in teaching sequences in which the main point is to elucidate the connections of mathematics with culture and society. As examples of such sequences one can mention the role of mathematics in painting and music, the scientific revolution in the 17th century, viewpoints on the role of mathematics in the societies of today.[5]

The new regulations had an immediate impact on the textbooks for the gymnasium, both because the textbook market is—and has always been—completely free, and because there is no time-consuming authorization of textbooks in Denmark. Mathematics textbooks are usually in one or two volumes for each of the three gymnasium years; some existing textbooks already had something on the history of mathematics in them, and in all cases the authors of most textbooks launched new editions without delay. Some of these new editions are in the main unaltered except for a short historical appendix in the back of each volume or a short historical section at the end of each chapter. Others have had one or two substantional chapters on history added to each volume, sometimes written

by a historian of mathematics. One quite new textbook by Flemming Clausen et al presents every mathematical theme with its history completely integrated into the exposition, and presents mathematics as an important part of human culture.[6] Independent of the textbooks, many books on diverse historical topics have been published for use in the gymnasium or for the teachers' background reading, also before the new regulations. The larger of these books are often anthologies or collections of sources; the smaller ones appear sometimes in a series of booklets from the same publisher. Some of these books are written by university faculty and are also used at the universities.[7]

There are about 1,500 mathematics teachers in the Danish gymnasium, all of them university-educated, all with master's degrees in mathematics or sometimes in physics or chemistry. Many of them will have met with the history of mathematics during their university years, since there is a long and unbroken tradition for both research in it and teaching of it in the mathematics departments of Danish universities. At Copenhagen University this goes back mostly to Hieronymus Georg Zeuthen (1839–1920) and Johan Ludvig Heiberg (1854–1928), both still remembered for their critical work on classical Greek mathematics. Zeuthen is remembered mainly for his understanding of the role of the constructions in Euclid's *Elements* as existence proofs, for his book on the conics in antiquity,[8] and for his two volumes on the history of mathematics.[9] Heiberg is remembered especially for his rediscovery in 1906 of the *Method* of Archimedes,[10] but also for his text editions in Greek of the extant works of Archimedes, of Euclid's *Elements,* and of other classical works, which are still standard.[11] The contemporary Danish translation by his student Thyra Eibe of his text of the *Elements* is now being reprinted, not the least for possible use in the gymnasium.[12] During his stay in Copenhagen in the 1930s, Otto Neugebauer added to the university's tradition of history of mathematics with his study of ancient Babylonian mathematics. At the four newer Danish universities, especially in Aarhus, it has also been possible to take courses in the history of mathematics. Therefore many gymnasium teachers have been well prepared to attend to this new aspect of mathematics teaching in their schools. Of course, universities have already registered an increased attendance among present students for courses in the history of mathematics, so one can hope that many of the future gymnasium teachers will be well prepared.[13]

Even before the new math history requirement, gymnasium teachers had a long-established practice of taking voluntary additional education courses arranged by their association, taught by university faculty, and supported financially by the Ministry of Education, and more courses on historical subjects are now included. Also, historical matters are discussed and papers on them are published both in the same association's newsletter, *LMFK-bladet,* and in the journal *Normat (Nordisk Matematisk Tidskrift),* which is published jointly by this association and its counterparts in Finland, Iceland, Norway, and Sweden, together with the mathematical societies in Denmark and these four countries. So the new idea of making the history of mathematics an established part of the teaching of mathematics in the Danish gymnasium may have a good chance of success.

In the folkeskole the situation with respect to the history of mathematics is quite different. Here 90% of all 7- to 17-year-olds are educated; the rest attend private schools which are privately owned but heavily subsidized by the state on condition of a certain conformity with the folkeskole. There are no nationwide official rules for what should be taught in the folkeskole; local authorities make their own decisions for the schools in their district. However, the Ministry of Education issues a set of guidelines— one for each subject—and most local authorities follow them rather closely.[14]

In the Ministry's guideline for teaching of mathematics in the folkeskole the history of mathematics is not mentioned explicitly, but in many ways it is in the air, and has been for a long time, as suggested by articles appearing here and there in many textbooks, both old and new.[15] One of the main reasons for this sympathy towards history is Poul la Cour's book *Historisk Matematik (Historical Mathematics).*[16] It was written in 1888 as a textbook for the author's own mathematics teaching at one of Denmark's folkehøjskoler (folk high schools), a special sort of historically-minded schools for young adults, in those days mostly from rural surroundings, but it was widely read by the general public and reprinted many times.[17]

In 1993, a new law for the folkeskole was enacted. In the law itself, history of mathematics is not mentioned, but in an accompanying provision entitled *Formål og centrale kundskabs- og færdighedsområder. Folkeskolens fag* (Goals and Central Areas of Knowledge and Proficiency. The Subjects of the Folkeskole), published by the Ministry of Education in 1994, the following passage occurs (on p. 55) in the section on mathematics (pp. 54–57): 'Elements of the history of mathematics must be included, and the importance of the subject for the development of the society must be illustrated.' It remains to be seen how this will affect the Ministry's coming guideline for the teaching of mathematics, and then the textbooks, and the teaching in the schools, and also the education and further education of teachers.

Mathematics is taught at all levels in the folkeskole. It is usual for a class to have the same mathematics teacher through the first six or seven years and often through all ten years. The same is true for other subjects as well.

Teachers in the folkeskole are supposed to be prepared to teach all subjects at all levels of the school, at least up to grade 6 or 7. Teachers have a four-year education beyond the gymnasium from what is called a seminarium. During the first year it is the same for all and covers all subjects, none of them deeply. Through the other three years half the time is devoted to two subjects chosen by the student (called line subjects) and the other half to generalities and to teaching practice. Nowhere in the regulations for the obligatory first year mathematics or for mathematics as a line subject is the history of mathematics emphasized. Thus, given the short time available, it is no surprise that, even if a teacher has taken mathematics as a line subject, that teacher will often not have a deep understanding of the subject or a broad knowledge of its history.

At many of the 24 seminariums, however, the position of mathematics in society —past, present, and future—and the history of mathematics itself, are more prominent than the regulations demand, which has something to do with the influence of Poul la Cour and his book. It is possible that something more definite might materialize in the new law for the seminarium which is under preparation. In any case it is safe to say that future teachers of mathematics in the folkeskole will not always be unaware that it really has a history, and that this might be of some importance in their teaching.

The same cannot be said of the current teachers in the folkeskole. There are around 53,000 of them, and a majority of them teach mathematics as one of their subjects, at least in the lower grades. But many— maybe most—of them consider themselves as teachers in a general sense and not as teachers of mathematics in a specialized sense, and they often see mathematics only as applied mathematics with home economy the only application, or the only application worth mentioning in school. They often seem to think that, if you can balance your cheque-book and fill in your income tax return and compute the area of your kitchen floor so that you can lay down your own vinyl tiles, then that is about as much as you need of mathematics education. These teachers are not interested in the deductive side of mathematics and not at all in the history of mathematics. So there is room for some pessimism.

But there is room for some optimism, too, because something is happening all the time: there is an active association of teachers of mathematics in the folkeskole, even if its members are of course only those teachers who see themselves as mathematics teachers. This association arranges meetings and short courses all over the country, and every year a well-attended four-day summer course. It publishes books and a journal called *Matematik,* and the history of mathematics can be seen at many places in this activity.

Denmark also has a tradition for a more institutionalized further education of teachers in the folkeskole. Since 1856 a special institution has existed for this purpose. It is called Danmarks Lærerhøjskole, literally Denmark's Teachers' High School but officially translated as The Royal Danish School of Educational Studies. The word "Royal" signals that it is a state institution. From a modest beginning it has grown in this century into a nationwide enterprise with its main branch in Copenhagen and branches in eight provincial towns around the country.[18] Every year as many as 25% of all teachers in the folkeskole attend a course in some subject at one of the branches of this teachers' university, as it might be called. There are short and longer courses; the typical one-year course meets once a week for 33 weeks, either three morning hours or three afternoon hours or a whole six-hour day. To make it more practicable and more attractive for teachers to attend such courses, their teaching load is reduced by nearly as many hours as they attend, without any reduction in their salary. Of course this last has its side effects: since the local school authorities have to pay for the extra teachers needed because there are always some teachers attending courses, they then also have to endorse the teachers' applications for the courses, making it more difficult for teachers to attend courses as often as they wish and to choose in complete freedom among the courses which are offered.

Since 1963 I have worked full-time at the mathematics department of the Copenhagen branch of this teachers' university. History of mathematics has always been present in my own and in many of my colleagues' mathematics courses, whatever the mathematical theme has been, and occasionally—in the last four or five years, every year—I have given courses also in the history of mathematics as a subject, either as a small one-year course (99 hours) or as a concentrated one-week, full-time course (28 hours), or as a special part (21 hours given as seven three-hour mornings) of a large (198 hours) one-year course. The last of these formats is the most usual one, since local school authorities often are not inclined to endorse an application from a teacher for a course solely in the history of mathematics, which they tend to consider as a luxury.

Of course it is difficult—or rather impossible—to present a comprehensive history of mathematics in 21 or 28 hours or even in 99 hours, especially since participants

in these courses often do not know much mathematics, and not much general history either. They are victims of our ahistoric times. In one of my courses it helped that one participant was a history teacher rather than a mathematics teacher. He would come up with highly pertinent observations and comments. In the 21-hour course the purpose must be to show especially that here is something of interest, something to return to, something to go on with. It is underlined that presented material is not meant to pass unadapted into the participants' own classrooms but to colour and maybe improve what goes on there.

About the content of these courses I will make just one remark: I always try to include non-Euclidean geometry. Most teachers of elementary mathematics have never heard about it, and it comes as something of a shock for them that mathematics is something separate from physics in that mathematical statements are not true in a straightforward physical sense. And it is strange—but I hope also enlightening —for them to get to know that mathematicians had to realize this acutely more than a hundred years ago, a few pioneers even earlier than that, and with the same initial incredulity as their own.

This ends my survey of the situation in Denmark with respect to the second of the three questions: what might be done to make the history of mathematics have its due place in the teaching of mathematics. Of course, we have in Denmark our special difficulties and special opportunities, and it may well be that I have painted too optimistic a picture: there are many obstacles, and things take time, also in Denmark. Worse than that: sometimes we make mistakes with the best of intentions, also in Denmark.

This brings me to the third of the three original questions: what can go wrong, and what should we try to avoid when we include history in the teaching of mathematics? Many mistakes occur because we who teach mathematics, at whatever level, are not all professional historians of mathematics. I am not one, and neither are teachers in school, and no one would insist that we should be. Therefore we have to rely on secondary or even tertiary sources, with the result that there is a risk that we misrepresent the history of mathematics—that we tell colourful but untrue historical anecdotes, that we reproduce historical reasonings with wrong points or with anachronistic twists, and so on.

A string of examples of such misrepresentations follows. They are drawn from all levels of mathematics education and are very different with respect to both scope and importance. I will, of course, be grateful for any criticism here—I may well have made mistakes in my presentation of these mistakes. I will give full references, but only for

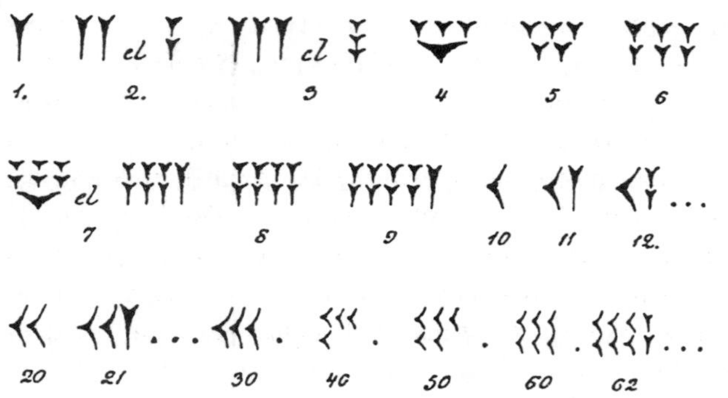

FIGURE 1

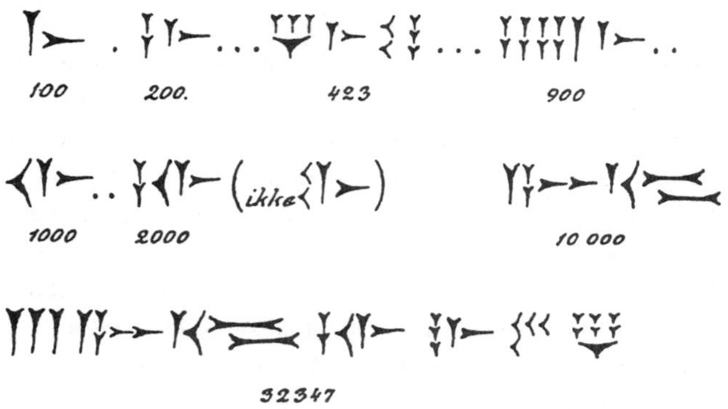

FIGURE 2

reasons of documentation, since I am chiefly concerned with pointing out different sorts of possible mistakes and not at all with blaming anyone for having made them. It can happen to any of us. To emphasize this I begin with an example from my own country from school mathematics.

An activity booklet for grade 8 from a new series of textbooks for grades 1–10 in the Danish folkeskole, entitled *Matematik 2000*, gives on page 3 a figure showing the Babylonian cuneiform numerals for one to 12, 21, 30, 40, 50, 60 and 62.[19] Some are given in more than one version, and all are drawn instructively and correctly except that, instead of one vertical wedge for 60, the figure shows six of the wedges which taken alone designate 10, and the same in the wedge-symbol for 62, as shown on Fig. 1. Now one of the main points of acquainting 15-year-olds with Babylonian numerals must be to give them a deeper understanding of the place-value system by showing them that our—or rather the Indians' and Arabs'—choice of base ten is arbitrary, in that we could just as well have used 60 as the Babylonians did, or another number. The reasons for choosing those two number bases can be illuminating. Another point must be to show students the interesting fact that we follow the Babylonian subdivision of the hour into 60 minutes and the minute into 60 seconds and similarly in subdivision of angles, starting with the angle in an equilateral triangle, but in some countries now also stopping there, because decimal subdivision of one degree has been introduced. The reasons for all this can help to open the pupils' eyes to ways of cultural influence. So by giving the wrong wedge-symbol for 60, the authors miss several important points and some good pedagogical opportunities.

Since Babylonian numerals are described in many books, one can wonder how this mistake could happen. I have no doubt: in the venerable book by Poul la Cour already mentioned, one finds a plate showing the same

cuneiform numerals—even with precisely the same different versions and the same errors. In fact, Fig. 1 is taken from la Cour's plate which also contains the strange cuneiform numerals shown in Fig. 2, and Egyptian, Greek and Chinese numerals. It would be interesting to know the exact origin of this plate—it also contains a picture of a Chinese merchant using his abacus, the suan-pan, with two amazed onlookers outside a window, a picture which has recently been reproduced with the information that it was made in San Francisco in the 1880s.[20]

In any case, it seems that the deeper reason for the mistake made by the authors of *Matematik 2000* is that they copied from an old book without checking its accu-

FIGURE 3

racy, which is in itself a mistake. Of course they may have used a newer edition of la Cour's book and believed it to be up-to-date, but the prefaces of all of them state that, in order to preserve the author's original tone, they are virtually unaltered since the first edition. As it came out 40 years before Neugebauer laid the foundations for a real understanding of Babylonian mathematics, one cannot reproach la Cour with his plate's being faulty. I have not been able to confirm another proposition: that around 1880 only some late Assyrian tablets with mainly economic, not mathematical, content had been deciphered, and that they had only such non-typical cuneiform numerals.

$$\begin{matrix} |\,|\,|\,|\,| \\ |\,|\,|\,|\,| \end{matrix} \;=\; \cap \qquad\qquad \cap \cap |\,|\,| \;=\; 23$$

FIGURE 4

The next example is from the well-known American journal on mathematics education, *Arithmetic Teacher.* In a proposal for work with ancient Egyptian numerals in grades 1–8, strange anachronisms spring to the eye. The equality sign introduced in 1557 by Robert Recorde[21] is placed between the ancient numerals and between them and our (Indo-Arabic) numerals, as shown in Fig. 4, and plus and minus signs also originating in European renaissance (with medieval ancestors) are also placed between Egyptian numerals, as shown in Fig. 5. Pupils are asked to write their birthdays and other dates from the Gregorian calendar, which was first proposed in 1582, with Egyptian numerals. Nothing is said about which calendar the Egyptians might have used.

Anachronisms aside, it is a real distortion to ask pupils, as is done here, to perform multiplications and divisions in our algorithms but with all the numbers written in Egyptian numerals. Our algorithms are evidently constructed for use with Indo-Arabic numerals. They can, of course, be used with other place-value systems, but not very well with number systems like the Egyptian. So when the author proposes that the teacher ask the pupils which number system is easier to use, he is not being fair to the ancient Egyptians. Their algorithms were based on repeated doublings and halvings, and one of the things one can achieve by introducing Egyptian numerals in school

mathematics is to show the pupils that algorithms different from ours can be efficient. Moreover, one can show them that the Egyptian algorithms are not essentially more quaint than ours, just different, and that, if they wonder at the Egyptian ones, they must also wonder at our own —which thereby maybe can be lifted up from being just dull routine. There is still more to be achieved, for the repeated doubling continues not only in what is sometimes called Russian peasants' multiplication and similar methods elsewhere, but it is also highly related to binary systems and therefore to what goes on inside the pupils' own hand calculators. The author of these activity proposals misses all this completely, so it seems that his only purpose for all the activities is to brighten up some (otherwise dull?) mathematics lessons by introducing some sort of curiosity, and that he has chosen Egyptian numerals more for being exotic than for historical reasons. Not only is history being distorted, but several educational opportunities are missed.[22]

Another example touching upon ancient Egypt is the following. In many books on history of mathematics it is told that Egyptian surveyors, who were called rope stretchers (among others by Democritus), systematically constructed right angles with the help of knotted ropes which could be stretched out to form Pythagorean (3, 4, 5) triangles with the knots at the vertices, as shown on Fig. 6. This has become the anecdote which textbook authors cannot resist; as one example among many, see the otherwise excellent new Danish textbook, mentioned above, by Flemming Clausen et al, where there even is a figure in which the knots do not coincide with the vertices of the triangle.[23]

It seems that this anecdote can be traced back to Moritz Cantor's classical *Geschichte der Mathematik,* in which it is presented as a conjecture based on two problems in the Berlin papyrus (from around 1300 B.C.), namely the following: The sum of the areas of two squares is 100, and the side of one is 3/4 of the side of the other; find the sides. And: The sum of the areas of two squares is 400, and the side of one is to the side of the other as 2 to 1; find the sides. The first problem is solved correctly in the papyrus: the sides are 6 and 8; and likewise the second: the sides are 16 and 12. Of course 100 and 400 are the areas of

$$\cap |\,|\,| \;+\; \cap \cap |\,|\,|\,| \qquad\qquad \cap \cap \cap \;-\; |\,|\,|\,|$$

FIGURE 5

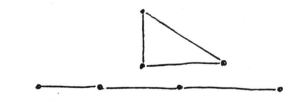

FIGURE 6

squares with sides 10 and 20, but nowhere in the papyrus are triangles with sides 6, 8, and 10 or 12, 16, and 20 mentioned. Nevertheless, Cantor conjectured that the Egyptians must have thought of these triangles and seen that they are right-angled, and that the rope stretchers could therefore have used them as described above.[24]

Now come two examples from a book on mathematics for the general public in which history is given a prominent place. The book, whose title translates as *The Phantastic Mathematics,* is written by Swedish journalist Kristin Dahl, helped by some Swedish mathematicians, and she and they have succeeded very well.[25] However, on page 33 one finds a figure showing a Babylonian multiplication table with double entry, "quite in agreement with the modern arrangement," says the accompanying text. No specialist I have consulted has ever seen a Babylonian multiplication table so arranged; they are always single entry tables. An example can be seen in Asger Aaboe's *Episodes From the Early History of Mathematics* (1964)[26] in a figure on page 7, which has the advantage that anybody can learn Babylonian numeration from it without any preparation at all, by real discovery.

The other example from Dahl occurs on page 35, where it is told how Archimedes approximated the circumference of a circle by looking first at a circumscribed square, then a circumscribed regular pentagon, hexagon, heptagon and so on until he got to a 96-gon. But of course he didn't! He started with a regular hexagon, and, since he was able, given a tangent line and an angle belonging to it, to calculate the tangent line belonging to half the angle, he could go directly from the hexagon to a regular polygon with twice as many sides: a twelve-gon (why should he then ever have bothered with a heptagon which moreover would have caused him a lot of trouble?). And with precisely the same method he could then go on to a 24-gon, to a 48-gon and to a 96-gon. Here we have a case of making history much more complicated than it really was, and much less elegant. Archimedes went from a hexagon to a 96-gon in just four jumps, and by using the same method four times—and he was completely aware that by using it once more he would have come to a 192-gon. Incidentally, the accompanying figures in Dahl show a circumscribed square, pentagon, and heptagon but no hexagon, and it mentions as Archimedes's approximation to π the number 3.14271 which has nothing to do with the upper bound of 31/7 which Archimedes found (and which is approximately 3.142857). Moreover, the book misses the chance to tell that Archimedes did not content himself with an approximation from above—that he also worked with inscribed regular polygons and was again able to go from a

hexagon to a 96-gon by using the same method four times, since he was able, given a chord with its angle, to calculate the chord belonging to half the angle—and of course he was completely aware that he could have gone on to a 192-gon, and further—and finally that he in this way found a lower bound of $3\frac{10}{71}$ (which is approximately 3.140845).[27] So the book misses Archimedes's greatness here, which is that he did not give an approximate value for π (or even believed that he had found it exactly) but gave an interval in which it lies, and a method by which this interval could be diminished further.[28]

One can only guess at the reasons for mistakes like the two I have here described from this book; maybe the author has just read somewhere or heard from someone that "the Babylonians had real multiplication tables" (correct) and that "Archimedes went all the way up to a 96-gon" (also correct) and then imagined the rest.

We have just seen an example where unnecessary complications were introduced in some historical mathematics. The opposite can also occur: someone takes a piece of exceedingly complicated historical mathematics and introduces it—with the best of intentions—into the teaching of some mathematical theme to make things easier; this can be done only by leaving out all difficulties. In the 1970s the (British) Schools Council Project produced a series of books for teachers, in which school mathematics for 11 to 16-year-olds was discussed. In one of these books, *From Graphs to Calculus* (1977) by Hilary Shuard and Hugh Neill, a chapter written by Rolph Schwarzenberger on "The Early Historical Development of Calculus" is included.[29] Here Bonaventura Cavalieri's "work on the area under the curve $y = x^p$ "—incidentally this is not what Cavalieri would have called it—is brought in on page 151 only to be dismissed on page 152 with a complaint that Cavalieri did not argue in the way the author would have wished him to do. Instead "he introduced the unfortunate and confusing notion of 'indivisibles'." But this notion is fundamental in Cavalieri's thought and central for his achievement. See Kirsti Andersen's "Cavalieri's Method of Indivisibles," where one also can read in the introduction that Cavalieri often has been misrepresented in this way:

> Cavalieri is well known for the method of indivisibles which he created during the third decade of the 17th century. The ideas underlying this method, however, are generally little known. This almost paradoxical situation is mainly caused by the fact that authors dealing with the general development of analysis in the 17th century take Cavalieri as a natural starting point, but do not discuss his rather special method in detail, be-

cause their aim is to trace ideas about infinitesimals. There has even been a tendency to present the foundation of his method in a way which is too simplified to reflect Cavalieri's original intentions.[30]

There are good reasons for trying to make introductory calculus intuitive, but it cannot be done in Cavalieri's name. What we call Cavalieri's Principle was for him not his point of departure but a main result proven by difficult arguments on indivisibles. By the way, Gottfried Wilhelm Leibniz's infinitesimals are not much easier to understand or to handle, even if they are quite different from Cavalieri's indivisibles.

The next example is one of the really great stories in the history of mathematics, the one about poor, miserable Évariste Galois writing down all his mathematical discoveries during the night before his fatal duel, time after time breaking off to scribble in the margin "I have not time, I have not time." I had—as many others have done —told this story many times through the years in courses both on algebra and on history of mathematics when in 1982 Tony Rothman in a well-argued paper told us that Eric Temple Bell had invented both the moving detail and much, much more.[31] Galois wrote those famous words, but only once, and in a marginal note in one of his papers, and in the following form, inclusive of the cool signature in parentheses:

> There are a few things left to be completed in this proof.
> I have not the time. (Author's note.)

Then Bell wrote it all down in his famous book, *Men of Mathematics,*[32] thereby creating the Galois legend, which has since been repeated by most writers and lecturers—and is still repeated despite Rothman's paper. Although there are other legends in Bell's book, it is recommended reading, but must be read with caution.

Here are two more examples of long-established truths which have been retold hundreds of times and have turned out to be not so true. Both are about Georg Cantor, one biographical and the other mathematical. Both can be found in David Burton's *Abstract Algebra* (1991)—even on the same page—a very good book by an author who has also written a fine history of mathematics.[33] On page 2 of Burton's *Abstract Algebra* one reads:

> One of Cantor's former professors at Berlin, Leopold Kronecker, considered Cantor a scientific charlatan, a renegade, and a "corrupter of youth". (As a result of continuing animosity such as this, Cantor suffered periods of mental break-

down and spent the last five years of his life in a sanatorium.)

According to newer biographies of Cantor, by Herbert Meschkowski, by Ivor Grattan-Guinness, by Joseph Warren Dauben, and by Walter Purkert and Hans Joachim Ilgauds, there exists no documentation at all for the words "as a result of"; on the contrary there is good evidence for Cantor's mental illness being congenital and having nothing to do with Kronecker's dislike for his mathematics, which was intense, or for his person, which may be mythical. It must be mentioned that the traditional explanation of Cantor's illness as being caused by the animosity of Kronecker and others persists. And even if E. T. Bell did not invent it, he certainly did much, in the last chapter of his above-mentioned *Men of Mathematics,* to spread it "with additional ill-founded embellishments of his own."[34]

And now for the second of the questionable truths about Georg Cantor, which can also be found on page 2 of Burton's *Abstract Algebra*:

> About the turn of the century . . . a number of entirely unexpected logical contradictions, termed "paradoxes", were discovered on the fringes of set theory. In his desire to produce a concept as general as possible, Cantor had allowed for any collection to be a set. He intuitively assumed that if one could describe a sensible property of objects, then one could also speak of the set of all objects that possess the property. This naive and unsystematic approach led to upsetting consequences. Certain "large" collections, such as the "set of all sets", had to be discarded to circumvent the troublesome paradoxes. The remedy was to construct a sufficiently restrictive system of postulates for deciding which objects were to be regarded as sets and what properties sets should have. The leader of the movement of rehabilitation was Ernst Zermelo (1871–1953), who published the first axiomatic treatment of set theory in 1908.

In the above-mentioned Cantor biographies, especially in the one by Purkert and Ilgauds, and before that in a 1986 paper by Purkert, it is argued that Cantor was well aware of these "paradoxes" from the beginning of his work with set theory,[35] and had even formed his original definition of a set, which appears in English translation in Burton in this form:

> By an aggregate (set) we are to understand any collection into a whole M of definite and distinct objects m of our intuition or thought[36]

so as to avoid the paradoxes, or better: he deliberately formed his definition so that it talks only of collections which can be handled by humans, and where paradoxes therefore will not occur. In the definition the word "collection" denotes rather a process than its product, and Cantor knew very well that the process cannot always be carried through; if it can, then the word "whole" denotes the product (all this is more evident in the original German words "Zusammenfassung zu einem Ganzen"). So Cantor was quite aware that we cannot speak for example of the set of all sets, and he distinguished between on one side finite and transfinite multitudes which we can consider as sets, and on the other side absolute multitudes which we cannot—only God can comprehend them. This is evident from several of his letters to David Hilbert and from many footnotes in his papers on set theory. The reason why Cantor put his reservations in footnotes must have been that he did not want to hinder the main argument of his text, but, when after his death his collected works were published, edited by Ernst Zermelo,[37] all footnotes were left out, and it seems that since then not many have read the originals.

The latest examples have had as their theme what I called the telling of colourful but untrue historical anecdotes. Other such anecdotes are those of the famous villains, for example Girolamo Cardano and Guillaume-Francois-Antoine de L'Hospital—they did not simply steal their results even if Niccoló Fontana (or Tartaglia) and Johann Bernoulli said so (Bernoulli prudently after the death of L'Hospital). Let me conclude with an example whose theme is what I called the reproducing of historical reasonings with wrong points or with anachronistic twists.

In textbooks for all levels of mathematics teaching and in monographs on number theory and sometimes in journals, one can find the theorem that there exist infinitely many primes proved "just as Euclid did it."[38] What one gets is an indirect proof along these lines: Let us assume that there are only finitely many primes, let them be $p_1, p_2, \ldots, p_n$; then the prime factors of the number $p_1 p_2 \cdots p_n + 1$ must all be different from $p_1, p_2, \ldots, p_n$ which were then after all not all the primes, and we have a contradiction. This is a correct proof but it is not Euclid's. Thomas Heath's translation of proposition IX.20 from Heiberg's Greek text goes like this:[39] Let A, B and C be some primes; I can show you one more, namely any prime factor of $ABC + 1$. This is really a constructive proof, and not at all indirect (there is a small indirect turn in Euclid's argument for any prime factor of $ABC + 1$ being different from A, B and C, but that is a minor item). But there is even more to this, because Euclid does not say: Given an arbitrary number of primes, I will come up with

one more. He takes three primes and shows us how to find a fourth, and then he leaves the rest to us. Most children—and many adults—will understand such a typical argument much better than a general one, so it seems much better to do it this way and then maybe ask: But are you quite content with this? In this way one may lay a foundation for a fruitful discussion of inductive reasoning.

A natural reaction to such a long row of examples might be to say: Yes! we must really guard against uncritically accepting misrepresentations, against committing such mistakes. But I think we must realize that we cannot guard ourselves completely in these matters, that there is no absolutely reliable intellectual safety net. This is an unusual situation for mathematicians to find themselves in. We are used to a certain feature of mathematics in which it differs, I believe, from all other subjects, namely this: We do not have to trust any authorities, we can always see for ourselves, each of us. The pupils do not have to rely on the teacher for the truth or falsity of what they are told; every one of them has the possibility of checking it. They can for example check if it is really true that there is no rational number whose square is two—and it is marvellous that they can do it by an argument which may go back to Aristotle. There is no other subject in which the pupils have this freedom. The same holds true all the way through: the teacher does not have to rely on the textbook author, and the textbook author does not have to rely on some next authority. This special feature of mathematics can even be exploited in teaching, as when I tell my students not to trust me completely when I wave my hands and talk of uniform convergence.

But all this changes as soon as we include history in our teaching—suddenly we are without the safety net. This has always been so for the history teacher, but we mathematics teachers are not used to it, and we are not always happy about it.

So far I have concentrated on the relations between the history and teaching of mathematics. It might be appropriate here also to touch on the relations between the history of mathematics and research in mathematics. I will do this by quoting Gian-Carlo Rota from his paper "Mathematics and Philosophy":

> In short, no mathematician will ever dream of attacking a substantial mathematical problem without first becoming acquainted with the history of the problem, whether the real history or an ideal history that a gifted mathematician might reconstruct. The solution of a mathematical problem goes hand-in-hand with the discovery of the inadequacy of previous attempts, with the enthusiasm

that sees through and does away with layers of irrelevancies inherited from the past which cloud the real nature of the problem. In philosophical terms, a mathematician who solves a problem cannot avoid facing up to the historicity of the problem. Mathematics is nothing if not a historical subject par excellence.[40]

Similarly, in *Proofs and Refutations* (1976) Imre Lakatos explicitly depicts mathematical research as an ongoing historical process.[41]

So the history of mathematics is not just a box of paints with which one can make the picture of mathematics more colourful, to catch the interest of students at their different levels of education; it is part of the picture itself. If it is such an important part that it will give a better understanding of what mathematics is all about, if it will widen horizons of learners, maybe not only their mathematical horizons, if it will make them wonder— then it must be included in teaching even if we are not always sure of the details. This can also be told, and the teacher must be ready to leave out a colourful detail if it turns out to be not so true as it used to be—even the best anecdote is a hindrance to understanding if it is not true.

If we include the history of mathematics in our teaching in this spirit, I think we can bear some risks with confidence.

Acknowledgements.   A draft of this paper was presented at the conference held by the International Study Group on the Relations Between History and Pedagogy of Mathematics in Toronto, August 12–14, 1992. The author wishes to thank the editor Ronald Calinger for important references and for useful advice on language and style.

References

1. Torkil Heiede, "Why Teach History of Mathematics?" *The Mathematical Gazette,* 76 (1992): 151–157.
2. Torkil Heiede, "Hvorfor undervise i matematikkens historie?" *Normat (Nordisk Matematisk Tidskrift),* 39 (1991): 153–161 & 192.
3. Robert L. Hayes, "History—A Way Back to Mathematics," *History and Pedagogy of Mathematics Newsletter,* 22 (1991): 10–12. Robert L. Hayes has for several years been teaching mathematics and basic numeracy courses to various adult groups in Australia. Many of his students came "with a deep-seated dislike of mathematics—generally due to poor teaching, lack of succes, and unpleasant memories during their primary and secondary school years," in which they had received "a sterile, polished, and clinical form of mathematics, with no impression of the feeling, pain and anguish connected with its development." A final quotation from Hayes's paper: "Mathematics should be taught as an exciting, dynamic part of human culture and all students should be allowed to share in its richness."
4. *Matematik. Bekendtgørelse og vejledende retningslinier* (Copenhagen: Undervisningsministeriet, Direktoratet for Gymnasieskolerne og HF, 1988).
5. The above-mentioned *Matematik. Bekendtgørelse. . .* (1988): 351 & 360.
6. A pioneering textbook containing much history even in its first edition, is Poul O. Andersen, Stig Bülow & Hans Jørgen Helms, *Matematik for gymnasiet,* 5 vols. (Copenhagen: Gyldendal, 1963–1966, new ed. by Poul O. Andersen, Stig Bülow & Torben Christoffersen, 1978–1979). A textbook with substantial historical chapters added in a newer edition is Steffen Jensen & Karin Sørensen, *Teori og redskab,* 5 vols. (Copenhagen: Christian Ejlers' Forlag, 1979–1983, 2nd ed. in 6 vols. with historical chapters by Tinne Hoff Kjeldsen, 1988–1990). The only textbook so far in which history appears prominently throughout, is Flemming Clausen, Poul Printz & Gert Schomacher, *Ind i matematikken,* 6 vols. (Copenhagen: Munksgaard, 1989–1994).
7. See for example the following: On Greek mathematics, Bent Hirsberg & Klaus Holth, *Tal og geometri* (Vejle: Forlaget Trip, 1982); on the history of vectors, Kirsti Andersen, *Hvor kommer vektorerne fra?* (Aarhus: Videnskabshistorisk Museums Venner, 1984); on the three classical problems and their ramifications through the history of mathematics, Jesper Lützen, *Cirklens kvadratur, vinklens tredeling, terningens fordobling: Fra oldtidens geometri til moderne algebra* (Herning: Forlaget Systime, 1985); on the Pythagoreans, Flemming Clausen & Jørgen Falkesgaard, *Tal & tanke: Pythagor`ernes verdens- og livsanskuelse* (Copenhagen: Munksgaard, 1986); a collection of sources from the history of equations "from Babel to Abel", with commentaries, Kirsti Andersen, ed., *Kilder og kommentarer til ligningernes historie* (Vejle: Forlaget Trip, 1986); a translation of excerpts from L'Hospital's *Analyse des infiniment petits pour l'intelligence des lignes courbes,* with commentaries, Guillaume-François-Antoine de L'Hospital, *Analyse af de uendeligt små størrelser til forståelse af kurver,* transl. and comm. Kirsti Andersen, (Aarhus: Videnskabshistorisk Museums Venner, 1988); the following two both on the history of π, Kurt Bøge, *Elementer af tallet π's historie* (Vejle: Abacus, 1991) and Torben Svendsen, *Bogen om π* (Herning: Forlaget Systime, 1992); on the determination of tangents seen historically, Jens Lund, *Tangentbestemmelse historisk set* (Copenhagen: Matematikl`rerforeningen, 1992); and on Greek geometry, especially on Euclid, Hans Fich, *Geometri uden tal* (Copenhagen: Gyldendal, 1992). The last four are booklets from four different series, one of them published by the association of mathematics teachers in the gymnasium. Note also that two of the books in the list are from a series published by a society connected with a science museum.
8. Hieronymus Georg Zeuthen, *Keglesnitsl`ren i Oldtiden* (Copenhagen: Videnskabernes Selskab, 1885), German transl., *Die Lehre von den Kegelschnitten im Altertum* (Copenhagen: Høst & Søn, 1886, repr. Hildesheim: Olms, 1966).
9. Hieronymus Georg Zeuthen, *Mathematikens Historie: Oldtid og Middelalder* (Copenhagen: Høst & Søn, 1893),

new ed. rev. by Otto Neugebauer, *Mathematikens Historie: Oldtiden* (Copenhagen: Høst & Søn, 1949), German transl., *Geschichte der Mathematik im Altertum und Mittelalter* (Copenhagen: Høst & Søn, 1896), and idem, *Mathematikens Historie: 16de og 17de Aarhundrede* (Copenhagen: Høst & Søn, 1903), German transl., *Geschichte der Mathematik im XVI und XVII Jahrhundert* (Leipzig: Teubner, 1903).

10. See the introduction to the *Method* in Thomas L. Heath, transl. and comm., *The Works of Archimedes* and *The Method of Archimedes* (Cambridge: Cambridge University Press, 1897 and 1912), reprinted in one vol. (New York: Dover, 1956).

11. Johan Ludvig Heiberg, ed., *Archimedes opera omnia*, 3 vols., (Leipzig: Teubner, 1880–1881, 2nd ed. 1910–1915), and idem (with H. Menge), *Euclidis opera omnia*, 7 vols., (Leipzig: Teubner, 1883–1896), and idem, *Apollonii Pergaei quae Graece extant*, 2 vols., (Leipzig: Teubner, 1890–1893), and idem, *Cl. Ptolemaei opera quae extant omnia*, 2 vols., (Leipzig: Teubner, 1898–1907).

12. Thyra Eibe, transl., *Euklids Elementer 1–13*, 6 vols., (Copenhagen: Gyldendal, 1897–1912, repr. of *1–4*, Vejle: Forlaget Trip, 1985).

13. Among the materials produced for students at the universities and also for additional education of gymnasium teachers, see for example the following: On mathematics and science in antiquity, Olaf Pedersen, *Matematik og naturerkendelse i oldtiden*, (Copenhagen: Akademisk Forlag, 1975); a collection of essays on many different topics from the history of mathematics, Kirsti Andersen et al, *Nogle kapitler of matematikkens historie*, 2 vols. (Aarhus: Matematisk Institut, Aarhus Universitet, 1979); on infinite series and their history, Lars C. Mejlbo, *Uendelige r`kker: En historisk fremstilling*, (Aarhus: Matematisk Institut, Aarhus Universitet, 1983), an anthology of sources and secondary literature from the history of calculus and analysis, Kirsti Andersen, Henk Bos & Jesper Lützen, eds., *Træk af den matematiske analyses historie: En antologi af kilder og sekund`r litteratur* (Aarhus: Institut for de eksakte videnskabers historie, Aarhus Universitet, 1987); a translation of Lobachevsky's *Geometrische Untersuchungen zur Theorie der Parallellinien*, N. I. Lobatjevskij, *Geometriske undersøgelser over teorien for parallelle linier*, transl. Lars C. Mejlbo, (Aarhus: Matematisk Institut, Aarhus Universitet, 1988); and a history of elementary geometry with many sources included in translation, also the one just mentioned, Lars C. Mejlbo, *Om den element`re geometris historie* (Aarhus: Matematisk Institut, Aarhus Universitet, 1989).

14. On the folkeskole, see Finn Br`strup Karlsen & Paulette Møller, *The Folkeskole: Primary and Lower Secondary Education in Denmark* (Copenhagen: The Ministry of Education, 1988). The Ministry's guideline for the teaching of mathematics is *Undervisningsvejledning for Folkeskolen 15: Regning/matematik* (Copenhagen: Undervisningsministeriet, 1976).

15. For example, Maya, Aztec, Egyptian, Babylonian, and Roman numerals are treated (together with numerals from sign language and the Braille and Morse alphabets) in an activity booklet for grades 6 or 7 in the folkeskole, published by the association of mathematics teachers in the folkeskole:

Lene Christensen & Jørgen Schiermacher, *Tal, tegn ... og underlige gerninger* (Nordby: Forlaget Matematik, 1991). Also, in a series of textbooks for the folkeskole now being published, Marianne Holmer & Svend Hessing, *Faktor Matematik* (Albertslund: Forlag Malling Beck, 1992–), one finds in the text for grades 8–9 short accounts, with portraits, of the following: in the section on geometry, Archimedes; in the section on algebra, Niels Henrik Abel, Arthur Cayley, Francois Viète and Évariste Galois; in the section on coordinate systems, René Descartes; in the section on probability, Jakob Bernoulli and Pafnuti Lvovitch Chebychev, and in the sections on numbers, Leopold Kronecker and Richard Dedekind.

16. Poul la Cour, *Historisk Matematik* (Copenhagen: P. G. Philipsen, 1888, later eds., Gyldendal and also Grønholt Pedersen, 1899, 1909, 1942 and 1962).

17. This is not the place to review the Danish folkehøjskole movement and its importance for adult education in Denmark and for Danish culture as a whole from around 1850. The reader may consult Steven M. Borish, *The Land of the Living* (Grass Valley, California: Blue Dolphin Press, 1991). In this study—by a sympathetic American anthropologist—the description of the folkehøjskole movement is interwoven with an account of our country's non-violent road from absolute monarchy to modern (or even post-modern) democracy since around 1780.

18. On this institution, see Ernst Larsen, *Danmarks L`rerhøjskole 1950–64. Udgivet i anledning af L`rerhøjskolens 125 års jubil`um*, and Torben Brostrøm, Ingrid Markussen & Kaj Spelling, eds., *Forskning for Folkeskolen— undervisning og videreuddannelse* (both Copenhagen: Danmarks L`rerhøjskole, 1981), and Harald Torpe, *Aldrig løses mand af l`re: Bidrag til Danmarks L`rerhøjskoles historie med hovedv`gt på tiden 1959–1977* (Copenhagen: Danmarks L`rerhøjskole, 1988).

19. Bent Dyrby, Esben Esbensen & Kim Foss Hansen, *Aktivitetsh`fte Tal 8* (Copenhagen: Gjellerup & Gad, 1988), an activity booklet from the textbook series *Matematik 2000*.

20. See Poul la Cour, *Historisk Matematik*, and John McLeish, *Number* (London: Bloomsbury, 1991), Swedish transl., *Matematikens kulturhistoria*, (Stockholm: Forum, 1992), the plates between pages 122 and 123, respectively 128 and 129; in this book the picture carries a reference to *Mary Evans Picture Library*.

21. Robert Recorde, *The Whetstone of Witte* (London: Jhon Kyngstone, 1557). The title page is reproduced om page 319 in vol. 1 of David Eugene Smith, *History of Mathematics*, 2 vols., (Boston: Ginn, 1923–1925, repr. New York: Dover, 1958), and the page on which the equality sign is introduced is reproduced on page 319, 2nd ed. page 324, of Carl B. Boyer, *A History of Mathematics* (New York: John Wiley, 1968, 2nd ed. rev. by Uta C. Merzbach, 1989).

22. Joseph N. Payne, "Ideas", *Arithmetic Teacher*, 34, No. 1 (1986): 26–32.

23. See for example page 81 in vol. 1 and page 288 in vol. 2 of David Eugene Smith, *History of Mathematics*, and page 6 in W.W. Rouse Ball, *A Short Account of the History of Mathematics*, (London: Macmillan, 4th ed. 1908, repr. New York: Dover, 1960), and page 5 in Stuart Hollingdale, *Mak-*

ers of Mathematics (London: Penguin, 1989), and page 115 in vol. 1 of Flemming Clausen et al, *Ind i matematikken.*

24. See pages 95–96 and 105–106 in vol. 1 of Moritz Cantor, *Geschichte der Mathematik,* 4 vols., (Leipzig: Teubner, 1880–1908, repr. New York: Johnson, 1965). See also pages 6 and 28–29 in B.L. van der Waerden, *Science Awakening* (Groningen: P. Noordhoff, 1954), page 24 in idem, *Geometry and Algebra in Ancient Civilizations,* (Berlin: Springer, 1983), and pages 161–162, 238 and 242 in Richard J. Gillings, *Mathematics in the Time of the Pharaohs,* (Cambridge, Massachusetts: The Massachusetts Institute of Technology Press, 1972, repr. New York: Dover, 1982).

25. Kristin Dahl, *Den fantastiska matematiken* (Stockholm: Fischer, 1991), Danish transl., *Den fantastiske matematik* (Copenhagen: Høst & Søn, in prep. 1993).

26. Asger Aaboe, *Episodes From the Early History of Mathematics* (New York: Random House & L.W. Singer, 1964), Danish ed., *Episoder fra matematikkens historie* (Copenhagen: Munksgaard, 1966, repr. Borgen, 1986).

27. See Archimedes, *Measurement of a Circle,* pages 91–98 in Thomas L. Heath's above-mentioned translation of Heiberg's *Archimedes.*

28. For a discussion of Heiberg's text of Archimedes's *Measurement of a Circle* (which is, as Heiberg was well aware of, incomplete), and a comparison with alternative versions preserved by the late commentators Pappus and Theon, see Wilbur R. Knorr, "Archimedes' Dimension of the Circle: A View of the Genesis of the Extant Text," *Archive for History of Exact Sciences,* 35 (1986): 281–324.

29. R. L. E. Schwarzenberger, "The Early Historical Development of Calculus," pages 145–155 in Hilary Shuard & Hugh Neill, *From Graphs to Calculus,* (Glasgow and London: Blackie, 1977).

30. Kirsti Andersen, "Cavalieri's Method of Indivisibles," *Archive for History of Exact Science,* 31 (1984/85): 291–367.

31. Tony Rothman, "Genius and Biographers: The Fictionalization of Évariste Galois," *The American Mathematical Monthly,* 89 (1982): 84–106.

32. Eric Temple Bell, *Men of Mathematics* (New York: Simon & Schuster, 1937, and London: Penguin, 2 vols., 1953), Danish transl. *Matematikkens Mænd* (Copenhagen: C.A. Reitzel, 1944). The Galois legend can be found on page 413 of the Penguin edition.

33. David M. Burton, *Abstract Algebra* (Dubuque, Indiana: Wm. C. Brown, 1991, and idem, *The History of Mathematics: An Introduction* (Newton, Massachusetts: Allyn & Bacon, 1985 and Dubuque, Indiana: Wm. C. Brown, 2d ed. 1991).

34. Herbert Meschkowski, *Probleme des Unendlichen: Werk und Leben Georg Cantors* (Braunschweig: Vieweg, 1967); Ivor Grattan-Guinness, "Towards a Biography of Georg Cantor," *Annals of Science,* 27 (1971): 345–391 and plates xxv–xxviii; Joseph Warren Dauben, *Georg Cantor: His Mathematics and Philosophy of the Infinite* (Cambridge, Massachusetts, and London: Harvard University Press, 1979), especially the introduction and chapter 12, "Epilogue: The Significance of Cantor's Personality;" and Walter Purkert & Hans Joachim Ilgauds, *Georg Cantor, 1845–1918* (Basel: Birkhäuser, 1987), in the series *Vita mathematica,* revised and augmented edition of an earlier book with the same title (Leipzig: Teubner, 1985), especially pages 50–53, 79–92 and 193–195 in the 1987 edition. The quotation mentioning Bell's "ill-founded embellishments" is from page 3 in Dauben's book.

35. Walter Purkert, "Georg Cantor und die Antinomien der Mengenlehre," *Bulletin de la Société Mathématique de Belgique,* 38 (1986): 313–327. In the Cantor biography by Dauben see especially chapters 10, "The Foundation and Philosophy of Cantorian Set Theory," and 11, "The Paradoxes and Problems of Post-Cantorian Set Theory," and in the biography by Purkert & Ilgauds (rev. ed.) see especially pages 147–159.

36. Burton, *Abstract Algebra* (1991), page 2.

37. Georg Cantor, *Gesammelte Abhandlungen mathematischen und philosophischen Inhalt* ed. by Ernst Zermelo, (1932, repr. Hildesheim: Olms, 1962).

38. See for example pages 3–4 in I. A. Barnett, *Elements of Number Theory* (Boston, Massachusetts: Prindle, Weber & Schmidt, 1969), pages 61–63 in David M. Burton, Elementary Number Theory (Dubuque, Indiana: Wm. C. Brown, 2d ed. 1991), pages 3–4 in Paolo Ribenboim, *The Book of Prime Number Records* (New York: Springer, 2d ed. 1989), page 3 in idem, *The Little Book of Big Primes* (New York: Springer, 1991), and Abner Shimony, "That There Exists No Greatest Prime," *Synthese,* 92 (1992): 313–314.

39. See page 412 in vol. 2 of Thomas L. Heath (transl.), *The Thirteen Books of Euclid's Elements,* 3 vols. (Cambridge: Cambridge Univeristy Press, 1908, 2d ed. 1926, repr. New York: Dover, n.d.).

40. Gian-Carlo Rota, "Mathematics and Philosophy: The Story of a Misunderstanding," *Humanistic Mathematics Network Newsletter* 6 (1991): 49–55.

41. Imre Lakatos, *Proofs and Refutations: The Logic of Mathematical Discovery* ed. by John Worrall and Elie Zaher, (Cambridge: Cambridge University Press, 1976). See also Teun Koetsier, *Lakatos' Philosophy of Mathematics: A Historical Approach* (Amsterdam: North-Holland, 1991).

Ethnomathematics: An Explanation

Ubiratan D'Ambrosio
Universidade Estadual de Campinas, Brazil

The Cultural Context of Mathematics and the Program Ethnomathematics

The history of mathematics needs to be examined in a broader context than is generally done today, so as to incorporate other forms of knowledge of natural phenomena. This is more than an academic exercise, since its implications for pedagogy are clear, especially if we refer to recent advances in cognitive science which show culture and cognition to be strongly related.[1] Although there have long been indications of a close connection between cognitive mechanisms and cultural environments, a reductionist tendency traceable to René Descartes has grown in parallel to the development of natural science. This tendency, which has dominated education until recently, implies culture-free cognition models. David Lancy's holistic recognition of the interpenetration of biology and culture (1983) suggests a fertile ground for research on culture and scientific cognition.[2] The connections of this position with education have been discussed with special regard to science and mathematics education.[3] This paper will discuss and help create a conceptual framework for ethnomathematics.

Ethnomathematics lies on the border between history of mathematics and cultural anthropology. It may be conceptualized as the study of techniques (*tics*) developed in different cultures (*ethno*) for explaining, understanding, and coping with (*mathema*) their physical and socio-cultural environments, which encompasses the study of scientific and technological development in direct relation to social, economic, and cultural contexts.[4] In recent years ethnomathematics has been drawing more and more attention. The disciplinary context in which we generally operate is a product of modern western science. Choosing ethnomathematics as a title for a general theory of knowledge recognizes the special place that mathematically related notions, such as size and dimension, quantity, and shape, as well as order, classification, and inferences occur in all cultures, although in different modes. If we regard ethnomathematics as related to modes of thought, ethnomathematics is broader than its meaning as given by most authors—authors who restrict it to practices of mathematical nature (such as counting and measuring) of non-literate cultures, which only recently have been recognized as structured forms of knowledge.[5]

The etymological exercise which led to the term ethnomathematics for this broader approach to knowledge was justified by the conceptual origin of the three components of the word ethnomathematics. Ethno refers to the cultural roots of a people, much broader than the racial component which shows cultural mixing everywhere. Mathema's origin goes back to Pythagoras and suggests explaining and understanding. Tics originates from techné, the same root that gave us art and technique. Thus, ethnomathematics is the art or technique of explaining, understanding, knowing and perceiving, which is rooted in cultural components.[6]

When we see practices that are typically scientific—practices such as counting, ordering, sorting, weighing and measuring performed in different ways in different settings—notions of a universality of methodology of mathematics and science must be seriously challenged. The discovery of this cultural variety of practices has encouraged further studies on the evolution of scientific concepts and mathematical practices within a richer cultural and anthropological framework.

To date this has been done only to a limited, and perhaps, timid extent. Raymond L. Wilder's *Mathematics as a Cultural System* (1981)[7] and Charles Smorinsky's "Mathematics as a Cultural System"[8] seem important attempts by mathematicians to pursue this appoach. In a survey paper Walter S. Sizer[9] restricts his analysis to preliterate people and effectively discusses ethnomathematics although without actually using that name. Others, like Marcia Ascher and Paulus Gerdes, restrict their studies to cultures which have no writing—cultures they call "pre-literate." In addition to these scholars, other anthropologists have begun to produce literature which bridges the approach between anthropologists and historians of mathematics and which constitutes an important step toward recognizing different modes of thought which lead to different forms of science.

Anton Dimitriu's extensive *History of Logic*[10] briefly describes Indian and Chinese logic as background for his historical study of the logic stemming from Greek thought. Many historians do not recognize the existence of different forms of mathematics and logic. But we know from other sources that such a basic concept as "number one" is quite a different concept in the Nyãya-Vaisesika epistemology: "number one is eternal in eternal substances, whereas two, etc. are always non-eternal" and from this follows a well-developed arithmetic.[11] Almost nothing is known about the assumptions underlying the Inca treatment of numbers. What is known through the study of the "quipus" represents a mixed qualitative-quantitative language.[12] The concept of experience, or the experimental method, must also be discussed. René Thom argues in favor of a Heracletian position, and he challenges what we might call the "experimental basis of scientific knowledge" in favor of theoretical reflection. This suggests the possibility of a new conceptualization for experience, hence for science. Details are given elsewhere.[13]

Academic and Practical Mathematics

Without considering ethnomathematics, which some might call "lower mathematics," René Thom recognizes since Greek antiquity two mathematics—the first based on understanding the world; the second acting upon it. These two mathematics correspond to this polarization: *la mathématique de l'intelligibilité,* which goes from the global to the local through a process of analysis, and *la mathematique de la maitrise,* which goes from the local to the global, through a process of extrapolation.[14] In his view of what he calls the essence of mathematics, Thom comes close to recognizing that both understanding and acting upon are manifestations of the same dialectical process of action, through which we constantly modify reality. Indeed, he sees "analytic continuation" as the algorithm that allows the transit between local and global. But this is the key concept in Newtonian fluxions and depends essentially on the choice of the parameter *time* structured associatively.[15]

During this historical moment two kinds of mathematics—one coming from the Greeks and the other a distinct ethnomathematics—started to be united into a single mathematics. This process increased the power of those who would master this knowledge, essential for the modernization of the world starting at the birth of modern science. The paradigms of this knowledge indeed turned out to be the paradigms of the new power structure and of the modes of production then emerging.

On the Nature of Science and Mathematics

These remarks ask for a broader interpretation of science and mathematics. Mathematics is more than ciphering and arithmetic, mensuration, and relations of planetary orbits. It is also classifying, ordering, inferring, and modelling. This broad range of human activities throughout history has been expropriated by the establishment, and formalized and codified by scribes and other academics. Essentials of these modes of thought are alive in culturally identified groups in marginality and constitute routines in their daily practices.

Using the broad conceptualization of science, we are able to identify several practices which are essentially scientific in nature. For this reason, I use ethnoscience, instead of athematics, physics, medicine, and so on of distinct groups. My concept of culture results from an hierarchization of behavior, from individual behavior through social behavior and leading to cultural behavior. It relies on a model of individual behavior based on the cycle...

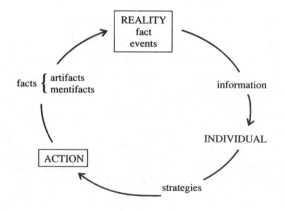

FIGURE 1

reality ⟶ individual ⟶ action ⟶ reality ... schematically described in Figure 1.

This holistic model simply assumes reality in a broad sense—natural, material, social, and psycho-emotional. Now, links are possible through mechanisms of information, including sensorial and memory (genetic and acquired) systems, which produce stimuli in the individual. Through reification and what H. Maturana calls autopoiesis[16] these stimuli foster strategies (based on codes and models), which generate action. Action introduces facts into this reality, both artifacts and "mentifacts." This neologism means all the results of intellectual action which are not material, such as ideas, concepts, theories, reflections and thoughts. These are added to reality and modify it. The concept of reification has been used by the sociobiologists as "the mental activity in which hazily perceived and relatively intangible phenomena, such as complex arrays of objects or activities, are given a factitiously concrete form, simplified and labelled with words or other symbols."[17]

If this is the basic mechanism through which strategies for action are defined, such action, whether through artifacts or mentifacts, modifies reality and produces additional information, which through a reificative process modifies or generates new strategies for action. This cycle is the basis for the theoretical framework for the ethnoscience concept.[18]

Individual behavior, which is homogenized through education, builds into societal behavior, which generates culture. Again, a scheme such as Figure 2, depicts culture as the strategy for societal action. Now the mechanism of reification is replaced by communication, while information is replaced by history, which affects society as a whole. I shall not go deeper into this theoretical framework for the concept of culture.

As mentioned above, culture manifests itself through jargons, codes, symbols, and ways of reasoning and inferring. Associated with these are practices such as ciphering and counting, measuring, observing, classifying, ordering, inferring, modelling, and so on, which are the essentials of ethnomathematics and ethnoscience. What we identify now as a mathematics has, obviously, these same essentials.

The basic question is the following: how "theoretical" can ethnoscience be? Scientific practices, like those mentioned in the end of the previous paragraph, are known to several culturally differentiated groups and are pursued in different ways from the western academic ways of doing them. These ways have been recognized by anthropologists and regarded as native science. For over a century anthropologists have studied the ways native peoples think. But what is the underlying structure of inquiry of these practices commonly classified as ad hoc?

I pose the following questions:
1. How do we pass from ad hoc practices and solution of single problems to methods?

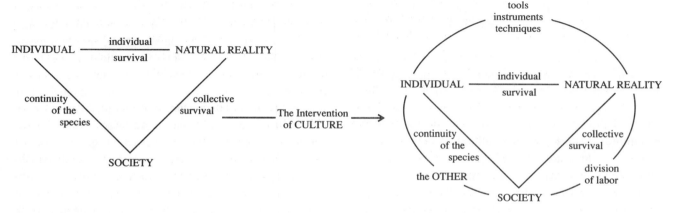

FIGURE 2

2. How do we pass from methods to theories?

3. How do we proceed from theories to inventions?

Although I will not go into details here, these three questions suggest a research program in the history of science and mathematics, which may look into these steps as the steps in building up scientific theories. Of course, this is intimately related to ethnomathematics.

The main methodological issue lies in the concept of history, in particular the history of science. The initial sentence in Enrico Bellone's excellent book *A World on Paper*:

> There is a temptation hidden in the pages of the History of Science—the temptation to derive birth and death of theories, the formalization and growth of concepts, from a scheme (either logical or philosophical) always valid and everywhere applicable... Instead of dealing with real problems, history would then become a learned review of edifying tales for the benefit of one philosophical school or another.[19]

Ideas which have been generated and developed in certain times and places can not be treated as universally applicable and valid. In dealing with science in general, and particularly with mathematics, one must understand the entire process of the generation, organization, transmission, institutionalization, and diffusion of knowledge to have any hope of incorporating them into the streams of history, or of integrating them into the cultural bases of society.

School systems in a democratic society should encourage critical acquisition of knowledge. Regrettably, systems in both aristocratic or oligarchic societies are structured to identify individuals through *filters* which distribute people into categories such as achievers and failures, successful and unsuccessful, winners and losers, those who will serve the aristocracy or oligarchy. Medals, titles, tests and exams, and other selective processes serve as filtering instruments. Managing these filtering systems was exercised sometimes by religious orders and court practices, by civil service, mainly in the Asian tradition, and in modern society, by military services.

The basic assumption in the cognitive sciences is that knowledge depends upon interactive dynamics involving the individual and his environment. Knowledge is generated through socio-emotional and cognitive steps originating in the interaction between an individual and the environment—natural, social, cultural—since birth. The dynamics of this interaction, mediated by communication and the resulting codification and symbolization, produces structured knowledge, which eventually becomes disciplines. These are expropriated by the power structure or the establishment and returned to the people, who in fact

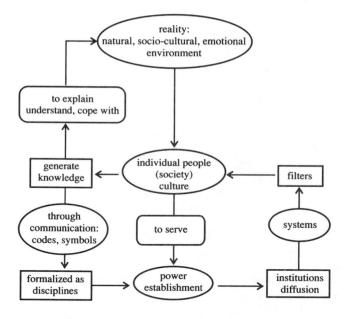

FIGURE 3

generated this same knowledge, by a diffusion process, impregnated with filters, structured as systems. Figure 3 clarifies this entire process.

The filtering is done so that the interests of the power structure are served. For aristocracies and oligarchies, this is done by preserving dynasties and parties or corporations in power. For similar reasons, in some civilizations science is controlled by a priesthood.

This model of generation, organization and transmission, and institutionalization of knowledge does not exclude systems as a basic factor in bridging the gap between individual, social, and cultural behavior. The first steps in moving from individual to social behavior come when a child discovers another. The first "others" to influence the child are those emotionally close to the child: parents, siblings, close relatives, neighbors, friends. All carry an emotional attachment. Only in the school does the child get involved with the other as a stranger, and the process of socialization advances. Schools functions as a step towards social behavior. More than functioning as a means of transmitting knowledge (which is also done through other systems, including the media), schools initiate conviviality with strangers in the diffusion of knowledge, which is essential in modern society. Other potential goals of school are to teach students to cooperate and to collaborate in studies aand to learn about the common good. This instruction is based on *interaction*. Every single child must master existing coping mechanisms and develop increasing confidence at distinct levels.

Knowledge is transmitted not only by school systems but also more generally by current media (printed materials—books and journals—and telecommunications). Some believe that the amount and variety of information make it impossible to use older teaching methods. The teacher must rely on the students and on fresh information which was sometime not available to the teacher. This new role for the teacher was discussed in a former paper of mine.[20]

Science and Mathematics in the Modern World

How did science come to be regarded as a large scale professional activity? As already mentioned, this situation evolved only since the early nineteenth century. Although scientists communicated through scientific periodicals, meetings and national and local associations from at least the seventeenth centuries. But only in the nineteenth century did the professionalization of the sciences transpire, and it occurred as the the differentiation of science into scientific fields progressed. Educating scientists, now professionals with specific qualifications in a specific subject, was done in universities or institutes, and mechanisms to qualify them for professional activities were developed. done in a specific subject, in universities or similar institutions, and mechanisms to qualify them for professional activities were developed. Standards of evaluation of credentials were developed. The University of Berlin established criteria for academic careers, a practice soon followed by universities all over he world. Demonstrated scientific knowledge became the basis for acquiring credentials to pursue professional activity. This same process, practiced in many strata of society at different levels of sophistication and depth, was expropriated by those responsible for professional accreditation. A power structure based on the authority to grant specific certification thus developed.

Consider the case of mathematics, the parallel development of a scientific discipline outside the established model of the profession. The example is the discovery of Dirac's delta function, which only about 20 years after being in full use among physicists, was expropriated and became a mathematical subject as the theory of distribution. This is part of the internal dynamics of knowledge vis-a-vis society and illustrate the left column of Figure 3.

There is a time lag between the appearance of new ideas in mathematics and the general acceptance of these ideas into mathematics, with their formalization. The case of the delta function is an example of what we call ethnomathematics in its broadest sense. Eventually, it becomes

mathematics in the style of thought recognized as such. In many cases, it never gets formalized, and the practice continues to be restricted to the culturally differentiated group which originated it. Schooling replaces these practices by equivalent practices which present a codified version available only to those educated. The same is true of scientific knowledge in general.

Those practices, which have not yet reached the level of science in the usual sense, or paraphrasing the terminology of Thomas S. Kuhn, which are not "normal science," are usually called ethnosciences. It is unlikely that these will generate revolutionary science in the Kuhnian sense. Ethnoscience evolves as a result of societal change, but new forms simply replace former ones, which disappear. The cumulative character of this form of knowledge cannot be recognized, and its status as a scientific discipline becomes questionable. Internal revolutions in ethnoscience, which result from societal changes as a whole, are not sufficiently linked to normal ethnoscience. Historical development is perhaps not recognizable. Consequently, ethnoscience is not recognized as a structured body of knowledge, but a set of ad hoc practices. To identify ethnoscience as a structured body of knowledge, it is essential to follow steps 1, 2, and 3 above, and this requires a new historiography and the development of methodological alternatives.

At present, there is a collection of examples and data on practices of culturally differentiated groups, which are identifiable as scientific and mathematical practices, hence ethnoscience and ethnomathematics. Efforts are under way to link those practices as a pattern of reasoning, a mode of thought. From cognitive theory and cultural anthropology, it may be possible to trace the origin of those practices. In this way, we may reach a systematic organization of these practices into a body of language.

Conclusion

Effective research on ethnomathematics requires both scholarly experience in mathematics and research methods to understand their context. This requires advanced anthropological research in the sciences together with social history of science, which aims at understanding mutual sociocultural, economic, and political factors in the development of science. Only these can go beyond thematic studies to shed light on the underlying ground which is needed to understand better the evolution of scientific knowledge.

History of science and mathematics also acquires a more global, holistic approach, by the consideration of methods, objectives and contents of scientific knowledge in solidarity with anthropological findings and social historical analysis.

It is a mistake only to analyze these components individually.

The distinction between areas of knowledge and disciplines is artificial and ideologically dangerous. To define research priorities in such a way as to incorporate national priorities into scientific practices is also difficult. This problem leads to a close of the theme of this paper by considering the relation of mathematical knowledge and ideology.

Ideology permeates every style of thought. It is possible to find relations between styles of dressing, housing, sports, and leisure. This includes the logic of structured knowledge as well. Science and mathematics result from some logic which underlies the ideological roots of western and oriental civilizations.

This paper has assumed a broad conceptualization of science and mathematics, which allows for examining apparently unstructured forms of knowledge. This results from a concept of culture which accepts a hierarchization of behavior, from individual through social behavior to cultural behavior. The model for the cycle is... $\longrightarrow$ reality $\longrightarrow$ individual $\longrightarrow$ action $\longrightarrow$ reality $\longrightarrow$ .... This model underlying science and mathematics allows for the inclusion of marginal practices of a scientific nature called ethnoscience. These common practices are impregnated with ideological overtones deeply rooted in the cultural texture of practitioners. Understanding of these ideological overtones is a main objective of the program ethnomathematcs.

Endnotes

1. See the important book by Geoffrey B. Saxe: *Culture and Cognitive Development. Studies in Mathematical Understanding* (Hillsdale, NJ: Lawrence Erlbaum Associates, Pub., 1991).

2. David F. Lancy: *Cross-Cultural Studies in Cognition and Mathematics* (New York: Academic Press, 1983).

3. Ubiratan D'Ambrosio: *Socio-Cultural Bases for Mathematical Education* (Campinas: UNICAMP, 1985) ; and Ubiratan D'Ambrosio : "Culture, Cognition and Science Learning", *Proceedings of the Conference on Science Education in the Americas (held in Panama, 1984)* (Washington: NSTA, 1986).

4. Ubiratan D'Ambrosio: "Science and Technology in Latin America During Discovery," *Impact of Science and Society,* 1977, 27, no.3: pp. 267–274.

5. The works of Paulus Gerdes, Marcia Ascher, Bill Barton, Eduardo S. Ferreira, and many others, including myself in my early writings on ethnomathematics, reveal such a focus on mathematics of other cultural groups.

6. Ubiratan D'Ambrosio: "On Ethnomathematics" *Philosophia Mathematica,* 1989, Second Series, vol.4, no. 1: pp. 3–14.

7. Raymond L. Wilder: *Mathematics as a Cultural System,* (Oxford: Pergamon Press, 1981).

8. Charles Smorynski: "Mathematics as a Cultural System," *Mathematical Intelligencer,* 1983 vol. 5, no. 1: pp. 9–15.

9. Walter S. Sizer: "Mathematical notions in preliterate societies," *The Mathematical Intelligencer,* 1991 , vol. 13, no. 4: pp. 53–60.

10. Anton Dimitriu: *History of Logic* (Kent: Abacus Press, 1977).

11. Karl H. Pooter (ed.): "Indian Metaphysics and Epistemology" *Encyclopedia of Indian Philosophies,* (Princeton N.J.: Princeton University Press, 1977) p. 119.

12. Marcia Ascher and Robert Ascher: *Code of the Quipu* (Ann Arbor: University of Michigan Press, 1981)

13. Ubiratan D'Ambrosio: (to appear), "Teoria da Catastrofes: Um Estudo em Sociologia da Ciencia."

14. René Thom: *Apologie du Logos* (Paris:Hachette; 1990); pp. 325–26.

15. The fact that analytic continuation is closely associated with associativity in the structure of time is studied in Ubiratan D'Ambrosio: *Dynamical Systems* and Huygen's *Principle Philosophia Mathematica,* 1972, vol. 9, p. 27–39.

16. See Humberto R. Maturana and Francisco J. Varela, *The Tree of Knowledge. The Biological Roots of Human Understanding,* (Boston: New Science Library, 1987).

17. Charles J. Lumsden and Edward O. Wilson: *Genes, Mind and Culture* (Cambridge: Harvard University Press, 1981).

18. This is explained in detail in Ubiratan D'Ambrosio: *Da Realidade 'a Acão. Reflexões sobre EducaÅão (e) Matemática* (2a ed.) (São Paulo: Summus Editorial, 1988).

19. Enrico Bellone: *A World on Paper* (Cambridge: MIT Press; 1980) p. 1.

20. Ubiratan D'Ambrosio: "Environmental Influences" *Studies in Mathematics Education,* vol. 4, Robert Morris ed., (Paris, UNESCO, 1985); pp. 29–46.

The Necessity of History in Teaching Mathematics

V. Frederick Rickey
Bowling Green State University

The first edition of collected works of Euclid, *Euclidis quæ supersunt omnia,* was published at Oxford in 1703 by the Scottish mathematician David Gregory (1659–1708). In 1683, before he received his degree, Gregory was appointed professor at Edinburgh to the same chair that had been vacant since its previous holder, his more famous uncle, James Gregory (1638–1675), had died. In 1691, David Gregory became Savilian Professor of Geometry at Oxford, and he held the chair until his death in 1708.

In the background of the etching on the title page of Gregory's Euclid is the Sheldonian Theater, an early design of Christopher Wren (1632–1723), the great architect who would be better known as a mathematician had it not been for the fire of London in 1666. In the foreground are a set square and pair of compasses, symbolizing the tools of the geometer. The frontispiece (next page) is even more interesting.[1] It shows a ship wrecked on the shores of Rhodes and three philosophers, the Socratic philosopher

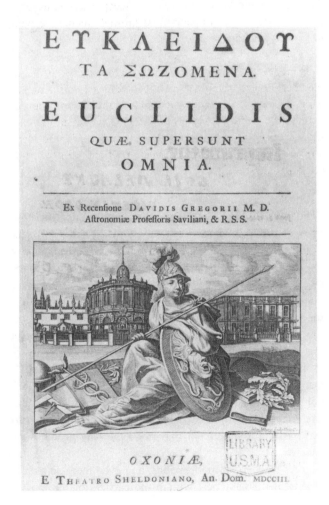

Aristippus Philosophus Socraticus, naufragio cum eiectus ad Rhodiensium litus animadvertisset Geometrica schemata descripta, exclamavisse ad comites ita dicitur: Bene speremus, Hominum enim vestigia video. Vitruv. Architect. lib. 6. Praef.

Aristippus (c. 435–356 BC) and two companions, who have made it safely to shore (curiously, they do not seem wet). They have come across several geometrical diagrams in the sand. As the Roman architect Vitruvius (first century, B.C.) tells the story, one of them exclaims "Bene speremus, hominum enim vestigia video"—We have nothing to fear, for I see the footprints of men.[2]

This aphorism of Vitruvius sets the tone for the point that I would like to argue here: Mathematics is the work of individuals. It is a discipline that has been developed by many people over the ages, some making great contributions, some making minor contributions, with the cumulative effect that mathematics has developed into a rich field that has had a significant impact on the way people view their world.

As teachers of mathematics, and even more so as historians of mathematics, we are the carriers of the mathe-matical culture. It is our solemn responsibility to transmit this culture to our students. It is not sufficient that we simply present the mathematical details. Our responsibility is much greater. We would be derelict in our responsibilities as teachers if we presented mathematics as a fully developed discipline, a discipline that seemingly appeared millennia ago in perfect form.

In the United States there are currently two big reform movements under way in mathematics education. One is the calculus reform movement, the move toward a lean and lively calculus, with the stress on understanding rather than manipulation. The other pushes teachers to follow the standards developed by the National Council of Teachers of Mathematics (published 1989). Both encourage substantial use of calculators and computers to teach mathematical ideas. Their use will eliminate much of the arithmetical and algebraic drill and rote from the classroom work of students. This will free up time in our classrooms for more substantial work. Unfortunately this means that teachers will no longer have to spend so much time on the parts of mathematics that they have become adept at *training* students to do. The hard part remains. Now we must deal with the *ideas* of mathematics. Ideally, we wish to give our students a deep and penetrating understanding of what mathematics is all about. This will be a new and substantial challenge to us, but I firmly believe that the careful and judicious use of the history of mathematics can help us to meet these challenges.

What is lost if the classroom teacher does not use history? Think about your own education. Only two of my teachers were inclined to make historical remarks, and both of them influenced me significantly. Those remarks helped me to understand the material better and helped me to see how it fit into the wider domain of mathematics. Without these two teachers my education would have been much poorer. I remember taking many courses in the theorem-proof style. Yes, I learned a lot from them, but now I realize that I could have learned much more. Everyone who does research in mathematics is somewhat aware of the history of their own field, especially the recent work in the field. It is important to know who proved what and when they did it and how it fits into the development of the field. Graduate advisors pass some of this on to their own doctoral students, but basically students are left to learn it on their own. This is a mistake. The situation must be corrected. Moreover, if this way of proceeding is beneficial at the research level, then it is all the more appropriate at lower levels. If we would show our students what a wonderful field mathematic is, then we would have more interested students who would take more mathematics courses. Eventually we would have a general population with a much better feel for what mathematicians do and why it is important.

What can we, as members of this International Study Group for the relations between History and Pedagogy of Mathematics, do to improve the situation? Here are three suggestions:

1. Convert your colleagues to our cause. When I present talks about using history in the classroom, I like to give very specific examples of how it can be done.[3] I like to think this does some good, but I get the awful feeling that what I say does not go beyond the audience of the day. They may use my ideas in their classroom, and I am pleased when they do, but I do not get the feeling that they are sharing them with their colleagues. The teachers who really need to learn about the many ways of using history in the classroom are the ones who do not attend meetings. The only way to reach them is to seek them out individually. So I implore you: talk to your colleagues, preferably the younger ones, and convince them to use history. Show them specific examples, give them a note you have written or one you picked up at a meeting and encourage them to try it out in their classroom. Then talk to them about how it worked and suggest other ideas for them to try. This will be a slow arduous process, but it is one way to make progress. Perhaps it is the only way.

2. Prepare and publish materials for the classroom teacher. We who know a lot of history and are really interested in it must do things to help the classroom teacher. The teachers in the schools obviously do not have time to do the historical research necessary to introduce history in the classroom. Let me give a specific example of how we can help the teachers. All of us know of the work of Muḥammad ibn Mûsâ al-Khwârizmî. We know how his name and the title of his book enriched our language by giving us the words 'algorithm' and 'algebra.' We know his work on the quadratic equation, and that he coined the phrase "completing the square" because it had a definite geometric meaning. We know that this geometric approach helps students to understand quadratic equations better. Yet when I ask undergraduates if they have seen this before, I usually get blank stares. When I ask a group of high school teachers if they know about this, far too many of them do not. The message is simply not getting out. We must do more to make teachers aware of how history can be used in their classrooms to motivate and teach their students.

At the very least we need to prepare bibliographies with indices which can tell teachers precisely where to find information on a given topic. But most importantly, we need to write short notes on specific classroom topics that explain all of the history and mathematics in detail, provide good graphics that the teacher can easily reproduce, and provide teaching tips that will convince the teacher of the benefit of proceeding historically. One way to do this is to work with the editors of our local mathematics newsletters, the ones that go directly to teachers. Write to them and volunteer to write and edit a column with historical teaching tips.

The *HPM Newsletter*[4] has become very popular and provides an excellent summary of what is happening in the world in the history of mathematics, but it does have one fault that we should correct. There is not enough material there for the teacher. What we need to do is to start sending Victor Katz, the editor, short articles about how we use history in the classroom to present specific ideas. I am not talking about general arguments like those that I am giving here, but very concrete ideas with all the material included that the harried teacher needs. He would be happy to publish them, and all of us would benefit greatly.

3. Participate in an email group. My final suggestion is something that will primarily benefit those of us with a serious interest in history and pedagogy of mathematics, but will also have an indirect benefit on our students and other teachers. We should use email extensively to communicate among ourselves. This is an excellent way to exchange information about good publications, to ask about references on a given topic, to publicize future meetings and to report on what happens at them, to let our colleagues know of historical notes we have written and how to obtain copies. This mode of communication will not replace the *HPM Newsletter*, for while many historians have access to electronic mail, most teachers don't; it is intended as a supplement. If you would like to join an email group dealing with the history of mathematics, send an email note to

`majordomo@maa.org`

consisting of the single line

`subscribe math-history-list`

and your name will be added to the list.

Enough of philosophy and preaching. Now let me enumerate the various ways that I have used history in the classroom:

To introduce a new topic
History of specific concepts
History of notation
Etymology of terms
Pictures of mathematicians
Biography: Identify every individual mentioned
Quotations by famous mathematicians
Anecdotes
Title pages from famous books
Problems from old textbooks
As a way to discuss advanced and modern topics
Historical errors

I would like to give examples of each of these ways of using history, for it is my firm belief that what we need is not talk about using history, but examples of how it has been successfully used in the classroom. But space is limited, so I have chosen to give one concrete example of how historical errors can be used to teach our students more about mathematics.

Karl Weierstrass (1815–1897) developed, over many years, a wonderful reputation as a teacher of advanced mathematics, so I would like to discuss one thing that we can learn from him. He had taught in several Gymnasia in East Prussia for fourteen years before being invited to join Borchardt, Kummer, and Kronecker at the University of Berlin in 1857. For several years before Weierstrass arrived in Berlin, Jacobi had been conducting a mathematics seminar at Königsburg—it was the first in the world—and, when Kummer and Weierstrass read about it, they adopted the idea and wrote to the minister of culture with their plan of improving teaching, including mathematics and elocution, via a seminar. Although Berlin was a research university, the mathematicians demanded that the students should learn how to be good teachers in the high schools and colleges.[5]

Naturally, they also desired to expand the mathematical experiences of their students by introducing them to problems on the frontiers of mathematics. Weierstrass knew that he had benefited from reading the masters—Euclid, Archimedes, and Euler—and he wanted to introduce his students to the works of these masters. Each year an area of mathematics was chosen to concentrate on and the students were given a series of problems so that they could discover for themselves what the field was all about. Weierstrass believed that the way to understand mathematics and to become a creative mathematician was to follow how the breakthroughs occur.

The Berlin mathematics seminar began April 6, 1860, with two functions, to create new mathematics and to improve teaching. They began with problems, proceeded to in-depth research, and used the seminar time to report on what they had done. There were only 12 students in the seminar at a time. One of them was Georg Cantor (1845–1918).

Weierstrass felt that a teacher and researcher had a responsibility to tell students about the mistakes that one had made. The Göttingen group gave the students actual mistakes and had them find them. In fact, they felt that you should not even tell the students that they are there. This is critical to the learning process, for there is nothing unusual about making errors. One key to becoming an expert in a field is to learn how to detect and correct the errors that you do make. The Berlin teachers showed the students how to go beyond the problems and to create new mathematics. Weierstrass was exceedingly important in all of this.

Let me give an example that I customarily present in a senior-level class on the foundations of mathematics for prospective secondary teachers which contains an introduction to Cantor's work on transfinite cardinals. It shows one way to use errors made by creative mathematicians in our classrooms.

Theorem. *There are precisely as many points on the line as there are in the plane.*

Proof. Since we have already proved (in class[6]) that there is a bijection between the points in the interval $(0, 1]$ and the whole real line $\mathbb{R}$, it suffices to show that there is a bijection between the square $(0, 1] \times (0, 1]$ and the interval $(0, 1]$. If we represent real numbers by decimals, then a point (x, y) in the square has the form:

$$(0.x_1 x_2 x_3 x_4 \ldots, 0.y_1 y_2 y_3 y_4 \ldots).$$

If we now map this pair of reals to the single real

$$0.x_1 y_1 x_2 y_2 x_3 y_3 x_4 y_4 \ldots$$

which is obtained by interlacing the digits of the two decimals representing the point, then we have the desired bijection. QED

In class, I discuss this map in sufficient detail so that the students are all convinced that it really is a one-to-one correspondence. The decimals represent reals in $(0, 1]$ so the map has the right domain and range. I make it plausible that this map is a bijection by holding up my hands and using the fingers on one hand to represent the digits of x, and the fingers on the other to represent y, and then I interlace my fingers just the way the map interlaces the digits of the decimals. Students seem to appreciate this manual representation of the map.

When all of my students are convinced that this proof is correct, I tell them some of its history. After Georg Cantor showed in 1874 (in the first paper on set theory) that there are more real numbers than there are natural numbers, he next looked for sets of greater cardinality. He guessed that there were more points in the plane than on the line, but was unable, after three year of work, to prove this result.[7] Cantor consulted with several leading mathematicians about the correctness of this result and they responded that it was obvious that there were more points in the plane than on the line. Finally, in 1877, Cantor found the proof that I have just presented. He was so amazed with the result that he wrote his friend Richard Dedekind (1831–1916), "Je le vois, mais je ne le crois pas."—I see it, but I don't believe it![8]

Cantor soon realized that his proof—intuitive that it was—was incorrect. The idea is a neat one, and correct in spirit, but the details are not correct, for the interlacing 'function' is not really a function. This is the problem I set my students: provide an example to show the interlacing is not a function and then show how the details of the specious proof can be altered so that it becomes correct. This is a hard question for them, and only the best students in the class can find a solution, but it forces all of them to really come to grips with understanding the proof. The main reason that I present this exercise is that it forces them to deal with the way we represent real numbers as decimals.[9] This is not a trivial detail—for every secondary teacher must have a solid understanding of the real number system.

The important part about this historical presentation of Cantor's mathematics is that the student learns more than if I had simply presented the correct proof right away. For one thing they would then be puzzled about why the complicated details were necessary, and it would take a good deal of time to explain that. With the historical approach the student learns how mathematics is really done. They learn that very good mathematicians make mistakes, and that mathematicians work hard at eliminating them before their work is published. By making an effort to find the error, the student makes an intellectual commitment to the problem and is thus prepared to accept and appreciate the details of the proof.

Cantor found a correct proof that used continued fractions (remember that he wrote his dissertation in number theory at Göttingen) and published it in 1878. This paper was delayed by Kronecker and was finally published through the intercession of Weierstrass. The result was a surprise to the mathematical community, for it showed that dimension was not determined by cardinality. This led to the development of a new field of mathematics called dimension theory.

So we see here another of the benefits of using history of mathematics in the classroom. It allows us to talk about more advanced and more recent results in mathematics where it would be impossible to present the mathematical details. This is one point where the physicists have done a good job, and, if we want to attract the best students into mathematics and to improve the image of mathematics in the mind of the general population, then we need to emulate the physicists and find a way of talking about recent results in layman's terms.

Perhaps the simplest way to take advantage of errors in the classroom is to utilize the mistakes that we teachers (and our students) make in class. Some are simply a slip of the chalk, but sometimes we get off on the wrong track in doing a problem or presenting a proof. There is a temptation to erase and start over, for this is the quickest way to salvage the situation. But it is not the best for the student. We should continue with the bad approach until the students see the folly of it. Then we should do our best to make them explain where we got off the track. This way the students will learn from our errors.[10]

Naturally this example of Cantor's can only be done with a more advanced class, but the same idea can be used in elementary classes. The book *Fallacies in Mathematics* (1961) by E. A. Maxwell—to cite just one specific example—is a collection of erroneous proofs that can be adapted to both the high school and college classroom. I encourage you to pay more attention to mistakes in your classroom.

Endnotes

1. A similar frontispiece appears in *Apollonii Pergaei conicorum libri octo et Sereni Antissensis de sectione cylindri et coni libri duo*, edited by Edmund Halley, Oxford, 1710. This frontispiece is reproduced in *The Greek Study of Curves*, p. 27, which is Unit 4 of the Open University course on the history of mathematics, prepared by John Fauvel, 1987. The difference between this frontispiece and the one in the 1703 Euclid is that the geometric diagram in the 1710 Apollonius relates to the conics. For years I thought this was the only difference. It was not until I visited the Bullitt Collection at the University of Louisville (see Richard M. Davit, "William Marshall Bullitt and his amazing mathematical collection," *The Mathematical Intelligencer*, vol. 11, Fall 1989, pp. 26-33) and was able to lay the two volumes side by side that I realized that the plate had been completely redrawn, but carefully copied, for the 1710 Apollonius. The engraver, M. Burghers, was a very talented individual, but I know nothing about him. It would be very interesting to know more about the frontispieces that are used (and occasionally reused) in mathematics texts. I would especially like to know about the allegories they contain.

2. After surviving the shipwreck, Vitruvius continues, Aristippus went to Rhodes where he discussed philosophy. The townspeople rewarded him with gifts, showing that they had a high regard for learning. When his fellow travellers departed for home, they asked if he had any advice for his countrymen. He asked them to say that children should be provided with "resources of a kind that could swim with them even out of a shipwreck." In other words, it's what's in your head that counts. See *Vitruvius. The Ten Books on Architecture*, translated by Morris Hicky Morgan, Harvard University Press, 1914; reprinted by Dover, 1960; see p. 167.

3. Today I am giving a more philosophical talk because I want to convince you—who are already among the true believers in using history in the classroom—that we are not doing enough.

4. If you do not receive this newsletter, write to the editor and ask to have your name put on the mailing list: Victor J. Katz, Department of Mathematics, University of the District of Columbia, 4200 Connecticut Ave. N.W., Washington, D.C. 20008 USA. Simpler still, send him email: vkatz@udcvax.bitnet

5. This is something that we need to pay more attention to today. We who teach graduate students must do all in our power to guarantee that they become good teachers. It is not sufficient that they simply be capable of doing research. There are not many research sinecures today. Moreover, we need to see that all graduate students take a course in the history of mathematics, for it will not only improve their teaching, but it will give them a much better idea of what mathematics is all about and will thus improve their research. We must work to insure that our departments do both of these things.

6. When presenting the related result $(0, 1] \sim (0, 1)$ in class, I invite a student to the board to construct the necessary bijection. We start with a function similar to the identity function. But what should $f(1)$ equal? When this is fixed by giving a value in the range space, say $1/2$, the revised function is not one-to-one. We need to redefine one value of f to fix this, namely $f(1/2)$, but then the function is still not one-to-one, so another value of f must be changed. This process of fixing one problem, but creating another, continues until the students see a pattern and can come up with a good bijection. Through this struggle, the students gain some insight into how mathematics is done. After they have constructed a good bijection, I show them the function that Cantor himself constructed. In fact, I show an overhead of Cantor's letter to Dedekind of June 25, 1877 with his lovely diagram of his function. This has the added benefit that the student sees that there is nothing unique about bijections between sets. For a copy, see Joseph W. Dauben, "Denumerability and dimension: the origins of Georg Cantor's theory of sets," *Rete,* 2, 105-134.

7. It is beneficial for our students to realize that good mathematicians often work on problems for many years before they are successful. I point this out to my students in the hope that it will encourage them to work on problems for more than ten minutes.

8. Curiously, although Cantor was writing to his friend in German on 29 June 1877, he broke into French for this one phrase. The letter has been published in E. Noether and J. Cavaillès (eds.), *Briefwechsel Cantor-Dedekind* (Actualité Scientifiques et Industrielles, no. 518. Paris, Hermann & Cie, 1937), p. 34. It also appeared in *Philosophie mathématique,* Jean Cavaillès (ed.), Paris, Hermann, 1962, p. 211. After the first of these was published, the original correspondence disappeared. Amazingly, it was located many years later by Clark H. Kimberling, while he was writing a paper on Emmy Noether (1882–1935). He sent a query to the lawyer who had probated her estate, and the correspondence turned up in the lawyer's files. See his "Emmy Noether," *American Mathematical Monthly,* 79(1972), 136–149.

9. To see that interlacing of decimals is not a function consider, for example, the images of the pair $(0.2333\ldots, 0.5000\ldots)$ and of the equal pair $(0.2333\ldots, 0.4999\ldots)$. This difficulty is avoided by agreeing to represent reals by infinite decimals that do not end in a string of 0s. But now there is another difficulty. The preimage of, say, .234040404... is not properly represented. A new trick is needed. The simplest one, apparently due to J. König (1849–1914), is to break a decimal into blocks of digits, each consisting of zero or more 0s followed by a nonzero digit (for example, 0.3|0002|5|01|003|... and then to interlace blocks of digits. For additional details and history, see Abraham A. Fraenkel's *Abstract Set Theory,* third edition, 1966, pp. 100–105.

10. Having listed quotations earlier as one way to use history in the classroom, let me quote from a letter of Newton to Nathaniel Hawes of 25 May 1694: "A Vulgar Mechanick can practice what he has been taught or seen done, but if he is in an error he knows not how to find it out and correct it, and if you put him out of his road, he is at a stand; Whereas he that is able to reason nimbly and judiciously about figure, force and motion, is never at rest till he gets over every rub." Perhaps you will note that it is this quotation that inspired the title of Richard S. Westfall's fabulous biography of Newton: *Never at Rest.*

Mathematical Masterpieces: Teaching with Original Sources

Reinhard C. Laubenbacher and David Pengelley
New Mexico State University

Our upper-level university honors course, entitled *Great Theorems: The Art of Mathematics,* views mathematics as art and examines selected mathematical masterpieces from antiquity to the present. Following a common practice in the humanities, for example in Chicago's Great Books program and St. John's College curriculum, we have students read original texts without any modern writer or instructor as intermediary or interpreter. As with any unmediated learning experience, a special excitement comes from reading a first-hand account of a new discovery. Original texts can also enrich understanding of the roles played by cultural and mathematical surroundings in the invention of new mathematics. Through an appropriate selection and ordering of sources, students can appreciate immediate and long-term advances in the clarity, elegance, and sophistication of concepts, techniques, and notation, seeing progress impeded by fettered thinking or old paradigms until a major breakthrough helps usher in a new era. No other method shows so clearly the evolution of mathematical rigor and abstraction.

The end result is a perception of mathematics dramatically different from the one students get from traditional courses. Mathematics is now seen as an evolving human endeavor, its theorems the result of genius struggling with the mysteries of the mathematical universe, rather than an unmotivated, ossified edifice of axioms and theorems handed down without human intervention. For instance, after reading Cayley's paper (see below) introducing abstract group theory, the student is much less bewildered upon seeing the axiomatic version so devoid of any motivation. Furthermore, Cayley makes the connection to the theory of algebraic equations, which a student might otherwise never become aware of. An additional feature of the method is that suddenly value judgments need to be made: there is good and bad mathematics, there are elegant proofs and clumsy ones, and of course plenty of mistakes and unsubstantiated assertions which need to be examined critically. Later follows the natural realization that new mathematics is being created even today, quite a surprise to many students.

To achieve our aims we have selected mathematical masterpieces meeting the following criteria. First, sources must be original in the sense that new mathematics is captured in the words and notation of the inventor. Thus we assemble original works or English translations. When English translations are not available, we and our students read certain works in their original French, German, or Latin. In the case of ancient sources, we must often depend upon restored originals and probe the process of restoration. Texts selected also encompass a breadth of mathematical

subjects from antiquity to the twentieth century, and include the work of men and women and of Western and non-Western mathematicians. Finally, our selection provides a broad view of mathematics building upon our students' background, and aims, in some cases, to reveal the development over time of strands of mathematical thought. At present the masterpieces are selected from the following.

Archimedes: The Greek method of exhaustion for computing areas and volumes, pioneered by Eudoxus, reached its pinnacle in the work of Archimedes during the third century BC. A beautiful illustration of this method is Archimedes's determination of the area inside a spiral. [10] An important ingredient is his summation of the squares of the terms in certain arithmetic progressions. As in all of Greek mathematics, even this computation is phrased in the language of geometry. Further advance toward the definite integral did not come until the Renaissance.

Omar Khayyam: The search for algorithms to solve algebraic equations has long been important in mathematics. After Babylonian and ancient Greek mathematicians systematically solved quadratic equations, progress passed to the medieval Arab world. The work of Arab mathematicians began to close the gap between the numerical algebra of the Indians and the geometrical algebra of the Greeks. Notable is Omar Khayyam's *Algebra* of the late 11th or early 12th century. Here he undertakes the first systematic study of solutions to cubics and writes: "Whoever thinks algebra is a trick in obtaining unknowns has thought it in vain. No attention should be paid to the fact that algebra and geometry are different in appearance. Algebras are geometric facts which are proved." In addition to a general discussion of his view of algebra, an excellent selection to read is his treatment of the cubic $x^3 + cx = d$, which is solved geometrically via the intersection of a parabola and a circle. [20]

Girolamo Cardano: The next major advance toward solving algebraic equations did not come until the sixteenth century with the work of Cardano and his contemporaries in Europe. During that time, ancient Greek mathematics was rediscovered, often via Islamic sources, and old problems were attacked with new methods and symbols. The general arithmetic solution for equations of degree three and four essentially awaited Cardano's seminal *Ars Magna* (1545), in which Khayyam's equation $x^3 + cx = d$ now receives a virtually algebraic treatment. [16] It is instructive to compare the two texts. The final chapter in the search for general solutions for the quintic or higher-degree equations was later written by Niels Abel and Evariste Galois.

Evangelista Torricelli: By the early seventeenth century the Greek method of exhaustion was being transformed into Cavalieri's method of indivisibles, the precursor of Leibniz's infinitesimals and of Newton's fluxions. One of the most astonishing results of this period was the discovery that an infinite solid can have finite volume. Torricelli, a pupil of Galileo, demonstrated by the method of indivisibles that the solid obtained by revolving a portion of a hyperbola about its axis has finite volume. [19]

Blaise Pascal: Closed formulae for sums of powers of consecutive integers such as $\sum_{i=1}^{n} i^m$ were already of interest to Greek mathematicians. For instance, Archimedes used them to determine areas, such as for the spiral above. After much effort over the centuries, Fermat in the early seventeenth century first recognized the existence of a general rule, and called this "perhaps the most beautiful problem of all arithmetic." [2] Shortly thereafter Pascal provided a recursive description of the formulae for sums of powers in an arithmetic progression in *Potestatum Numericarum Summa (Sommation des Puissances Numériques)*. [15] As Pascal mentions, these formulae are connected to the continuing development of integration techniques at the time.

Jacques Bernoulli: Improving on the work of Pascal (which he apparently was not aware of), Bernoulli, in the late seventeenth and early eighteenth centuries, provided the first general analysis of the polynomial expressions giving the sums of powers. His *Ars Conjectandi* (1713) noticed surprising patterns in the coefficients, involving a sequence of numbers now known as the Bernoulli numbers. [18] Today these Bernoulli numbers are important in many areas of mathematics, such as analysis, number theory, and algebraic topology.

Leonhard Euler: The eighteenth century was dominated by applications of the calculus, many of them provided by Euler, who was a master in working with infinite series. His *De Summis Serierum Reciprocarum* contains a variety of results on sums of reciprocal powers, including a recursive analysis of $\sum_{i=1}^{\infty} \frac{1}{i^{2m}}$. [8] Euler's computations are examples of the general formula

$$\sum_{i=1}^{\infty} \frac{1}{i^{2m}} = \frac{(-1)^{m+1} B_{2m} 2^{2m-1} \pi^{2m}}{(2m)!},$$

which involves Bernoulli numbers and can be derived directly from the above text of Bernoulli.

Sophie Germain: The early nineteenth century saw the beginnings of modern number theory with the publication of Gauss' *Disquisitiones Arithmeticae* in 1801. Efforts to prove Fermat's Last Theorem contributed to the development of sophisticated techniques by mid-century. Before then, however, the only progress toward a general solution, beyond confirmation of the conjecture for exponents five and seven (three and four were confirmed by Fermat and Euler), was provided by Sophie Germain. She developed a general strategy toward a complete proof, and used the theorems she proved along the way to resolve Case I of Fermat's Last Theorem for all exponents less than 100. Sophie Germain never published her work. Instead, a part of it appeared in 1825, in a supplement to the second edition of A. M. Legendre's *Théorie des Nombres*, where he credits her in a footnote.[1] [11]

Nicolai Lobachevsky: From its beginning, Euclid's parallel postulate [9] was controversial. Attempts to prove it from the others led to the nineteenth-century discovery that it is independent of the rest, allowing for other geometries. Lobachevsky, the co-discoverer of non-Euclidean geometries along with Gauss and Bolyai, made several attempts at gaining the attention of the mathematical world with his ideas. In 1840 he published the very readable book *Geometrische Untersuchungen zur Theorie der Parallellinien*, laying out the foundations of hyperbolic geometry. [1]

Gotthold Eisenstein: In addition to Fermat's Last Theorem, another driving force in the development of number theory was the Quadratic Reciprocity Theorem and the study of higher reciprocity laws. The theorem, discovered by Euler and restated by Legendre in terms of the symbol now bearing his name, was first proven by Gauss. The eight different proofs Gauss published, for what he called the Fundamental Theorem, were followed by dozens more before the end of the century, including four given by Gotthold Eisenstein in the years 1844–45. His article *Geometrischer Beweis des Fundamentaltheorems für die quadratischen Reste* [7] gives a particularly elegant and illuminating geometric variation on Gauss' third proof. [12,13]

William Rowan Hamilton: After extended efforts, Hamilton's attempts to define a multiplication on three-dimensional vectors led to his flash of insight in 1843 that this was possible if one allowed vectors of dimension four. Selections from his book *Elements of Quaternions* give an interesting account of his geometric view of the quaternions. [17] They provided one of the first

important examples of a non-commutative number system, thus spurring the development of abstract algebra.

Arthur Cayley: By the mid-nineteenth century, group structures had emerged implicitly in several branches of mathematics, for instance in modular arithmetic, the theory of quadratic forms, as permutations in the work of Galois on algebraic equations, in the work of Hamilton on the quaternions, and in the theory of matrices. In his paper *On the Theory of Groups, as Depending on the Symbolic Equation* $\theta^n = 1$ Cayley was the first to investigate the abstract concept of a group and began the classification of groups of a given order in a purely abstract way. [4]

Richard Dedekind: In an attempt to explain calculus better to his students, Dedekind constructed the real numbers through what is now known as Dedekind cuts, from which their continuity can be deduced rigorously. Together with Cantor's equivalent construction, this work represents the culmination of the century-long effort to arithmetize analysis. In 1872 he published these ideas in the celebrated *Stetigkeit und die Irrationalzahlen*. [6]

Georg Cantor: Mathematics was changed forever toward the end of the nineteenth century by Cantor's bold embrace of the infinite. Selections from his readable *Beiträge zur Begründung der Transfiniten Mengenlehre* develop the foundations of his theory of transfinite numbers, or what we today call ordinals and cardinals. [3]

John Conway: A vast generalization of a Dedekind cut, combined with ideas from game theory, led Conway in the 1970s to create, with a single construction, the so-called surreal numbers, an enormous number system containing both the real numbers and Cantor's ordinals. Chapter 0 of his book *On Numbers and Games* provides a delightful introduction to his surreal world. [5] Conway's work is one of the rare examples of very recent mathematics that is deep but can be read with minimal background.

In our experience students find the study of original sources fascinating, especially when combined with readings in the history of mathematics. The benefits for instructors and students alike are a deepened appreciation for the origins and nature of modern mathematics, as well as the lively and stimulating class discussions engendered by the interpretation of original sources. As part of their assignment students complete a research project on a topic of their choice, with the only constraint that it be part mathematical and part historical. Other assignments focus for the most part on mathematical points in the sources and related topics.

---

[1] A more comprehensive evaluation of Germain's work on Fermat's Last Theorem, based on her original manuscripts, can be found in [14].

After using the traditional lecture approach for some time, we discovered the amazing effectiveness of a combination of two pedagogical devices: the "discovery approach"; and extensive writing. The discovery method assumes that students should discover the mathematics for themselves. Hence, for each source we briefly provide the historical and mathematical context, alert the students to any difficult points in the text, and then stand by to answer questions while they work through the source in pairs. A wrap-up discussion lets everyone share his or her understanding of the material, and any remaining difficulties are resolved. This method generates tremendous enthusiasm and a genuine sense of discovery. Strikingly, we see that this method also leads to a deeper understanding of the sources than the lecture approach achieves.

The students write frequently and about every aspect of the course: the mathematical details of the sources, their historical context, lecture notes, thoughts jotted in the throes of problem solving, and their own ideas about the process that creates mathematics. This writing experience leads to a more comprehensive view of the great theorems we study as well as a much better grasp of the mathematical details in their proofs.

For students in the sciences, engineering, and mathematics education, our course provides both a broad and humanistic view of mathematics, and for many students it is a breath of fresh air within the traditional mathematics curriculum. For mathematics majors the course is an enriching capstone for their entire undergraduate experience.

References

1. Roberto Bonola, *Non-Euclidean Geometry*, Dover, New York, 1955.
2. Carl Boyer, "Pascal's formula for the sums of powers of the integers," *Scripta Mathematica*, **9** (1943), 237–244.
3. Georg Cantor, *Contributions to the Foundations of the Theory of Transfinite Numbers, Article I*, Dover, New York, 1895.
4. Arthur Cayley, *The Collected Mathematical Papers of Arthur Cayley*, vol. 1, 423–424; vol. 2, 123–132, Cambridge University Press, Cambridge, 1889.
5. John Conway, *On Numbers and Games*, Academic Press, London, 1976, pp. 3–14.
6. Richard Dedekind, *Essays on the Theory of Numbers*, Dover, New York, 1963, pp. 1–27.
7. F. Gotthold Eisenstein, "Geometrischer Beweis des Fundamentaltheorems für die quadratischen Reste," *Journal für die Reine und Angewandte Mathematik (Crelle's Journal)*, **28** (1844), 246–248 and Taf. II: Figs. 1,2.
8. Leonhard Euler, *Opera Omnia*, Teubner, Leipzig & Berlin, 1924–25, Series I, vol. 14, 73–86.
9. T. L. Heath (ed.), *The Elements*, Dover, New York, 1956, vol. I, pp. 153–155.
10. T. L. Heath, *The Works of Archimedes*, Dover, New York, pp. 107–109, 176–182.
11. Adrien M. Legendre, *Sur quelques objets d'analyse indeterminée et particulièrement sur le théoreme de Fermat*, Second Supplément (Sept. 1825) to *Théorie des Nombres*, Second Edition, 1808.
12. Reinhard Laubenbacher and David Pengelley, "Eisenstein's Misunderstood Geometric Proof of the Quadratic Reciprocity Theorem," *College Mathematics Journal*, **25** (1994), 29–34.
13. Reinhard Laubenbacher and David Pengelley, "Gauß, Eisenstein, and the 'Third' Proof of the Quadratic Reciprocity Theorem: Ein Kleines Schauspiel," *Mathematical Intelligencer*, **16** (1994), 67–72.
14. Reinhard Laubenbacher and David Pengelley, "Here is What I Have Found": Sophie Germain's Forgotten Number Theory Manuscripts, in preparation.
15. Blaise Pascal, *Oeuvres de Blaise Pascal*, Kraus Reprint, Vaduz, Liechtenstein, 1976, vol. 3, 341–367.
16. David E. Smith, *Source Book in Mathematics*, Dover, New York, 1959, pp. 203–206.
17. Ibid., pp. 677–683.
18. Ibid., pp. 85–90.
19. Dirk J. Struik, *A Source Book in Mathematics, 1200–1800*, Princeton Univ. Press, Princeton, 1986, pp. 227–230.
20. J. J. Winter and W. "Arafat, The Algebra of Umar Khayyam," *Journal of the Royal Asiatic Society of Bengal*, **41** (1950), 27–78.

A History-of-Mathematics Course For Teachers, Based on Great Quotations

Israel Kleiner
York University

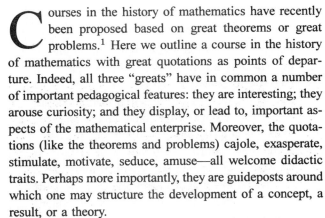

ourses in the history of mathematics have recently been proposed based on great theorems or great problems.[1] Here we outline a course in the history of mathematics with great quotations as points of departure. Indeed, all three "greats" have in common a number of important pedagogical features: they are interesting; they arouse curiosity; and they display, or lead to, important aspects of the mathematical enterprise. Moreover, the quotations (like the theorems and problems) cajole, exasperate, stimulate, motivate, seduce, amuse—all welcome didactic traits. Perhaps more importantly, they are guideposts around which one may structure the development of a concept, a result, or a theory.

At York University, "History of Mathematics" is one of two required courses in an In-Service Master's Program for mathematics teachers offered in the Department of Mathematics and Statistics. The Program, which has been especially designed for teachers, attempts to give students a broad overview of major mathematical fields and issues, to expand their horizons and deepen their understanding of mathematics, to teach them relatively elementary mathematics from a relatively sophisticated point of view, and to broaden their perspective on the mathematics they teach, so that they can better judge what to emphasize in their teaching and why to emphasize it.

The effective teaching of mathematics requires more than a sound command of the subject matter. In his classic *Mathematical Discovery* (1981), George Pólya explains:

> To teach effectively a teacher must develop a feeling for his subject; he cannot make his students sense its vitality if he does not sense it himself. He cannot share his enthusiasm when he has no enthusiasm to share. How he makes his point may be as important as the point he makes; he must personally feel it to be important.[2]

Wise counsel! The history of mathematics can increase teachers' enthusiasm for the subject, promote a sense of its importance, even greatness, and encourage students to ask "why?" in addition to "how?" All these are among the objectives of this course. The following two quotations by Charles Edwards and Otto Toeplitz, respectively, express some of these sentiments well:

> Although the study of the history of mathematics has an intrinsic appeal of its own, its chief raison d'être is surely the illumination of mathematics itself. For example, the gradual unfolding of the integral concept—from the volume computations of Archimedes to the intuitive integrals of Newton and Leibniz and finally the definitions of Cauchy, Riemann and Lebesgue—cannot fail

to promote a more mature appreciation of modern theories of integration.[3]

Regarding all these basic topics in infinitesimal calculus which we teach today as canonical requisites..., the question is never raised, "Why so?" or "How does one arrive at them?" Yet all these matters must at one time have been goals of an urgent quest, answers to burning questions, at the time, namely, when they were created. If we were to go back to the origins of these ideas, they would lose that dead appearance of cut-and-dried facts and instead take on fresh and vibrant life again.[4]

The focus of the course is on mathematical ideas—their origin and evolution. But the ideas are presented in the context of the mathematics which sustains the ideas. The historical context provides the motivation often lacking in the schools. It also provides an opportunity to review or to do some new mathematics with the students, to fill some gaps in their mathematical knowledge.

To come back to the ideas. The biggest idea of all is undoubtedly the nature of mathematics. I put it to my students as a question: "What is mathematics?" This is the $64,000 question, which I do not intend to answer because I do not know the answer. But raising the question is important. Teachers taking this course have been studying, doing, and teaching mathematics for many years, but have probably reflected little on what mathematics is. I do not suggest that this question should be constantly on their minds, but they should have thought about it at least once in their mathematical careers. To paraphrase the conclusion to the preface of Paul Halmos' *Naive Set Theory* (1960), I give the following advice to my students in connection with the question "What is mathematics?": Think about it, try to assimilate it, but don't worry too much about it.[5]

The question "What is mathematics?" is a question in the philosophy of mathematics. But it cannot be addressed without an understanding of the history of mathematics. In general, I heartily endorse Lakatos's remark (paraphrasing Immanuel Kant) that

The history of mathematics, lacking the guidance of philosophy, [is] blind, while the philosophy of mathematics, turning its back on the most intriguing phenomena in the history of mathematics, [is] empty.[6]

I begin to address the question "What is mathematics?" by setting down various definitions and descriptions of the subject given over the years. Here are some:

Mathematics is the study of number and form (Anon).

It is not of the essence of mathematics to be conversant with the ideas of number and quantity. (Boole)[7]

Mathematics is an art. (Anon)

Mathematics is the 'Queen of the Sciences'. (Gauss)[8]

The profound study of nature is the most fertile source of mathematical discoveries. (Fourier)[9]

It is true that Fourier had the opinion that the principal object of mathematics was public use and the explanation of natural phenomena; but a philosopher like him ought to know that the sole object of the science is the honor of the human spirit, and that under this view a problem of [the theory of] numbers is worth as much as a problem on the system of the world. (Jacobi)[10]

Mathematics is the science which draws necessary conclusions. (Benjamin Peirce)[11]

The essence of mathematics lies in its freedom. (Cantor)[12]

Mathematics, in its widest signification, is the development of all types of formal, necessary, deductive reasoning. (Whitehead)[13]

Logic merely sanctions the conquests of the intuition. (Hadamard)[14]

You will note that I have arranged these quotations in more or less opposing pairs. This may at first seem confusing and paradoxical to the students. But confusion and paradox should be seen not as impediments to learning but rather as opportunities for clarification. Although each of the questions merits considerable discussion, at this point their role is to arouse the students' curiosity and stimulate their interest. The quotations also give students an indication of the subtlety and complexity of the question "What is mathematics?"

But one cannot deal with these quotations in an historical vacuum. So I next give the students a traditional, *very concise*, chronological account of the history of mathematics. The idea is to discuss with students some characterizing features of various historical periods, and thus to give them a brief panoramic view of selected main currents of mathematical thought through the ages (e.g., pre-Greek mathematics, the mathematical aspects of the Greek "miracle", the mathematics of the Renaissance).[15] We can then return to the quotations and discuss them more meaningfully.

So let us reconsider the first pair of quotations. The first, "Mathematics is the study of number and form", gives me the opportunity to discuss conjectured origins of mathematics, be they utilitarian or ritualistic;[16] to talk about the relation of number to form, for example, their coexistence in early Greek mathematics, severed by the supposed "crisis of incommensurability"; and to raise the question of whether my students are familiar with mathematics *not* dealing with number or form. This question leads to the second quotation, "It is not of the essence of mathematics to be conversant with the ideas of number and quantity," by Boole. Boole's "heretical" view of mathematics (espoused in 1847) was shared by others in the nineteenth century. For example:

Pure mathematics is the theory of forms. (Grassman, 1844)[17]

Mathematics is concerned only with the enumeration and comparison of relations. (Gauss, 1851)[18]

[Mathematics is] purely intellectual, a pure theory of forms, which has for its objects not the combination of quantities or their images, the numbers, but things of thought to which there could correspond effective objects or relations, even though such a correspondence is not necessary. (Hankel, 1867)[19]

Mathematics is the science which draws necessary conclusions. (Peirce, 1870)[20]

Grassman, Hankel, and Peirce, not to speak of Gauss, were leading mathematicians of the nineteenth century, but the perspective expressed in these quotations was then a minority view. To most mathematicians of that time the subject was firmly anchored in "real" entities.[21]

Let us now consider the second pair of quotations given in our list of opposed pairs: "Mathematics is an art" and "Mathematics is the 'Queen of the Sciences'." This pair provides the opportunity to discuss aspects of mathematics shared by the sciences and the arts, and to suggest that mathematics possesses characteristics of both.[22] Assuming science is discovered and art created, the question arises: Is mathematics discovered or created? This brings us face to face with foundational issues of platonism and formalism.[23]

Let me set aside the other three pairs of descriptions of mathematics. The important moral for the students to learn from all of these apparently contradictory pairs of quotations is that they are not mutually exclusive but complementary. Each gives new insights into mathematics, and together they illustrate its many facets. But not only are these pairs of quotations not mutually exclusive, they are

far from exhaustive. Here are several others, to bring home that point.

The laws of Nature are written in the language of mathematics ... the symbols are triangles, circles and other geometrical figures, without whose help it is impossible to comprehend a single word .... (Galileo)[24]

Galileo had much to do with the supplanting, in the seventeenth century, of theology by mathematics as the queen of the sciences.

The science of mathematics presents the most brilliant example of how pure reason may successfully enlarge its domain without the aid of experience. (Kant)[25]

An eighteenth-century view by one of the foremost philosophers of the Enlightenment.

No mathematician can be a complete mathematician unless he is also something of a poet. (Weierstrass)[26]

There was undoubtedly "poetry" in the mathematics of Karl Weierstrass.

Mathematics is not the art of computation, but the art of minimal computation. (Anon)

Certainly not the average person's view of mathematics.

In mathematics ... we find two tendencies present. On the one hand, the tendency toward *abstraction* seeks to crystallize the *logical* relations inherent in the maze of materials ... being studied, and to correlate the material in a systematic and orderly manner. On the other hand, the tendency toward *intuitive understanding* fosters a more immediate grasp of the objects one studies, a live *rapport* with them, so to speak, which stresses the concrete meaning of their relations. (Hilbert)[27]

So much for Hilbert the formalist!

The constructs of the mathematical mind are at the same time free and necessary. The individual mathematician feels free to define his notions and set up his axioms as he pleases. But the question is, will he get his fellow mathematicians interested in the constructs of his imagination. We cannot help the feeling that certain mathematical structures which have evolved through the combined efforts of the mathematical community bear the stamp of a necessity not affected by the accidents of their historical birth. Everybody who looks at the spectacle of modern algebra will be struck by this complementarity of freedom and necessity. (Weyl)[28]

A perceptive statement, indeed, about the nature of mathematics.

Finally, here is perhaps the most recent definition of mathematics:

Mathematics is what mathematicians do. (Anon)

Mathematicians give definitions of mathematics and discuss the nature of mathematics, but they also experiment, visualize, discover, compute, invent, prove, model, apply, and classify. The preceding quotation embodies some of these recent thoughts on the nature of mathematics. The related philosophy of mathematics, given formal expression within the last two decades, is called quasi-empiricism.[29]

Having "settled" the big idea—the nature of mathematics—I can get on with discussing "lesser" things, such as the evolution of a concept, result, or theory. There are many possibilities here, of course. Two "must" topics for teachers are noneuclidean geometry and the mathematical infinite. We begin with the former. It is a fascinating story spanning more than two millennia, and it has fundamental implications in mathematics, philosophy, physics, and pedagogy. Hilbert gives us a possible entrée with his statement:

Every mathematical discipline goes through three periods of development: the *naive*, the *formal*, and the *critical*.[30]

I tell my students that the evolution of a mathematical idea often proceeds in *four* stages: discovery (or invention), use, understanding, and justification. But regardless of whether there are two, three, or more levels, the point to stress is that, when it comes to the evolution of mathematical ideas, the big bang theory rarely applies.

The following are several quotations around which one can structure some of the major issues in the evolution of noneuclidean geometry.

You must not attempt this approach to parallels. I know this way to its very end. I have traversed this bottomless night, which extinguished all light and joy of my life. I entreat you, leave the science of parallels alone ... I thought I would sacrifice myself for the sake of the truth. I was ready to become a martyr who would remove the flaw from geometry and return it purified to mankind. I accomplished monstrous, enormous labors; my creations are far better than those of others and yet I have not achieved complete satisfaction ... I turned back when I saw that no man can reach the bottom of the night. I turned back unconsoled, pitying myself and all mankind.

I admit that I expect little from the deviation of your lines. It seems to me that I have been in these regions; that I have travelled past all reefs of this infernal Dead Sea and have always come back with broken mast and torn sail.

The ruin of my disposition and my fall date back to this time. I thoughtlessly risked my life and happiness—*aut Caesar aut nihil*.[31]

A wonderfully evocative quotation, from Wolfgang Bolyai, a friend of Gauss, to his son János, one of the inventors of noneuclidean geometry. Mathematical passions!

The assumption that the sum of the three angles is less than 180° leads to a curious geometry, quite different from ours [the Euclidean], but thoroughly consistent, which I have developed to my entire satisfaction, so that I can solve every problem in it with the exception of the determination of a constant ...

The theorems of this geometry appear to be paradoxical and, to the uninitiated, absurd; but calm, steady reflection reveals that they contain nothing at all impossible .... All my efforts to discover a contradiction, an inconsistency, in this non-Euclidean geometry have been without success ....

I do not fear that any man who has shown that he possesses a thoughtful mathematical mind will misunderstand what has been said above, but in any case consider it a private communication of which no public use or use leading in any way to publicity is to be made. Perhaps I shall myself, if I have at some future time more leisure than in my present circumstances, make public my investigations.[32]

Unmistakably Gauss, in an 1824 letter to Franz Taurinus, who had also been working on the theory of parallels.

Mathematical discoveries, like springtime violets in the woods, have their season which no human can hasten or retard.[33]

The season was upon them, and Wolfgang Bolyai admonished his son János to publish his discoveries in noneuclidian geometry lest others claim priority. This raises an interesting question: If mathematical discoveries have their season, what role does the individual play in the development of mathematics? For example, are the near misses of Galileo on the infinite and of Saccheri on noneuclidean geometry due to the fact that the "season" for these two areas of mathematics had not yet arrived?[34]

> If geometry were an experimental science, it would not be an exact science. It would be subjected to continual revision .... *The geometrical axioms are therefore neither synthetic a priori intuitions nor experimental facts. They are conventions.* Our choice among all possible conventions is guided by experimental facts; but it remains *free,* and is only limited by the necessity of avoiding every contradiction, and thus it is that postulates may remain rigorously true even when the experimental laws which have determined their adoption are only approximate. In other words, *the axioms of geometry* (I do not speak of those of arithmetic) *are only definitions in disguise.* What then are we to think of the question: Is Euclidean Geometry true? It has no meaning. We might as well ask if the metric system is true and if the old weights and measures are false; if Cartesian coordinates are true and polar coordinates false. One geometry cannot be more true than another; it can only be more convenient.[35]

This is Henri Poincaré's famous statement about the newly emergent view of the nature of geometry and its relation to the physical world.[36]

To finish the geometry segment, here is a celebrated quotation which makes it starkly clear how axiomatics of geometry à la Euclid differs from axiomatics of geometry à la Hilbert:

> It must be possible to replace in all geometric statements the words point, line, plane by table, chair, mug (Hilbert).[37]

Surely Euclid and his contemporaries would have found this view shocking!

The next segment focuses on the infinite in mathematics.

> No other question has ever moved so profoundly the spirit of man; no other idea has so fruitfully stimulated his intellect; yet no other concept stands in greater need of clarification than that of the infinite. (Hilbert)[38]

We are still far from having clarified the concept of the infinite, nearly a century after Hilbert's challenge.

Here are several more quotations which help focus the discussion.

> [The difficulties in the study of the infinite arise because] we attempt, with our finite minds, to

discuss the infinite, assigning to it those properties which we give to the finite and limited; but this ... is wrong, for we cannot speak of infinite quantities as being the one greater or less than or equal to another.[39]

Galileo, resigned, following an attempt to compare for size the positive integers and their squares.

> I see it, but I don't believe it.[40]

This is Georg Cantor's expression of bewilderment, conveyed in a letter to Richard Dedekind, following his proof that the real numbers and the complex numbers have the same cardinality.

> Later generations will regard set theory as a disease from which one has recovered. (Poincaré)[41]

> No one shall expel us from the paradise which Cantor has created for us. (Hilbert)[42]

Who said there is no democracy in mathematics! Of course, the idea of "democracy" in mathematics is hard for students to accept, but it is a much more common phenomenon than might appear.[43]

Although Cantor's set theory is standard fare, its implications for the students are far from standard. Here are some:

(a) The whole need not be greater than its parts.[44]
(b) Infinity comes in different sizes.[45]
(c) There are "arithmetics" in which the additive and multiplicative cancellation laws, the commutative laws of addition and multiplication, and one of the two distributive laws fail (cardinal and ordinal arithmetic).[46]
(d) One can have two equally consistent mathematical theories contradicting one another (Cantorian and non-Cantorian set theories).[47]
(e) "Simple" assumptions can have formidable consequences (the axiom of choice as the assumption and the Banach-Tarski paradox as a consequence).[48]

We conclude the discussion of the infinite in mathematics with the following quotation by Weyl.

> Mathematics has been called the science of the infinite. Indeed, the mathematician invents finite constructions by which questions are decided that by their very nature refer to the infinite. This is his glory.[49]

This is one of the paradoxes about mathematics which make the subject so alluring.

As a final topic for consideration, it is important to give high school teachers a sense of at least *some* twentieth-century developments in mathematics. Among other things, this will demonstrate that mathematics has not stopped growing and prospering. The quotations below provide entry points into a number of central ideas of the mathematics of this century, including foundational issues, Gödel's work, and the role of the computer. The first is Bertrand Russell's provocative, perhaps facetious, description of mathematics.

> Mathematics is the subject in which we do not know what we are talking about nor whether what we are saying is true.[50]

This one raises the teachers' eyebrows.

> The great edifice of mathematics was shown to be like an enormous inverted pyramid delicately balanced upon the natural number system as a vertex.[51]

This quotation, from Howard Eves's *Great Moments in Mathematics* (1981), recalls the arithmetization of analysis in the late nineteenth century and points to a useful insight which students should be aware of—a latter-day pythagoreanism.[52]

Before I state the next quotation, I need a definition of religion (attributable to the contemporary mathematician De Sua):

> Religion is any discipline whose foundations rest on an element of faith, irrespective of any element of reason which may be present.[53]

Now the quotation, also from De Sua:

> Mathematics is the only branch of theology possessing a rigorous demonstration of the fact that it should be so classified.[54]

De Sua is referring here to Gödel's revolutionary work. An awareness of Gödel's ideas should be part of every student's mathematical culture. Here is another way of saying what De Sua says:

> Gödel gave a formal demonstration of the inadequacy of formal demonstrations (Anon).

The next quotation is from Jean Dieudonné:

> Now... the basic principle of modern mathematics is to achieve a complete *fusion* [of] 'geometric' and 'analytic' ideas (Dieudonné, 1977).[55]

For examples of the "fusion" of algebra and geometry, note such areas of mathematics as algebraic geometry, algebraic topology, topological algebra, and diophantine geometry, as well as the use of metric notions in number theory (p-adic numbers), of topology in algebra (the Zariski topology), and of algebra in geometry (Klein's Erlangen Program). Some sense of this unity-in-diversity of mathematics can and should be conveyed to students. For example:

(a) To solve $x^2 + y^2 = z^2$ (nontrivially) in integers is to find the points with rational coordinates on the unit circle $u^2 + v^2 = 1$.[56]

(b) To prove the nonconstructibility with straightedge and compass of the three Greek classical construction problems one must resort to abstract algebra.[57]

(c) The only known proof of the fact that in a finite projective plane Desargues's theorem implies Pappus's theorem involves showing that a finite division ring is a field.[58]

No account of twentieth-century mathematics is adequate if it does not mention the computer. Here, then, is a quotation by Lynn Steen which fills the bill admirably:

> The intruder has changed the ecosystem of mathematics, profoundly and permanently.[59]

Steen has in mind here not so much the use of computers in mathematics as the way in which computers have affected the direction of mathematics: its problems, its methods, and mathematicians' conception of their subject.

A final comment. The history of mathematics can be studied chronologically, thematically, topically, and biographically. I have used in this course elements of each approach. The quotations have played a pivotal function in all of them. It is perhaps not inappropriate to conclude this paper with several more quotations—historical and pedagogical.

> The Divine intellect indeed knows infinitely more propositions [than we can ever know]. But with regard to those few which the human intellect does understand, I believe that its knowledge equals the Divine in objective certainty.... (Galileo)[60]

> I have had my results for a long time, but I do not yet know how to arrive at them. (Gauss)[61]

> If I only had the theorems! Then I should find the proofs easily enough. (Riemann)[62]

> The utmost abstractions are the true weapons with which to combat our thought of concrete fact. (Whitehead)[63]

> God exists since mathematics is consistent and the devil exists since we cannot prove the consistency. (Weyl)[64]

> Education is that which remains when one has forgotten everything learned in school. (Einstein)[65]

> Being a language, mathematics may be used not only to inform but also, among other things, to seduce. (Mandelbrot)[66]

To teach creatively is not to cover, but to uncover the syllabus. (Bowden)[67]

References

1. For the former see the preceding essay, as well as: William Dunham, *Journey Through Genius: The Great Theorems of Mathematics* (New York: John Wiley & Sons, 1990). For the latter consult Israel Kleiner, "Famous Problems in Mathematics: An Outline of a Course", *For the Learning of Mathematics*, 1986, **6**(1): 31–38. Also Reinhard Laubenbacher and David Pengelley, "Great Problems of Mathematics: A Course Based on Original Sources", *American Mathematical Monthly*, 1992, **99** : 313–317.

2. George Polya, *Mathematical Discovery* (New York: John Wiley & Sons, 1981).

3. Charles Edwards, *The Historical Development of the Calculus* (New York: Springer-Verlag, 1979), p. vii.

4. Otto Toeplitz, *The Calculus: A Genetic Approach* (Chicago: The University of Chicago Press, 1963), p. v.

5. See Paul Halmos, *Naive Set Theory* (New York: Springer-Verlag, 1960). His statement is: "Read it, absorb it, and forget it."

6. Imre Lakatos, *Proofs and Refutations* (Cambridge: Cambridge University Press, 1976), p. 2.

7. George Boole, *An Investigation of the Laws of Thought* (1854; rpt. New York: Dover, 1951), p. 12.

8. Eric T. Bell, *Mathematics : Queen and Servant of Science* (London: G. Bell & Sons, 1952), p. 1.

9. Morris Kline, *Mathematical Thought from Ancient to Modern Times* (New York: Oxford University Press, 1972), p. 1036.

10. Kline, *Mathematical Thought*, p. 813.

11. Benjamin Peirce, "Linear Associative Algebras", *American Journal of Mathematics*, 1881, **4** : 97–215, on p. 97.

12. Kline, *Mathematical Thought*, p. 1031.

13. Alfred N. Whitehead, *A Treatise on Universal Algebra* (New York: Hafner, 1960), p. vi.

14. Kline, *Mathematical Thought*, p. 1026.

15. See, for example, Dirk Struik, *A Concise History of Mathematics*, 4th ed. (New York: Dover, 1987).

16. See Abraham Seidenberg, "The Origin of Mathematics", *Archive for History of Exact Sciences*, 1978, **18** : 301–342.

17. Michael Crowe, *A History of Vector Analysis* (Notre Dame: University of Notre Dame Press, 1967), p. 65.

18. Eric T. Bell, *The Development of Mathematics*, 2nd ed. (New York: McGraw-Hill, 1945), p. 211.

19. Kline, *Mathematical Thought*, p. 1031.

20. Peirce, "Associative Algebras", p. 97.

21. See, for example, Victor Katz, *A History of Mathematics* (New York: Harper Collins, 1993), or Kline, *Mathematical Thought*.

22. See Armand Borel, "Mathematics: Art and Science", *The Mathemataical Intelligences*, 1983, **5**(4): 9–17.

23. See Philip Davis and Reuben Hersh, *The Mathematical Experience* (Boston: Birkhäuser, 1981), or Nicolas Goodman, "Mathematics as an Objective Science", *American Mathematical Monthly*, 1979, **86**: 540–551.

24. Kline, *Mathematical Thought*, pp. 328–329.

25. Immanuel Kant, *Critique of Pure Reason*, transl. by M. Müller, 2nd ed. (New York: Macmillan, 1927).

26. Eric T. Bell, *Men of Mathematics* (New York: Simon & Schuster, 1937), p. 432.

27. David Hilbert and S. Cohn-Vossen, *Geometry and the Imagination* (New York: Chelsea, 1952), p. iii.

28. Hermann Weyl, "A Half-Century of Mathematics", *American Mathematical Monthly*, 1951, **58**: 523-553, on pp. 538–539.

29. See Thomas Tymoczko, *New Directions in the Philosophy of Mathematics* (Boston: Birkhäuser, 1986).

30. Reinhold Remmert, *Theory of Complex Functions* (New York: Springer-Verlag, 1989), p. 240.

31. Herbert Meschkowski, *Noneuclidean Geometry* (New York: Academic Press, 1964), pp. 31–32.

32. Harold Wolfe, *Introduction to Non-Euclidean Geometry* (New York: Holt, Rinehart & Winston, 1945), pp. 46–47.

33. Bell, *Development of Mathematics*, p. 263.

34. See Raymond Wilder, *Evolution of Mathematical Concepts: An Elementary Study* (New York: John Wiley & Sons, 1968).

35. Henri Poincaré, *Science and Hypothesis* (New York: Dover, 1952), pp. 49–50.

36. See Marvin J. Greenberg, *Euclidean and Non-Euclidean Geometries: Development and History*, 2nd ed. (San Fransisco: W.H. Freeman & Co., 1980).

37. Hermann Weyl, "Axiomatic Versus Constructive Procedures in Mathematics", *The Mathematical Intelligencer*, 1985, **7**(4): 10–17, on p. 14.

38. Eli Maor, *To Infinity and Beyond* (Boston: Birkhäuser, 1987), p. vii.

39. Galileo Galilei, *Two New Sciences*, tr. by H. Crew and A. de Salvio (New York: Dover repr., 1954), p. 31.

40. Kline, *Mathematical Thought*, p. 997.

41. Kline, *Mathematical Thought*, p. 1003.

42. Kline, *Mathematical Thought*, p. 1003.

43. See Israel Kleiner and Nitsa Movshovitz-Hadar, "Aspects of the Pluralistic Nature of Mathematics", *Interchange*, 1990, **21** : 28–35.

44. See Maor, *Infinity*.

45. See Rudy Rucker, *Infinity and the Mind* (Boston: Birkhäuser, 1982).

46. See Halmos, *Set Theory*.

47. See Paul Cohen and Reuben Hersh, "Non-Cantorian Set Theory", *Scientific American*, 1967, **217** (Dec.): 104–116.

48. See Stan Wagon, *The Banach-Tarski Paradox* (Cambridge: Cambridge University Press, 1985).

49. Weyl, "Axiomatic", p. 12.

50. Kline, *Mathematical Thought*, p. 1196.

51. Howard Eves, *Great Moments in Mathematics (After 1650)* (Washington, D.C. : The Mathematical Association of America, 1981), p. 132.

52. Recall the pythagorean dictum that "all is number".

53. Frank De Sua, "Consistency and Completeness: A Resumé", *American Mathematical Monthly*, 1956, **63**: 295–305, on p. 305.

54. De Sua, "Consistency", p. 305.

55. Jean Dieudonné, "Should We Teach Modern Mathematics?" *American Scientist*, 1973, **61** (Jan./Feb.): 16–19, on p. 19.

56. See John Stillwell, *Mathematics and Its History* (New York: Springer-Verlag, 1989), p. 4.

57. See Joseph Gallian, *Contemporary Abstract Algebra*, 2nd ed. (Lexington, Mass. : D.C. Heath & Co., 1990).

58. See Leonard Blumenthal, *A Modern View of Geometry* (San Francisco: W.H. Freeman & Co., 1961).

59. Lynn Steen, "Living With a New Mathematical Species", *The Mathematical Intelligencer*, 1986, **8**(2) : 33–40, on p. 34.

60. Galeleo Galilei, *Dialogues Concerning the Two Chief World Systems*, tr. by S. Drake, 2nd ed. (Berkeley: University of California Press, 1970), p. 103.

61. Lakatos, *Proofs and Refutations*, p. 9.

62. Lakatos, *Proofs and Refutations*, p. 9.

63. Morris Kline, *Mathematics in Western Culture* (New York: Oxford University Press, 1953), p. 466.

64. Kline, *Mathematical Thought*, p. 1206.

65. Albert Einstein, *Ideas and Opinions* (New York: Crown Publ., 1954), p. 63.

66. Benoit Mandelbrot, *Fractals: Form, Chance, and Dimension* (San Francisco: W.H. Freeman & Co., 1977), p. 20.

67. M.M. Schiffer and Leon Bowden, *The Role of Mathematics in Science* (Washington, D.C. : The Mathematical Association of America, 1984), back jacket.

Measuring an Arc of Meridian

**Marie Françoise Jozeau
and Michéle Grégoire**

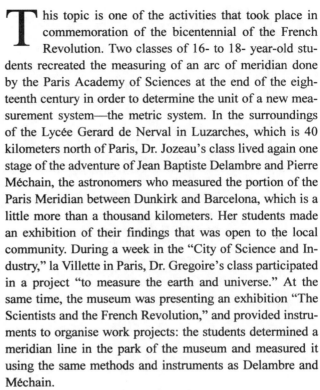

This topic is one of the activities that took place in commemoration of the bicentennial of the French Revolution. Two classes of 16- to 18- year-old students recreated the measuring of an arc of meridian done by the Paris Academy of Sciences at the end of the eighteenth century in order to determine the unit of a new measurement system—the metric system. In the surroundings of the Lycée Gerard de Nerval in Luzarches, which is 40 kilometers north of Paris, Dr. Jozeau's class lived again one stage of the adventure of Jean Baptiste Delambre and Pierre Méchain, the astronomers who measured the portion of the Paris Meridian between Dunkirk and Barcelona, which is a little more than a thousand kilometers. Her students made an exhibition of their findings that was open to the local community. During a week in the "City of Science and Industry," la Villette in Paris, Dr. Gregoire's class participated in a project "to measure the earth and universe." At the same time, the museum was presenting an exhibition "The Scientists and the French Revolution," and provided instruments to organise work projects: the students determined a meridian line in the park of the museum and measured it using the same methods and instruments as Delambre and Méchain.

This project was an outside the curriculum activity for both classes, and was prepared by classroom exercises in geometry and trigonometry. Here is a brief account of the historical context leading to the metric system and a review of principal student activities in this measuring of an arc of meridian.

1. From early units of measure to the metric system

From antiquity, there was a need for common units and yardsticks for measuring as long-distance trade was developing. Many measures differed from one area to another. Standards were kept inviolable in churches, temples, and palaces. By 789 the emperor Charlemagne suggested the "kingsfoot" as a universal yardstick ($\approx$ 0.3248m). This unit was adopted by Latin monastic building orders, but other communities used other units. In the eighteenth century more than 800 differents units were still in use. The units differed for measuring forest, meadows, fields, or vineyards, or different types of fabric, firewood, timber, charcoal, or coal, or to weigh different types of fruit, or to weigh cereal or salt. As science developed amid an *esprit géométrique* calling for greater precision, and central states evolved, pressure grew to establish common measures.

Understanding and converting one unit into another one was a type of exercise for younger stu-

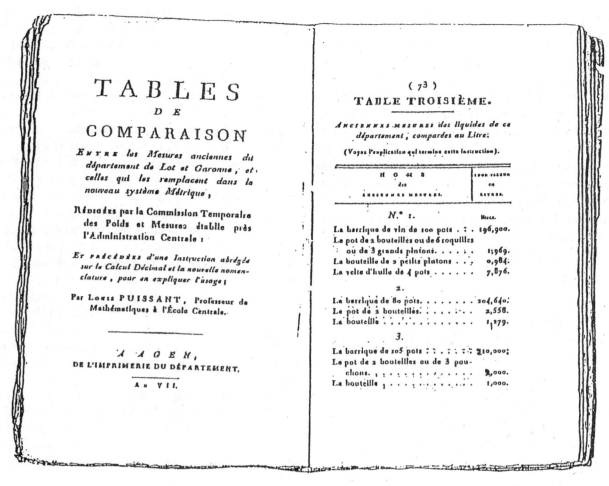

FIGURE 1

dents, who studied comparison tables from the French Revolution period. Figure 1 shows an example of such tables, in order to convert different capacity units into liters.

In 1670, Gabriel Mouton, a curate of the city of Lyons, suggested as a common unit the "milliare," which is the 60th part of one degree of the meridian, and is still the nautical mile ($\approx$ 1,852km). Despite many such attempts, until the French Revolution the political will to unify the measuring system was less strong than the self-interest of lords or corporations who profited from the muddle. At the beginning of the revolutionary period the people wrote "cahiers de doléances" which asked that all types of grain be measured with the same unit, in order not to harm the poor.[1]

In 1790, the assembly deputies asked the Paris Academy of Sciences to determine a measurement system that could be accepted by the whole word, without any particularity linked to any situation on the earth. French mathematicians and physicists, among them Lagrange, Laplace, Monge, Condorcet, and Lavoisier,[2] chose as length unit the ten-millionth part of one quarter of a terrestrial meridian, and they called it Meter. This definition is inspired by the egalitarian spirit of the French Revolution: "under every human being's feet goes a meridian and all meridians are equal." The weight unit was defined by the help of the length unit as the weight of a 0.1 m sided cube of distilled water at the melting ice temperature. The decimal division of units was adopted.

---

[1] the cahiers de doléance are complaints books, written by local populations in every district.

[2] Lavoisier was asked to determine the weight unit, and was secretary and treasurer of the committee.

2. The length of the meridian before 1789

Measuring the earth—its radius or circumference or meridian—had been done earlier. In the 4th century BC, Aristarchus, author of a heliocentric system, estimated the circumference of the earth at 400000 stades ($\approx$ 63000 km) and Archimedes (3th century BC) at 300000 stades. But how? Eratosthenes of Alexandria (200 BC) computed an early measure of the earth. He measured the distance between Alexandria and Syene (today Asswan), both cities situated approximately on the same meridian, measured the angle Φ, the difference of their latitudes, and calculated the circumference of the earth (see Figure 2). His conclusion was equivalent to 39,000 km.

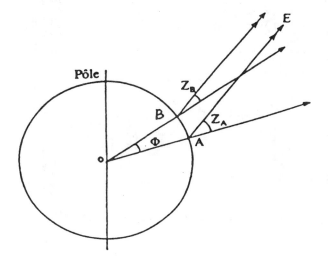

FIGURE 2
The method for measuring an arc of meridian. The angles Z_A and Z_B are the zenithal distances of a same star E. For Erathostenes, E is the sun and $Z_A = 0$, (Asswan is on the tropic of Cancer and the observation is done at the summer solstice).

$$\Phi = Z_A - Z_B.$$

Measuring the earth was a task of cartographers, but no precise measure of a meridian was tried in the western world before 1669. Jean-Baptiste Colbert, a prime minister of Louis XIV, requested the members of the new Academy of Sciences to draw a map of the French Kingdom. The astronomer Jean Picard was put in charge of measuring the distance corresponding to one degree of the Paris meridian, between Sourdan and Malvoisine (see Figure 3). Picard was personally congratulated by Louis XIV, allthough the conclusion of his observations was that France's territory was shrinking (see Figure 4).

La mesure de la Terre

PICARD .1671 - planche II

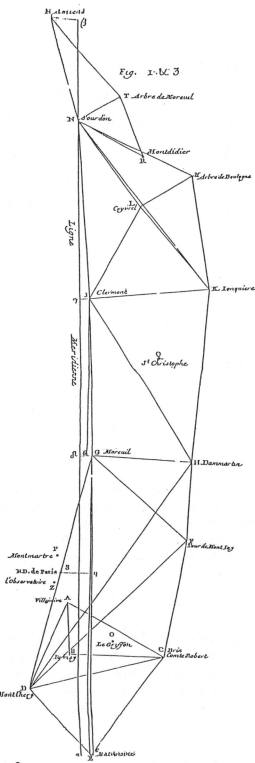

FIGURE 3
The measure of the meridian between Sourdan and Malvoisine.

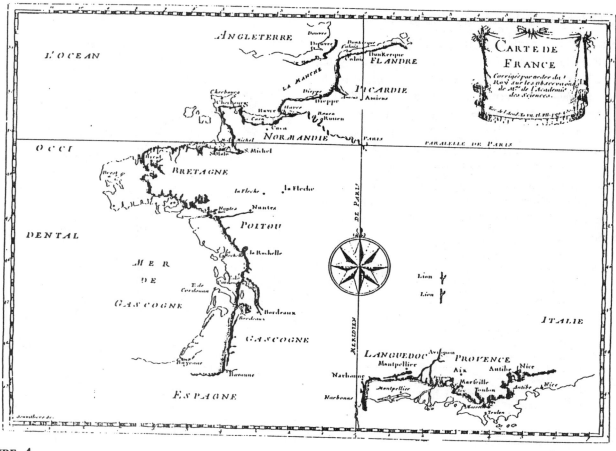

FIGURE 4
Map of France (1684).

(1684)

Picard applied the triangulation method invented around 1617 by Willebrord Snel, the Dutch mathematician and cartographer, who had worked all his life to draw a map of his country. As lengths are more difficult to measure than angles, Picard measured a more or less 11-km-long base (the distance Juvisy–Villejuif) and then the angles of a chain of thirteen triangles. With the help of the triangulation formula

$$\frac{a}{\sin A} = \frac{b}{\sin B} = \frac{c}{\sin C},$$

he could calculate the sides of the triangles and their projections on the meridian line. The sides of the triangles (except ABC) were around 40 km long.

When our students learn this triangulation formula, they are asked to read the page of Picard's report, which appears on Figure 5, and to calculate the distance GE.

Figure 6 presents the instrument used by Picard to measure angles: a "quarter circle" which radius is about one meter. The precision was about 5 seconds of a degree.

Picard's conclusion was 57060 toises by degree, which is about 111.210 km. The actual length of one degree of a meridian is 111.111... km[3] New measures were done in the eighteenth century by different members of the Italian astronomer family, the Cassinis, who directed the Paris observatory, and the last one in 1740, by the French astronomer Nicolas Louis de La Caille.

3. The measure of the meridian by Delambre and Méchain

Since the measures of La Caille had to be verified to have an acceptable definition of the meter, Delambre and Méchain were asked to do it.

Their operations lasted from 1792 to 1799, their path bristling with all sorts of difficulties: beside natural difficulties, uneven ground (especially in the Pyrenees) and bad

---

[3] The length of the toise Picard used was about 1.949m.

34   *Mefure de la Terre,*

qui ne donnoient les minutes que de fix
en fix, ils n'ont pas laiffé d'approcher de
la juftefle autant qu'il étoit néceffaire ,
pour faire voir qu'on ne s'étoit pas trom-
pé aux conclufions.

I. TRIANGLE ABC.
Pour connoitre le côté A C.

C A B............54°....4'....35".
A B C............95.....6....55.
A C B............30...48....30.
    A B......5663..*Toifes de mefure actuelle.*
Donc A C....11012..*Toifes 5 pieds.*
Et B C....8954..*Toifes.*

II. TRIANGLE ADC.
Pour D C & A D.

D A C............77°....25'....50".
A D C............55........10.
A C D............47....34....0.
    A C....11012..*Toifes 5 pieds.*
Donc D C....13121..*Toifes 3 pieds.*
Et A D......9922..*Toifes 2 pieds.*

III. TRIANGLE DEC.
Pour D E & C E.

D E C............74°....9'....30".
D C E............40...34....0.
C D E............65...16...30.
    D C....13121..*Toifes 3 pieds.*
Donc D E....8870..*Toifes 3 pieds.*
Et C E....12389..*Toifes 3 pieds.*

par M. l'Abbé Picard.   35

IV. TRIANGLE DCF.
Pour D F.

D C F............113°....47'....40".
D F C............33....40......0.
F D C............32....32....20.
    D C....13121.....*Toifes 3 pieds.*
Donc D F....21658.....*Toifes.*

Notez que dans ce quatriéme triangle, l'angle D F C a été augmenté de
10", qui manquoient à la fomme des
trois angles.

V. TRIANGLE DFG.
Pour D G & F G.

D F G............92°....5'....20".
D G F............57....34......0.
G D F............30....20....40.
    D F......21658..*Toifes.*
Donc D G....25643...*Toifes.*
Et F G....12963...*Toifes 3 pieds.*

Enfuite de ces cinq triangles, il a été
facile de conclure la diftance G E entre
Malvoifine & Mareuil , fans fuppofer
aucune nouvelle Obfervation.

FIGURE 5

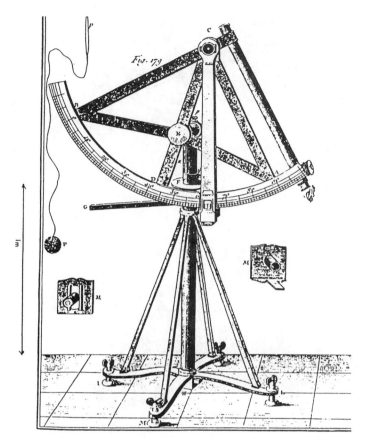

FIGURE 6

FIGURE 7
J. B. Delambre and P. Méchain.

weather, Méchain got ill, and was temporarily paralyzed;
there was a war between France and Spain; the astronomers
were suspected of being spies, of being counterrevolution-
aries; they were jailed, their measuring signals were often demolished before they could finish their observations.
Their adventures were told by Denis Guedj, in a novel "La
Méridienne" (1987).

Since the first precise measures done by Picard, the
observation instruments had been improved: in 1752, Tobias

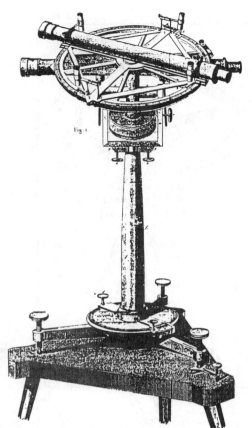

FIGURE 8
Repetitive circle of Borda.

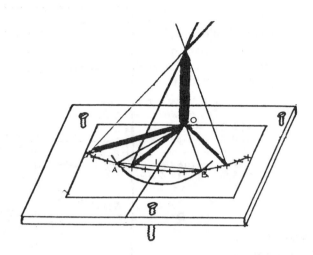

FIGURE 9
To determine a meridian line.

Mayer, a German astronomer, repeated the same observation many times without using every time the zero division to lessen errors. Following his idea, Delambre and Méchain ordered special "repetition circles," which had been built first in 1784 by Lenoir. It had been used for measuring the junction between the Greenwich and the Paris meridians (see Figure 8).

At the City of Science and Industry, students used theodolites, a more modern version of this instrument, and measured three triangles along the meridian of a metallic structure in the park. The students had also to determine a meridian line, as follows:

> Draw a line joining the extremities of the shadows of a stick put in O, every 10 minutes during 2 hours in the middle of the day (Figure 9). This line cuts a circle centered in O in two points A and B. At this points, shadows are equal: the corresponding instants are symmetrical in relation to the solar midday. The meridian joins O to the middle I of $[AB]$.

At the same period, in the lycée Gerard de Nerval, students drew a meridian line in the schoolyard under the direction of the physics teacher. The lycée stands near Saint-Martin du Tertre, one of the observation stations from where Delambre focused on Clermont, on the Panthéon, on Saint-Christophe and on Dammartin (see Figure 10).

The students took part in the commemorative event organized by Denis Guedj all over France. From school to school, from Dunkirk to the Pyrenees, students recognized the observation stations and focused with theodolites.

Both of our groups studied excerpts of Delambre's report "Base du système metrique décimal," like the one on Figure 11.

Let us comment on the meaning of the different lines of Figure 11.

a) The first line expresses that *36* sightings have been made to measure the first angle. The reading gives *3041.629 grades*; the 1/36 th of it is *84.48969 grades*, which, transformed into degrees, is *76° 2′ 26″ 6*. The measures were repeated. On the second line, *18* sightings were made; the result is *1520.817 grades*; the 1/18th of it is *84.48986 grades*, which gives again *76° 2′ 27″ 06*. The arithmetic mean of the two measures is taken on the third line. The first measures were done by Delambre and then by an engineer helping Delambre, named Bellet *(D. puis B.)*; the second done by an astronomer named Le Français, using the first repetition circle *(F. n° 1)*. On the sixth line, a third serie of measures of the same angle is done by Delambre with the first repetition circle (20 sightings).

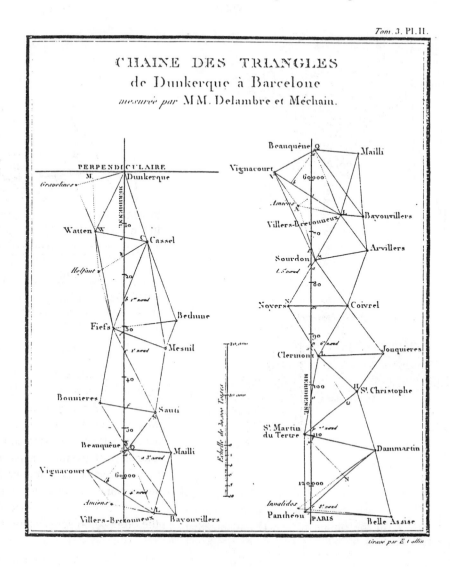

FIGURE **10**

Chain of triangles
between Dunkirk and Paris.

FIGURE **11**

Entre le Panthéon et Dammartin.

36	3041 6629	84 48969	= 76° 2' 26"6	D. puis B.
18	1520 817	84 48986	= 76° 2' 27"06	F. n° 1.

$$76° \ 2' \ 26"83$$
$$+ \ 2"48$$
$$76° \ 2' \ 29"31$$

$r = 0'21181 \quad y = 6° \ 56' \ 53"$

20	1689 792	84 4896	= 76° 2' 26"30	D. n° 1.

$r = 0'37153 \quad y = 10° \ 51' \ 10"$ $\quad + \ 3"98$

$$76° \ 2' \ 30"28$$

Première série . . . . 76° 2' 29"08
Seconde série . . . . 76° 2' 29"34
Moyenne . . . . 76° 2' 29"63
$\qquad\qquad\qquad\qquad\qquad + \ 1"20$
Horizon . . . . 76° 2' 30"83

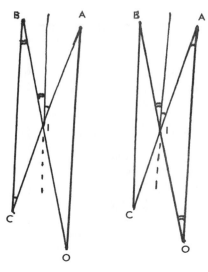

FIGURE 12

$$ACB = AIB - CBO$$

$$AIB = AOB + OAC$$

$$ACB = AOB + OAC - CBO$$

more simply $C = O + A - B$. In the triangle OCB,

$$\frac{\sin B}{OC} = \frac{\sin O}{BC};$$

hence $\sin B = r\dfrac{\sin y}{G}$.

b) On lines 4 and 7, a correction called *Center reduction* is done. The observation instrument can not often be put at the center C of the station, which is the spot that will be focused on from the next station, if C is the top of a steeple for example.... Delambre calculated the difference between the observation angle AOB and the ideal angle ACB. The principle of the method is explained in Figures 12 and 13.

c) On the last lines another correction is done called *Horizon reduction*: When the observed angle c is not situated on an horizontal plane, Delambre calculated the projection C of this angle onto the horizontal plane (see Figure 14), with the use of the formula that Legendre, the mathematician, has established at the express demand of Delambre. This formula was obtained with the help of spherical trigonometry.

d) The sphericity of the earth: Picard's calculations do not take sphericity into account. In a following stage of their calculations, which does not appear on this page of the report, Delambre and Méchain did take sphericity into

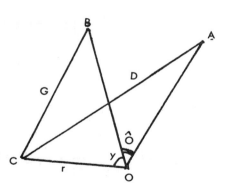

FIGURE 13

In the triangle AOC

$$\frac{\sin A}{OC} = \frac{\sin COA}{AC}$$

hence $\sin A = r\dfrac{\sin(O + y)}{D}$. The angles A and B are very small, and measured in seconds; hence

$$A \approx \frac{r\sin(O + y)}{D} \times \frac{1}{\sin 1''}$$

$$B \approx \frac{r\sin y}{G} \times \frac{1}{\sin 1''}$$

$$ACB \approx O + \frac{r\sin(O + y)}{D\sin 1"} - \frac{r\sin y}{G\sin 1''}$$

account. Here was the opportunity to deliver a course about the sphere, even to explain a bit of spherical trigonometry, to determine which circle is the shortest from one point to another one.

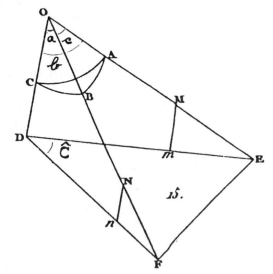

FIGURE 14

$$\sin^2 \frac{1}{2}C = \frac{\sin\dfrac{a - b + c}{2}\sin\dfrac{b + c - a}{2}}{\sin b \sin a}.$$

4. Conclusions

In November 1798, the two astronomers presented their conclusions to a committee of scientists from many different European nations which correspond today to Italy, Spain, Portugal, Switzerland, Denmark, Holland... who, in 1799, approved their definition of the meter and the kg. Hence, yardsticks started being built. Joseph Dombey, the botanist, was sent by the young French republic to bring the meter and the kg to the United States Congress in Philadelphia. But his vessel was attacked by pirates, and Dombey was put into slavery. Nobody knows where the meter and the kg diasppeared. Later on the political situation changed many times, and with it support for the metric system. Bonaparte was in favour of the metric system, but, as an emperor, he authorized the use of ancient measures. With the restoration of the monarchy in 1815, these ancient measures prevailed in France. In 1837, however, the metric system became compulsory in France. In 1869, an international meter commission rallied more than 30 nations and the platinum irridium French yardsticks were adopted. Between 1960 and 1983, the meter is defined as 1650763.73 wave lengths in the vacuum of the radiation produced by the transition between two levels of krypton 86. In 1983, the meter is defined as the distance covered by the light in the vacuum in 1/299792458 second.

Bibliography

Delambre J.B., *Base du système métrique décimal ou Mesure de l'arc du méridien compris entre les parallèles de Dunkerque et Barcelone exécutée en 1792 et années suivantes,* Suite de Mémoires de l'Institut, 3 vol., Paris, 1806–1807–1810.

Guedj D., *La révolution des savants.* Découvertes Gallimard, Paris, Gallimard, 1988

Legendre A.M., *Eléments de Géométrie,* Paris, Didot, 1794.

Picard J., *La mesure de la terre,* 1671 in *Recueil de plusieurs traitez* vol. 2, Académie royale des Sciences, Paris, 1676.

Observatoire de Paris, *Une mesure révolutionnaire: le mètre.* Paris, 1988.

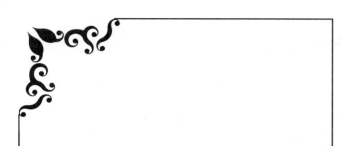

From Egypt to Benjamin Banneker: African origins of false position solutions

Beatrice Lumpkin
Malcolm X College,
Chicago City Colleges

Benjamin Banneker (1731–1806) is one of the most appealing figures in the history of mathematics. Inspired by the Banneker role model, mathematics teachers have organized the Benjamin Banneker Association, an NCTM affiliate, to advance the mathematics education of African American students. Although he himself was never a slave, Banneker was the son and grandson of Africans kidnaped and brought to Maryland into slavery and who later won their freedom. In those years the slave power was expanding and his freedom remained uncertain. For freedom's sake, he dedicated his almanacs to the cause of equality and peace. He wanted to show by example that African Americans were as intellectually gifted as any other people.

Banneker was one of the earliest, practicing mathematicians and astronomers in the United States. Primarily self-educated, he said he "had advanced to false position through his own efforts."[1] Banneker continued a tradition begun in Africa thousands of years earlier. This paper traces the development of the method of false position from its uses in ancient Egypt, the Hellenistic world, and medieval Islam, to its transmission to Europe and subsequent appearance in mathematics texts of Banneker's time.

Yet the scope of his genius is still little known.[2] He spent most of his life doing hard physical labor on an isolated farm. As a child in the 1730s he had the rare good fortune to learn to read. The rest he learned on his own. At the age of 22 Banneker became famous for building a wood clock, making all the parts himself with only a borrowed pocket watch as a model. He designed the clock to strike the hours, and it kept good time for 54 years. The clock worked until his house burned on the day of his funeral, destroying all his papers except a manuscript journal and his printed almanacs that were not in the house. The knowledge that Banneker gained from designing his clock was very useful in his later work as an astronomer.

After he built the clock, Banneker's reputation continued to grow, based on his ability to solve and create the mathematical puzzles that were popular in that day. People sent him puzzles from all over the colonies and later from the new republic. Still, until age 57, he spent most of his time as a farmer. At that time, a neighbor, who was leaving the surveying business, loaned Banneker some books and astronomy instruments.[3] Without a teacher and with only a few semesters of elementary schooling in his childhood, Banneker taught himself the algebra, geometry, logarithms, trigonometry, and astronomy needed to become an astronomer. He also learned on his own how to use a compass, sector, and other instruments to make astronomical predictions, including that of eclipses. One book that he

279

studied was written in Latin, which indicates the breadth of his self-study.[4]

When Banneker late in life gained access to astronomy texts, he worked his way through them and in little over one year became a practicing astronomer. Gaining confidence, he began to calculate an ephemeris for an almanac for the Baltimore, Washington, D.C., and Philadelphia area. Locally published almanacs contained ephemerides which adapted information from the observatory of Greenwich, England to local latitude and longitude. Local information included times of sunrise and sunset, moonrise and moonset, planetary positions, times of eclipses, tides, and so on.

Based on Banneker's work on his almanac, he was appointed an astronomer on the team of surveyors that drew up the outline for the new nation's capital, Washington, DC. Banneker was appointed because he was one of the few in the country capable of doing such work. Charles Leadbetter, author of an astronomy book that Banneker studied, wrote that knowledge of astronomy in London was, "so rare,... not one of 20,000 hath attained to it."[5] Knowledge of astronomy was even rarer in the new United States. Banneker's work so impressed Thomas Jefferson, then Secretary of State, that he wrote Banneker that he was sending a copy of the almanac to the Paris Academy of Sciences.

How did Banneker keep his mind so sharp during the decades he spent working his farm? He loved to solve the mathematical puzzles that were popular at that time. His manuscript journal shows that he was also a keen observer of nature, and he used a scientific approach to search for rational explanations. For example, even at the risk of his personal safety, he made the following observation:

> August 27, 1797. Standing at my door I heard the discharge of a gun, and in about 4 or 5 seconds of time after the discharge, the small shot came rattling about me, one or two of which struck the house, which plainly demonstrates that the velocity of sound is much greater than that of a common bullet.

In his famous letter to Thomas Jefferson dated August 19, 1791, Banneker wrote (in his beautiful handwriting and in the spelling and of that day) that he:

> long had unbounded desires to become Acquainted with the Secrets of nature. I have had to gratify my curiosity herein thro my own assiduous application to Astronomical Study, in which I need not to recount to you the many difficulties and disadvantages which I have had to encounter.[6]

Early use of the False Position Method

In Banneker's time, false position solutions were widely employed. The method was variously called rule of false, rule of position, double position, and scales (a type of diagram). A book by Daniel Adams, *The Scholar's Arithmetic or Federal Accountant,* written before 1801, gives a good summary of the method:[7]

> Position is a rule which, by false or supposed numbers, taken at pleasure, discovers the true one required. It is of two kinds, Single and Double.
>
> Single Position is the working with one supposed number, as if it were the true one, to find the true number.
> RULE: 1. Take any number and perform the same operations with it as are described to be performed in the question. 2. Then say, as the sum of the errors is to the given sum, so is the supposed number to the true one required.

Examples of the "single position" method are found in the ancient Egyptian Rhind Mathematical Papyrus copied by the scribe A'hmosè. It dates back to before 1800 BCE. Two examples from problems 26 and 24 have been used by the Chicago Public Schools to teach proportional thinking and to honor the African contribution. Problem 26 asks: "A quantity and 1/4 of the quantity are added and the sum is 15. What is the quantity?" In modern notation, $x + x/4 = 15$.

Let 4 be the trial solution value. This gives $4 + \frac{1}{4}(4) = 5$. But the sum should be 15. A correction factor is needed. Since $\frac{15}{5}$ is 3, 3 is the correction factor for the "false" or trial value. Multiply the trial solution value of 4 by 3 to get the correct solution 12. This gives $12 + \frac{1}{4} = 15$ or $12 + 3 = 15$. This method, stated as a proportion, is $5 : 15 = 4 : x$.

A similar procedure was used for RMP 24, "the sum of a quantity and its $\frac{1}{7}$ is 19." The scribe naturally assumed the quantity was 7. But $7 + \frac{1}{7}(7)$ is 8, not 19. The correction factor is $\frac{19}{8}$. Since Egyptians reduced fractions other than $\frac{2}{3}$ to unit fractions, $\frac{19}{8}$ was written as $2 + \frac{1}{4} + \frac{1}{8}$. Multiplied by the assumed value 7, the desired quantity is 16 and $\frac{1}{2}$ and $\frac{1}{8}$.[8]

The scribe, A'hmosè, also used false position to solve a problem involving arithmetic series:

> Divide 100 loaves among five persons so there is a common difference. Also, the sum of the highest three must be 7 times the sum of the lowest two.

A'hmosè started by taking the common difference to be $5\frac{1}{2}$, a happy choice, and assuming a value of 1 for the lowest share. The shares would be: 1, $6\frac{1}{2}$, 12, $17\frac{1}{2}$, and 23.

The sum of the three highest, $23 + 17\frac{1}{2} + 12$, is 7 times the sum of the two lowest, $1 + 6\frac{1}{2}$. But the series sum is 60, not 100, as required. As many times as 60 is multiplied to give 100, so each term of the series must be multiplied. The correction factor is $\frac{100}{60}$ or $\frac{5}{3}$ and the desired shares are: $1\frac{2}{3}$, $10\frac{5}{6}$, 20, $29\frac{1}{6}$, $38\frac{1}{3}$.

Was the trial value of $5\frac{1}{2}$ for the common difference just a lucky choice? The translator of the A'hmosè papyrus, Arnold Chace, thinks not, and shows how A'hmosè could have found this value. He believes that the Egyptians experimented with different common differences, starting with 1, 2, and so on. They assumed that the first term was 1.

Series	Sum of smallest two − $\frac{1}{7}$ Sum of largest three
1, 2, 3, 4, 5	$3 - 1\frac{5}{7} = 1\frac{2}{7}$
1, 3, 5, 7, 9	$4 - 3 = 1$
1, 4, 7, 10, 13	$5 - 4\frac{2}{7} = \frac{5}{7}$

Each time the common difference increases 1 unit, there is a decrease of $\frac{2}{7}$ in the difference between the sum of the smallest two and $\frac{1}{7}$ the sum of the largest three. How much must the common difference be increased to make the two values equal? Divide $1\frac{2}{7}$ by $\frac{2}{7}$ to get $4\frac{1}{2}$. Add $4\frac{1}{2}$ to 1 for a common difference of $5\frac{1}{2}$.[9]

The false position method also appears in other Egyptian papyri, for example Kahun LV,3 and the Berlin 6619 papyrus.[10] The Berlin papyrus problems involve a system of two equations, one of second degree. That problem asks for two squares such that the sum of the two areas equals a third square of area 100 cubits, and one square has a side $\frac{3}{4}$ the length of the side of the other square. What are the lengths of the sides?

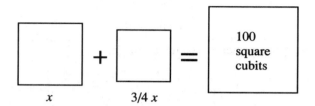

x 3/4 x

FIGURE 1

"Always take a square of side 1," the scribe recommends for the false position solution. Then the squares have assumed lengths of 1, and $\frac{3}{4}$ cubit, and the sum of the assumed squares is $1\frac{9}{16}$ square cubits. To find the side of the square, the scribe takes the square root of $1\frac{9}{16}$ square cubits and finds it to be $1\frac{1}{4}$ cubits. But the required sum is 100, with side 10. Divide 10 by $1\frac{1}{4}$ to get the correction factor, 8. The required sides are 8 and 6 cubits.[11] The correction factor was not the ratio of the areas but the ratio of the square roots.

Incidentally, ancient Egyptians sometimes solved equations by the same methods we use today. The equation in problem 19 of the Moscow Papyrus is solved by transposing, collecting similar terms, and then division by the coefficient of the variable.[12] False position methods appeared to be favored through the millennia, in Egypt and other parts of the world, notably in China.[13] Bringing multicultural examples of false position into the classroom can encourage experimentation and proportional thinking, approaches that are encouraged in the NCTM Curriculum and Professional Standards for the 1990s.[14]

Ancient Egyptian mathematics never died but fused with the Hellenistic at Alexandria and, later, with medieval Islamic mathematics at Cairo. Mathematics advanced in part through the blending of cultures—the ancient Egyptian and Babylonian with the Greek and the Indian and the Chinese. False position was a favored method in Africa throughout these periods, for example, in the work of Diophantus (c. 250) and Abu Kamil (born 850). Diophantus's false position solutions for problems 15 and 37 of his *Arithmetica* Book IV were so beautifully developed that in the early 1900s Thomas Heath and Gino Loria urged their inclusion in modern textbooks.[15]

In Book IV, problem 15, Diophantus asked for three rational numbers such that the sum of any two, multiplied by the third, is a given number. He gives his method in the traditional Egyptian manner, using specific but representative values of 35, 27 and 12 for the "given numbers." In modern general notation the problem could be stated:

$$(x + y)z = a$$
$$(y + z)x = b$$
$$(x + z)y = c$$

The form of the problem, as Diophantus stated it, was:

$$(\text{first} + \text{second})(\text{third}) = 35$$
$$(\text{second} + \text{third})(\text{first}) = 27$$
$$(\text{first} + \text{third})(\text{second}) = 32$$

Solution: Let the third number be the unknown; let's call it "z". Then $(\text{first} + \text{second}) = 35/z$. Assume a "false" value, such as $10/z$ for the first number. Then the second $= 25/z$, and their product will be $250/z^2$. Substitute these values in the second and third equations:

$$(250/z^2) + 10 = 27$$
$$(250/z^2) + 25 = 32$$

The result of assuming $10/z$ for the first number produces equations that are inconsistent. (Subtracting the top from the bottom equation results in the false statement $15 = 5$.) To obtain consistent equations, $35/z$, the sum of the first and the second numbers, must be split into two parts whose difference is $32 - 27$ or 5. These parts are $15/z$ and $20/z$. Try an assumed value of $15/z$ for the first number. Then the second number is $20/z$, and their product is $300/z^2$. Substitution in the original second and third equations results in:

$$(300/z^2) + 15 = 27$$
$$(300/z^2) + 20 = 32.$$

On transposing and collecting terms, each of these equations becomes $300/z^2 = 12$ and $z = 5$. Substitution gives $(3, 4, 5)$ as the solution. Negative solutions were not considered. Heath believed that Diophantus was working from a general solution. Heath presents the Diophantus solution in a general form and shows that the choice of constants 35, 27, and 32 fills the requirements of the general solution.[16]

The work of Diophantus precedes that of another Alexandrian mathematician, Hypatia (died 415). Hypatia, the earliest female mathematician whose name we know, wrote a commentary on Diophantus. Paul Tannery believes that it is only through Hypatia's commentary that Diophantus's work has survived because all other copies were lost.[17] Hypatia's commentary may have been used by Abu Kamil (born 850) and known as "the Egyptian Calculator" in Cairo. His work was so close to Diophantus that Jacques Sesiano believes Abu Kamil must have had original sources written in Greek.[18] Abu Kamil's *Algebra* extends the work of al-Khwarizmi both technically and theoretically. He included more proofs and worked with higher-degree equations and irrational coefficients.[19] Among his now lost works is a "Book on the Two Errors" or Double False Position. In his *Algebra* Abu Kamil solves equations by false position:

> One says that 10 is divided into three parts, and if the small one is multiplied by itself and added to the middle one multiplied by itself, it equals the large one multiplied by itself. And when the small is multiplied by the large, it equals the middle multiplied by itself.[20]

In modern symbols,

$$x + y + z = 10 \qquad x < y < z$$
$$x^2 + y^2 = z^2$$
$$xz = y^2$$

Abu Kamil assumes that $x = 1$. Then $z = y^2/1$ from equation 3, and substitution in the second equation gives $1 + y^2 = y^4$. Abu Kamil uses the quadratic formula to get:

$$y^2 = \frac{1 + \sqrt{1 + 4}}{2}$$

$$y = \sqrt{\frac{1}{2} + \sqrt{1\frac{1}{4}}}$$

$$x + y + z = 1 + \sqrt{\frac{1}{2} + \sqrt{1\frac{1}{4}}} + \left(\sqrt{\frac{1}{2} + \sqrt{1\frac{1}{4}}}\right)^2$$

$$= 1\frac{1}{2} + \sqrt{1\frac{1}{4}} + \sqrt{\frac{1}{2} + \sqrt{1\frac{1}{4}}}$$

but should $= 10$.

So the correction factor for $x = 1$, is:

$$\frac{10}{1\frac{1}{2} + \sqrt{1\frac{1}{4}} + \sqrt{\frac{1}{2} + \sqrt{1\frac{1}{4}}}}$$

Undaunted, Abu Kamil simplified the radicals and found the correction factor needed to multiply times the assumed value of 1 to get the correct value. The correct value is $x = 5 - \sqrt{\sqrt{3,125} - 50}$, approximately 2.57. A similar procedure is followed for the other variables.

The work of medieval Islamic mathematicians came into Europe during Moorish rule of Spain and through trade routes. During trading visits to North Africa before 1202, Leonardo of Pisa (Fibonacci) studied Islamic mathematics and brought this knowledge back to Italy. Fibonacci made extensive use of Abu Kamil's work. Abu Kamil's translators, Mohammed Yadegari and Martin Levey, wrote that: "There is proof that Leonardo used dozens of Abu Kamil's problems in his algebra. From *On the Pentagon and Decagon*, Leonardo used seventeen of its twenty problems carrying over the exact number facts."[21] Fibonacci

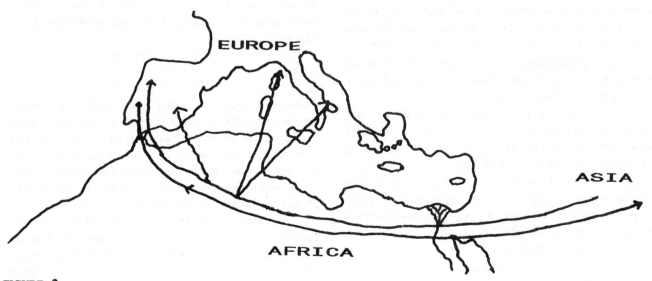

FIGURE 2
Major trade routes and dissemination of Islamic learning.

also borrowed from al-Karaji, a mathematician in Baghdad who wrote a book *The Marvellous* (ca. 1020). In turn, al-Karaji had been influenced by Abu Kamil.[22]

It is not surprising that Leonardo's mathematics often used false position solutions. He called this method "el chatayam," a corruption of the Arabic expression for double false, "hisab al-khataayn." Chapter 13 of Leonardo's *Liber Abaci* is devoted to the rule of false position.[23]

With false position Fibonacci solves some classic problems.[24] Instead of the "frog in the well," Leonardo had a lion in a pit, 50 handbreadths deep. Every day, the lion climbed 1/7 handbreadth up and fell back 1/9 handbreadth. In how many days would the lion climb out of the pit?

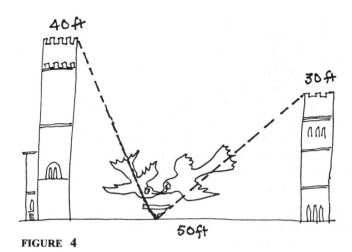

FIGURE 4

FIGURE 3

Leonardo's false assumption was a convenient value, 63 days, the least common denominator of $\frac{1}{7}$ and $\frac{1}{9}$. But in 63 days the lion would make only 2 handbreadths' progress whereas 50 handbreadths were needed. Then 2 : 50 :: 63 : x. Leonardo's result was 1,575 days. However, on day 1,572 the lion would be out and would not go back into the pit (unless hunger had dulled his senses after so many years in the pit!)

Another problem, known also from Chinese and Indian sources, involved two birds flying from the tops of two towers, at the same time, at equal speeds, to a fountain below. They arrive at the same time. How far from the towers is the fountain if one tower is 40 feet high, the other tower 30 feet high, and they are 50 feet apart at the base? (Find paths of equal lengths.)

Fibonacci tried two different values in a double false position solution. Stated in modern form, with x for the unknown, x_1 and x_2 for the two "false" assumptions, the formula for the correct value is:

$$x = \frac{x_2 f(x_1) - x_1 f(x_2)}{f(x_1) - f(x_2)}.$$

First he assumed the fountain was 10 feet from the 40-foot-high tower. By the right triangle theorem, the squares of the paths of the birds (the hypotenuses) are found to differ by 800. Then he tried a value of 15 feet from the 40-foot-high tower but still got a difference of 300. Fibonacci used the errors resulting from the two "false" values to show how much closer the second guess came to the solution. In this problem:

Guess	Error
10	800
15	300
18	0[25]

The double false position formula also gives the distance as 18 feet.

$$\frac{15(800) - 10(300)}{800 - 300} = 18.$$

Check: $18^2 + 400^2 = (50 - 18)^2 + 30^2$.

Banneker's journal includes a similar problem. A 60-ft. ladder, placed between two walls, reaches a 37-ft high window on one wall. Tilted the other way, without moving the bottom, the ladder reaches a window 23 ft. high on the other wall. How far apart are the walls? We do not know if Banneker used a false position solution. The distance can be found by direct application of the right triangle (Pythagorean) theorem which gives the distance between the walls as the sum of 47.23 ft + 55.42 ft = 102.65 ft.

An Islamic technique for double false position, called the method of the scales, appeared in early European mathematics books such as Humfrey Baker's *Wellspring of Science,* first printed in London in 1568. Baker must have been pleased with his efforts because he quoted an old maxim, "Where good wine is to sell, there needs no garland be hanged out." He describes the Rule of Falsehood or False Position: "It teacheth to find out the true nu(m)ber that is demaunded. And this of all y vulgar rules which are in practise is the most excellent."[26] His first example employs single false position:

> I have delivered to a banker, a certaine summe of pounds in money to have of him by the yeare simply, 6 li (livres or pounds B.L.) upon the 100 li. And at the end of 10 years, he payed me 500 li for all, both principal and gaine, I demand how much was the principall summe that I delivered him at the first.

> Suppose that I delivered y first 200 li, do make but 320 li, and I must have 500 li. Thus you see that I have three terms for the Rule of three; the one which shall contain the Question, the other two which I have formed artificially, which are 200 and 320: in such sort, that 320 outh to have such proportion to 200 as 500 hath unto the number that I seeke; that is to say, unto the true principal sum, then must I have recourse unto the Rule of three, after the sort, saying. If 320 li become of 200 li, of how much shall come 500 li. I do multiply 500 by 200, and they are 10,000, the which I must divide by 320 li. Thereof commeth 312 li 1/2, which is the summe that I delivered at the y first. And thus this Rule has some co'gruence with the Double rule of three.[27]

Double position is used for an inheritance problem in Baker's book, a problem probably derived from Islamic scholars, along with the scales diagram for false position.

> A man lying at the point of death" bequeathed 100 duckets to three friends. The second got twice the bequest of the first, abating 8 duckets. The third got three times the first less 15 duckets. How much did each get?

FIGURE 5

Baker assumed the first had 30. Then the other two would have received 52 and 75 for a sum of 157, 57 too much. Next he assumed the first had 24. This assumption gives a sum of 24 + 40 + 57 or 121, an error of 21. The guesses of 30 and 24, and the resulting errors of 57 and 21, are entered into the scales formula. As Baker said:

> "Multiply crosswise, 30 (which is the first position) by 21 which is the second error = 630. Likewise multiply 24 (which is the second position) by 57 which is the first error and I find 1365." Then Baker reasoned, errors alike (same sign) so subtract: 1368 − 630 = 738, the dividend, and 57 − 21 = 36, the divisor. The dividend 738, divided by 36, is 201/2.
> Answer: 20 1/2, 33, 46 1/2

$$\frac{630}{30\ 57}$$
$$\frac{24\ \ 21}{1368\ 36}$$
$$\frac{630}{738}$$

Evidently rules for signs created trouble, which does not surprise modern high-school algebra students. To aid in applying the formula, Baker offered a poem:

> The signs both like, subtraction do require
> And unlike signs, addition will desire.[28]

False Position in Banneker's Time

The method Baker used appears in such textbooks published in Banneker's time as Daniel Adams's *The Scholars Arithmetic,* and John Bonnycastle's *Scholar's Guide to Arithmetic,* 1780. Bonnycastle gave an algebraic rationale, stating that the rule of double position, applies only if: $r : s :: x - a : x - b$, where r and s are the errors resulting from substituting the assumed values a and b. He then shows algebraically that this proportion is equivalent to:

$$x = \frac{rb - sa}{r - s}.[29]$$

On Banneker's use of false position, documentation is sparse. Unfortunately, only one manuscript journal escaped the fire that consumed his house the day of his funeral. His printed almanacs show only results, not the calculations, Only one additional problem that Banneker had composed in verse was recorded by a friend after Banneker's death.

Banneker's Manuscript Journal includes a puzzle that he attributed to "Ellicott, Geographer General," Major Andrew Ellicott. Banneker gave the correct answer but does not show his method:

> Divide 60 into four Such parts, that the first being increased by 4, the Second decreased by 4, the third multiplyed by 4, the fourth part divided by 4, that the Sum, the difference, the product and the Qutient shall be one and the Same Number—
> Ans. first part 5.6 increased by 4 ( ) 9.6
> Second part 13.6 decreased by 4 ( ) 9.6
> third part 2.4 Multiplyed by 4 ( ) 9.6
> fourth part 38.4 divided by 4 ( ) 9.6 [30]

I think Banneker used two methods inherited from African and Asian mathematicians, working backwards from the end result and false position. Select any value for the end result, say 10. Working backwards,

first part is 6

second part is 14

third part is 10/4

fourth part is 40

Their sum is 62 1/2, but should have been 60.

$60 : 62 1/2 :: x : 10$. The correction factor $60/(62 1/2)$ is 0.96, and $10(0.96) = 9.6$, the result that Banneker gave in his answer.

The Hare and the Hound

In the Manuscript Journal this poem-puzzle was written in Banneker's beautiful script:

> When fleecy skies have Cloth'd the ground
> With a white mantle all around
> Then with a grey hound Snowy fair
> In milk white fields we Cours'd a Hare
> Just in the midst of a Champaign
> We set her up, away she ran,
> The Hound I think was from her then
> Just thirty leaps or three times ten
> Oh it was pleasant for to see
> How the Hare did run so timorously
> But yet so very Swift that I
> Did think she did not run but Fly
> When the Dog was almost at her heels
> She quickly turn'd and down the fields
> She ran again with full Career
> And 'gain she turn'd to the place she were
> At everyturn she gain'd of ground
> As many yards as the greyhound
> Could leap at thrice, and She did make,
> Just Six, if I do not mistake
> Four times She Leap'd for the Dogs three

> But two of the Dogs leaps did agree
> With three of hers, nor pray declare
> How many leaps he took to Catch the Hare.
> Just Seventy two I did Suppose,
> An Answer false from thence arose,
> I Doubled the Sum of Seventy two,
> But still I found that would not do,
> I mix'd the Numbers of them both,
> Which Shew'd so plain that I'll make Oath,
> Eight hundred leaps the Dog did make,
> And Sixty four, the Hare to take.

$$4 : 72 :: 48$$

$$\frac{48}{576}$$
$$\frac{288}{4/3456}$$

864 ans.

Double false position is Banneker's choice of method, with 72 and double 72 or 144 as assumed values. "Mixing the Numbers" may refer to the double false-position formula.

Trying 72, the hare would have made $(4/3)(72) = 96$ leaps to cover ground equivalent to $(2/3)(96) = 64$ hound leaps. Or simply, the hare covers 8/9 the distance that a hound covers in the same time. Adding the hare's head start of 30, plus the 3 hound leaps the hare gained at each turn (18 gained on 6 turns) the hound would need $64 + 48$ leaps to catch the hare. So 72 hound leaps is $72 - (64 + 48)$, or 40 hound leaps too few.

Also, 144 hound leaps, as Banneker said, will not do. It is 32 too few. The additional 72 hound leaps reduced the error by 8. To reduce the error of 32 to zero, add $4 \times 72 = 288$ to the guess of 144 for a total of $288 + 144 = 432$ hound leaps. Unfortunately, 432 is only half of Banneker's answer.

Plugging in the trial values and resulting errors in the double position formula also gives 432.:

$$\frac{72 \times 32 - 144 \times 40}{32 - 40} = 432,$$

The method used for a similar problem in the *Treviso Arithmetic* also yields 432,[31] as does the algebraic method of Nicholas Saunderson.[32] However, if Banneker's handwriting is read as numeral 22, instead of script "Six" for the number of hare turns, then the formula indeed gives 864 hound leaps. Silvio Bedini, a biographer of Banneker, checked the manuscript and confirmed his reading as "Six." The microfilm copy held by the Benjamin Banneker Association was too faint to confirm either value. The Rule of

Three (4 : 72 :: 48 :) that Banneker uses may be the clue to his interpretation of the poem, and which values were used.

The Cooper and the Vintner

This Banneker poem-puzzle was written down after his death by a friend, Charles W. Dorsey. The poem reads:

A cooper and vintner sat down for a talk,
Both being so groggy that neither could walk;
Says cooper to vintner, "I'm the first of my trade,
There's no kind of vessel but what I have made,
And of any shape, sir, just what you will,
And of any size, sir, from a tun to a gill."
"Then,"says the vintner, "you're the man for me.
Make me a vessel, if we can agree.
The top and the bottom diameter define,
To bear that proportion as fifteen to nine,
thirty-five inches are just what I crave,
No more and no less in the depth will I have,
Just thirty-nine gallons this vessel must hold,
Then I will reward you with silver or gold,—
Give me your promise, my honest old friend."
"I'll make it tomorrow, that you may depend!"

So, the next day, the cooper, his work to discharge,
Soon made the new vessel, but made it too large,
He took out some staves, which made it too small,
And then cursed the vessel, the vintner, and all.
He beat on his breast, "By the powers"he swore
He never would work at his trade any more.
Now, my worthy friend, find out if you can,
The vessel's dimensions, and comfort the man!"[33]

Solutions. The puzzle fixes the height of the barrel and the ratio of top and bottom diameters, of the barrel, leaving only the diameters to be found. The curvature of the sides is not stated, but some standard shape may have been understood. Since the puzzle speaks of a vintner, it seems reasonable to assume that the barrel was intended to hold wine. Gallons were not, and still are not uniform measures of volume. A wine gallon contains 231 in³, but a gallon of ale or beer contains 282 in³. For 39 gallons of wine, the volume would be 9009 in³, but 39 gallons of ale would contain 10,998 in³.

Adams's textbook,[34] written a few years after Banneker's Almanac, gives a formula for "gauging a cask." His formula for the volume, given the end diameters and the height, allows for curvature of the staves. A mean di-

ameter in inches is found; then the formula for the volume of a cylinder is used with Adams's simplification of $4/294 = \pi/231$ (to 4 places).

If Ds = diameter of the small end,
DL = diameter of the large end.
L = the length or height
The mean diameter, $D_{av} = D_s + (2/3)(D_L - D_s)$
Volume (wine gallons) = $(D_{av})^2(L)/294$.

If Banneker used this formula, he could have used false position to find the dimensions. Assume a large diameter of 15 inches; then the small diameter would be 9 inches, and the average diameter would be:

$$D_{av} = 9 + (2/3)(15 - 9) = 13 \text{ inches.}$$

$$\text{Volume} = (13)2(35)/294 = 20.12 \text{ gallons}$$

but should be 39 gallons. The correction factor is not the ratio of the volumes, but the ratio of the square roots: $\sqrt{39}$: $\sqrt{20.12} = 1.3922$. (Using the square roots here is similar to the procedure in the Egyptian Berlin papyrus problem of the sum of two squares.) Multiply 15 inches, the assumed diameter, by the correction factor, 1.3922, to get the desired large diameter, 20.88 inches. The small diameter would be $(9/15)(20.88) = 12.53$ inches.

In 1854 Benjamin Hallowell of Alexandria presented a proposed solution to a meeting of the Maryland Historical Society.[35] The Hallowell solution of 24.745 inches and 14.8476 inches for the diameters follows from taking the shape to be a truncated cone, with straight barrel sides made of flat staves. Hallowell appears to have used ale gallons of 282 inches3, rather than wine gallon measure of 231 inches³. For classroom discussion, this is a good opportunity to bring in the importance of standardization in the sciences and industry.

The volume of a truncated cone can be found by completing the cone, then subtracting the volume, or cutting off, the smaller cone at the apex. This is just the type of 3-dimensional thinking encouraged by the NCTM Curriculum Standards.

First find the height of the cutoff cone. A vertical section through the completed cone shows two similar triangles with corresponding sides proportional:

$$\frac{y}{3x} = \frac{35 + y}{5x}.$$

So $y = 52.5$ inches is the height of the cutoff cone. Height of the large cone is $52.5 + 35 = 87.5$ inches. The volume of a cone is

$$V = \frac{1}{3} \times \text{height} \times \text{area of the circular base}$$

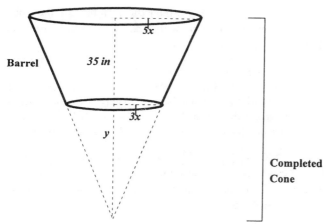

Barrel

35 in

5x

3x

y

Completed
Cone

FIGURE 6
Volume of truncated cone-shaped barrel.

or

$$V = \frac{1}{3}\pi r^2 h.$$

Volume of completed cone is

$$\frac{1}{3}(\pi)(5x)^2(87.5) \text{ inches}^3.$$

Volume of cutoff cone is

$$\frac{1}{3}(\pi)(3x)^2(52.5) \text{ inches}^3.$$

Volume of barrel is:

volume of completed cone − volume of cutoff cone.

39 ale gallons = 10,998 inches^3.

$$V = 10,998 = \frac{1}{3}\pi(5x)^2(87.5) - \frac{1}{3}\pi(3x)^2(52.5)$$

$$= 1795.9438x^2$$

$$x = 2.4746 \text{ in.}$$

The diameters are twice the radii of $3x$ and $5x$. Diameters, $6x$ and $10x$, = 14.84778728 inches and 24.74631213 inches, which may be rounded off to 14.8478 inches and 24.7463 inches.

Another problem from Banneker's Manuscript Journal was probably solved by trial-and-error, not by false position:

A, B and C, discoursing about their ages, Says A, if from double the Cube Root of B's age, double the biquadrate root of C's age betaken the remainder will be equal to the Sursolid Root of my age, Says B, the Square root of my age is equal to one fourth part of A's, and Says C,

the Square root of my age is one more than the Square root of B's, Required their several Ages—
A's) (32 The Sursolid Root of which is 2
B's) (64 The Cube of Root of which is 4
C's) (81 The biquadrate root of which is 3 [36]

Sursolid refers to the fifth root and biquadrate to the fourth root. Using the given letters to represent their ages, we get three equations with radicals.

$$2\sqrt[3]{B} - 2\sqrt[4]{C} = \sqrt[5]{A}$$

$$\sqrt{B} = (1/4)A$$

$$\sqrt{C} = 1 + \sqrt{B}$$

In Banneker's time, it was often assumed that the solutions are rational numbers, perhaps integers. Since the situation requires a fifth root, there are few values that could refer to a person's age and still have an integral fifth root. Consider fifth powers of integers:

$$1^5 = 1$$

$$2^5 = 32$$

$$3^5 = 243 \text{ (not a valid human age)}$$

Try A's age = 32 years. Then $\frac{1}{4}$ of A's age is 8. For the second equation, if $\sqrt{B} = 8$, then $B = 64$. From the third equation, we get: $\sqrt{C} = 1 + \sqrt{64}$. Then $\sqrt{C} = 9$. Therefore, $C = 81$. Test these values in the first equation:

$$2\sqrt[3]{64} - 2\sqrt[4]{81} = \sqrt[5]{32}.$$

In conclusion, I agree with Gino Loria[37] that teaching false position is valuable beyond its historical interest. Not only does it teach proportional thinking; it also encourages experimentation. Selecting a false value and then seeing how it works in the problem is somewhat like introducing a perturbation into the equations for a physical system. The results yield a more dynamic understanding of how the system operates. In this respect, Benjamin Banneker's experimental approach for the natural sciences may have guided his work with false position solutions. The African background of false position methods also adds to its interest in classrooms.

Some historians wonder why medieval mathematicians valued false position so highly. Perhaps this disparagement of false position results from judging one time period by the standards of another. In time to come, our present method of solving equations may be little more than a historical curiosity, since hand-held computers will do it much faster and more reliably. But then, as now, principles behind the old methods will still be worthwhile subjects for study.

References

1. At the New Orleans HPM meeting (1991) Florence Fasanelli quoted Benjamin Banneker, that he "had advanced to double position through his own efforts." My statement, that the false position method came from ancient Africa, led to a suggestion by Fasanelli for this paper on the African roots of Banneker's mathematics.
2. T.F. Mulcrone, S.J., "Benjamin Banneker, pioneer Negro mathematician," *The Arithmetic Teacher,* 54 (1961) 32.
3. Silvio A. Bedini, *The Life of Benjamin Banneker,* Rancho Cordova, CA: Landmark Enterprises, 77f.
4. Mulcrone, op. cit., 34.
5. Charles Leadbetter, *A Treatise on Eclipses of the Sun and Moon,* London: Wilcox, 1731, 1.
6. Banneker Astronomical Journal, repository Maryland Historical Society. This journal is referred to in the literature as "Banneker's Manuscript Journal."
7. Daniel Adams, *The Scholar's Arithmetic or Federal Accountant,* Keene, NH: John Prentiss, 1807 (written before 1801) 200–201.
8. Chicago Public Schools, *Algebra I Framework,* 1991, 96–97.
9. Arnold B. Chace, *The Rhind Mathematical Papyrus,* Reston, VA: National Council of Teachers of Mathematics (NCTM), 1986, 12.
10. Richard J. Gillings, *Mathematics in the Time of the Pharaohs,* Cambridge, MA: MIT Press, 1972, 156–62.
11. Ibid., 161–62.
12. Ibid., 157.
13. George G. Joseph, *The Crest of the Peacock, Non-European Roots of Mathematics,* London: Tauris, 1991, 158.
14. NCTM, *Professional Standards for Teaching Mathematics,* Reston, VA: NCTM, 1991, 26.
    Curriculum and Evaluation Standards for School Mathematics, Reston. VA: NCTM, 1989, 5, 70.
15. Thomas Heath, *Diophantus of Alexandria,* New York: Dover, reprint 1964, originally 1910, 175.
16. Ibid.
17. Thomas Heath, *A History of Greek Mathematics,* vol 2. New York: Dover, 1981, originally 1921, 449.
18. Jacques Sesiano, *Books IV to VII of Diophantus' Arithmetic in the Arabic Translation, attributed to Qusta ibn Luqa,* New York: Springer Verlag, 1982, 9–10.
19. Marshall Clagett, in *The Algebra of Abu Kamil in a Commentary by Mordecai Finzi,* tr. and commentary by Martin Levey, Madison: University of Wisconsin, 1966, ix.
20. Martin Levey, ed., *The Algebra of Abu Kamil in a Commentary by Mordecai Finzi,* 186–192.
21. Mohammed Yadegari and Martin Levey, *Abu Kamil's On the Pentagon and Decagon,* Tokyo: History of Science Society of Japan, Supplement 1, 1971, 1.
22. Charles Gillispie, ed., *Dictionary of Scientific Biography,* vol. 4, Scribners, New York, 1971, 611.
23. David E Smith, *History of Mathematics,* vol. 1, New York: Dover, 1958, orig 1923, 216.
24. Joseph and Frances Gies, *Leonard of Pisa,* New York: Crowell, 1969, 68.
25. Ibid., 71.
26. Humfrey Baker, *The Wellspring of Science,* London, 1612, 181.
27. Ibid., 182.
28. Ibid. 184.
29. John Bonnycastle, *The Scholar's Guide to Arithmetic,* J. Johnson, London, 1780. This information was supplied by F. Fasanelli in her paper on "Benjamin Banneker," presented at the International study group on the relationship of History and Pedagogy of Mathematics (HPM) at the national NCTM convention, April 1991, in New Orleans.
30. Banneker, Manuscript Journal.
31. Frank Swetz, *Capitalism and Arithmetic, the New Mathematics of the 15th century,including the full text of the Treviso Arithmetic,* translated by David E. Smith, LaSalle IL: Open Court, 1987.
32. Nicholas Saunderson, *The Elements of Algebra,* Cambridge, England: University Press, 1741, 145.
33. J. Saurin Norris, Sketch of the Life of Benjamin Banneker from notes taken in 1836, Maryland Historical Society, 1854, 20.
34. Adams, op. cit., 205.
35. Norris, op. cit., 20.
36. Silvio Bedini now agrees with me that Banneker listed C's age as 81. On page 327 of Bedini's biography, *The Life of Benjamin Banneker,* this value is given as 84.
37. Quoted in Heath, *Diophantus,* 178.

Mary Everest Boole (1832–1916): An Erstwhile Pedagogist for Contemporary Times

Karen Dee Ann Michalowicz
Langley School
McLean, Virginia

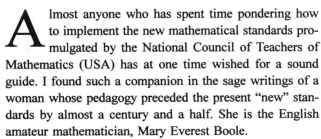

Almost anyone who has spent time pondering how to implement the new mathematical standards promulgated by the National Council of Teachers of Mathematics (USA) has at one time wished for a sound guide. I found such a companion in the sage writings of a woman whose pedagogy preceded the present "new" standards by almost a century and a half. She is the English amateur mathematician, Mary Everest Boole.

The question is how could Mary Boole's "modern" pedagogy be ignored for so long? How could her name be absent from the history of mathematics education? To answer this question we must look at her life in light of the Victorian era in which she lived. This paper is not simply the biography of a remarkable woman mathematics educator; it also includes selections from her pedagogical ideas and the philosophy and psychology from which her ideas were formed. It is hoped that Mary Boole's writings will give teachers ideas and inspiration.

Born in the village of Wickwar, Gloucestershire, England, in 1832, Mary Boole lived during an era in which women had almost no opportunity to pursue higher studies. Further, she stood in the shadow of her uncle, Sir George Everest, a Surveyor General of India during the early nineteenth-century after whom Mt. Everest was named, and her husband, George Boole, the famous algebraist and logician. Biographies about mathematicians give nominal attention to Mary. In fact, H.J. Mozans, an anagrammatic pen name for Rev. John A. Zahn, a contemporary of Boole, and a recognized biographer of famous women of science, does not even mention her. This is unusual because Mozans found space for Theona, the wife of Pythagoras, about whom less is known. It appears that Mary Everest Boole is a mathematical pedagogist who wrote about and encouraged the use of "hands-on" activities and of concrete materials described today as manipulatives. She is also an early advocate of "cooperative learning" in mathematics education.

The story begins more than a century ago in England. A lovely daughter was born to Rev. Thomas Roupell Everest, rector of Wickwar, and Mary Ryall whose brother was Vice-President of Queen's College, Cork. Following his long illness, the Reverend Everest moved his family to Poissy, France, close to Paris, to be near Samuel Hahnemann, the father of holistic medicine. Hahnemann prescribed a strict routine. Not only the father, but the whole family and the family staff were "forced" to participate.[1] The regimen included frigid baths, a special diet, and long vigorous walks before breakfast, especially on cold days. For eleven years, the Everest family followed Hahnemann's experiment, which worked.

During her time in France, daughter Mary became bilingual. Indeed, she considered French her first language, the language in which she would think and form ideas. Mary's first school experiences were in a day school and with a daily governess. Her mother also gave her children "hopelessly dreary" lessons.[2] Mary spoke of her mother as being very intelligent but unintellectual. A gentleman, "Monsieur Deplace," was hired to tutor the children for two hours every day in French and arithmetic. Deplace[3] so impressed Mary that years later she was to write about his extraordinary ability to teach.[4]

One lesson Mary writes about is her problem in arithmetic with the "double rule of three."[5] Deplace told her nothing at the beginning, but he asked her a succession of questions. She was to write down each answer, as he asked the question. Then he asked her to analyze his questions and to review her answers to his questions. In this way, step by step, she perceived the answer to her complicated problem. Mary writes that Deplace was educated in the French school of mathematics. Although she doesn't explain her statement, it is known that during the eighteenth century and early nineteenth century most French mathematicians were not associated with universities. Many were private teachers or tutors.[6] One can speculate that Deplace must have been well versed in the Frenchman Jean-Jacque Rousseau's "child-centered" philosophy, as described in his book, *Emile* (1762), which encouraged individual thought. The orientation of Deplace not only helped Mary in her own study of mathematics, but also had an impact on her mathematics pedagogy. A common thread throughout her writings is that teachers should elicit understanding from students through questioning and journal-notation. Deplace's method of questioning and journal-writing remains excellent pedagogy for mathematics educators today.

A low point in Mary's stay in France came when she was ten, in 1842, after she overheard her father and an unidentified English friend speak about Cambridge University as a center of great mathematics and about Charles Babbage. After the conversation her father told her mother that he did not know what he was going to do with his children. His son had the interests and disposition he would expect of his daughter, and his Mary, if she could go to Cambridge "would carry everything before her . . . But what could a girl do learning mathematics?"[7] Before this conversation, Mary had not realized that universities were not open to women. This incident was her "triple shock." Her first was that she would not be able to attend a university; second, that she might "never go to Cambridge and meet Mr. Babbage," and finally that, for the accident of being a girl, she would be deprived of learning mathematics.[8]

When she was 11, Mary's family returned to England. During her teen years she met her father's many friends, among them John Herschel and Charles Babbage. She became her own mathematics tutor using the family texts. Other gifted women likewise learned by self-instruction. The same thing happened with the French number theorist, Sophia Germain, and the Russian, Sonya Kovalevskaya, a differential-equation analyst. During her teen years, Mary's illustrious uncle returned from India. He spent many hours sharing with her his knowledge and traveling experiences. George Everest even asked to adopt young Mary.

While on a visit in 1850 to her uncle, John Ryall, Vice-President and Professor of Greek at Queen's College in Cork, Mary became acquainted with George Boole. During the following two years they frequently corresponded about mathematics and science. A deep friendship developed. In 1852, George visited the Everest family. At this time he became tutor, confidant, and adviser to Mary. For several years, they continued their correspondence about mathematics. In 1855, Mary's father died. George, concerned about her welfare, proposed marriage. After a short engagement, they married. Mary was 23, and George, 40.[9]

George and Mary Boole took up residence at Queen's College in Cork County, Ireland, where George was a professor. Theirs was a close successful marriage in which Mary shared in her husband's interests. At his encouragement, she attended her husband's lectures, an unheard of practice at that time; and she advised her husband as he developed his comprehensive work on differential equations.[10]

By all accounts, the nine years that George and Mary shared together were happy family years. During this time five daughters were born to the Booles. The girls brought George great delight. Even though he spent only a short time with them, he was always involved in their discipline and education.[11]

The happiness of family life did not carry over to George's work. Mary reports that his work was arduous, and college politics taxed his mental strength. Mary continually worried about his mental and physical health. On November 24th, 1864, George Boole, in his forty-ninth year, walked to a lecture in the pouring rain. He lectured in wet clothes, caught a fever, and died two weeks later. Thus, by the time she was 32, Mary Boole was the widowed , almost penniless, mother of five daughters, including an infant of six months.

After George Boole's death, Mary moved back to England and was offered a position at Queen's College in London. The college trained women to become governesses. Although Queen's College did offer college-level courses

SERIES II   **BOOLE CURVE-SEWING CARDS**   No. 4.

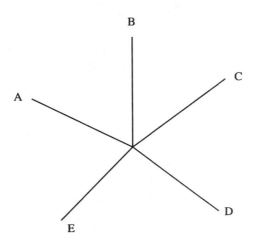

This may be treated in several ways; e.g., the
line marked A may be worked against that
marked C. C against E. E against B. B against
C., etc. Or A against B. B against D. D
against A, and then C against E, etc.

G.P. & S., Ltd.

Knots and finishings on this side.

LONDON

FIGURE 1

to women, college degrees were not offered to women in
England during the middle of the nineteenth century. Neither was it permitted for women to hold college faculty
positions. Instead, Mary was a librarian at Queen's College, which was considered a "respectable" position for a
woman. Mary also tutored children in mathematics.

Mary enjoyed teaching. Her ideas about mathematical
instruction included using natural materials, such as twigs,
shadows, and stones to develop children's understanding
of and excitement in mathematics. As children did things
physically (manipulatively), she believed that the unconscious understanding of many fundamental mathematical
ideas grew. Unfortunately, this pedagogy appears to have
made no impact on nineteenth-century English elementary
education. Mary only gained some recognition in the United
States in the 1940s. Ralph W. Tyler, a leader in the progressive schools movement and chairman of the Department of
Education of the University of Chicago, said: "Her conceptions of child psychology and of learning as well as her
understanding of the psychological nature of mathematics
and science, make her a pioneer in this generation as in
the last."[12] He distributed copies of *The Preparation of*

the Child for Science written by Mary Boole in 1904 to
many of his colleagues, but her work did not have much of
an impact on mathematics education in the United States.
Perhaps Mary Boole's ideas, which approximate ideas presented in the new mathematics standards, will generate
some interest a century after their publication.

To encourage students to explore their environment
and discover concepts, Mary invented curve-stitching
cards, called Boole cards, which became popular in England for a time.[13] Figure 1 is a replica of a Boole Card.[14]

Unfortunately, as Mary's ideas faded, so did the use
of the Boole cards. Today, one can find ideas similar to
Mary's in teaching geometry with string art and string design, notably in Victoria Pohl's excellent book, *How To
Enrich Geometry Using String Designs* (1986).

Although Mary was not permitted to teach at Queen's
College, she still inspired students of mathematics. One
should not forget that Mary had studied under her husband, George Boole. She organized "Sunday Night Conversations" during which she and other students discussed
her husband's mathematics, Darwinian theory, philosophy,
and psychology. These soirees were popular. Years later, a

former student wrote Mary to say that she thought that the Sunday night experiences were to amuse, not teach. However, the student found after leaving college that she had received from Mary a power which allowed the student to think for herself and discover what she needed to know.[15]

Mary considered herself a mathematical psychologist who throughout her life tried to understand how people learn mathematics using their conscious and unconscious minds and their bodies (manipulating things). She never tired of using her ideas and techniques with her students.[16]

Mary wrote several books, most of which were published years after they were written. It is said that her first book, *The Message of Psychic Science* (1883), was suppressed for fifteen years. It would hardly be considered controversial today. Its pages reveal her genius. Again, the question is asked: why has her pedagogy remained unknown for so long? Why did others not realize her brilliance? This question may be answered by examining the English times in which she lived.

Although Victorian England experienced many changes during the so-called second industrial revolution, some characteristics about elementary education in England were not typical of the western world. First of all, "while Germany and France were making some progress in connection with popular (national) education, England's attitude seemed to be one of indifference."[17] There were a number of reasons for this. Perhaps one of the easiest to understand reasons was the fact that the "Lockean" spirit of individualism was so characteristic of English education: "Each man to his own means," was frequently quoted.[18] John Stuart Mill said in 1859, "A general State education is a mere contrivance for molding people to be exactly like one another; and as the mold in which it casts them is that which pleases the predominant power of government."[19] In addition Herbert Spencer, an educational reformer and theorist of the nineteenth century, stood in strong opposition to national education. He and like-minded Englishmen couldn't stop the national education movement, but they did slow its progress.[20]

Nonetheless, the Victorian era produced many "experiments" in elementary education. These experiments supposedly incorporated Continental ideas. But the theories and methods of Rousseau, Pestalozzi, Herbart, Froebel, and Spencer were paid only lip service. The two chief models of education in Victorian England, the monitorial school and the infant school movements, did not really demonstrate "child-centered" sensory education.[21]

The model which survived into the mid-nineteenth century was the "monitorial system," which Andrew Bell, an Anglican clerk, described in his monograph in 1797. In this model, younger students were taught by older student monitors. It was much like the blind leading the blind. In this way one teacher could be in charge of as many as 1,000 students. Joseph Lancaster, a Quaker schoolmaster, opened a school for the London poor which was an instant hit.[22] The monitorial model spread everywhere. It was even popular in Continental Europe and the United States. It provided education of a sort to many students at a low cost.[23]

The second model, the infant school movement was founded by Robert Owen. Owen was a successful factory owner who believed that "infants of one class could easily be formed into men of another class."[24] Beginning with the three-year-old, the infant school conceived by Owen sought to have students reach a better place in human society through education. Owen was an admirer of Johann Heinrich Pestalozzi and Philippe Emanuel Fellenberg. The child was not to be hampered with books, but was to be encouraged by objects discreetly displayed around him. Unfortunately, Owen's political views of free thought and socialism led to his ouster from the school he founded. Samuel Wilderspin replaced him. Wilderspin again made books central. As a result, infant school education became classical and was conducted by rote, with little emphasis on discovery and observation. Wilderspin's much admired version of infant education spread widely.[25] Later, the pedagogy of Froebel influenced the infant school movement. Still, the British infant school remained more formal, emphasizing reading and other academic skills rather than the "kindergarten" ideals found in Europe.[26]

During the 1830s, Pestalozzian methods were introduced into England. Pestalozzi stressed three main principles: reducing subjects to their simplest structure and then proceeding from the simple to the complex; the use of the object lesson; and oral (not textbook) instruction. English Pestalozzianism centered on object lessons for children. Collections of objects of all sorts were be available for children to observe and describe. This led to student generation of problems and their solutions.[27] In Pestalozzi's words: "Observation is the absolute basis of all knowledge."[28] Unfortunately, the object was for observation only and not for manipulation. Often, the teachers themselves did not know enough about the object to encourage their students to describe the object fully. For example, an object lesson on polyhedra would not benefit any student if the teacher didn't know something about polyhedra.[29] Observation, not discovery and manipulation, was emphasized.

An inspection of the noteworthy selection of nineteenth-century British elementary arithmetic textbooks in the Artemus Martin Collection of The American University in Washington, DC, suggests that the Pestalozzian philosophy was rarely found in popular British mathematics textbooks. Rev. T.V. Short in his respected mathematics pedagogical work, *Hints on Teaching Vulgar and Decimal Fractions* (1840) was found to write in the Preface:

> To those who are at all acquainted with the usual method of teaching arithmetic in England, it can be no wonder that the mass of children so taught are not fond of calculations. The rules are utterly unintelligible. . . . The right way of teaching arithmetic is to teach the principles analytically, and to go on to the practice.

Fifty years later, Isaac Pitman in *The Avon Practical Arithmetic for Junior Pupils* (1893) hadn't changed the message:

> The shortest and surest road to a knowledge of arithmetic is by constant practice in solving exercises . . . a large number of exercises is needed.

Pedagogies which influenced English education were still designed and promoted by men. A woman would have difficulty generating interest in her pedagogical ideas at a time in history when she was allowed to teach only at the lowest level, not in universities or centers for teacher education. There was no academic forum for a woman's thought during this time period.

Although Mary Boole's endeavors were directed primarily toward clarifying her husband's *Law of Thought* for her lay readers, she wrote on many other matters which were considered controversial: the occult, eastern philosophy, evolution, and animal rights. Hebrew and Indian thought interested her at a time when anti-semitism and prejudice toward "non-Christian" thought was growing. She wrote about the conscious and the unconscious before the work of Sigmund Freud was widely read, much less accepted. She wove psychological observations into her husband's mathematical work. Because of her psychological approach, her integrity as a scientist and mathematician was questioned. Perhaps this is why Mozans did not mention her in his book, *Women in Science* (1913), in which he had written about Ada Lovelace and Mary Sommerville, both of whom had known Mary. These appear to be some of the reasons that the pedagogical works of Mary Boole were not recognized.

George and Mary Boole left an additional legacy to scholarship and the academic community through their progeny. Their daughter Alicia Boole Scott was interested in space and dimension. She coined the word "polytope" to describe the convex regular solids in four dimensions. In her later years, she met and collaborated with H.S.M. Coxeter, the famous geometer.[30] Daughter Lucy Boole became a professor and the head of Chemical Laboratories at the London School of Medicine for Woman. Daughter Ethel Lilian Boole Voynich was a prolific novelist, romantic, and music lover. *The Gadfly* (1897) is among her works.

Alicia's son, Leonard Boole Scott (1892–1963) , a physician, was a pioneer in the treatment of tuberculosis. He also invented a portable X-ray machine. The son of Margaret Boole Taylor, Geoffrey Ingram Taylor, was a notable early twentieth-century mathematical physicist. He received a number of honorary doctorates and awards from around the world. The grandson of Mary Ellen Boole Hinton , Howard Everest Hinton, is considered one of the eminent twentieth-century entomologists. Several of the descendants of George and Mary also exhibited interest in and were involved with social causes.[31]

The present National Council of Teachers of Mathematics (USA) *Curriculum and Evaluation Standards for School Mathematics* (1989) advises that elementary students should be actively involved in doing mathematics in an environment where children can explore, develop, test, discuss, and apply ideas.[32] Teachers are encouraged to use journals in mathematics class to give students an opportunity to reflect on their learning. Moreover, children learn from one another as they communicate in groups. The following words from Mary Boole's works on teaching arithmetic and on teaching science seem to have been driven by the NCTM standards instead of preceding them by a century. One can find an analogous presentation of Mary Boole's views in D.G. Tahta's compilation (1972) published by England's Association of Teachers of Mathematics. Each heading below has a phrase in parenthesis which compares it to contemporary pedagogy.

Mary Boole's Words

Working in Groups (Cooperative Learning)

Hardly ever are there two children working on quite the same problem; therefore there can be no competition as to who answered best or quickest. At the end of the lesson, the whole class are in possession of some information which no member of it possessed before. It is not something told them by the teacher; it is often quite new to the teacher. The skill of the teacher is shown, not by the knowledge which she imparts, but by the manner in which she utilizes the thinking power of the children for the purpose of finding out what she does not yet know.[33]

Only dead mathematics can be taught where the attitude of competition prevails: living mathematics must always be a communal possession.[34]

Children have a right also to a share in that still higher and purer delight, self-effacing communal Research. The opportunity for it should be provided in school. For if taken, under judicious supervision, by a group approximately equal in age and attainments, it is ... invigorating in itself.[35]

It will be found a good rule to make all work which is properly intellectual, all which involves serious thought, either individual or communal; and to reserve competition for those portions of time which are devoted to the acquiring of skill, accuracy and speed, by practice in what the pupils already understand perfectly.[36]

The stimulus of competition, when applied at an early age to real thought-processes, is injurious both to nerve-power and to scientific insight.[37]

"Hands-on," Discovery Learning (Using Manipulatives, Concrete Learning, Discovery)

Adopt the custom of introducing children to each mathematical principle by some easy examples, in play, several months before it will be used in the work of school problems.[38]

Each object (in school) is catalogued as intended to teach this, or to prove that, or to illustrate so-and-so; many ... seem to have no idea that it may be well to let a child have things and handle them, without any one talking, and find out what things have to say.[39]

It is desirable that children should sometimes be free to experiment under varied and accidental conditions ... and learn by making mistakes with some one about ... to whom they can apply when puzzled or discouraged.[40]

Receptivity cannot be generated by early teaching of a subject mixed up with the use of its appropriate technical machinery; but only by suggesting the new ideas by means of objects already familiar to the child's eye and touch.[41]

When he begins to do sums on paper, let him still, for a considerable period, do each addition, subtraction, etc., first in counters; and then, while these are still on the table, work out the same sum on paper.[42]

By training the hand to trace out Nature's action, we train the unconscious mind to act spontaneously in accordance with Natural Law; and the unconscious mind, so trained, is the best teacher of the conscious mind.[43]

First comes the education of the senses. From the time when an infant begins to stroke the cat ... have geometric solids as ornaments or toys, so that the senses of sight and touch may actually develop in contact with true type-form.[44]

At early stages the needle and thread has many advantages over any other implement yet devised; a child can ornament cards by setting long straight stitches in a way which causes beautiful curves to grow under his hands without his knowing why or how, and without any pattern being set for him. ... The beauty of some of the designs is unquestionable; and there can be no second opinion about the value of the method, as training, from the point of view of geometry as well as from that of art. What is not quite so obvious at first sight is its bearing on the training of the unconscious mind for science. Without the slightest intellectual strain it puts the children through that normal sequence of orderly attention to classification and detail interspersed with nodal points of synthesis which may be called the very breathing-rhythm of the scientific discoverer.[45]

Anything which he (the teacher) intends to prove should never be stated; children should be led up to find it out for themselves by successive questions.[46]

Geometry

If children of twelve are to learn what Euclid wrote for advanced men, children of three should be acquiring the subconscious physical experiences which lads in Greece picked up in the course of nature and by the accidental help of architecture and statuary.[47]

The geometric education may begin as soon as the child's hands can grasp objects. Let him have, among his toys, the five regular solids and a cut cone. Let eyes and hands be early accustomed to recognize the difference between natural form (that which evolves without human aid), artificial form (that which is made by man in response to a need from without, as a table, a tool or a toy), artistic form (that which is created by man in response to the senses of beauty within mathematical sense).[48]

As soon as the hands can hold steadily compasses and set-square the child should be encouraged both in copying diagrams..and in inventing others for himself. It is desirable that, before any systematic teaching of mathematics begins, the compass, set-square, and ruler marked in fractions of an inch should be as familiar implements as the fork and spoon.[49]

Mathematical Imagination
(Critical and creative thinking)

The teacher who would educate the mathematical imagination of pupils must begin by cultivating his own.[50]

The cultivation of the mathematical imagination should include not only its development but its orderly and systematic exercise. A child is too often made to pass from a particular case which suggests a law to other particular cases which require an application of it, without a sufficient amount of ... analysis of the law and in tracing out its results exhaustively. Between the time when a child handles an actual cube, cuts sections, etc., and the time when he comes, among his ordinary geometrical exercises, to problems requiring him to draw the elevation of a cube cut in some particular way, there is a period when he finds it useful, and very delightful, to go through a set of processes in imagination and to express them in his own words (Author's editorial note, this is the time for thinking and journal writing.)[51]

Mak(ing) mind pictures, is not intended to supersede the usual permitted intervals between lessons; they are an integral part of the Logic lesson itself.[52]

Inventing Ones Own Formulas
(Discovering patterns and functions)

Never let yourself get fixed ideas that numbers will not come right unless your sum is set or shaped in a particular way. Have a way in which you usually do a particular kind of sum, but do not let it haunt you.[53]

The applier of Mathematics ... should be able, when occasion requires, to ... invent a new formula for himself. He cannot begin to learn how to do this, straight away; he should have had from the first, the habit of seeing through formulae and notations; of watching them coming into being, of helping to construct them.[54]

It should be understood from the first that no such thing exists as a right method of performing any operation in elementary mathematics; because all rightness, and I may add all mathematicalness, depends essentially on getting each operation performed by two methods; the first, a roundabout one, which represents and registers, the conscious action of the mind during the process of discovery; the second a short method which condenses the roundabout one, assists in stowing its results away in the memory, and facilitates the using of them sub- consciously.[55]

Interdisciplinary Activities in Math and Using Mathematics History (Connections)

Again, the teaching of Arithmetic is much facilitated, if, besides actual exercises on 0 and 1, and on very simple fractions, logical exercises ... deal with questions of Art, or such simple portions of History, Ethics, or social relations, as come fairly within the scope of a child's intelligence. Each operation of Arithmetic may find its analogue in one of these borderland exercises.[56]

Arithmetic seems to some people dry and unbeautiful. ... If we had sympathy with the struggles and labours of others, Arithmetic would be easier to understand and pleasanter to learn than many children find it. ... But my own experience is that the dramatic introduction of the savage element (referring to teaching about the counting of early man, the use of mathematics history) ... has a wonderful effect in conquering the apathy of which so many teachers complain, and making Arithmetic real and living.[57]

Journal Writing (Communications)

They (children) should begin early the practice of entering certain kinds of results in a book. This book may be divided into two parts: anything specially needful to remember, if found out by the children, should be entered in one part; if told them by the teacher, in the other.[58]

Algebra

It is also desirable, for many reasons, that children should be accustomed to use letters for unknown quantities at an early age. It assists the imagination in keeping a clear distinction between a quantity in itself and the effect on the quantity of a certain operation or group of operations ... those who are to learn Algebra should not be exposed to the mental violence of being introduced to Algebraic notation for the first time, just when they will have to begin to learn the actual manipulation of unknown quantities by processes properly Algebraic.[59]

The questions so often put by parents, "At what age do you think my child had better begin Algebra?" ... For the majority of children the really right time for beginning any department of mathematics never comes, because the really right time is the time when the right condition of development has been attained; and for nine children out of ten it never exists at all.[60]

Everybody ought to be able to make Algebras; and the sooner we begin the better. It is best to begin before we can talk; because, until we can talk, no one can get us into illogical habits; and it is advisable that good logic should get the start of bad.[61]

Arithmetic itself is always best taught, when it is taught on methods which constitute a sound preparation for the study of the higher Algebra.[62]

This paper concludes with a passage from the preface of Mary Boole's *Lectures on the Logic of Arithmetic* (1903). Her words could have easily been written in recent years. Furthermore, she was to note, "In America more children fail in arithmetic than any other subject."[63] What Mary agonized about ninety years ago might be worse today.

Teachers of such subjects as Electricity complain of the difficulty of getting pupils to apply what they know of Mathematics, at whatever level, to the analysis and manipulation of real forces. It is not that the pupil does not know enough of Arithmetic, or Algebra, or the Calculus, as the case may be, but he too often does not see, and cannot be got to see, how to apply what he knows. . . . Some faculty has been paralyzed during his school life; he lacks something of what should constitute a living mathematical intelligence. . . . Whatever skill he may have acquired in the manipulation of those notations and formulae which he has been taught to use, he knows hardly anything about the manner in which such things come into being.

This paper has presented the life and pedagogy of a remarkable woman. Maybe some readers will be as inspired by Mary Everest Boole as I have been.

Reference Notes

1. Eleanor M. Cobham, *Mary Everest Boole* (Ashingdon, England, C.W. Daniel Company, 1951) pp. 16–17.
2. Eleanor M. Cobham, ed. , "A Child's Idyll," *Mary Everest Boole—Collected Works* (Ashingdon, England: C.W. Daniel Company, 1931) p. 1506.
3. Nowhere in Mary Boole's writings nor in her biography is Monsieur Deplace's first name given.
4. Cobham, ed., "A Child's Idyll," *M. Boole—Collected Works,* p. 1505–1506.
5. The rule of three relates to solutions using proportions, i.e., $a:b=c:d$. An example of a double rule of three problem can be found in a homemade cipher-book of Sebastian Graff from 1769: "If 14 men in 7 Days make 38 poles of a Ditch how many men will it take to make 98 poles in 4 Days."
6. Carl B. Boyer, *A History of Mathematics,* Second Edition (New York, NY: John Wiley & Sons, Inc., 1968, 1989, 1991) p. 467.
7. Cobham, ed., "A Child's Idyll," *M. Boole—Collected Works,* p. 1521.
8. Ibid., p. 1522.
9. Desmond MacHale, *George Boole—His Life and Work* (Dublin, Ireland: Boole Press Limited, 1985), pp. 108–109.
10. MacHale, *George Boole,* pp. 219–220.
11. Ibid., pp. 158–159.
12. Cobham, *Mary Everest Boole,* p. xi.
13. Teri H. Perl and Joan M. Manning, *Women, Numbers and Dreams* (Newton, MA: The WEEA Publishing Center, 1982) pp. 75, 85.
14. Dick G. Tahta, *A Boolean Anthology* (London, England: Association of Teachers of Mathematics, Nelson) p. 36.
15. Perl and Manning, *Women, Numbers and Dreams,* p. 73.
16. Ibid., *Women, Numbers and Dreams,* p. 81.
17. Edward J. Power, *Main Currents in the History of Education,* Second Edition (New York, NY: McGraw-Hill Book Company,1962, 1970) p. 527.
18. Power, *Main Currents,* p. 530.
19. Adolphe E. Meyer, *An Educational History of the Western World,* Second Edition (New York, NY: McGraw Hill Book Company, 1965, 1972) p. 286.
20. Power, *Main Currents,* p. 529.
21. John Lawson and Harold Silver, *A Social History of Education in England* (London: Methuen & Co Ltd, 1973), p. 281.
22. Meyer, *An Educational History of the Western World,* 1965, p. 281.
23. Ibid., 1972, pp. 293–295.
24. Ibid., 1965, p. 279.
25. Ibid., 1972, pp. 290–292.
26. Henry G. Good, and James D. Teller, *The History of Western Education,* third edition. (London: The Macmillan Company, 1969), p. 409.
27. Power, *Main Currents,* p. 493.
28. Michael R. Heafford, *Pestalozzi* (London: Methuen & Col, Ltd, 1967), p. 53.
29. Good and Teller, *The History of Western Education,* p. 497.
30. MacHale, *George Boole,* pp. 262, 263.
31. Ibid., p. 262–276.
32. *Curriculum and Evaluation Standards for School Mathematics* (Reston, VA: NCTM, 1989) pp. 68–69.
33. Cobham, ed. "The Crank," *Mary Everest Boole—Collected Works,* p. 1196.
34. Cobham, ed., "Mistletoe and Olive: An Introduction for Children to the Life of Revelation," *Mary Everest Boole—Collected Works,* p. 1009.
35. Cobham, ed., "The Preparation of The Child for Science," *Mary Everest Boole—Collected Works,* p. 892–893.
36. Ibid., p. 893.
37. Ibid., pp. 892–893.
38. Cobham, ed., *Mary Everest Boole—Collected Works,* Preface, p. vi.
39. Cobham, ed., "The Preparation of The Child for Science," *Mary Everest Boole—Collected Works,* p. 895.
40. Ibid., p. 898.

41. Ibid., p. 916.
43. Ibid., p. 921.
43. Ibid., p. 902.
44. Ibid., 905.
45. Ibid., p.906.
46. Mary E. Boole, *Lectures On the Logic of Arithmetic,* (Oxford: Clarendon Press, 1903), p. 16.
47. Cobham, ed., "The Preparation of The Child for Science," *Mary Everest Boole—Collected Works,* p. 904.
48. Mary E. Boole, *The Preparation of the Child for Science,* (Oxford: Clarendon Press, 1904) p. 110.
49. Ibid., p. 114.
50. Ibid., p. 98.
51. Ibid., pp. 121–122.
52. Cobham, ed., "Lectures On the Logic of Arithmetic," *Mary Everest Boole—Collected Works,* pp. 813–814.
53. Cobham, ed., "Philosophy and The Fun of Algebra," *Mary Everest Boole—Collected Works,* p. 1239.
54. Boole, *Lectures On the Logic of Arithmetic,* pp. 9–10.
55. Ibid., p.101.
56. Ibid., p. 15
57. Ibid., p. 29.
58. Ibid., p. 17.
59. Ibid., pp. 69–70.
60. Cobham , ed., "The Preparation of the Child for Science," *Mary Everest Boole—Collected Works,* pp. 912–913.
61. Cobham, ed., "Philosophy and The Fun of Algebra," *Mary Everest Boole—Collected Works,* p. 1233.
62. M. Boole, *Lectures On the Logic of Arithmetic,* p. 141.
63. Cobham, ed., *Mary Everest Boole—Collected Works,* p. vi.

III

Integration of History
with Mathematics Teaching

Origins and Teaching of Calculus

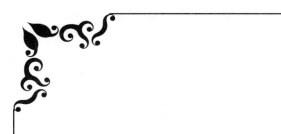

Pupils' Perception of the Continuum

Peter Bero
Comenius University

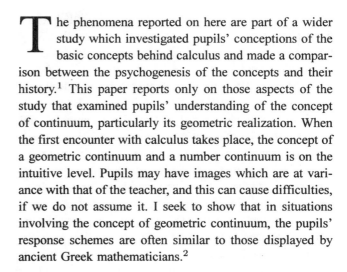

The phenomena reported on here are part of a wider study which investigated pupils' conceptions of the basic concepts behind calculus and made a comparison between the psychogenesis of the concepts and their history.[1] This paper reports only on those aspects of the study that examined pupils' understanding of the concept of continuum, particularly its geometric realization. When the first encounter with calculus takes place, the concept of a geometric continuum and a number continuum is on the intuitive level. Pupils may have images which are at variance with that of the teacher, and this can cause difficulties, if we do not assume it. I seek to show that in situations involving the concept of geometric continuum, the pupils' response schemes are often similar to those displayed by ancient Greek mathematicians.[2]

A Glimpse at the History of the Concept

Historically, we can trace two important conceptions of the continuum, which were essential in the development of calculus. These are the continualist and the atomic conceptions, which have appeared in history as a counterpoint—two different views, used sometimes simultaneously, in order to make use of advantages from both, and sometimes treated as completely incompatible approaches.

What are the main characteristics of these conceptions? The acceptance of the possibility of infinite divisibility in the continuum is a distinguishing feature of the continual conception, and the existence of indivisibles is a distinguishing feature of the atomic conception. The acknowledgement of the infinite divisibility emphasizes abstract homogeneity of the continuum, in contradiction to the indivisibles which emphasize a discrete essence of the continuum.

So far it is not clear whether atomism did originate in Greece or not, but Pythagorean geometry of the fifth century B.C., based on the discrete number concept, may be taken as a beginning of the atomistic conception in Greek mathematics. According to the early Pythagoreans, a line segment consists of points, whose quantity can be determined by a finite number. This directly entails a distinguishing idea of Pythagorean mathematics—that of commensurability of every two segments. According to this, a proportion of every two segments must be expressed by two natural numbers. The discovery of the existence of incommensurable geometric magnitudes, for example an edge and a diagonal of a square, necessitated a thorough reexamination and the consequent fall of this idea.

The ancient continuum conception profited by the infinite divisibility of a straight line segment, but simultane-

ously employed a point, that is, according to Euclid, something possessing no parts and having zero length. Every line was treated as consisting of infinite number of these indivisible points.

The consistency of such an approach was questioned by Zeno in his famous paradoxes of plurality. Here he divided all magnitudes into dimensionless magnitudes and assumed: the sum of an infinite number of arbitrarily small (positive) magnitudes must be infinite, and the sum of even an infinite number of dimensionless magnitudes must be zero. Zeno used these assumptions as a basis for the following dilemma: If a line segment is resolved into an aggregate of infinitely many elements, then two cases are possible: either these elements are of positive length and the aggregate of them is of infinite length, or the elements are of zero length and then their aggregate is of zero length.[3] In other words: If points have no dimension, then even an infinite number of them would not form figures having dimensions, and, if they have finite dimension, then the infinite number of points would form an infinite magnitude.

The atomism of Democritus and Leucippus emerged as a response to Zeno's doctrine.[4] They believed that everything was composed of indivisible parts, "ameros," which are indivisible because of their smallness and absence of parts. They are not, however, infinitely small; they are just imperceptible.

By putting together two or more indivisible parts, we can arrive at an elementary straight line. Elementary straight lines can be compounded in two ways: a) by piling a very big (but not an infinite) number of them on each other, an elementary plane will be formed; b) by putting them next to each other, a straight line will be formed. When piling the elementary planes on each other, the physical minimum of substance—an atom—will be formed. The atom is indestructible because of its solidity, though it has parts.

An important feature of Democritus's theory is an existence of only one infinite number—the number of indivisibles in the whole universe. There exists a limit, after which the number of indivisibles seems infinite, but it is given only by our sensuous perception. In fact, there is only a finite number of indivisibles in each object.

We can argue against atomism, but it is a very fruitful approach in mathematics. Atomism provides a possibility to build hypotheses and to calculate. Archimedes, for example, worked in this way.[5]

A renaissance of atomism in mathematics started late in the sixteenth century in works of Galileo Galilei, Johannes Kepler, and Bonaventura Cavalieri. These scholars, unlike ancient Greek mathematicians, did not analyze the structure of the continuum; each of them made up only some calculus based on atomism and used it to solve different tasks.[6]

Atomism showed a continuing strength, when Isaac Newton and Gottfried W. Leibniz independently discovered calculus (17th century). Disregarding the final versions of their theories, they stemmed from atomism, but their calculus lacked logical foundations.

In providing satisfactory foundations for calculus, Augustin Cauchy, Bernhard Riemann, and Karl Weierstrass displaced atomism from established mathematics. This displacement lasted only until nonstandard analysis appeared in the mid-twentieth century. Since then atomism has been basic again.

Research Project

Children in Slovakia first attend eight grades at an elementary school and then (approximately a third of them) four grades at a gymnasium. Styles of teaching mathematics at these two schools are very different. The difference lies mainly in precision and structure. Mathematics at the gymnasium is mathematics of definitions, theorems, and proofs, unlike the elementary school, where those are rare.

In 1984/1985 I interviewed 40 pupils from 1 to 18 years old and nine of their teachers, with the aim of identifying the current pupils' ideas of the continuum. These interviews and my teaching experience were a base for the following questionnaire:

1. What does a straight line consist of?
2. What does a segment consist of?
3. Has a point any dimension at all?

The questionnaire clearly concentrated only on the issue of the structure of the geometric continuum which underlies in my opinion, all mathematics taught in elementary school and at gymnasium.

The questionnaire was administered in grades 5–8 at elementary schools and in grades 1–4 at gymnasiums, 20 pupils in each grade, which totals 160 pupils. In case of a vague response, the pupil was interviewed. In order to avoid pages of tables, results are arranged by the structure of responses for two large groups—pupils from elementary school and pupils from gymnasium. The abbreviation ES for an elementary school, G for gymnasium, B-6 for a boy from the sixth grade of an elementary school, G-III for a girl from the third grade of a gymnasium, and so on.

Responses to Question One

Practically all answers to Question One were to the effect that a straight line consists of points. In their responses pupils used some other concepts. The most often used were:

point .... 156 pupils (76–ES, 80–G)
infinity .... 100 pupils (51–ES, 49–G)
set .... 62 pupils (30–ES, 32–G)
line .... 32 pupils (26–ES, 6–G)

In Slovak we have completely different words for a line and a straight line: the word "čiara" for a line, and the word "priamka" for a straight line.

In addition to saying a line consists of the points, pupils added other attributes:

a) Number of points. This attribute was used by 128 pupils (65–ES, 63–G). They spoke of "number of points," "a certain number of points," and gave similar expressions. In an interview following the questionnaire I found out that all pupils, except eight of them, were thinking of an infinite number of points. The most typical response was: B-II: "A straight line consists of the infinite number of points."

b) Way of arrangement of points. Forty-eight pupils (39–ES, 9–G) utilized this attribute. Nearly all pupils employed the expression "tightly to each other." Their typical response was:

B-6: The straight line consists of the points, which are tightly to each other.

c) Dynamic description. Dynamic description expresses a dynamic understanding of a straight line; it expresses, as a matter of fact, a process of an origin of a straight line (by an increase of the amount of points). The description was used by 40 pupils (21–ES, 19–G). They typically stated:

B-6: A straight line is an amount of points, whose number is always increasing.

G-6: The straight line consists of a certain number of points perpendicular to each other and going to the infinite.

d) (Un)boundedness. Twenty-four pupils (13–ES, 11–G) made use of this attribute. They implicitly compared a straight line to certain things with boundaries. This can be seen from answers of one pupil to the first and second question.

B-6: A straight line consists of points, which succeed one another and have no points as boundaries.

The line segment consists of the points, which lie after each other and have two points as boundaries.

e) Beginning–end. This attribute was referred to by 20 pupils (12–ES, 8–G). As in the previous example, it is a transfer of a finite experience.

G-8: A straight line consists of an infinite number of points. It is a line without a beginning and without an end, it has no points as boundaries."

During this inquiry into the pupils' perception of the structure of continuum, I found the points a) and b) to be the most important. The absence of any links between children's response and their drawing experience as well as the absence of any signs of materialistic grasp was surprising.

Responses to Question Two

All the pupils think that the line segment consists of points. The most recurring concepts:

point .... 160 pupils (80–ES, 80–G)
set .... 41 pupils (17–ES, 24–G)
line .... 26 pupils (19–ES, 7–G)

The pupils added other characteristics.

a) Number of points. Used by 122 pupils (80–ES, 42–G). All elementary school pupils referred to the finite number of points. Only one pupil from a gymnasium did likewise.

b) Boundedness. Used by 86 pupils (57–ES, 31–G). All the responses were similar to this response:

B-II: A line segment is a set of points edged by two end-points.

c) Beginning–end. This characteristic was employed by 22 pupils (19–ES, 3–G). This probably reflects the influence of drawing pictures. While teachers at elementary schools emphasize the marking of the end-points of the line segment, at gymnasiums it is usual to abandon it. A typical response here was:

G-8: A line segment is an edged line with one starting point and one ending point.

d) Way of arrangement of points. The characteristic was employed by 35 pupils (33-ES, 2-G). A typical response on this matter was:

B-7: A line segment is a set of points, which are next to each other and are edged with two points.

Compared with the first question, this attribute occurs less frequently. An explanation may be provided by this response:

G-7: A line segment is a set of points arranged in the same way like in a straight line, except that the line segment has its boundaries, has its beginning and end.

In these responses, unlike those to the first question, the influence of a material model of the line segment—the picture of it—is stronger. The dynamic description, often in the case of a straight line, is not encountered at all.

Responses to Question Three

An expressive difference appeared between the elementary school pupils and the gymnasium pupils. It is interesting that practically all pupils from the elementary schools presented reasoned responses, whereas none of the gymnasium pupils reasoned their responses.

The responses to Question Three can be divided into three groups: a) **A point has a length** was the response of 85 pupils (76–ES, 9–G). b) **A point has no dimensions** was the response of 67 pupils (3–ES, 64–G). c) **I do not know** was the response of 8 pupils (1–ES, 7–G).

The most frequent argument in group a) was like this:

B-6: A point has a dimension, because it forms a straight lines and other figures we can measure.

Different dimensions were acceptable for eight pupils. For example,

G-8: In my opinion, the point has a dimension. It can be small or big.

Other pupils used emotive language such as

B-7: It has, but it is very small, invisible particle.

In the subsequent interview all but one answer coincided, that the point is circular (spherical)-shaped. Only one maintained that the point is cross-shaped.

The responses in both group b) and c) were not commented on. Pupils just stated briefly "A point has no dimensions," and "I do not know."

Elementary school pupils were intensely involved in responding to this third question. Approximately half of the responses were corrected from the negative to the positive response. For example,

B-7: A point has no dimension, for it is immeasurable. (This was crossed out.) A point has di-

mension, for it forms line segments and other figures which are measurable.

Analysis and Conclusion

The questionnaire results show that for practically all pupils both a straight line and a segment consist of points. I expected such a result for gymnasiums' pupils, who are taught this explicitly, as well as that the point has no dimensions. Elementary school pupils are not taught this, and I therefore expected at least minimal variety of opinions. It seems however, that expressions like "a point of a segment," "a point on a straight line" have been implicitly building up the idea of a line consisting of points.

I consider it important to know which parts of the mental structure of a pupil are active when the concept of continuum is evoked. Therefore the research instrument was intentionally designed for the purpose of probing for additional characteristics which pupils used without being asked to do so: a way of arrangement of points, boundedness, beginning–end, and the dynamic of the origin of a straight line. Here it might be interesting to parallel their views with the Euclidean understanding of a straight line.

First consider the elementary school pupils. Analyzing responses to the second and third questions we see, that 76 (out of 80) pupils consider a straight line to consist of a finite number of points that have finite dimensions. In the interview 67 pupils said that the point is divisible.

G-7: We can halve the point into two halves, or into two even smaller points.

Eight pupils were afraid of practical difficulties:

G-8: I can halve the point but it is improbable that I succeed.

On the basis of these responses, I think that the opinions of the pupils of basic schools are parallel to those of Democritus and Pythagoras. An interference between concepts of infinity and boundedness causes problems. The pupils do not imagine an infinite number of points in a bounded object—a line segment. Probably their idea that a point has dimensions intervenes here. Statements like "because shapes having dimensions consist of points," could have been taken from Zeno's considerations. They intuitively feel, but are not aware of, Zeno's antinomies.

Patterns in the responses of gymnasium pupils are entirely different. False ideas are practically absent. Only one (the last grade) considered a line segment to consist of the finite number of points, and only nine pupils (from all grades) thought of the point as having dimensions. It may seem strange, but I do not believe this is a salutary fact. Why? In the first grade of gymnasium, pupils are offered "right answers" when they are taught an axiomatic approach to geometry. They learn that a straight line consists of points, a point has no dimensions, and, because these are primitive concepts, we do not analyze them anymore. The established versions have to be used, and further teaching does not provide pupils with opportunities to think over their own ideas of concepts. But this does not mean that pupils replace their own ideas by official ones. This may lead to the situation that, when we teach pupils the calculus of the late 20th century, as we usually do, their ideas of continuum are actually on the level of the third century B.C.

Pupils should be made aware of their mistakes, but concurrently they have to be given enough time to think them over. They should be urged to deliberate on their own conceptions. The topic of the continuum as such is inspirational for pupils, as responses to the third question shows. Frequent occurrence of corrected answers reveals an interest from pupils, and we can again find an inspiration from history. In this case Zeno's antinomies are very engaging, provocative and effective for pupils.[7]

References

1. Peter Bero, "Basic Concepts of Calculus", (in Slovak) unpublished thesis, 1987, University of Comenius, Bratislava.
2. Related issues are addressed in the papers: David Tall, "Understanding the calculus," *Mathematics Teaching,* 1985, 110: 49–53; P. Tsamir, D. Tirosh, "Students Awareness of Inconsistent Ideas about Actual Infinity," to appear in Proceedings of PME XVI, 1992, New Hampshire; E. Fishbein, D. Tirosh, and P. Hess, "The Intuition of Infinity," *Educational Studies in Mathematics,* 1979, 10: 3–40; John Monaghan, "Adolescents' Understanding of Limits and Infinity," unpublished thesis, 1986, University of Warwick.
3. A broad philosophical analysis of Zeno's paradoxes is given in Adolf Grünbaum, *Philosophical Problems of Space and Time,* (Dordrecht, 1974).
4. S.J. Lurje, *Infinitesimals in ancient atomism,* (in Russian), (Moskva-Leningrad, AS USSR, 1935).
5. Charles H. Edwards, Jr., *The Historical Development of the Calculus,* (New York: Springer-Verlag, 1979).
6. Dirk J. Struik, (1969), *A Source Book in Mathematics, 1200–1800,* (Cambridge, MA: Harvard University Press, Cambridge, 1969); C.H. Edwards, *The Historical Development of the Calculus.*
7. Milan Hejný, "Achilles and tortoise as a problem," (in Slovak), *Matematika a Fyzika ve Skole,* 1978, 2: 102–105.

Historical Motivation for a Calculus Course: Barrow's Theorem[1]

Martin E. Flashman
Humboldt State University

Calculus plays a central role in most college mathematics programs in the United States. Though calculus is described frequently as the motivation for many topics taught in earlier courses, little is done in pre-calculus courses to pursue the problems of tangency and area. Consequently students begin the study of calculus with a minimal background concerning its central problems and results. They are certainly unaware that these problems were studied vigorously before the development of calculus by Isaac Newton and Gottfried Leibniz.

Beginning students generally believe that calculus problems have been studied only with the methods of calculus. They may also assume that the creators of calculus—Newton and Leibniz—came upon the methods and results simultaneously without any prior historical developments. Students may even conclude that only one method exists for solving these problems and that the tools they learn in the calculus course are both the beginning and the end of the story.

Since 1987, I have tried to improve the historical background of my calculus students at the course's beginning by discussing some of the mathematics done prior to Newton and Leibniz. In the first section describing the course's content I introduce Isaac Barrow's Theorem relating the tangent and area problems. This result serves as a thematic prelude for the course while showing some of the geometric style and ingenuity that was background for Newton and Leibniz.[2] By exploring these historical problems and results in the first week of the course, students gain greater appreciation of the significance of the fundamental theorem of calculus to the development of calculus as a mathematical theory.

Sir Isaac Barrow (1630–1677) stated the theorem we will consider in his *Lectiones Geometricae* (*Geometrical Lectures,* 1670). Although not always connected to Barrow's name, with a little effort this theorem can be recognized as a geometric version of the fundamental theorem of calculus in its derivative form. (See later in this paper.) The following presentation explains the content of the theorem by introducing both the tangent and the area problems geometrically. Algebraic formulations are included as appropriate for the reader familiar with calculus. Though expressed in somewhat more modern terms, the theorem illustrates the mathematics just prior to Newton and Leibniz, and the conceptual background and symbolic practices that may have kept others from finding a calculus earlier.[3] The theorem can be used also to initiate discussion of the connection of concepts to problem solutions, the desire to produce algorithmic solutions, and the role of symbols and algebra in describing and resolving problem situations.

By recognizing symbolic and conceptual differences in approach between the work of Barrow and current views based on functions, the student can see that mathematics has evolved gradually rather than as a sudden revolution traced only to the genius of a few individuals. This human aspect of mathematics is important for a course that frequently becomes a confusing collection of definitions, theorems, techniques, and applications.

After students are exposed to Barrow's Theorem early in a calculus course, they can grasp its relation to the calculus at later stages of the course. This illustrates how key elements of the theory fit together. The calculus of derivatives appears in part as a systematic approach to the tangent problem using the algebraic representation of curves that graph functions. Differential equations can be interpreted in terms of rates and motion as well as tangents to curves. The area problem can be connected to solving differential equations using concepts of estimation and accumulation of change. With a connection to differential equations, the area problem can be related to the tangent problem later in the course. With this early treatment, students are prepared for modern statements of the fundamental theorem of calculus along with more analytic proofs that complement the geometric argument given by Barrow.

Here then is a slightly edited excerpt from The Sensible Calculus Program materials introducing calculus. Included at the end is a proof based on Barrow's own geometric argument and some exercises to help students experience some of the thinking that might have resulted from applications of Barrow's Theorem before the full development of the calculus.

What is Calculus?

There are too many books with titles using the term "Calculus" and courses with the same name to believe that you have not heard this word. But few of these books tell you why the subject is called calculus and precisely what it is. While I, like other authors, may fail to answer these questions adequately, in this section I shall tell you something about what calculus is and about what mathematics was just before calculus was invented.

Quite simply any "calculus" is a systematic method for determining a result, for arriving at a conclusion or answer. In this sense there are many calculi, such as the calculus of propositional logic, the calculus of set operations, the calculus of probabilities, etc. But when someone talks about "The Calculus," be it "differential" or "integral" calculus or the calculus of infinite series, the reference is usually to versions of the calculus of Isaac Newton (1642–1727) and

Gottfried Leibniz (1646–1716). Without going into greater detail here, let me say that calculus is concerned with *problems of analyzing change* and determining such things as the line tangent to a curve, the area of a planar region enclosed by a curve, and estimates and exact solutions to many related problems.

Several great mathematicians before Newton or Leibniz had studied these questions, from Euclid and Archimedes in antiquity to Descartes, Fermat, and Pascal to name only a few mathematicians living within the century before Newton. Newton and Leibniz differed from those who had worked on these topics before them in the general overview they achieved. Their approaches, developed independently[4], solved the problems *using systematic techniques of calculation* that depended primarily on the *algebraic description of the problem*. With these techniques, the user may avoid repetitive conceptual analysis of each different curve and formula. The conceptual analysis was summarized in proofs of calculus rules. This was the achievement of the new calculus. The results for problems contained in works by Newton and Leibniz may not have been new, but the methods were revolutionary in their generality.

Consider briefly two of these problems, the *tangent* problem and the *area* problem. You may have discussed tangent lines in a geometry course that gave a direct method for drawing a line that touches a circle at only one specific point, P. [See Figure 1.] This line is called the *tangent line* to the circle at P. The construction of this line draws a radius from the center of the circle O to P and then constructs at P a line that is perpendicular to the line OP. This last constructed line is the desired tangent line.

Archimedes discussed a similar problem for a parabola (and the spiral as well)[5]. The problem was to find a method for drawing a line that touches a parabola at only one spe-

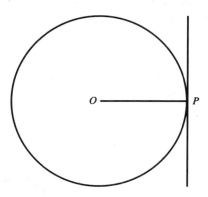

FIGURE 1

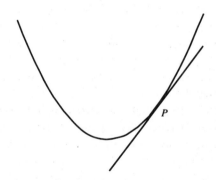

FIGURE 2

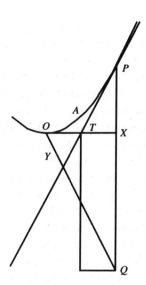

FIGURE 4

cific point, P. [See Figure 2.] This line is called the *tangent line* to the parabola at P. The *general tangent problem* is to find a method for drawing a line tangent to a curve at any specified point.

The *area* problem is perhaps more familiar as a measurement problem today, but its early treatment was also geometric. For example, the area problem for a right triangle seeks a method for constructing a rectangle with area equal to that of a given right triangle. [See Figure 3.] Here the geometric solution is simple. Bisect one leg of the right triangle. The rectangle formed by the other leg and one of the resulting bisected segments has the desired area.

Once measurement and algebra are added to the tools with which we analyze the area problem, we arrive at a familiar formulation of this last result. [See Figure 4.] If the legs of the triangle have lengths a and b then the area A of the triangle (and the rectangle) is $\frac{1}{2}ab$. If we use x instead of a for the length of one leg and mx instead of b for the other leg, then we have an equation that expresses the relation between A, x, and m:

$$A = \tfrac{1}{2}xmx = \tfrac{1}{2}mx^2.$$

One of the most important results of calculus, now referred to as the *fundamental theorem of calculus,* connects

the tangent problem to the area problem. The result was known to Isaac Barrow, Newton's predecessor on the faculty at Cambridge, and is described as *Barrow's Theorem.* A slight paraphrasing of Barrow's statement follows.

Theorem.   *Suppose the curve A has the length of the segment XP equal to the area of the planar region enclosed by the curve Y, the X-axis, and line segment XQ. (See Figure 5.) If the point T is chosen on the X-axis so that TX times XQ is equal to XP, then the line TP is tangent to the curve A at the point P.*

Barrow's theorem certainly connects the two problems under discussion. To solve the tangent problem for the area curve A, we need only solve the area problem related to the curve Y.

The result can be stated in slightly more modern terms by using the concept of the slope of a line. If the line tangent to A at the point P has a slope, m_{tan}, then Barrow's

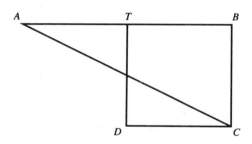

FIGURE 3

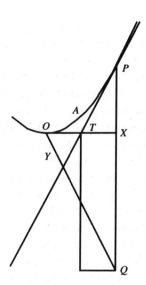

FIGURE 5

Theorem says this slope is the value of the length of the segment XQ.

$$m_{\text{tan}} = PX/TX = TX \cdot XQ/TX = XQ.$$

In the current calculus language of functions, derivatives, and integrals, the result is: If $y = f(x)$ is a continuous function and

$$A(t) = \int_0^t f(x)/\,dx,$$

then $A'(t) = f(t)$.

Note that algebra is missing from the original statement (and proof) of Barrow's theorem. Only the geometric content appeared. When Newton and Leibniz arrived at algebraic versions of this result, its utility became apparent. Both Newton and Leibniz developed algebraic methods for a calculus of tangents with fluxions and differentials respectively. With algebra they applied Barrow's theorem to explore methods for a calculus that systematically resolved problems of area, volume, arc length, etc.

What then is the calculus of Newton and Leibniz? Briefly again, calculus is a *conceptual framework for solving problems such as those just mentioned that are suitably posed in the language of analytic geometry and algebra.*

Here are some problems frequently covered in discussions of calculus over the centuries. These problems and many others of more current interest will be solved as we progress in our study of calculus.

1. The Tangent Problem.   Determine the line tangent to a given curve at a given point. (Also, define precisely the concept of "tangent.")

For example, determine the line in the plane tangent to the circle with equation $X^2 + Y^2 = 25$ at the point $(-4, 3)$.

2. The Velocity Problem.   Determine the instantaneous velocity of a moving object. (Also, define precisely the concept of "instantaneous velocity.")

For example, determine the instantaneous velocity of an object moving on a straight line at time $t = 5$ seconds when its position at time t seconds is $t^2 - 6t$ meters from a given point P.

3. Extremum Problems.   Determine the maximum and minimum values of a dependent variable. For example, when $Y = X^2 - 6X$, determine any maximum and minimum values for the dependent variable Y as X is allowed to vary over real numbers between 0 and 10.

4. The Area Problem.   Determine the area of a planar region enclosed by a suitably defined curve. For example, determine the area of the planar region enclosed by the X-axis, the lines $X = 2$, $X = 5$, and the parabola with equation $Y = X^2 - 6X$.

5. The Arc Length Problem.   Determine the length of a suitably defined curve. For example, determine the length of the parabola with equation $Y = X^2 - 6X$ between the points $(0, 0)$ and $(6, 0)$.

6. The Tangent-Curve Problem Reversed.   Determine a curve so that the tangent to the curve at any point on the curve is predetermined by some description depending on the point's position in the plane. For example, determine a curve in a coordinate plane so that the slope of the tangent to the curve at the point (a, b) is $a + b$.

7. The Position Problem (The Velocity Problem Reversed).   Determine the position of an object moving on a straight line from knowledge of its initial position and instantaneous velocity at every instant. For example, determine the position of an object moving on a straight line at time $t = 5$ seconds knowing its initial position is P on the line and its instantaneous velocity at time t is precisely $t^2 - 6t$ meters per second.

Barrow's Geometric Proof

Theorem.   (See Figure 6.) *Suppose curve A has (the number measuring) length PX equal to (the number measur-*

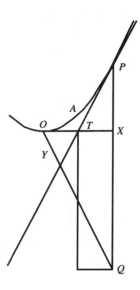

FIGURE 6

ing) the area of the planar region enclosed by the curve Y, the X-axis, and the line segment XQ. If point T is chosen on the X-axis, so that TX times XQ equals PX, then the line TP is tangent to the curve A at the point P.

Proof. Barrow shows that TP is tangent to the curve A at the point P by demonstrating that for any other point R on the curve A the line TP does not pass through R. (See Figure 7.)

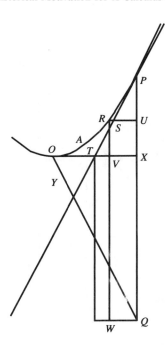

FIGURE 8

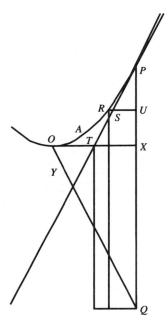

FIGURE 7

For convenience assume that as X increases, the length of XQ increases as well and that R occurs before P on A. Draw a line through R parallel to the X axis which will meet TP at the point S and the line XP at the point U.

It suffices to show that $R \neq S$. Triangle TPX is similar to triangle SPU, so

$$QX = PX/TX = PU/SU.$$

Multiplying by SU gives $QX \cdot SU = PU$.

Draw a line through R parallel to PXQ which meets the X axis at V and the curve Y at W so that the length of RV is equal to the area of the planar region enclosed by the curve Y, the X-axis, and the line segment WV. (See Figure 8.) Now $RV = UX$ and since

$$PX = PU + UX,$$

PU has a length equal to the area of the region enclosed by the line segments VW, VX, XQ, and the curve Y between Q and W. But this region is contained in a rectangle determined by VX and XQ, so PU is less than the product $QX \cdot VX$. Thus we can conclude that $QX \cdot SU < QX \cdot VX$ and therefore $SU < VX$. But $VX = RU$, so $SU < RU$ showing that $R \neq S$ as claimed.   EOP.

Many symbols and phrases have declared the end of a proof. Traditional mathematics has used the abbreviation Q. E. D. standing for the Latin clause "quod erat demonstrandum," which translates into "what was to be shown." This text signifies the end by EOP, meaning "end of proof." That seems simple and sensible enough.

Exercises

In these problems assume that Barrow's Theorem is true to find the slope of lines tangent to parabolas arising from quadratic functions of the form $Y = CX^2$ where C is a constant.

1. Suppose the right triangle in Figure 9 has side OX of length 5 on the X-axis and vertical side XQ of length 10. Above it is a sketch of the area curve A related to the line Y which is the hypotenuse of this triangle.

a. If $OX' = x$, $X'Q' = q$, and $X'P' = p$, express q and p as functions of x. [Remember that $X'P'$ is the

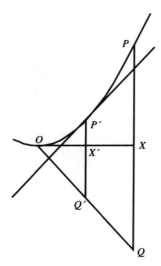

FIGURE 9
Not drawn to scale.

area of the region enclosed by the X-axis, the line segment $X'Q'$, and the line segment OQ'.]

   b. When $x = 3$, find the position of T so that $P'T$ is a line tangent to the curve A at P'.

   c. Find the slope of the line tangent to the curve A at P', when $x = 3$.

   d. Find the slope of the line tangent to the curve A at P', when $x = 1, 2,$ and 4.

   e. Find the slope of the line tangent to the curve A at P', when $x = a$.

2.   Suppose the right triangle given in Figure 9 has side OX of length 10 on the X-axis and vertical side XQ of length 5. Above it is the area curve A related to the line Y, which is the hypotenuse of this triangle.

   a. If $OX' = x$, $X'Q' = q$, and $X'P' = p$, express q and p as functions of x. [Remember that $X'P'$ is the area of the region enclosed by the X-axis, the line segment $X'Q'$, and the line segment OQ'.]

   b. When $x = 5$, find the position of T so that $P'T$ is a line tangent to the curve A at P'.

   c. Find the slope of the line tangent to the curve A at P', when $x = 5$.

   d. Find the slope of the line tangent to the curve A at P', when $x = 1, 2,$ and 4.

   e. Find the slope of the line tangent to the curve A at P', when $x = a$.

3.   Draw a right triangle with side of length 8 on the X-axis and vertical side of length 8 and label a point on the

X-axis X and the appropriate point Q on the hypotenuse and O on the X-axis for Barrow's theorem.

   a. If $OX = x$, $XQ = q$, and p is the area of the triangular region enclosed by the X-axis, the line segment XQ and the line segment OQ, express q and p as functions of x.

   b. Sketch the area curve A related to the line Y, which is the hypotenuse of this triangle.

   c. When $x = 5$, find the position of T so that PT is a line tangent to the curve A at P.

   d. Find the slope of the line tangent to the curve A at P, when $x = 5$.

   e. Find the slope of the line tangent to the curve A at P, when $x = 1, 2, 3,$ and 4.

   f. Find the slope of the line tangent to the curve A at P, when $x = a$.

4.   Draw a right triangle with side of length L (> 1) on the X-axis and vertical side of length M and label a point on the X-axis X and the appropriate point Q on the hypotenuse and O on the X-axis for Barrow's theorem.

   a. If $OX = x$, $XQ = q$, and p is the area of the triangular region enclosed by the X-axis, the line segment XQ and the line segment OQ, express q and p as functions of x.

   b. Sketch the area curve A related to the line Y, which is the hypotenuse of this triangle.

   c. When $x = 1$, find the position of T so that PT is a line tangent to curve A at P.

   d. Find the slope of the line tangent to the curve A at P, when $x = 1$.

   e. Find the slope of the line tangent to the curve A at P, when $x = a$.

5.   Use Barrow's theorem to show the slope of the line tangent to the graph of $Y = CX^2$ at the point (a, Ca^2) where $a > 0$ is $2Ca$. [Hint: Relate this curve to a right triangle with legs of length L and $2CL$.]

Final Comments

After students have been exposed to Barrow's theorem early in the calculus course, they are reminded of its relation to the calculus at later stages of the course to emphasize how the key elements of the entire theory fit together. The calculus of derivatives is seen in part as a systematic approach to the tangent problem using the algebraic representation of curves that graph functions. Differential equations are interpreted with the language of rates and motion as well as tangents. The area problem is connected to the problem of solving differential equations through the use of estimations. Thus the area problem is again related to the tangent problem, leading to contemporary statements and proofs of Barrow's theorem as the fundamental theorem of calculus.

Reference Notes

1. . A draft of this paper was presented at the quadrennial International Study Group on The Relations Between History and Pedagogy of Mathematics Meeting at Victoria College in Toronto, Canada, August 12–14, 1992.
2. The author is developing materials as part of the "Sensible Calculus Program: Articulating Precalculus and Calculus" at Humboldt State University that will discuss these problems and solutions earlier in the precalculus curriculum. See Martin E. Flashman, "A Sensible Calculus," *The UMAP Journal,* 1990, 11:93–96.
3. Readers may find excerpts of Barrow's work in D.J Struik, ed., *A Source Book in Mathematics, 1200–1800* (Cambridge: Harvard University Press, 1969). Other discussions of Barrow's Theorem may be found in Margaret E. Baron, *The Origins of the Infinitesimal Calculus* (Oxford: Pergamon Press, 1969), C.H. Edwards, *The Historical Development of the Calculus* (New York: Springer, 1979), and Victor J. Katz, *A History of Mathematics* (New York: HarperCollins, 1993).
4. See A.R. Hall, *Philosophers at War* (Cambridge, Cambridge University Press, 1980).
5. See "The Quadrature of the Parabola" and "On Spirals" in T.L. Heath, ed., *The Works of Archimedes* (New York: Dover, 1953).

The History of the Concept of Function and Some Implications for Classroom Teaching

Manfred Kronfellner
Technical University of Vienna

Some Important Steps in the Historical Development of the Function Concept

Implicit use of functions. The first roots of an implicit use of functions can be found in antiquity, actually in tabular form, as in the Babylonian tablet Plimptom 322, or as curves, such as conic sections, quadratrix, the Archimedian spiral, and others. One of the earliest steps toward an explicit concept of function, however, was made by the French mathematician and bishop of Lisieux, Nicole Oresme (1323?–1382). His doctrine of the latitudes of forms described graphically the dependency of two magnitudes—velocity and time—by using horizontal and vertical lines. By the seventeenth century, the increasing use of variables enabled mathematicians to describe such dependencies by formulas.

The first explicit definitions of a function. The first explicit definition of the function concept was given by James Gregory in 1667.[1] But this definition was lost sight of. In a 1673 manuscript, the Saxon scholar Gottfried Wilhelm Leibniz (1646–1716) used the word "function" ("functio").[2] After his friend Johann Bernoulli adopted Leibniz's terminology for a "function of x" in 1698, Bernoulli's student Leonhard Euler (1707–1783) first systematically investigated the theory of functions in *Introductio in analysin infinitorum* (2 vols, 1748). Euler stated:

> A function of a variable value is an analytical expression, which is composed of the variable value and constant values.[3]

A similar definition had appeared earlier in a paper of Johann Bernoulli published in 1718, which uses the notation íx .[4] Neither Euler nor Bernoulli made a distinction between "function" and "value of a function"; furthermore, there was no claim for uniqueness. As an example of an ambiguous function Euler mentioned expressions containing square roots. Mathematicians did not yet speak about domain and range, and we learn from another quotation that Euler subsumed under his concept of function only continuous functions, for functions are—in Euler's eyes—given not only by analytical expressions but also by a "curve drawn with the free hand."[5]

The domain interval and distinction between function and its description. Works leading towards the domain interval soon emerged. For the differential equation of the vibrating string

$$\frac{\partial^2 y(x,t)}{\partial t^2} = a^2 \cdot \frac{\partial^2 y(x,t)}{\partial x^2}$$

the French mathematician Jean d'Alembert (1717–1783) found the following solution:[6]

$$y(x,t) = \frac{1}{2} \cdot \psi(at + x) - \frac{1}{2}\psi(at - x)$$

(where ψ is an arbitrary function fulfilling certain conditions depending on the length of the string and on initial conditions). His Swiss colleague Daniel Bernoulli (1700–1782) found another solution:[7]

$$y(x,t) = \sum_{n=1}^{\infty} a_n \cdot \sin\frac{n\pi x}{\ell} \cdot \cos\frac{n\pi ct}{\ell}.$$

Comparing these two solutions, Euler argued that d'Alembert is more general, because his solution is arbitrary outside the interval given by the length of the string whereas Bernoulli's function solving the above differential equation is—because of periodicity—already fixed outside the interval $[0, \ell]$ by the values inside.

The French mathematician Joseph Fourier (1768–1830) realized that the assumption that the variables in an expression can be substituted by every number is a restriction of the concept of function. He wrote:

> We see by this that we must admit into analysis functions which have equal values, whenever the variable receives any values whatever included between two given limits even though on substituting in these two functions, instead of the variable, a number included in another interval, the results of the two substitutions are not the same.[8]

This insight led to the concept of the domain of a function, but at this time only intervals, and not more general sets, were considered.

Previously, a function was essentially identified with its description. Joseph Fourier discovered that a function can be described by different expressions. For example, the function

$$f(x) = \sum_{n=1}^{\infty} \frac{\sin((2n + 1)x)}{2n + 1}$$
$$= \begin{cases} 1 & \forall x \in]2k\pi, (2k + 1)\pi[, \ k \in \mathbb{Z} \\ -1 & \forall x \in](2k - 1)\pi, 2k\pi[, \ k \in \mathbb{Z} \\ 0 & \forall x = k\pi, \ k \in \mathbb{Z} \end{cases}$$

can be described in the interval $(0, \pi)$ by the series above and by $x \mapsto 1$. In this sense Peter Lejeune Dirichlet (1805–1859) gave a similar function :

$$f(x) = \lim_{n\to\infty} \lim_{k\to\infty} \left(\cos^2(n!\pi x)\right)^k = \begin{cases} 1 & \forall x \in \mathbb{Q} \\ 0 & \forall x \in \mathbb{R}\backslash\mathbb{Q} \end{cases}$$

Dirichlet's function also shows that continuity is not a self-evident characteristic of a function. This insight made clear that "function" is a more general concept than previously assumed. Hermann Hankel (1839–1873) called it "Dirichlet's concept of function" to distinguish this general concept from the previous narrower concepts. He defined:

> A function is called y of x, if each value of the variable magnitude x within a given interval corresponds to a certain value of y, no matter whether y depends on x in the whole interval according to the same law or not, no matter whether this dependency can be expressed by mathematical operations or not.[9]

Mappings, Functions and Relations. In the nineteenth century considerations of new mathematical objects (such as vectors, matrices, functionals), the subsequent development of new branches (for example topology, measure theory, or functional analysis) and further investigations of the concept of number, especially the real numbers, led to generalizations of the concept of function. Richard Dedekind (1831–1916) gave the following definition.

> A mapping ϕ of a system S is a law, which assigns to each element s of S a certain thing, which is called image of s and which is written $\phi(s)$.[10]

In this definition the domain is already arbitrary, and there is a distinction between the function (mapping) and its value (image). Still, this definition does not contain the concept of the range of the function. This concept and its special symbol was first used by Georg Cantor (1845–1918).

In their definitions Dirichlet, Hankel, Dedekind, and Cantor avoided phrases like "analytical expression", but still used such words as "law", "dependency", and "correspondence". The question how to define these terms rigorously leads to the definition of a function as a special relation (A, B, G). But, if $G \subseteq AKB$ is an infinite set, we need a "method" to describe this set. The question of which methods are admitted or possible, leads to problems in the field of philosophy of mathematics and to axiomatic set theory. This development shows that the concept of function can be widely reduced to set theory, another example for the importance of the concept of a set as one of the most basic concepts of mathematics.

II. Some Implications from the Historical Development

In the historical development of concepts we can often observe similar patterns.

Implicit use before explicit definition. In the history of mathematics we can often see that prior to an explicit definition of a concept, it has long been used implicitly. This "implicit phase" often yields the "material," which after an explicit introduction of the concept can be structured and considered from a more general point of view.

Changeability of mathematical concepts. The present day definition of a concept is thus the result of a long process. But this result is only provisional and most likely liable to future alternations.

Understandable modification of concepts. Modifications of the function concept did not drop from the blue sky; new formulations arose from concrete problems.

Generalization also means a changing the point of view. Questions which led to a modification of a concept were often of a different kind than previous problems: In the case of the function concept we observe a reflexive shift towards foundations of mathematics.

Adequate level of generality and exactness. Although using elementary concepts in their area, mathematicians from Euler to Leibniz to Cantor produced impressive mathematics. Their formulations of concepts, especially the degree of exactness and generality were adequate to their problems.

Increasing generality often leads to loss of aspects[11]. In a set-theoretical definition of a function the aspect of dependency (given by a formula), the aspect of a curve, of continuity (or even smoothness), of monotony, are no longer recognizable.

III. Some Consequences for Classroom Teaching

One of my fundamental aims in teaching is to present mathe- matics as a developing science. I assume that aspects which were recognized in the historical development before other ones are often easier to understand by pupils than aspects which were seen afterwards. Therefore my teaching follows the main ideas of the historical development as often as possible. In the case of the concept of function, my course consists of the following steps.

Step1: Formulas as a preparation of the function concept. First, before an explicit definition of the function concept, pupils should deal with various types of formulas and their graphical representation to develop a feeling for dependen-

cies, correspondences, monotony, and others. (See section II of this paper: Implicit use of the function concept)

Step 2: Working with provisional definitions. Dealing with the amount of examples in the first step pupils should be guided step by step to an explicit definition of the concept of real function, working with provisional versions. One possible sequence of such versions might be:

Version 1: A function is a correspondence (or dependency) between numbers.

In other words:

Version 2: If to every (real) number x there corresponds a (real) number y, then we call this correspondence a function.

The question of whether $\sqrt{4} = 2$ or $\sqrt{4} = \pm 2$ (see section II: understandable modifications of concepts) leads to

Version 3: If to every (real) number x there corresponds a unique (real) number y, ...

The consideration of functions like $f(x) = \sqrt{x}$ and the question $f(-4) = ?$ (see section II: understandable modifications of concepts) leads to the concept of the domain of a function:

Version 4: If to every (real) number $x \in A \subseteq \mathbb{R}$ there corresponds a unique $y \in \mathbb{R}$, ...

The range of a function is only needed for the investigation, whether there exists an inverse function of a given function. This yields the following definition.

Version 5: If to every (real) number $x \in A \subseteq \mathbb{R}$ there corresponds a unique $y \in B \subseteq \mathbb{R}$, this correspondence is called a real function from A to B.

Step 3: The generalization from real functions to arbitrary functions. Considerations of formulas with more variables, for example, $V = V(r, h) = r^2 \pi h$, or correspondences like $a \mapsto a$ or $P \mapsto (x, y)$ lead to the generalization from real functions to arbitrary functions.

Version 6: If to every $x \in A$ there corresponds a unique $y \in B$ (A, B arbitrary sets), this correspondence is called a function from A to B.

Step 4: Function as a special relation. A final step could be to discuss how to define the term "correspondence," which motivates the definition of the relation concept.

IV. Comments

Applications of the function concept are possible at all levels: most of the functions modeling simple problems in physics, life sciences, or economy, in particular functions of the type $f(x) = kx + d$, $p(x) = c \cdot x^n$, $N(t) = N_o a^t$, can be used definitely after the second version and possibly—implicitly—before the first version, and inverse functions (for example logarithmic functions) after the fifth version. Various types of problems arising in dealing with different types of functions yield modifications of definitions. (See section II: understandable modifications of concepts)

In my opinion it is nonsense to give the definition of a function as a special relation followed by simple exercises dealing with linear functions. This general definition is not a tool to be applied to such calculations (see section II: adequate level of generality and exactness) but a good opportunity to talk about mathematics, its fundamentals, and its development (see section II: change of point of view).

After my courses my students do not have higher abilities in solving simple problems than those being taught in the traditional way. But various discussions with my students have led me to the conviction that they have got a lively impression of mathematics as a developing science.

A final remark: The awareness of the history of mathematics can make the teacher more tolerant. Nearly 20 years ago I became angry when a pupil did not distinguish between a function and the value of the function or said,

$$f : x \mapsto \begin{cases} x + 1 & \text{for } x < 0 \\ 2 - x & \text{for } x \geq 0 \end{cases}$$

are *two* functions. Now having learned that Leibniz or Euler would probably have begun answering in a similar way, I realize that these pupils seem to be grappling with the germ of the idea in a potentially fruitful way.

References

1. Morris Kline, *Mathematical Thought from Ancient to Modern Times* (New York: Oxford University Press, 1972), p. 339.
2. A. P. Youshkevitch, "The Concept of Function up to the Middle of the 19th Century," *Archive for the History of Exact Sciences* 16 (1976/77), pp. 37–85, on p. 56.
3. Leonhard Euler, *Einleitung in die Analysis des Unendlichen*, vol. 1 (Berlin: Matzdorff, 1788) (trans from latin by Johann Andreas Christian Michelsen), p. 3.
4. Kline, *Mathematical Thought*, p. 340.
5. Youshkevitch, "The Concept of Function," p. 68.
6. Kline, *Mathematical Thought*, p. 507.
7. Ibid., p. 508.
8. Jerome H. Manheim, *The genesis of point set topology* (London: Pergamon Press, 1964), p. 49.
9. Oskar Becker, *Grundlagen der Mathematik in geschichtlicher Entwicklung* (Frankfurt/Main: Suhrkamp, 1975), p. 222.
10. Richard Dedekind, *Was sind und was sollen die Zahlen* (Braunschweig: Vieweg, 1911), p. 6.
11. Roland Fischer, Günther Malle, *Mensch und Mathematik* (Mannheim, Wien, Zürich: Bibliographisches Institut, 1985), p. 156.

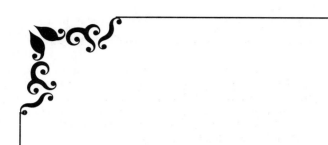

Integration in Finite Terms: From Liouville's Work to the Calculus Classroom of Today

SIU Man-Keung
University of Hong Kong

There is a question which a college mathematics teacher always wants to have an answer for but is afraid students may ask: why can't e^{x^2} (or $\sin x/x, \ldots$) be integrated? To be precise, why are their indefinite integrals not elementary functions? This esoteric topic of integration in finite terms is seldom explained in class. This article attempts to outline its development from Joseph Liouville's papers in the 1830s to its revival after almost a century in the work of Joseph Fels Ritt and subsequent authors. Pedagogically, an upsurge of interest in recent years, due to progress in symbolic computation, leads some to query: should students learn integration rules?

Introduction

At least some 30 years ago, in a beginning course in calculus there was a plethora of exercises regarding indefinite integrals. To some this may seem an elegant art or an amusing game, but to many this presents a source for anxiety and failure! It is not unusual to see some fairly artificial-looking integrals such as

$$\int \log(\cos x)\tan x\,dx;$$

this particular one happens to be $-[\log(\cos x)]^2/2$ (plus a constant), obtained through substituting a new variable for $\cos x$. However, the less artificial-looking integral

$$\int \log(\cos x)\,dx$$

cannot be found using a similar means. Adding to one's perplexity is the fact that a similar integral,

$$\int \cos(\log x)\,dx,$$

can be found by applying the technique of integration by parts twice with some deftness to yield the answer $x[\sin(\log x) + \cos(\log x)]/2$ (plus a constant). Textbooks sometimes include the first and sometimes also the third integral in their exercises but omit the second integral. In this era of computer software students often ask: why are there integrals that a machine cannot handle?[1] This is perhaps a question to which every teacher in calculus wants to know the answer but is afraid students may ask! When we are forced into a corner—when we are confronted with a difficult integral produced by a student out of the blue and wish to impress upon the student that integration is an art—we wield that typical counterexample,

$$\int e^{x^2}\,dx \quad \text{or} \quad \int \frac{e^x}{x}\,dx,$$

and announce that there is no answer in closed form. But what sense will students make of that? For instance, they know that integration by parts yields the answer $(x-1)e^x$ (plus a constant) for the integral

$$\int xe^x \, dx;$$

what difference does it make to have x dividing e^x instead of x multiplying e^x? An analogous situation occurs when a student wants to trisect an angle only to receive the reply that in general this cannot be done, but the question remains: why not and under what condition? Incidentally, there is a more contextual analogy between these two questions of "impossibility". (See the last section.) In 1835 the integral

$$\int \frac{e^x}{x} \, dx$$

was produced by the French mathematician Joseph Liouville (1809-1882) as an example of a celebrated theorem which now bears his name. It is one of the earliest examples of an integral which cannot be expressed in finite terms.

This article examines the story of integration in finite terms from Liouville to modern times, including some of its related developments and its pedagogical implications in teaching calculus. In particular, the availability of readily accessible software on symbolic computation compels us to ask whether students still need to go through the pleasure or torture, depending upon one's inclination, of integration rules. Looking at this page in history may shed some light on its answer. Throughout this article, "integral" is taken to mean indefinite integral, also known to some as antiderivative or primitive. (This article is the text of a talk given at the 7th International Congress of Mathematical Education at Québec City in August 1992.)

Section One
What Happened in the Nineteenth Century?

Liouville is usually called the founder of the theory of integration in finite terms. In a series of papers published between 1833 and 1835, he investigated the question of determining whether a given indefinite integral can be expressed as a finite expression involving only algebraic, logarithmic, exponential, trigonometric, or inverse trigonometric functions.[2] From 1839 to 1841 he treated the similar question for certain ordinary differential equations.[3] An important theorem (which will be stated in this section later), now named after him, was proved by him in 1834.[4] Most of his subsequent work is based upon this theorem.

However, history seldom, if ever, proceeds in a linear manner, and mathematical development has its root in tradition. In this case, the predilection for certain types of curves had long been a tradition with the ancient Greeks, as pointed out in the following passage in René Descartes' *La Géométrie*:[5]

> "The ancients were familiar with the fact that the problems of geometry may be divided into three classes, namely, plane, solid and linear problems. This is equivalent to saying that some problems require only circles and straight lines for their construction, while others require a conic section and still others require more complex curves. I am surprised, however, that they did not go further, and distinguish between different degrees of these more complex curves, nor do I see why they called the latter mechanical, rather than geometrical."

Descartes singled out among these complex curves those whose "relation must be expressed by means of a single equation",[6] that is, those curves that are graphs of a polynomial equation $f(x,y) = 0$, and to classify them according to the degree of the corresponding polynomial. He disregarded the other complex curves that cannot be so expressed. Later, Isaac Newton, Gottfried Wilhelm Leibniz, and others called the former type algebraic curves and the latter type transcendental curves. Both were accepted as genuine mathematical objects. However, while Newton felt no qualm in resorting to infinite series, Leibniz preferred to reduce transcendental expressions to certain elementary but finite forms.[7] He once discussed the possibility of reducing the quadrature problem to that of the hyperbola and the circle, or in terms of functions, of representing an integral by algebraic, logarithmic, trigonometric functions and their inverses.[8] In the eighteenth century, largely due to the influence of Leonhard Euler's *Introductio in analysin infinitorum* (2 volumes, 1748), the prominent roles of *elementary* functions were established. Leibniz's question became the problem of integration in finite terms. At about the same period, interest in the problem arose in another quarter, namely, computation of the so-called elliptic integral.[9] A typical example is the integral

$$\int \frac{1 - k^2 x^2}{\sqrt{1 - x^2}\sqrt{1 - k^2 x^2}} \, dx$$

in finding the perimeter of an ellipse. Nobody at the time could compute such integrals.

Johann Bernoulli first answered Leibniz's question. In *Acta Eruditorum* for 1702 he integrated some rational functions by the method of partial fractions and asserted that the integral involves only trigonometric or logarithmic functions.[10] By 1750 this theorem was an acknowledged fact, although a definitive proof involving factorization of a polynomial awaited a rigorous proof of the Fundamental Theorem of Algebra supplied by Carl Friedrich Gauss half a century later. For the benefit of exposition in this article, it is instructive to cast this theorem in a language familiar to the calculus classroom of today.

A rational function is a function of the form $P(x)/Q(x)$ where $P(x)$, $Q(x)$ are polynomials with real coefficients and $Q(x) \neq 0$. To determine integrals of rational functions it suffices to compute the integrals

$$\int \frac{dx}{(ax+b)^m} \quad (a \neq 0),$$

$$\int \frac{dx}{(ax^2+bx+c)^m} \quad (a \neq 0, \triangle = b^2 - 4ac < 0),$$

and

$$\int \frac{x\,dx}{(ax^2+bx+c)^m} \quad (a \neq 0, \triangle = b^2 - 4ac < 0).$$

It turns out that

$$\int \frac{dx}{ax+b} = \frac{1}{a}\log(ax+b),$$

$$\int \frac{dx}{(ax+b)^m} = -\frac{1}{(m-1)a(ax+b)^{m-1}} \quad \text{for } m > 1,$$

$$\int \frac{dx}{ax^2+bx+c} = \frac{2}{\sqrt{-\triangle}}\tan^{-1}\left[\frac{2ax+b}{\sqrt{-\triangle}}\right],$$

$$\int \frac{dx}{(ax^2+bx+c)^m} = \frac{2ax+b}{(m-1)(-\triangle)(ax^2+bx+c)^{m-1}}$$
$$+\frac{2(2m-3)a}{(m-1)(-\triangle)}\int \frac{dx}{(ax^2+bx+c)^{m-1}}$$
$$\text{for } m > 1,$$

$$\int \frac{x\,dx}{ax^2+bx+c} = \frac{1}{2a}\log(ax^2+bx+c) - \frac{b}{2a}\int \frac{dx}{ax^2+bx+c},$$

and

$$\int \frac{x\,dx}{(ax^2+bx+c)^m} = \frac{bx+2c}{(m-1)\triangle(ax^2+bx+c)^{m-1}}$$
$$+\frac{(2m-3)b}{(m-1)\triangle}\int \frac{dx}{(ax^2+bx+c)^{m-1}}$$
$$\text{for } m > 1.$$

If we work in the domain of complex numbers, then we can dispense with inverse trigonometric functions because these can be represented in terms of the logarithmic function, and we can rephrase the answer as:

$$\int \frac{P(x)}{Q(x)}\,dx = V(x) + C_1 \log U_1(x) + \cdots + C_n \log U_n(x),$$

where $V(x), U_1(x), \ldots, U_n(x)$ are **rational functions** and $C_1, \ldots, C_n$ are constants. The statement of Liouville's ba-

sic theorem resembles this expression. In order to describe his result, it is first necessary to see what Liouville meant by "finite explicit functions", or, in modern day terms, "elementary functions".

A function $y = f(x)$ is called **algebraic** if it is a root of a polynomial equation, that is, $y^n + A_{n-1}(x)y^{n-1} + \cdots + A_1(x)y + A_0(x) = 0$ for some positive integer n and rational functions $A_{n-1}(x), \ldots, A_1(x), A_0(x)$. (Example: $y = \sqrt{1+x^2}$.) Logarithms and exponentials of algebraic functions are called transcendental monomials of the first kind. (Example: $y = \log(1+x^2)$.) A function that is not algebraic but is an algebraic function of x and transcendental monomials of the first kind is called a transcendental function of the first kind. (Example: $y = \sqrt{1+e^x}$.) Logarithms and exponentials of transcendental functions of the first kind are called transcendental monomials of the second kind. (Example: $y = \log\sqrt{1+e^x}$.) A function that is not algebraic nor transcendental of the first kind but is an algebraic function of x and transcendental functions of the first kind and transcendental monomials of the second kind is called a transcendental function of the second kind. (Example: $y = (1+x^2)e^{(1+x)} + \log\sqrt{1+e^x}$.) In this way Liouville defined recursively transcendental **functions of the nth kind**, and he called all functions defined in this way **finite explicit functions**. In 1834 he proved the following result:[11]

> **Liouville's Theorem.** *Let y be an arbitrary algebraic function of x. If the integral $\int y\,dx$ is expressible in finite explicit form, then*
>
> $$\int y\,dx = t + A\log u + B\log v + \cdots + C\log w$$
>
> *where $A, B, \ldots, C$ are constants and t, u, v, $\ldots$, w are algebraic functions of x.*

With this theorem Liouville could establish that certain elliptic integrals are not expressible in finite explicit form,[12] a topic which drew much attention at the time and which started Liouville's interest in the theory of integration in finite terms. In 1835 he generalized his theorem to the following form:[13]

> **Liouville's Generalized Theorem.** *Let y and z, etc. be functions of x, which satisfy differential equations of the form $\frac{dy}{dx} = p$, $\frac{dz}{dx} = q$, etc., where p and q are algebraic functions of x, y, and z, etc. Further, let P be an algebraic function of x, y, and z, etc. If $\int P\,dx$ is expressible in finite explicit form, then*
>
> $$\int P\,dx = t + A\log u + B\log v + \cdots + C\log w$$

where $A, B, \ldots, C$ *are constants and* $t,$ $u,$ $v, \ldots, w$ *are algebraic functions of* $x,$ $y,$ *and* $z,$ *etc.*

With this generalized theorem, Liouville could demonstrate that certain integrals of the form $\int ye^x \, dx$, among them the integral $\int \dfrac{e^x}{x} \, dx$, are not expressible in finite explicit form.[14]

Actually, several eighteenth- and early nineteenth-century mathematicians had mentioned or even claimed to have proved results of the same nature. Some of these anticipated Liouville's ideas and might well have inspired his work.[15] Among early investigators were Alexis Fontaine (1764) and Marie-Jean Marquis de Condorcet (1765). About Fontaine, Liouville commented that the method was "in reality nothing but a laborious groping whose least fault is its disheartening length"; about Condorcet he commented that the "theorems lack demonstrations and some of them lack exactness".[16] As to the impossibility of expressing certain elliptic integrals in finite explicit form, Pierre Simon Laplace (1812) first claimed to possess a proof, but he did not publish rigorous proofs of the theorems he claimed to have found. The real contender to Liouville's priority as the founder of the theory of integration in finite terms is Niels Henrik Abel, who wrote on the subject about 1823.[17] Unfortunately, this paper of Abel met a fate worse than that of his famous Paris Academy Memoir on elliptic functions (1826)—the latter at Jacobi's insistence was published in 1841, fifteen years after Abel submitted it to the Paris Academy and twelve years after his death,[18] but the former paper seems to be completely lost.[19] By piecing together clues from other papers of Abel, Jesper Lützen is of the opinion that Abel had most of the ideas needed for a more systematic exposition of the theory of integration in finite terms, but, because of Abel's early death the job was left for Liouville. Although Liouville did not know of Abel's contributions when he began his investigation and so was not directly inspired by ideas of Abel, he made ample use of them after having learned of Abel's contributions.[20]

After this work on indefinite integrals, Liouville turned to solutions of ordinary differential equations in finite terms. The complexity of the problem can be seen from the length of elapsed time between setting himself the task in 1834 and the publication of his first paper on this topic in 1839,[21] more than double the period he needed to develop his theory of integration in finite terms. Although the three papers he produced fell short of his original ambitious project, they contained beautiful results. The last paper in 1841 concluded his published work in the subject. In it he answered an age-old problem about the Riccati equation:

the equation $\frac{dy}{dx} + ay^2 = bx^m$ can be solved by quadratures only for $m = -4n/(2n \pm 1)$ where n is a positive integer.[22]

As for indefinite integrals, the general problem of Liouville remained unanswered: "Given a finite explicit function of x; how does one determine in a finite number of steps whether its integral is also a finite explicit function? If the answer is in the affirmative, how does one compute its integral?" It remained unanswered until 1970 when Robert H. Risch rounded off the problem by giving such an algorithm.[23]

Section Two
What Happens in the Twentieth Century?

On March 30, 1834, Liouville wrote in his notebook that "we must begin to collect the material for a great work entitled *Essai sur la théorie de l'intégration des formules différentielles en quantités finis*".[24] He listed the content of the first part of this projected book, which however never materialized. His work on the solution of the Riccati equation in finite terms (1841) concluded his published work in the subject, and the subject more or less disappeared for nearly a century! In some sense, the comprehensive work which Liouville never published found its realization in the book *Integration in Finite Terms: Liouville's Theory of Elementary Methods* (1948) by the American mathematician Joseph Fels Ritt.[25]

Between the work of Liouville and Ritt there were activities going on in the field, mainly in Russia. This work was referred to in an appendix to *The Integration of Function of a Single Variable* by the British mathematician Godfrey Harold Hardy in 1905.[26] Hardy's book recreated interest in this near forgotten subject, and the Russian school began to add to Liouville's theory. Interestingly, while Hardy's approach was more function–theoretic than Liouville's original work, that of the Ukraine-born Swiss mathematician Alexander Ostrowski in 1946 was more algebraic than Liouville's original work.[27] This algebraic approach using the notion of field extension pointed to the way of extracting the algebraic ingredients of the investigation, thereby furnishing a simpler and more general treatment by which the original problem was eventually solved.

For thirty years, almost up to his death in 1951, Ritt produced a series of papers and books, including the 1948 classical account just mentioned, which gave impetus to the subject and opened up a new field christened "differential algebra" (by Ritt's student and successor Ellis R. Kolchin).[28] This trend in differential algebra which deals with differential equations, is covered in the next section. Let us continue with the trend, which deals with indefinite integrals. Although Ritt was at heart an analyst, he promoted the algebraic outlook of Ostrowski and (in Kolchin's words) "made a great effort to meet the algebraist half way".[29] In 1968 Maxwell Rosenlicht published the first purely algebraic exposition of Liouville's theory on functions with elementary integrals,[30] and in 1970 Robert Risch furnished an algorithm for solving the general problem.[31] Rosentlich's approach can be regarded as the algebraic approach, which had gradually developed out of the initial analytic approach in Liouville's work, pushed to its extremes.

In the language of abstract algebra, we define a **differential field** to be a field F, together with a **derivation** on F, i.e., a map of F into itself, usually denoted by $a \mapsto a'$, such that $(a+b)' = a' + b'$ and $(ab)' = a'b + ab'$ for all a, b in F. The constants of F, i.e., all elements c in F such that $c' = 0$, form a subfield of F. If a, b are elements of the differential field F, a being nonzero, we call a an **exponential** of b and b a **logarithm** of a if $b' = a'/a$. By a **differential extension field** of a differential field F, we mean a differential field which is an extension field of F whose derivation extends the derivation on F. An **elementary** extension field of F is a differential extension field of F which is of the form $F(t_1, \ldots, t_N)$ where, for each $i \in \{1, \ldots, N\}$, the element t_i is either algebraic over the field $F(t_1, \ldots, t_{i-1})$, or a logarithm or an exponential of an element of $F(t_1, \ldots, t_{i-1})$. We can now state the theorem proved by Rosentlich:[32]

> **Liouville's Theorem.** *Let F be a differential field of characteristic zero and $\alpha \in F$. If the equation $y' = \alpha$ has a solution in some elementary differential extension field of F having the same subfield of constants, then there are constants $c_1, \ldots, c_n \in F$ and elements $v, u_1, \ldots, u_n \in F$ such that*
>
> $$\alpha = v' + c_1(u_1'/u_1) + \cdots + c_n(u_n'/u_n).$$

By choosing F to be the field $\mathbb{C}(z, e^{g(z)})$, the field of complex rational functions of z with $e^{g(z)}$ adjoined, Rosentlich recovered from the theorem above a criterion due to Liouville: If $f(z), g(z)$ are rational functions of z, $f(z)$ being nonzero and $g(z)$ being non-constant, then $\int f(z)e^{g(z)}dz$ is elementary (that is, contained in some elementary extension field of $\mathbb{C}(z)$) if and only if $f = a' + ag'$ for some rational function $a = a(z)$ in $\mathbb{C}(z)$.[33] An equivalent formulation for the equality is that the integral is of the form $a(z)e^{g(z)}$ for some rational function $a(z)$. Let us

apply the criterion to $\int \frac{e^z}{z} dz$ for which the equation to look at is $1/z = a' + a$, which has no solution in $\mathbb{C}(z)$ (by comparing the order of poles). Hence $\int \frac{e^z}{z} dz$ is not elementary. At this point it is desirable to offer an answer to the query about integration in finite terms for students in a calculus class. We will use this same example, but try to refrain from introducing the language of differential field extension and to bypass the employment of knowledge about order of poles. We take as our starting point Liouville's criterion: The integral $\int f(z)e^{g(z)}dz$ is elementary if and only if it is of the form $a(z)e^{g(z)}$ where $a(z)$ is a rational function. This effectively lands us back on the familiar ground of polynomials. Suppose $\int \frac{e^z}{z} dz$ is elementary, then $\int \frac{e^z}{z} dz = a(z)e^z$ for some rational function $a(z)$. Differentiating both sides and using the Fundamental Theorem of Calculus, we have $1/z = (P/Q)' + (P/Q)$ where $a(z) = P(z)/Q(z)$ with $P = P(z)$, $Q = Q(z)$ being polynomials with no common factor and $Q \neq 0$. After differentiating P/Q and simplifying terms, we obtain

$$Q(Q - zP' - zP) = -zPQ'. \qquad (\#)$$

Since $Q \neq 0$, Q has a zero α of positive multiplicity m. We now divide our discussion into two cases: (i) $\alpha \neq 0$ and (ii) $\alpha = 0$. Suppose $\alpha \neq 0$. Since P, Q have no common factor, $P(\alpha) \neq 0$. Hence α is a zero of multiplicity $m-1$ of the polynomial on the right-hand-side of $(\#)$, but α is a zero of multiplicity at least m of the polynomial on the left-hand-side of $(\#)$. This is a contradiction. Suppose $\alpha = 0$; then we can write $Q = z^m R$ for some polynomial $R \neq 0$ which has no common factor with P and $R(0) \neq 0$. Equality $(\#)$ becomes

$$R(z^m R - zP' - zP + mP) = -zPR'. \qquad (\#\#)$$

By choosing a zero $\beta \neq 0$ of R of positive multiplicity, we can repeat the former argument to $(\#\#)$ for R, P instead of Q, P to obtain a contradiction. Hence $\int \frac{e^z}{z} dz$ is not elementary.[34]

These new developments in the early 1970s, coupled with the advent of computers since the 1960s, led to rapid progress in symbolic integration, which in turn stimulated research in the theory of integration in finite terms and its related topics. In a survey titled *Symbolic integration: The stormy decade*, written in 1971, Joel Moses said,[35]

> "In the beginning of the decade [1960s] only humans could determine the indefinite integral to all but the most trivial problems. The techniques used had not changed materially in 200 years. People were satisfied in considering the problem as requiring heuristic solutions and a good deal

of resourcefulness and intelligence. There was no hint of the tremendous changes that were to take place in the decade to come. By the end of the decade computer programs were faster and sometimes more powerful than humans, while using techniques similar to theirs. Advances in the theory of integration yielded procedures which in a strong sense completely solved the integration problem for the usual elementary functions."

Then in another survey titled *Symbolic integration—The dust settles?*, written in 1979, A.C. Norman and J.H. Davenport said,[36]

"... the last decade [1970s] has seen a great deal of consolidating work, with experimental programs being refined into practical tools and abstract mathematical techniques reduced to workable algorithms."

The titles of these two articles indicate substantial advance from 1971 to 1979. Much is still happening in this field today.

Section Three
Related Work: Galois Theory of Differential Equations

A related problem to integration in finite terms is that of solutions of differential equations. Liouville made some progress but stopped in the early 1840s. Subsequent developments in this direction are not as closely related to Liouville's theory, and their extensions and ramifications are so diversified and dynamic that their discussion falls outside the scope of this article. Let us just look at some highlights to appreciate how various mathematical strands are woven into a grand mathematical tapestry.

The Norwegian mathematician Sophus Lie conceived and carried out a much broader programme of application of group theory to differential equations. A rich variety of ideas and problems contributed to Lie's creation of a theory of continuous groups in the winter of 1873–74.[37] In subsequent years Lie developed his theory thoroughly in a series of books and articles.[38] The theory of Lie groups and Lie algebras, as the theory has come to be known, is today a fundamental part of mathematics which is in touch with a host of mathematical areas and applications,[39] but its original inspirational source was the field of differential equations. Like a beginning student in calculus today, mathematicians around the mid-nineteenth century saw the art of solving differential equations as a variety of special techniques. The profound insight Lie had was that these special techniques are subsumed under one general procedure

based on the invariance of the solutions of the differential equation under a continuous group of symmetries. To study these continuous groups, Lie made the fundamental step of assigning to each continuous group through "infinitesimal transformations" a corresponding vector space with a multiplication which is "anti-associative", thus switching the problem to the study of a more manageable object. From these notions come what we call today Lie groups and Lie algebras.

Lie tried to assign continuous groups to differential equations in the same spirit as in Galois's work on algebraic equations, although perhaps he did not have a full understanding of Galois's work.[40] He proved that those equations which correspond to solvable continuous groups have solutions by quadratures. Lie's theory of differential equations was popular, and its exposition even found its way into the curriculum of many universities. For instance, it was presented in the popular famous texts *Cours d'Analyse de l'École Polytéchnique* by Camille Jordan (1887) and *Traité d'Analyse de la Faculté des Sciences de Paris* by Émile Picard (1891-96).[41] However, the topic faded after the global, abstract formulation of Lie groups and Lie algebras championed by Élie Cartan gained dominance. Later emphasis on numerical solutions after the advent of computers further diminished the attractiveness of Lie's original scheme. Only much later in the twentieth century was interest in that idea rekindled when mathematicians and physicists sensed the significant role played by symmetry.[42]

A more refined "Galois theory of differential equations" was that proposed by Émile Picard (1883, 1887) and Ernest Vessiot (1891, 1892) for homogeneous linear ordinary differential equations. In 1948 Kolchin wrote his seminal paper, *Algebraic matric groups and the Picard-Vessiot Theory of homogeneous linear ordinary differential equations*, and placed the theory in its natural setting, the Ritt theory in differential algebra.[43] By studying what he meant by a Picard-Vessiot extension and a Liouvillian extension of a differential field, he characterized those differential equations which are solvable by quadratures. A self-contained clear exposition of this theory was provided by Irving Kaplansky.[44] Kolchin's work opened up the theory of linear algebraic groups and pushed forth research in differential algebra started by Ritt.[45] An account of modern differential Galois theory was given by Michael Singer recently.[46] In recent years there is an upsurge of interest in effective algorithms in differential algebra because of advances in symbolic computation on a computer.[47]

Section Four
Morals of the Story

Toni Kasper remarked at the conclusion of his succinct account of integration in finite terms:[48]

> "Risch makes the interesting suggestion that some features of his algorithm are suitable for presentation to calculus students. No calculus text at present provides this material, an omission that not only leaves the story of finite elementary integration incomplete, but deprives the calculus student of some valuable insights."

I am more interested in the last clause, and have a broader but less technical aim in mind. In the second and third sections of this article, I attempt to embellish the story with pertinent mathematical pointers to suggest a possible way of bringing this esoteric topic, seldom explained in class, into the calculus classroom. The notion of field extension is admittedly too advanced for an ordinary calculus class. However, with tactful exposition it is possible to at least get the general idea across,[49] just as it is possible to explain to a high-school class the impossibility of trisecting a general angle by straight edge and compasses—in some sense the two problems bear analogy in that they are both (in Kaplansky's words) "pre-Galois" theories which involve only basic properties of differential fields and ordinary fields respectively.[50] In the fourth section of this article, I attempt to exhibit several rich strands of ideas which are related to the topic and which develop into fundamental parts of mainstream mathematics. With carefully worked out embellishment these ideas can be introduced into relevant courses on a more advanced level to enhance understanding. Such use of history has been pointed out by Frederick Rickey who said:[51]

> "... we can *talk* about mathematical ideas that are too hard to present in detail in class. The results are still important and of interest, even if the proofs cannot be given. Black holes, quarks, DNA, and plate tectonics are things that we have all heard about and understand in a general way, even though few of us know the technical details. This is a lesson that we had better learn from the physical scientists: Popular presentations of scientific ideas attract students to the field, and leaves the general public with warm feelings towards it."

To conclude I ask a more general question: What can we learn from the page of history we unfolded in the preceding three sections?

1. There is a time to everything. The saying is true of mathematical development. Liouville devoted eight years to the study of the theory of integration in finite terms and achieved significant accomplishments, but he discontinued this line of research in the early 1840s. The line was picked up by Ritt after almost a century, and Ritt became the principal prophet and practitioner in the field of differential algebra that grew out of it. His student Kolchin in turn became the leader in this field. Although Kolchin's work significantly influenced related areas, the original interest in integration and differential equations was more or less confined to a small group around Kolchin in the 1960s and the early 1970s. Then came an upsurge of interest in symbolic computation in which the work of Rosentlich, Risch, and others played an important role. Today active research is going on in this field with journals, conferences, and special interest groups devoted to the subject.[52] How can we explain such ups and downs? In the case of Liouville, technical difficulties which seemed insurmountable at the time might have convinced him that there was little hope for a complete solution of the problem, that is, a general algorithm to decide which integrals are finite explicit functions, and an extension of the theory to the case of differential equations. But an even greater disappointment and discouragement might have come from the relatively little impact his theory had in his own time. Other mathematicians watched passively with a general attitude "of approval and indifference".[53] The main reason why Liouville's theory did not appeal to his contemporaries (but did appeal to mathematicians after a century) is the algebraic aspect of the techniques which did not fit into the mathematical community of the time. In the case of the second revival of interest, advances in computer science are the moving force.

2. Although skill is needed in technological advance, the underlying theory is of primary importance. The rapid development of computer algebra with the accompanying rekindled interest in the theoretical aspect is a good illustration of this blending of skill and theory. Another illustration can be found in the story of Lie's original intention to apply his continuous transformation groups to study differential equations analogous to Galois's work on algebraic equations. For a period it was a popular topic that even found its way into the university curriculum, but then fell into oblivion and lay dormant for nearly half a century. The last two decades, however, witnessed a new surge of interest and much research activity in this field by both physicists and mathematicians. The motivation does not lie with the skill of solving the differential equation—

high speed computers and techniques in numerical analysis can handle the job in a more efficient way—but with the description of symmetry and invariance of the differential equation and hence also that of the real objects modelled by the differential equation.

3.   In connection with the two points just stated, we can now attempt to answer the question: Should students learn integration rules? In a paper of the same title as the question above, Bruno Buchberger proposed a didactical principle which he named "White-Box/Black-Box Principle for using symbolic computation software in math education".[54] A rough summary of the principle says that students should understand an area X as a "white-box" at the stage when area X is new to them, and should use computation software in area X as a "black-box" at the stage when area X has been thoroughly studied by them. History informs us that skills and algorithms do not come from nowhere and that skills and algorithms, though useful and important, are means rather than ends. In studying the process whereby skills and algorithms are obtained, we gain insights and understanding of the subject. The most useful "methods" are actually "theorems." Although it will be unwise not to use the "black-box" when it is readily available and when it can enhance learning, a "black-box only" approach can be disastrous to mathematics education and even to the future development of mathematics. An understanding of the theory of integration in finite terms in its historial context can perhaps convince students that calculus is not just a cookbook of recipes but is in itself a beautiful subject with a close relationship to exciting modern development.

Appendix

For readers who seek after mathematical details to complete the story, my recommendation is Rosenlicht's article in the *American Mathematical Monthly*.[55] In this Appendix I want to give just enough details to convey the flavour of the topic, if only to show the basic elementary aspect of it—partial fractions used by Bernoulli. Let us try to see how we arrive at Liouville's criterion. Readers who do not wish to use the language of differential field extension may think of F as the field of rational functions and t as an exponential function, as in the situation of the proof of Liouville's criterion. (The technique employed appears throughout the theory, including the proof of Liouville's Theorem.[56])

Lemma.   *Let F be a differential field of characteristic zero, and $F(t)$ a differential field extension of F having*

the same subfield of constants, with t transcendental over F. If $t'/t \in F$, then for any $h(t) \in F[t]$ of positive degree, $\big(h(t)\big)' \in F[t]$ is of the same degree, and is a multiple of $h(t)$ only if $h(t)$ is a monomial.

Proof.   To prove the first assertion we need only consider the leading term so that we may assume $h(t) = at^n$ with $a \neq 0$ and $n > 0$. Suppose $t' = bt$ with $b \in F$. Since $(at^n)' = a't^n + nat^{n-1}t' = (a' + nab)t^n$ and $a' + nab \neq 0$ (or else at^n, being a constant, is in F and t is thus not transcendental over F), we see that $(at^n)'$ is of degree n. To prove the second assertion, suppose $\big(h(t)\big)' = dh(t)$ with $d \in F$ and $h(t)$ contains at least two monomial terms $a_m t^m$, $a_n t^n$ ($a_m \neq 0, a_n \neq 0, m \neq n$). By comparing coefficients we see that $a'_m + ma_m b = da_m$, $a'_n + na_n b = da_n$. Hence $(a_m t^m / a_n t^n)' = 0$ so that $a_m t^m / a_n t^n$, being constant, is in F, and t is thus not transcendental over F. Therefore $h(t)$ must be a monomial.   Q.E.D.

Proposition (Liouville's Criterion).   *Let $f(z)$, $g(z) \in \mathbb{C}(z)$, $f(z)$ being nonzero and $g(z)$ being nonconstant. Then $\int f(z)e^{g(z)}dz$ is contained in some elementary extension field of $\mathbb{C}(z)$ if and only if $f = a' + ag'$ for some a in $\mathbb{C}(z)$.*

Proof.   Put $F = \mathbb{C}(z)$ and $t = e^{g(z)}$. Note that t is transcendental over F and $t'/t = g' \in F$. By Liouville's Theorem we have

$$ft = v' + c_1(u'_1/u_1) + \cdots + c_n(u'_n/u_n) \qquad (*)$$

where $c_1, \ldots, c_n \in \mathbb{C}$ and $v, u_1, \ldots, u_n \in F(t)$, if $\int f(z)t(z)\,dz$ is contained in some elementary extension field of $\mathbb{C}(z)$. We are going to show that $v, u_1, \ldots, u_n$ must be of a very special form for the terms on the right side to add up to a polynomial in $F[t]$. By factoring each u_i as a power product (negative exponents allowed) of irreducible elements of $F[t]$ and using logarithmic derivatives if necessary, we may assume the $u_i's$ which are not in F are distinct monic irreducible elements of $F[t]$. We then expand v into partial fractions so that it is a sum of an element of $F[t]$ plus various terms of the form $k(t)/\big(h(t)\big)^r$ where $h(t)$ is a monic irreducible element of $F[t]$, r a positive integer, and $k(t)$ a nonzero element of $F[t]$ of degree less than that of $h(t)$. Thanks to the lemma, $h(t)$ does not divide $\big(h(t)\big)'$ if it is neither an element in F nor the monomial t. Suppose $h(t)$ occurs as some $u_i(t)$, then the fraction u'_i/u_i is in lowest term. Look at the maximal $r > 0$ for which $k(t)/\big(h(t)\big)^r$ occurs in $v(t)$. Then $\big(v(t)\big)'$ will consist of various terms having $h(t)$ in the denominator at most r times plus $-rk(t)\big(h(t)\big)'/\big(h(t)\big)^{r+1}$.

Note that the last fraction is in lowest term since $h(t)$ does not divide $k(t)\big(h(t)\big)'$. But since the right side of $(*)$ contains a fraction, this contradicts the fact that the left side is a polynomial in $F[t]$. Hence $h(t)$ cannot appear in the denominator of the partial fraction expansion of $v(t)$ and $h(t)$ cannot be any of the $u_i(t)$. The conclusion is: in $(*)$ each $c_i(u_i'/u_i)$ is an element of F, and v is of the form $\sum b_j t^j$ for j ranging over some set of integers with $b_j \in F$. Hence $f = b_1' + b_1 g'$ (since t is transcendental over F). Set $a = b_1 \in F = \mathbb{C}(z)$. Conversely, if $f = a' + ag'$ for some $a \in \mathbb{C}(z)$, then, by setting $t = e^{g(z)}$, we see that $(at)' = ft$ so that $\int f(z)e^{g(z)}dz = \int ft\,dz = at = ae^{g(z)}$, which is elementary.   Q.E.D.

References

1. See V. Frederick Rickey, "In praise of Liouville" in *Use of History in the Teaching of Calculus*, to appear in the MAA Notes Series. Availability of software in symbolic integration transforms this question from an academic query, which perhaps in the past occurred only to the more mathematically-minded, to a practical query, which can now occur to any average beginner in calculus. On the other hand, Robert D. Richtmyer proposed in an MAA invited address, "A Special Curriculum for Exceptional Students and Integration in Finite Terms" (January, 1993 at San Antonio, Texas) that the topic of integration in finite terms, which is customarily omitted from the undergraduate curriculum, will provide much stimulation for talented students in mathematics. See also (1) A.D. Fitt and G.T.Q. Hoare, "The closed-form integration of arbitrary functions", *Mathematical Gazette*, 1993, 77: 227–236, (2) E.A. Marchisotto and G.-A. Zakeri, "An invitation to integration in finite terms", *College Mathematics Journal*, 1994, 25: 295–308.

2. A detailed account is given by Jesper Lützen in Sections 14–41 of Chapter IX of *Joseph Liouville, 1809–1882, Master of Pure and Applied Mathematics* (New York-Berlin-Heidelberg: Springer-Verlag, 1990).

3. See Lützen, *Joseph Liouville*, Sections 42–51 of Chapter IX.

4. Joseph Liouville, "Sur les transcendantes elliptiques de première et de seconde espèce, considérées comme fonctions de leur amplitude," *Journ. Ec. Polyt.*, 1834, 14(23 cahier): 37–83.

5. René Descartes, *La Géométrie*, 1637. The translation is taken from René Descartes, *The Geometry of René Descartes*, translated by David Eugene Smith and Marcia L. Latham (1925; rpt. New York: Dover, 1954), p. 40.

6. Ibid., p. 48.

7. Lützen, *Joseph Liouville*, p. 353.

8. Ibid. p. 353. Leibniz discussed it in two letters of March 30, 1675, and July 12, 1677, to Henry Oldenburg. See Gottfried Wilhelm Leibniz, *Der Briefwechsel von Gottfried Wilhelm Leibniz Mit Mathematikern*, edited by C.J. Gerhardt (Berlin, 1899), pp. 248–249.

9. See Section 4 of Chapter 19 in Morris Kline, *Mathematical Thought From Ancient to Modern Times*. (New York: Oxford University Press, 1972).

10. Ibid., p. 411.

11. See Footnote 4 and Sections 26–32 of Chapter IX in Lützen, *Joseph Liouville*.

12. See Footnote 4 and Section 28 of Chapter IX in Lützen, *Joseph Liouville*. Liouville could establish that elliptic integrals of the first kind and second kind are not expressible in finite explicit form.

13. Joseph Liouville, "Mémoire sur l'intégration d'une classe de fonctions transcendantes", *Journ. Reine Angew. Math.*, 1835, 13: 93–118.

14. See Section 33 of Chapter IX in Lützen, *Joseph Liouville*.

15. See Sections 3–13 of Chapter IX in Lützen, *Joseph Liouville*.

16. See Footnote 4, pp. 37–38.

17. See Sections 7–13 of Chapter IX in Lützen, *Joseph Liouville*.

18. See Chapter 20 in Øystein Øre, *Niels Henrik Abel, Mathematician Extraordinary* (Minneapolis: University of Minnesota Press, 1957; rpt. New York: Chelsea, 1974).

19. Ibid., pp. 63–76.

20. See Section 13 of Chapter IX in Lützen, *Joseph Liouville*.

21. Joseph Liouville, "Mémoire sur l'intégration d'une classe d'équations différentielles du second ordre en quantités finies explicites", *Journal de Mathématique pures at appliquées*, 1839, 4: 423–456; *Comptes Rendus de l'Académie des Sciences*, 1839, 9: 527–530.

22. Joseph Liouville, "Remarques nouvelles sur l'équation de Riccati", *Journal de Mathématique pures et appliquées*, 1841, 6: 1–13; *Comptes Rendus de l'Académie des Sciences*, 1840, 11: 729. See Section 51 of Chapter IX in Lützen, *Joseph Liouville*.

23. Robert H. Risch, "The problem of integration in finite terms", *Trans. Amer. Math. Soc.*, 1969, 139: 167–189; Robert H. Risch, "The solution of the problem of integration in finite terms", *Bull. Amer. Math. Soc.*, 1970, 76: 605–608. See also Joel Moses, "Symbolic integration: The stormy decade", *Comm. Asso. for Computing Machinery*, 1971, 14: 548–560, and A.C. Norman, "Integration in finite terms", in *Computer Algebra: Symbolic and Algebraic Computation*, edited by Bruno Buchberger, George E. Collins, Rüdiger Loos (New York-Berlin-Heidelberg: Springer-Verlag, 1982; 2nd edition, 1983), pp.57–69. A review on the algorithm with successive improvement is discussed in Manuel Bronstein, "Symbolic integration: Towards practical algorithms", in *Computer Algebra and Differential Equations*, edited by E. Tournier (New York: Academic Press, 1990), pp.59–85.

24. Lützen, *Joseph Liouville*, p. 392.

25. Joseph Fels Ritt, *Integration in Finite Terms: Liouville's Theory of Elementary Methods* (New York: Columbia University Press, 1948).

26. Godfrey Harold Hardy, *The Integration of Functions of a Single Variable* (Cambridge: Cambridge University Press, 1905; 2nd edition, 1916; rpt. 1958). It appeared as no.2 in the Cambridge University Tract in Mathematics and Mathematical Physics in 1905.

27. Alexander Ostrowski, "Sur l'intégrabilité élémentaire de quelques classes d'expressions", *Comment. Math. Helv.*, 1946, 18: 283–308.

28. See the Preface, p.iii, in Joseph Fels Ritt, *Differential Algebra, AMS Colloquium Publications, vol.33* (Providence: Amer. Math. Soc., 1950; rpt. New York: Dover, 1966).

29. See the Preface in Ellis Robert Kolchin, *Differential Algebra and Algebraic Groups* (New York: Academic Press, 1973).

30. Maxwell Rosenlicht, "Liouville's theorem on functions with elementary integrals", *Pacific J. Math.*, 1968, 24: 153–161.

31. See Footnote 23.

32. See Footnote 30, and pp.968–970 in Maxwell Rosenlicht, "Integration in finite terms", *Amer. Math. Monthly*, 1972, 79: 963–972.

33. The proof is included as an appendix to this article.

34. Frederick Rickey did a similar thing for the integral $\int e^{-x^2} dx$ in his article. See Footnote 1.

35. See p. 548 in Joel Moses, "Symbolic integration: The stormy decade", *Comm. Asso. for Computing Machinery*, 1971, 14: 548–560.

36. See p. 398 in A.C. Norman, James Harold Davenport, "Symbolic integration—The dust settles?" in *Symbolic and Algebraic Computation (EUROSAM '79)*, edited by Edward W. Ng, *Springer Lecture Notes in Computer Science, no.72* (New York-Berlin-Heidelberg: Springer-Verlag, 1979), pp. 398–407.

37. See Thomas Hawkins, "Line geometry, differential equations, and the birth of Lie's theory of groups", in *The History of Modern Mathematics, Volume I: Ideas and Their Reception*, edited by David E. Rowe and John McCleary (New York: Academic Press, 1989), pp. 275–327; and S.S. Demidov, "The study of partial differential equations of the first order in the 18th and 19th centuries", *Archive for History of Exact Sciences*, 1982, 26:325–350.

38. Sophus Lie (with collaboration of Friedrich Engel), *Theorie der Transformationsgruppen, 3 volumes* (Leipzig: Teubner, 1888–1893).

39. See Johan G. Belinfante, Bernard Kolman, Harvey A. Smith, "An introduction to Lie groups and Lie algebras, with applications", *SIAM Review*, 1966:11–46; "An introduction to Lie groups and Lie algebras, with applications. II: The basic methods and results of representation theory", *SIAM Review*, 1968: 160–195.

40. See p. 294 in Hawkins, Footnote 37.

41. See p. 219 in Isaak Moiseevich Yaglom, *Felix Klein and Sophus Lie*, translated from Russian by Sergei Sossinsky, edited by Hardy Grant and Abe Shenitzer (Basel-Boston: Birkhäuser, 1988).

42. Ibid., p. 108. See also Introduction, p.xvi in Peter J. Olver, *Applications of Lie Groups to Differential Equations* (New York-Berlin-Heidelberg: Springer-Verlag, 1986).

43. Ellis Robert Kolchin, "Algebraic matric groups and the Picard-Vessiot theory of homogeneous linear ordinary differential equations", *Ann. of Math.*, 1948, 49: 1–42.

44. Irving Kaplansky, *An Introduction to Differential Algebra* (Paris: Hermann, 1957; 2nd edition, 1976).

45. See Preface in *Contributions to Algebra: A Collection of Papers Dedicated to Ellis Kolchin*, edited by Hyman Bass, Phyllis J. Cassidy, Jerald Kovacic (New York: Academic Press, 1977), and Ellis Robert Kolchin, *Differential Algebra and Algebraic Groups*, (New York: Academic Press, 1973).

46. Michael F. Singer, "An outline of differential Galois theory", in *Computer Algebra and Differential Equations*, edited by E. Tournier (New York: Academic Press, 1990), pp. 3–57.

47. This upsurge of interest is borne out by books like *Computer and Mathematics*, edited by Erich Kaltofen, Stephen M. Watt (New York-Berlin-Heidelberg: Springer-Verlag, 1989), and *Computer Algebra and Differential Equations*, edited by E. Tournier (New York: Academic Press, 1990).

48. Toni Kasper, "Integration in finite terms: The Liouville theory", *Math. Magazine*, 1980, 53: 195–201.

49. See the illustration on $\int \frac{e^z}{z} dz$ and also the Appendix.

50. See p. 5 in the second edition of Kaplansky, *An Introduction to Differential Algebra*; see also p.414 in Lützen, *Joseph Liouville*.

51. See Footnote 1.

52. See Introduction in *Computer Algebra: Symbolic and Algebraic Computation*, edited by Bruno Buchberger, George E. Collins, Rüdiger Loos (New York-Berlin-Heidelberg: Springer-Verlag, 1982; 2nd edition, 1983).

53. See p. 422 in Lützen, *Joseph Liouville*.

54. Bruno Buchberger, "Should students learn integration rules?", *SIGSAM Bulletin*, 1990, 24: 10–17.

55. Maxwell Rosenlicht, "Integration in finite terms", *Amer. Math. Monthly*, 1972, 79: 963–972.

56. Ibid. pp. 968–970.

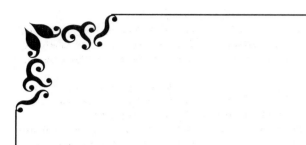

How Many People Ever Lived?

Jim Tattersall
Providence College

One of the main objectives for a mathematics teacher should be to have students ponder stimulating problems. This paper concerns a provocative exercise, which combines history, sociology, natural science, philosophy, and mathematics. In calculus, after the exponential law of growth and decay[1] is introduced, a number of intriguing problems may be posed concerning its various applications, for example, Isaac Newton's law of cooling,[2] safe and effective dosage,[3] and patterns for the spread of a disease.[4] While death is never a popular subject in a mathematics class many students are interested in careers in the actuarial sciences and a demographic exercise that generates considerable interest is determining how many people ever lived.

Historically, one of the first to attempt to determine the number of people who ever lived on the earth was done by Reverend Ezra Stiles (1727–1795), a Connecticut congregationalist, who studied law and theology at Yale and later became its president. Stiles participated in the global attempt to determine the earth-sun distance using data from his observations of the 1769 transit of Venus taken at Newport, Rhode Island.[5] Nine years earlier in a scriptural calculation he estimated that 120 billion souls existed in the approximately 6,000 years since the time of Adam and Eve.[6] It is interesting to see how Stiles's early approximation compared to some modern estimates. The demographers Wilhelm Winkler,[7] Edward Deevey,[8] Nathan Keyfitz,[9] and Annabelle Desmond[10] have estimated that the total human population of the earth up to the 1960s was respectively 500 billion, 110 billion, 100 billion, and between 77 and 96 billion people. The large variance in their answers can be attributed to the markedly different assumptions each made. Winkler started with two people 600,000 years ago and assumed a uniform growth rate until the world's population reached 3 billion in 1960. Some partitioned their data into several distinct parts. Both Deevey and Desmond saw three fundamental epochs in human development: cultural, agricultural, and scientific-industrial, while Keyfitz used four unclassified divisions. In order to estimate the total population in each region, Deevey used logarithmic growth curves, while Desmond and Keyfitz used exponential growth curves. Moreover, Keyfitz, using a semibiblical approach took into account the discoveries in East Africa of *Australopithecus* by Raymond Dart in 1924 and *Zinjanthropus boisei* by Mary Leakey in 1959 to commence his calculation with two humans 2 million years ago.

For our estimate, we assume a simple exponential growth model $N_i = N(t_i) = ae^{bt_i}$, where a and b are constants and N_i represents the population at a given time t_i.[11] In order to derive a formula for the number of people

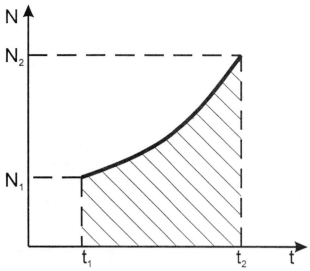

FIGURE 1

who lived during a given interval of time, say $t_2 - t_1$, we must first determine the area under the curve $N(t) = ae^{bt}$ over the interval $[t_1, t_2]$.

Note that

$$\frac{N_2}{N_1} = e^{b(t_2 - t_1)},$$

so $\ln(N2) - \ln(N1) = b(t2 - t1)$. Hence,

$$b = \frac{\ln(N_2) - \ln(N_1)}{t_2 - t_1}$$

and the total number of people-years, P, over the period $[t_1, t_2]$ is given by

$$P = \int_{t_1}^{t_2} ae^{bt}dt = \left(\frac{a}{b}\right)\left(e^{bt_2} - e^{bt_1}\right)$$
$$= \frac{(N_2 - N_1)(t_2 - t_1)}{\ln(N_2) - \ln(N_1)}.$$

Furthermore, if L denotes the average life expectancy during the period from t_1 to t_2, then $\frac{P}{L}$ represents the total number of people who lived during the period from t_1 to t_2.

There are two types of estimations that are crucial to obtaining a reliable figure for the total number of people who ever lived, namely the world's population for specific times in the past and the life expectancy between those given times. Deriving an accurate estimate of the world's population for a particular year in the distant past is practically impossible. Relatively reliable estimates and censuses of population are a recent phenomena and are limited to but a few countries. Still some estimates of the world's population before the twentieth century, albeit open

to question, are available. Students are advised to consult demographic yearbooks published by the United Nations[12] or almanacs to obtain reliable estimates for the world's population for the twentieth century. In order to obtain pre-twentieth century population estimates students select data from Table 1, which was compiled from several sources: demographer Alexa Carr-Saunders,[13] Deevey,[14] director of the Geochronometric Laboratory at Yale,[15, 16] Desmond[17] of the Population Reference Bureau in Washington, DC, Keyfitz[18], former Andelot Professor of Sociology and Demography at Harvard, a popular author, James Trager,[19] and Walter F. Willcox[20] of the National Bureau of Economic Research. Students are encouraged to consult Fernand Braudel's *Civilization and Capitalization*[21] for an informative discussion on the accuracy of statistical population estimates for earlier centuries. For convenience, I have used the expressions "$m(n)$" to denote "$m \cdot 10^n$" where m is a real number and n is a positive integer and "$-m$" to denote the year "m B.C." Hence, $-2(6)$ represents 2,000,000 B.C. Except for the 2*, which represents two people, all figures are given in millions.

The starting point of human population is still a subject of heated debate, and this issue is extremely important for our estimate. We begin by discussing Christian attempts to establish a chronology based on the *Bible*. In the seventeenth century the study of biblical chronology was a very respected field of research. Such early attempts to establish the origin of people on the earth drew upon the time of the origin of the earth itself. Isaac Newton considered his chronological work as among his most important contributions. Theologians first debated the notion that the earth had a finite past. This was before the study of fossils[22] and the rise of geology in the early nineteenth century, so their estimates are poor. In 1647 based on his biblical research, James Lightfoot (1602–1675), Master of St. Catherine's College, Cambridge, determined that the world began early on the morning of September 17, 3928 BC.[23] In 1650, *Cooper's Chronicle*[24] gave 5221 years as an estimate for the age of the earth.

However, the most well-known estimate of the age of the earth was done in 1656 by Archbishop James Ussher (1581–1656) of Armagh, Northern Ireland. Following a literal interpretation of the *Bible,* Ussher calculated that the world began on Sunday evening, October 23, 4004 BC.[25]

Scientific attempts to determine the age of the earth in the eighteenth century focused mainly on trying to determine how long it took the oceans to become salty or how long it took the molten earth to cool down. Compte de Buffon (1707–1788) calculated that it would take about 75,000 years for a molten earth to cool down but was uneasy with

YEAR	Carr-Sa	Deevey	Desmond	Keyfitz	Trager	Willcox
−2 (6)				2*		
−1 (6)		.125				
−3 (5)		1				
−25000		3.34				
−10000					3	
−8000		5.32		5	5.3	
−4000		86.5			85	
−3000					100	
−50				300		
1		133				
1000			275			
1100			306			
1200			348			
1300			384			
1400			375			
1500			446			
1600			486			
1650	545	545				470
1750	728	728		800		906
1800	906	906				
1825				1000		
1850	1171					1191
1900	1608	1610				1571
1950		2400				

TABLE 1

his estimate that about 3 million years was needed for the sedimentation phenomena to occur.[26] According to the theory of uniformitarianism, which asserted that the natural order of the past was uniform with that of the present and which was propounded by the Scottish geologists James Hutton (1726–1797) and Charles Lyell (1797–1875), the earth had to be several hundreds of millions of years old.

Undoubtedly, these estimates added a new human conception to time and history. The most accepted estimate was made in 1956 by Clair Patterson of Cal Tech. He used radiocarbon dating and determined from samples of the earth's crust that our planet was at least 4.5 billion years

old.[27] Students who wish to learn more about such historical attempts are directed to Brent Dalrymple's *The Age of the Earth.* [28]

After some discussion on how long it would have taken for a primordial atmosphere to acquire its present concentration and on the fossil record itself students usually assume as an initial reference point for their estimate that humans first inhabited the earth from 1 to 3 million years ago. Having determined a starting point and several point estimates for the world's population, it is now necessary to get a reliable estimate of the life expectancy for the periods between those point estimates. One of the earliest estimates

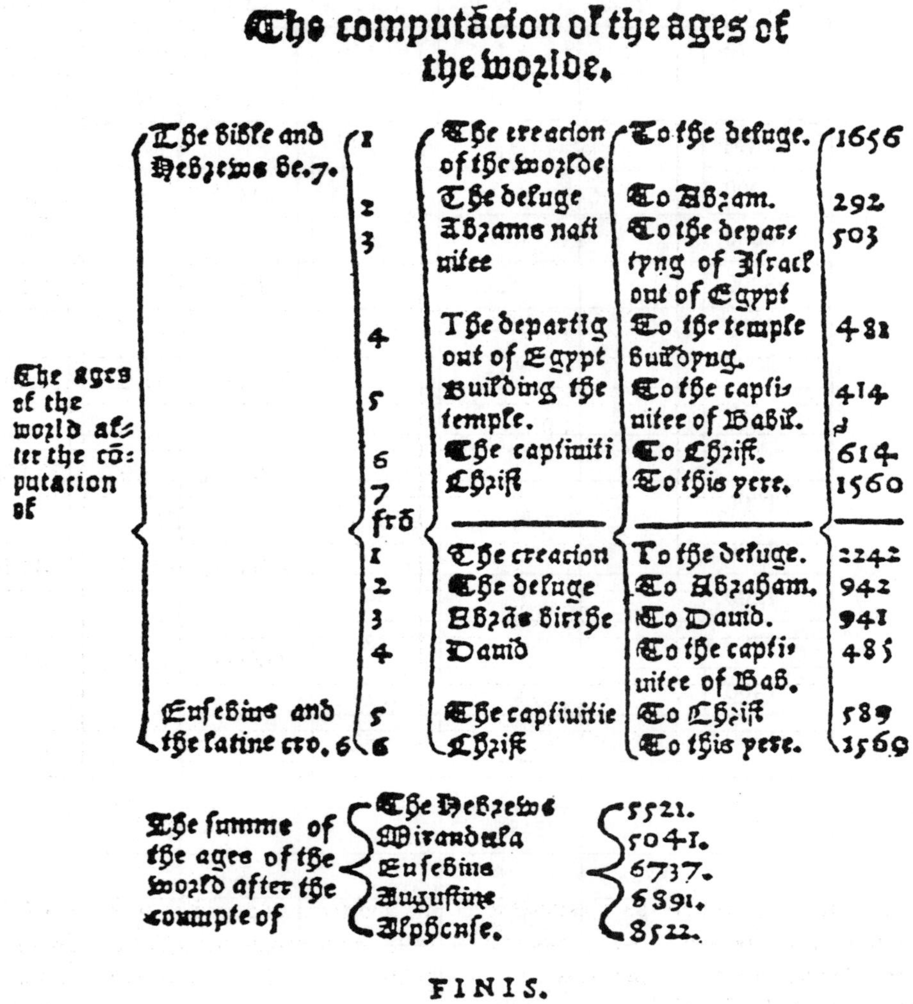

The computacion of the ages of the worlde.

The Bible and Hebrewes &c.7.

The ages of the worlde after the computacion of

Eusebius and the Latine cro. 6.

	The creacion of the worlde	To the deluge.	1656
2	The deluge	To Abram.	292
3	Abrams natiuitee	To the departyng of Israel out of Egypt	503
4	The departig out of Egypt Building the temple.	To the temple buildyng.	481
5		To the captiuitee of Babil.	414
6	The captiuiti	To Christ.	614
7	Christ	To this yere.	1560

1	The creacion	To the deluge.	2242
2	The deluge	To Abraham.	942
3	Abrds birthe	To Dauid.	941
4	Dauid	To the captiuitee of Bab.	485
5	The captiuitie	To Christ	589
6	Christ	To this yere.	1569

The summe of the ages of the worlde after the compte of

The Hebrewes — 5521.
Mirandula — 5041.
Eusebius — 6737.
Augustine — 6391.
Alphonse. — 8522.

FINIS.

FIGURE 2

of life expectancy was done by Edmund Halley (1656–1742), who drew up one of the first mortality tables and laid the foundations for the study of annuities.[29] The town of Breslau, Germany had hired Halley. His work ultimately led to the development of life insurance companies.[30] One of the earliest studies in historical demography was undertaken by William Macdonell,[31] when the statistician Karl Pearson suggested to him that one of the best ways to analyze the mortality rates in ancient Roman times was to gather data from tombstones inscriptions in Rome and in the surrounding provinces. In 1960 Durand using a sample of Macdonell's data, consisting of males who died between the ages of 15 and 42, obtained a life expectancy at birth for the population of the Roman Empire of between 25 and 30 years.[32] In an earlier study, Bessie Ellen Richardson used data from over 2000 Greek tombstones ranging from Classical Greece to the early Middle Ages to estimate that the average life expectancy in the sample was 29.4 years.[33] In 1947, using an osteoarcheological study, Juvenal Angel estimated that the average life expectancy in the Neolithic and Early Bronze Age was 32.1 years, for Classical Greece, 38.1 years, for the Roman Era, 36.8 years, for the Byzantine period, 33.7 years, and over the whole period in question, 35.4 years.[34]

However, in 1989 Mirko Grmek pointed out that, if Angel's study was restricted to only cranial specimens, which dated from 3500 B.C. to A.D. 1300, the mean life expectancy for Europe would have been 27.4 years.[35] Even if we account for the fact that there was a shortage of skulls of children and old people, Grmek believed that this figure is closer to the true mean life expectancy for that period. Furthermore, Grmek estimated the life expectancy in the

YEAR (t)	POPULATION (N)	$\ln(N)$	P	L	P/L
−2(6)	2	0.69			
			6.76(11)	24(9)	28.2 (9)
−8000	5(6)	15.43			
			5.76(11)	27	21.3 (9)
−50	3 (8)	19.5			
			5.94(11)	35	17.0 (9)
1348	4.7(8)	19.9			
			0.5 (11)	40	1.26(9)
1600	4.9(8)	20			
			1.33(11)	40	3.32(9)
1800	9.1(8)	20.63			
			1.46(11)	45	3.24(9)
1900	16 (8)	21.1			
			2.66(11)	50	5.32(9)
1992	53.3(8)	22.39			
					79.6 (9)

TABLE 2

Mesolithic era was 21.5 to 25 years. For their estimates, Keyfitz assumed a uniform life expectancy of 25 years, and Deevey calculated that human life-expectancy was 25±5 years from Neolithic times to 1900 and 45 years for the twentieth century.

Table 2, illustrates the results gleaned from a typical student exercise where exponential growth curves have been used to estimate population size. The data was taken from the work of Keyfitz and Desmond, and the life expectancies from Macdonell, Richardson, and Angel.

The figure for the world's population of 5.33 billion found in the *1992 Information Please Almanac* is 6.7% of the 79.6 billion people who ever lived. This is quite a contrast to 1973, when the world's population of 3.86 billion people constituted only 4% of the number of people who ever lived.

The Swiss-born mathematical genius Leonhard Euler (1707–1783) was interested in exponential population growth problems and included such exercises in his *Introduction to the Analysis of the Infinite*.[36] Students are required to solve similar exercises to conclude their demographic project. They are reminded of the religion teacher who once asked a class how they knew they were going to

die and the astute student who replied that it was because so far most people have.[37] According to the calculation in Table 2 the student was correct, but for how long will that be the case? Assuming an exponential rate of population growth, $N = N_o e^{bt}$, and 40 years for the world's population to double, hence $b = 0.017329$, it is a straightforward exercise to calculate the number of years it will take for the world's population to constitute 50% of the number of people who ever lived. We first set

$$\frac{5.33 \times 10^9 + p}{79.6 \times 10^9 + p} = \frac{1}{2}$$

to determine that we need $p = 68.94 \times 10^9$ more people. Substituting, $N = 74.27 \times 10^9$ and $N_o = 5.33 \times 10^9$, and b into $N = N_o e^{bt}$, and, solving for t, we find that in $t = 152$ years the world's population of 74.27×10^9 will constitute half the people that ever lived.

Using the exponential law, assuming that the present growth rate continues, and a figure of 1.497×10^8 square meters for the surface (land) area of the earth students determine the time when each person will have but one square meter of earth to stand on. Currently there are approximately 37 people per square kilometer of the earth's land (as opposed to 16 in 1960). If the world's population continues to grow exponentially at the same rate as in the period from 1900 to 1992, then in approximately 566 years each man, woman, and child will have but one square meter of earth to stand on. Nevertheless, there may well be other factors that we have not taken into account. For example, demographer Edward Wrigley has noted[38] that the water flea *Daphnia* shows a steady decline in fertility rate with increased crowding which may be the case with humans.

Students who have a good grasp of the calculus and are interested in learning more about mathematical demographics are encouraged to consult Keyfitz[39]. Further, Fernand Braudel's *Capitalism and Material Life*[40] contains relevant material on demography and its intimately interconnected economic and social structure. Jacqueline Kasun's *Consequences of Rapid Population Growth in Developing Countries*,[41] Rafael Salas's *Reflections on Population*,[42] and the United Nation's *The War against Population*[43] all discuss the population problem facing the world today, and such material can help foster germane nonmathematical discussions related to this demographic project.

Today the world's population is doubling at such a high rate that this rapid growth adversely effects the world's standard of living. It is incredible to think that such growth may continue unabated before some form of population self-balancing is accepted. Perhaps the change will be slight or perhaps catastrophic, which is not a very comforting thought with which to leave your class.

ACKNOWLEDGEMENT.   The author wishes to thank his students and the editor for their assistance and helpful suggestions for improving the paper.

Endnotes

1. If $f(t)$ represents the amount of a substance present at time t, then the exponential law requires that the rate of change of the amount present be proportional to the amount present or $f'(t) = k \cdot f(t)$.

2. Newton's law of cooling assumes that at any given time t the rate of change of the temperature of a body is proportional to the difference between the temperature of the body $T(t)$ and the constant temperature T_M of the surrounding medium or $T'(t) = k[T(t) - T_M]$.

3. For such problems, we assume that the concentration of a drug in the blood stream dissipates exponentially, and we determine at what time intervals and at what level a patient should be given a dosage which is both safe and effective.

4. If $p(t)$ represents the proportion of a population that has a disease at time t and c is a contagion factor, then under a variety of initial conditions we investigate solutions to the differential equation $p'(t) = c \cdot p(t)[1 - p(t)]$.

5. The ancient Greeks made several attempts to determine the earth-sun distance, culminating in Ptolemy's estimate of 1,210 earth-radii or about 5% of the actual distance. See Anton Pannekoek, *A History of Astronomy* (London: Allen and Unwin, 1961), p. 156.

6. Patricia Cohen, *A Calculating People: The Spread of Numeracy in Early America* (Chicago: University of Chicago Press, 1982), p. 112.

7. Wilhelm Winkler, Wieviele Menschen haben bisher auf der Erde gelebt? *International Population Conference, Vienna, 1959* (Vienna: Union Internationale pour l'Etude Scientifique de la Population, 1959), pp. 73–76.

8. Edward S. Deevey, Jr. "The Human Population," *Scientific American*, 1960, 203:194–204. Reprinted in the *Scientific American Resource Library: Readings in the Social Sciences*, 2 vols., Vol. 1 (San Francisco: Freeman, 1969), pp. 78–86.

9. Nathan Keyfitz, "How Many People Have Lived on the Earth?" *Demography*, 1966, 3:581–582.

10. Annabelle Desmond, "How Many People have ever Lived on Earth?" *The Population Crisis and the Use of World Resources* (Bloomington: Indiana University Press, 1964), pp. 27–46.

11. In particular, e denotes the base of the natural logarithm function $\ln(x)$.

12. Reliable recent population estimates can be found the *Demographic Yearbooks* published by the United Nations Department of Economic and Social Affairs Statistical Office, New York.

13. Alexa Carr-Saunders, *World Population, Past Growth and Present Trends* (Oxford: Clarendon, 1936), p. 42.

14. Deevey, "The Human Population."

15. Deevey's post-1650 data comes from Wladimir and Emma Woytinsky's, *World Population and Production: Trends and Outlook* (Twentieth Century Fund, 1953).

16. Deevey has taken into account the devastating effects of the Black Death during the late Middle Ages as have Michael Gonzales and William Carr in "Impact of the Black Death (1348–1405) on World Population: Then and Now," *Mathematics Teacher*, 1986, 79:92–94. Philip Zeigler, *The Black Death* (New York: John Day, 1969) has estimated that this plague caused a 21 percent decrease in the world's population from 1348 to 1400.

17. Desmond, "Population Crisis."

18. Nathan Keyfitz, "How Many People Ever Lived?" *Some Illustrative Examples of the Use of Undergraduate Mathematics in the Social Sciences* (Washington: CUPM-MSSB, 1976).

19. James Trager, *The People's Chronology* (New York: Holt, Rinhart, & Winston, 1979), pp. 2–4.

20. Walter Willcox, Population of the Earth. *International Migrations*, 2 vols. Vol. 2, Part I (New York: National Bureau of Economic Research, 1931).

21. Fernand Braudel, *Civilization and Capitalism in the 15th–18th Century*, 3 vols. Vol. 1: *The Structures of Everyday Life* (New York: Harper and Row, 1989), pp. 31–51.

22. Xenophanes (c 570–c 480 B.C.) made the first correct interpretation of fossils, see George Sarton, *Introduction to the*

History of Science, 3 vols. Vol. 1 (Washington: Carnegie Institute, 1927), p. 73.

23. James Lightfoot, *Harmony, Chronicle, and Order of the Old Testament* (London: Cotes, 1647).

24. Thomas Lanquett, *Cooper's Chronicle* (London, 1560), p. 378.

25. James Ussher, *Annals of the World* (London: Tyler, 1658).

26. Jacques Roger, "Georges-Louis LeClerc, Compte de Buffon," *Dictionary of Scientific Biography* (New York: Scribner's Sons, 1970), 2:576–582.

27. Clair Patterson, "Age of Meteorites and the Earth," *Geochimica et Cosmochimica Acta* 1956, 7:230–237.

28. Brent Dalrymple, *The Age of the Earth* (Stanford: Stanford University Press, 1991).

29. For a detailed account of the origins of actuarial science, see Oscar Sheynin, "Early History of the Theory of Probability." 1977, *Isis* 17:201–259.

30. Edmund Halley, "An Estimate of the Degrees of the Mortality of Mankind, Drawn from Curious Tables of the Births and Funerals at the City of Breslau; with an Attempt to Ascertain the Price of Annuities upon Lives," *Philosophical Transactions of the Royal Society,* 1694, 17:596–610.

31. Edward Deevey, "The Probability of Death," *Scientific American,* 1950, 186:58–60.

32. John Durand, "Mortality Estimates from Roman Tombstone Inscriptions," *American Journal of Sociology,* 1960, 45:365–373.

33. Bessie Richardson, *Old Age Among the Ancient Greeks* (Baltimore: Johns Hopkins, 1933).

34. Juvenal Angel, "Length of Life in Ancient Greece," *Journal of Gerontology,* 1947, 2:18–24.

35. Mirko Grmek, *Diseases in the Ancient Greek World* (Baltimore: Johns Hopkins, 1989).

36. Leonhard Euler, *Introduction to the Analysis of the Infinite* (New York: Springer-Verlag, 1988), p. 85–88; *Introductio in analysin infintorum* (Lausanne: Bosquet, 1748) trans. by John Blanton.

37. Michael Deakin. "We, the Living," *Function, 1990,* 14:42–46.

38. Edward Wrigley, *Population and History* (New York: McGraw-Hill, 1969), p. 39.

39. Nathan Keyfitz, *Applied Mathematical Demography.* New York: Wiley, 1977).

40. Fernand Braudel, *Capitalism and Material Life 1400–1800* (New York: Harper and Row, 1973), pp. 2–20.

41. Jacqueline Kasun, *Consequences of Population Growth in Developing Countries* (New York: Taylor & Francis, 1991).

42. Rafael Salas, *Reflections on Population* (New York: Pergammon, 1985)

43. *United Nations, The War Against Population: The Economic and Ideology of Population Control* (San Francisco: Ignatius, 1988).

Notes on Contributors

William Aspray is the director of the IEEE Center for the History of Electrical Engineering and a member of the graduate faculty in history at Rutgers University. He writes on the history of computing, electronics, mathematics, and high technology in the twentieth century. His address is IEEE-CHEE, 39 Union Street, Rutgers University, New Brunswick, NJ 08903. Dr. Aspray's phone number is 908-932-1066, and his e-mail address is w.aspray@ieee.org.

Evelyne Barbin, Maître de Conferences in epistemology and history of mathematics in the IUFM of Creteil, conducts research at the IREM of Paris-Nord (University of Paris 13). Her major areas of research are the history of mathematical proof and 17th-century mathematics. One of her papers appeared in the *Proceedings of the Eves Conference* (1991), and she has edited the IREM (Institut de Récherche sur l'Enseignement des Mathématiques) volume, entitled *Histoires de problemes, histoire des mathématiques* (1993). Dr. Barbin's address is IREM, Université Paris 7, 2, place Jussieu, 75251 Paris Cedex 05, France. Her phone number is 44 27 53 83/53 84, and her fax number is 44 27 56 08.

Peter Bero is a lecturer in mathematics education at Comenius University in Bratislava, Slovakia. His research areas are the history of mathematics, particularly calculus; the use of history in teaching; and the theory of concept development. His publications include "Intuitive Awareness of Dependence Relationships and Understanding of Mathematical Functions" in *Research in Education* (1993) and *Historical Approach in the Teaching of Mathematics* (1993). Dr. Bero's address is KZDM, MFF, Univerzita Komenskeho, Mlynska dolina, 842 15 Bratislava, Slovakia. His phone number is (+42 7) 720 003, and his e-mail address is bero@fmph.uniba.sk.

Ronald Calinger is an associate professor of history at The Catholic University of America. His main research area is the history of the mathematical sciences during the 18th-century Continental Enlightenment, especially the work of Leonhard Euler. He recently edited and wrote the historical sections of *Classics of Mathematics* (Prentice-Hall, 1995), and his latest article, "Leonhard Euler: the First St. Petersburg Years (1727–1741)," will appear in *Historia Mathematica* 23 (1996): 1–46. Dr. Calinger's address is Department of History, The Catholic University of America, Washington, D. C. 20064. His phone number is 202-319-5484, and his fax number is 202-319-5569. His e-mail address is calinger@cua.edu.

Roger Cooke is a professor of mathematics at the University of Vermont. He specializes in multiple trigonometric series and the history of mathematics. He is currently finishing *Issues in the History of Mathematics,* to be published by PWS Kent, and has recently published "Descriptive Set Theory and Uniqueness of Trigonometric Series, 1870–1985," *Archive for History of Exact Science* 45 (1993): 281–334. Dr. Cooke's address is Department of Mathematics and Statistics, University of Vermont, Burlington, VT 05401-1455. His telephone number is 802-656-4335, and his fax numbers are 802-656-2552 or 802-862- 9763. His e-mail address is cooke@emba.uvm.edu.

Zarko Dadić, professor of the history of science at the University of Zagreb, is the head of the Institute of the History and Philosophy of Science of the Croatian Academy of Sciences and Arts. He is also a member of the International Academy for the History of Science. His principal research area is the history of mathematics and astronomy, particularly in Croatia. His chief publications are a two-volume history of the exact sciences in Croatia and a biography of Rudjer Boskovic. Professor Dadić's address is Institute of the History and Philosophy of Science, Ante Kovacica 5, 41000 Zagreb, Croatia. His fax number is 385-01-449378.

Ubiratan D'Ambrosio is professor emeritus at the Universidade Estadual de Campinas in Brazil. His chief research and teaching areas are the history of mathematics and mathematics education. Two recent publications are *Metodos da Topologia* [*Methods of Topology*] (1994) and "Ethnomathematics, the Nature of Mathematics and Mathematics Education," *Mathematics, Education and Philosophy: An International Perspective,* ed. Paul Ernest, London: The Falmer Press, 1994, pp. 230–242. Professor D'Ambrosio's address is Rua Peixoto Gomide 1772 apt. 83, 01409-002 Sao Paulo, SP Brasil. His telephone and fax number is 55-11-280-0266.

John Fauvel is senior lecturer in mathematics at the Open University, Milton Keynes, UK. He has edited several books in the history of mathematics, including *The History of Mathematics: A Reader* (1987), *Let Newton Be!* (1988), and *Möbius and His Band* (1993). He is currently working on the history of mathematics at Oxford. Mr. Fauvel is the immediate past President of the British Society for the History of Mathematics and the current Chair of the International Study Group on the History and Pedagogy of Mathematics. His address is Faculty of Mathematics, The Open University, Walton Hall, Milton Keynes, MK7 6AA, UK. His e-mail address is j.g.fauvel@open.ac.uk.

Martin Flashman is a professor of mathematics at Humboldt State University, where he has taught since 1981. His research interests range from philosophy of mathematics to geometry and game theory, to sensible reform of the calculus curriculum. Dr. Flashman is book and media editor for *UME Trends,* in which he writes the column "Noteworthy Books and Such." His address is Department of Mathematics, Humboldt State University, Arcata, CA 95521. His phone number is 707-826-4950, and his e-mail address is flashman@axe.humboldt.edu.

Andrew Goldstein is a curator at the IEEE Center for the History of Electrical Engineering. His major research is in the history of post-war electronics. Recently he has studied the careers of Gordon Teal, Jack Avins, and John R. Pierce. Dr. Goldstein's address is IEEE-CHEE, 39 Union Street, Rutgers University, New Brunswick, NJ 08903. His phone number is 908-932-1066, and his e-mail address is a.goldstein@ieee.org.

Judy Grabiner, the Flora Sanborn Pitzer Professor of Mathematics at Pitzer College in Claremont, California, also teaches in a Claremont-wide program in science, technology, and society. Her primary research area is eighteenth- and nineteenth-century mathematics. Professor Grabiner has recently authored *The Calculus as Algebra: J. -L. Lagrange, 1736–1813* (1990) and "Descartes and Problem-Solving," *Mathematics Magazine,* April 1995, pp. 83–97. Her current research focuses on Colin Maclaurin, and her newest teaching enterprises are liberal-arts mathematics courses on "Mathematics, Philosophy, and the 'Real World'" and "Mathematics in Many Cultures." Professor Grabiner's office address is Pitzer College, 1050 North Mills Avenue, Claremont, California 91711-6110. Her phone number is 909-607-3061, and her e-mail address is jgrabiner@pitzer.claremont.edu.

Michèle Grégoire, mathematics teacher, Lycée Lavoisier, Paris 5 , is in charge of continuing education in history of mathematics and researcher, group M.A.T.H., IREM (Institute for Research in Mathematical Education) of University Paris-7 Denis Diderot, 2 place Jussieu 75005 Paris France. fax (33) 1 44 27 56 08. She studies history of perspective, and her publications include *Comment mesurer la pyramide?* in IREM, *Histoires de Problèmes, histoire des mathématiques,* ELLIPSES, Paris, 1993, *Histoires de pyramides,* Mnémosyne N° spécial I, IREM Paris-7 1994.

Torkil Heiede, who obtained his doctorate at Copenhagen University, has been at the Royal Danish School of Educational Studies since 1963. A professor of mathematics, he is coauthor of a book on set theory and has written three volumes on groupoid theory. Since 1966 he has been a co-editor of *Normat* and has been active in the Con Amore Problem Group. Dr. Heiede has taught courses on the history of mathematics and has published on its place in mathematical education at all levels. A recent article is "Why Teach History of Mathematics?" in *The Mathematical Gazette,* 76, 1992: 151–157. Dr. Heiede's address is Royal Danish School of Educational Studies, Emdrupvej 115B, DK-2400 Copenhagen NV, Denmark. His phone number is +45 3966 3232 ext. 2694 and FAX number is +45 3969 6626. Dr. Heiede's e-mail address is dkibmlh5@ibmmail.com.

Susann Hensel is an assistant professor at the Institute for the History of Medicine, Science, and Technology at the University of Jena. Her main research areas are the history of engineering education, in particular in the mathematical sciences, and the interactions of mathematics with engineering science in the 19th and 20th centuries, especially for Germany and the USA. She is the coauthor of *Technik im 19. Jahrhundert in Deutschland* (1989). Dr. Hensel's address is Friedrich-Schiller-Universitat Jena, Institut fur Geschichte der Medizin, Naturwissenschaft und Technik, Ernst-Haeckel-Haus, Berggasse 7, D-07745 Jena, Federal Republic of Germany. Her telephone number is (49) 641-632129, and her fax number is (49) 641-632345.

Gavin Hitchcock is a senior lecturer in mathematics at the University of Zimbabwe. His research interests lie in general topology and history of mathematics; he is also concerned with the enlivening of mathematics teaching by means of posters, models, games, plays, and history. Two recent articles are "Forgetting and Recalling in Topology" (The concept of excess structure in historical and heuristic perspective), *Colloquia Mathematica Societatis Janos Bolyai, 55, Topology* Pecs (Hungary), 1989, pp. 307–319 and "The Grand Entertainment: Dramatizing the Birth and Development of Mathematical Concepts," *For the Learning of Mathematics* 12, 1 (1992): 21–27. Dr. Hitchcock's address is Department of Mathematics, University of Zimbabwe, P. O. Box MP 167, Mount Pleasant, Harare, Zimbabwe. His telephone number is 303211, ext. 1801, and his fax number is 263- 4-732828. His e-mail address is gavin@maths.uz.zw.

Jens Høyrup is a reader at Roskilde University in Denmark. His work in the history of mathematics is primarily concerned with the cultural and conceptual history of pre-modern mathematics, in particular the conceptual structure of Babylonian mathematics and its links with general scribal culture. Recent publications include *In Measure, Number and Weight* (1995) and "The Babylonian Cellar Text BM 85200 + VAT 6599." Dr. Høyrup's address is Department of Languages and Culture, University of Roskilde, P. O. Box 260, DK-4000 Roskilde, Denmark. His telephone number is (45) 46 75 77 11, and his fax number is (45) 46 75 49 05.

Barnabus Hughes, who received his doctorate in the history of mathematics from Stanford University, is a professor of secondary education at California State University, Northridge. His major research areas are teacher training, problem solving, and history of medieval/renaissance mathematics. His extensive publications include critical editions of the Latin translations of al-Khwarizmi's *Al-jabr.* Professor Hughes' address is California State University–EDUC, Northridge, CA 91330. His fax number is 818-885-4737.

Hans Niels Jahnke is a privatdozent at the Institute for Didactics of Mathematics at the University of Bielefeld. His research interests include history and philosophy of 19th-century mathematics, history of education, and phylogenesis/ontogenesis of mathematical concepts. Dr. Jahnke's latest book is *Mathematik und Bildung in der Humboldtschen Reform* (1990), and a recent article is "Algebraic Analysis in Germany, 1780–1840: Some Mathematical and Philosophical Issues," *Historia Mathematica* 20 (1993): 265–284. His address is Institut für Didaktik der Mathematik, Universität Bielefeld, Postfach 100131, 33501 Bielefeld, F. R. Germany. Dr. Jahnke's phone number is 521-106-5056, and his e-mail address is njahnke@post.uni.bielefeld.de.

Marie Françoise Jozeau, mathematics teacher, Lycée G. de Nerval, Luzarches, is in charge of teacher training and continuing education, Academy of Versailles, and a member of the group M.A.T.H. (Mathematics: an approach through historical texts) of the IREM (Institute for Research in Mathematical Education) of University Paris-7 Denis Diderot, 2 place Jussieu 75005 Paris France. fax (33) 1 44 27 56 08. Her research stresses the history of geodesy in the 19th century in Europe, and her address is: 12 Clos des Gâtines, 95270 Luzarches FRANCE. Her publications include *La mesure de la terre au 19ème siècle; nouveaux instruments, nouvelles méthodes,* in *La Mesure, Instruments et philosophies,* dirigé par J.C. Beaune, coll Milieux, Champ Vallon, Seyssel, 1994

Victor J. Katz is a professor of mathematics at the University of the District of Columbia. During academic year 1994-95 he was visiting mathematician at the Mathematical Association of America. His major research area is the use of the history of mathematics in teaching mathematics. Two recent publications are *A History of Mathematics: An Introduction* (1993) and "Ideas of Calculus in India and Islam," *Mathematics Magazine,* 68 (1995): 163–174. Since 1989, he has also edited the *Newsletter* of the International Study Group on the Relations between History and Pedagogy of Mathematics. His phone number is 202-274-5374, and his fax number is 202-224-5399. His e-mail address is vkatz@maa.org.

Peggy Aldrich Kidwell, whose doctorate is from Yale University, oversees the mathematics collections at the National Museum of American History. She studies the history of mathematical instruments and the history of astrophysics. Recent publications include *Landmarks in Digital Computing: A Smithsonian Pictorial History* (with Paul Ceruzzi, 1994) and "Ideology and Invention: The Calculating Machine of Ramon Verea," *Rittenhouse* 1995, 33–41. Dr. Kidwell's address is MRC636, National Museum of American History, Smithsonian Institution, Washington, D. C. 20560. Her office phone number is 202-357-2392.

Israel Kleiner is a professor in the Department of Mathematics and Statistics at York University. His scholarly interests are the history of mathematics and its use in teaching. Two recent articles are "The Roots of Commutative Algebra in Algebraic Number Theory," *Mathematics Magazine* 69 (1995): 3–15 and "The Role of Paradoxes in the Evolution of Mathematics," (with N. Movshovitz-Hadar), *American Mathematical Monthly* 10 (1994): 963–974. Dr. Kleiner's address is Department of Mathematics and Statistics, York University, North York, Ontario M3J IP3. His telephone number is 416-736-5250, and his e-mail address is kleiner@vml.yorku.ca.

Wilbur R. Knorr is professor of history of science at Stanford University, where he teaches history of science, cosmology, and mathematics. His research has been primarily on ancient Greek mathematics and its transmission in the Arabic and Latin Middle Ages, on which he had published four books, including *Ancient Tradition of Geometric Problems* (1986, Dover reprint 1993), and *Textual Studies in Ancient and Medieval Geometry* (1989), as well as forty articles. His current projects include a survey of the work of Archimedes. Dr. Knorr's address is Program in History of Science, Stanford University, Stanford, CA 94305-2024. His phone number is 415-723- 2760, and his fax number is 415-723-3235.

Manfred Kronfellner, an assistant professor of mathematics at the Technical University of Vienna, conducts research on the didactics of mathematics, especially incorporating history of mathematics in the teaching of mathematics, and computer-algebraic systems (DERIVE) in mathematical education. His publications include "Das Prinzip der Linearisierung," in *Mathematica Didactica* 2 (1979), vol. 1: 1–32 and *Geschichte der Mathematik im Unterricht: Möglichkeiten und Grenzen* (1994). Dr. Kronfellner's address is Technical University of Vienna, Institut fur Algebra und Diskrete Mathematik, A-1040 Vienna, Wiedener Hauptstrasse 8-10, Austria. His phone number is (0)222/58801/5449, and his e-mail address is mkronfel@email.tuwien.ac.at.

Reinhard Laubenbacher is an associate professor of mathematics at New Mexico State University. He conducts research on algebraic K-theory, module theory, and computational algebra/algebraic geometry. Together with David Pengelley, he created two honors mathematics courses based on the study of original historical texts. One of his recent publications is "Gauss, Eisenstein, and the 'Third' Proof of the Quadratic Reciprocity Theorem...," in *The Mathematical Intelligencer* 16 (1994): 67–72. Dr. Laubenbacher's office address is Department of Mathematics, New Mexico State University, Las Cruces, NM 88003, USA. His e-mail address is reinhard@nmsu.edu.

Beatrice Lumpkin, associate professor emeritus of mathematics at Malcolm X College of the Chicago City Colleges, is a consultant on multicultural curriculum materials for Educational Equity Services of the Illinois State Board of Education and the Portland, Oregon, and Chicago school districts. She is the co-author of *Multicultural Science and Math Connections* (1995) and author of these children's books about Egyptian mathematics: *Senefer, Young Genius in Old Egypt,* and *Senefer Hatshepsut*. Dr. Lumpkin's address is 7123 S. Crandon Avenue, Chicago, Illinois 60649. Her phone number is 312-684-4553.

Karen Dee Michalowicz, the Upper School Mathematics Chair at the Langley School in McLean, Virginia, is one of the State of Virginia's teacher trainers for its systemic reform in mathematics education. Karen is also a feature writer for the National Council of Teachers of Mathematics journal, *Mathematics Teaching in Middle School*. In 1992 she was Virginia's State Mathematics Teacher of the Year, and in 1994, a national Presidential Awardee in Mathematics. Mrs. Michalowicz's address is 5855 Forest Drive, Falls Church, VA 22041. Her phone number is 703-820-1889, and her e-mail address is karendm@aol.com.

David Pengelley, an associate professor of mathematics at New Mexico State University, investigates homotopy theory and has co-written *Student Research Projects in Calculus*. With Reinhard Laubenbacher, he created two honors mathematics courses based on the study of original historical texts. They are completing a companion text of annotated original sources entitled *The Evolution of Mathematical Ideas: A Study of Original Sources.* Dr. Pengelley won the 1993 Distinguished Teaching Award from the Southwestern Section of the Mathematical Association of America. His office address is Department of Mathematics, New Mexico State University, Las Cruces, NM 88003, USA. His e-mail address is davidp@nmsu.edu.

V. Frederick Rickey is Distinguished Teaching Professor at Bowling Green State University in Ohio. Although educated as a logician, he is currently interested in the history of mathematics, especially the history of the calculus. His article "Isaac Newton: Man, Myth, and Mathematics," *The College Mathematics Journal,* 18(1987) 362–389 won the Pólya Prize. More recently he published "My favorite ways of using history in teaching calculus," in *Learn from the Masters!*, MAA, 1995. Prof. Rickey's phone number is 419-372- 7452; his fax number is 419-372-6092; and his e-mail address is rickey@math.bgsu.edu.

David E. Rowe teaches at the Johannes Gutenberg-Universität Mainz, where he heads the *Arbeitsgruppe Geschichte der Mathematik und der exakten Naturwissenschaften.* His major research interests are the history of geometry and nineteenth-century mathematics. With Karen H. Parshall, he coauthored *The Emergence of the American Mathematical Research Community, 1876–1900* (1994). Professor Rowe is the immediate past editor of *Historia Mathematica.* His address is FB-17 Mathematik, Universität Mainz, 55099 Mainz, Germany. His phone number is 49-6131-392837, and his fax number is 49-6131-394659. His e-mail address is rowe@mat.mathematik.uni-mainz.de.

Man-Keung Siu, who obtained his doctorate in algebraic K-theory from Columbia University, is a reader in mathematics at the University of Hong Kong. He has published in the fields of algebra, combinatorics, applied probability, mathematics education, and history of mathematics. The Chinese Mathematical Society selected his book *Mathematical Proofs* (1990, in Chinese) as one of seven outstanding books in mathematical exposition in 1991. Dr. Siu's office address is Department of Mathematics, University of Hong Kong, Hong Kong. His phone number is 852-2859-2258, and his e-mail address is mathsiu@hkucc.hku.hk.

Frank Swetz is professor of mathematics and education at the Pennsylvania State University-Harrisburg, where he has chaired the Mathematical and Computer Sciences Program. His efforts to humanize mathematics teaching and learning have taken him into studies in the history of mathematics and ethnomathematics. His two latest books are *From Five Fingers to Infinity: A Journey Through the History of Mathematics* (1994) and *Learning Activities from the History of Mathematics* (1994). Dr. Swetz's address is The Pennsylvania State University-Harrisburg, Middleton, PA 17057. His phone number is 717-948-6086.

James J. Tattersall is a professor of mathematics at Providence College in Rhode Island. His major research areas in the history of mathematics are Lucasian professors at Cambridge University and American mathematics in the age of Jefferson. Other research interests are number theory, combinatorics, astronomy, and Sherlock Holmes. His publications include "Who put the 'C' in A. -T. Vandermonde?" *Historia mathematica* 15 (1988): 361–377, and "Nicholas Saunderson: The Blind Lucasian Professor," *Historia mathematica* 19 (1992): 356–370. Dr. Tattersall's address is Department of Mathematics and Computer Science, Providence College, Providence, RI 02918. His telephone number is 401-865-2468, and his e-mail address is np180026@brownvm.brown.edu.

Bernard Orion Williams is the product development manager at R&D Publications, which print technical books and journals for the computer industry. His research covers early electronic devices, information design, and human factors of complex systems. His publications include *Computing with Electricity, 1935–1945* (1984). Dr. Williams' address is R&D Publications, Inc., 1601 West 23rd Street, Suite 200, Lawrence, Kansas 66046. His phone number is 913-841-1631, and his e-mail address is berney@rdpub.com.

Index

A

Aaboe, Asger, 238, 243
Aachen, 171, 196
Aachen resolution, 194
Aarhus, 233
Abacist, 107, 110, 111
Abacus, 197, 198, 236
    String, 94
Abel, Niels Henrik, 158, 159, 161, 164, 167,
    173, 178, 179, 181, 241, 242, 324, 329
Abel's theorem, 181, 189
Abelian
    functions, 153, 161, 165, 183, 185, 188,
      189
    integrals, 180, 182, 183
Abolition, 125, 126
Abramsom, Norman, 223, 228
Absolute value, 138
Abstraction, 178
Abu Bakr, 49–58, 60, 62–65
Abu Kamil ibn Aslam, 48, 100, 282–284, 289
Academic freedom, 154, 163
Academy, Plato's, 14, 154
Acceleration, 133, 135
Accumulating, 46
Ackerman, M., 195
Acta Eruditorum, 322
Acta Mathematica, 171
Adamo, Marco, 88, 96
Adams, Daniel, 281, 285, 289
Additions, 47
Adelman, Len, 217, 223, 228
Adler, M., 188, 189
Adrion, Richard, 230
Advanced Resemnarch Projects Agency
    (ARPA), 211
Aerodynamics, 176
Africa, 282
African Americans, 279
Ahmad, M. M., 62
Ahmad, Salah, 105
Air Force, US, 213 215 216, 222, 226, 227
    Office of Scientific Research, 210
Akkadian, 46, 47, 57, 61, 64
Aleph, 100
Alexandria, 69, 271, 282
Algebra, al-Khwarizmi's, 107, 108, 111
Algebra, 1–7, 8–10, 37–40, 48, 50, 57, 58, 60,
    108, 111, 118, 131–143, 148, 161, 165, 173,
    177, 180, 243, 253, 266, 279, 297, 299, 309
    abstract, 259
    Babylonian, 45–48, 53, 57
    Chinese, 94
    differential, 324
    geometric, 6–8, 15
    matrix, 176
    symbolic, 36, 46, 121

Algorithm, 33, 62, 100, 108, 138, 140, 146,
    147, 150, 216, 217, 218, 226, 227, 237,
    253, 309, 325–327
    bucket-brigade, 217
    rod, 92–94
Al-jabr, Islamic, 45, 48, 50–52, 56, 57, 59, 62,
    63, 111
Al-Karaji, 51, 286
Al-Khalil ibn Ahmad, 100
Al-Kharki, 63
Al-Khwarizmi, 48–51, 56, 57, 59, 62, 63, 107,
    108, 253, 283
Almagest, 81, 82
Alma mater, 155
Al-Ma'mun, 48
Almanac, 280
Almohade dynasty, 102
Al-Nasir, Mohammed ibn Ya'kub, 102
Alphabet, Arabic, 102
Al-Samaw'al, 101, 105
Altenstein, Karl, 155
Althoff, Friedrich, 174
Altitude, 69–71, 74
Ambrose, S. E., 205
American Philosophical Society, 6, 14, 201
American University, 295
Ameros (indivisibles), 304
Amey, D. C., 206
Amherst College, 203
Amnesty International, 125
Amsterdam, 119
Analogy, 12, 126, 131, 183, 217, 327
Analysis, 117, 122, 192, 239, 266
    algebraic, 115, 117, 118, 145–152
    complex, 169, 171
    Diophantine, 147
    elementary, 173
    foundations of, 169
    geometrical, 17, 118, 121
    image, 228
    mathematical, 153, 164, 167, 187, 188
    nonstandard, 304
    numerical, 209, 210, 213, 226
    real, 168
    tensor, 176
Analytic geometry, 36
Analytical Mechanics (1788), 125, 133
Analytical-synthetical method, 119, 121
Ancien Régime, 191
Ancient mathematics, 6–9
Andersen, Kirsti, 80, 81, 84, 238, 242
Andersen, Paul O., 241
Anderson, Alexander, 118, 119, 122, 123
Anderson, Phil, 216
Andrews, A. H., 199
Ang Tian-Se, 95, 96

Angel, Juvenal, 337
Angle, 18–20
Angled, acute-, 115, 122
    obtuse-, 115
    right-, 115, 116
Answers, 141
Anthropology, 246, 247, 249
Antiderivative, 322
Antimathematical movement, 193, 194
Antinomies, Zeno's, 306
Antiquity, Greek, 246
Apex, 116
Aphorism, 252
Apollonii Pergaei Inclintationum libri duo, 118
Apollonius of Perga, 6, 115–120, 255
Apollonius Gallus, 123
Apollonius redivivus . . ., 116–118, 120
Appending, 46
Application of areas, 6, 15
Applied mathematics, 195
Approximation, 180
Approximation method of convergence, 78
Arabic, 51, 102
Arbitur, 155
Arc, 90, 137, 312, 314
    of meridian, 269–277
Arch, 135
Archeion, 95
Archibald, Raymond C., 206
Archimedes, 6, 8, 22, 29, 63, 67–69, 71–75,
    78–84, 86, 137, 233, 238, 242, 243, 254,
    258, 260. 261, 271, 304, 311
Archive for History of Exact Sciences, 7, 8, 14,
    15, 84–86, 150, 173, 175, 330
Area(s), 136, 149, 310, 311
    surface, 198
Arendt, F., 84
Aristarchus, 271
Aristippus, 252
Aristocracy, 248
Aristotle, 8, 75, 83, 86, 117, 240
Arithmetic, 167, 168, 172, 246, 265, 299
Arithmetic Teacher, 237
Arithmetica, 59, 282
*Arithmetical Classic of the Gnomon and the
    Circular Paths of Heaven,* 88
Arithmeticon, 199
Arithmetics, 145
Arithmetization of analysis, 193
Arley, Niels, 61
Army, US, 216
Arnauld, Antoine, 19–22
Arndt, 163
Arrebala, 95
Arrow, 101
Arrow, Kenneth, 217